The Safe Handling of Chemicals in Industry

Volume 2

The Safe Handling of Chemicals in Industry

P. A. CARSON M.Sc., Ph.D., A.M.C.T., C. Chem., F.R.S.C., M.I.O.S.H
Safety Liason Manager, Unilever Research, Port Sunlight
C. J. MUMFORD B.Sc., Ph.D., D.Sc., C. Eng., M.I. Chem. E.
Senior Lecturer in Chemical Engineering, University of Aston

Volume 2

Copublished in the United States with
John Wiley & Sons, Inc., New York

Longman Scientific & Technical
Longman Group UK Limited
Longman House, Burnt Mill, Harlow,
Essex CM20 2JE, England
and Associated Companies throughout the world

Copublished in the United States with
John Wiley & Sons, Inc., 605 Third Avenue, New York, NY 10158

First published 1988

British Library Cataloguing in Publication Data
Carson, P. A.
Safe handling of chemicals in industry
1. Hazardous substances – Safety measures
I. Title II. Mumford, C. J.
363.1′79 TP149

ISBN 0-582-00304-0

Library of Congress Cataloging-in-Publication Data
Carson, P. A., 1944–
Safe handling of chemicals in industry / P. A. Carson, C. J. Mumford.
p. cm.
Includes index.
ISBN 0-470-20886-4 (Wiley)
1. Hazardous substances – Safety measures. I. Mumford, C. J.
II. Title.
T55.3.H3C37 1988
604.7 – dc 19

Set in Linotron 202 10/12 pt Times

Printed and bound in Great Britain
at the Bath Press, Avon

Contents

Preface

This text is intended to assist in the safe use of chemicals whether in the laboratory or factory, in offices or workshops, in schools and colleges, on the land, or elsewhere. While it should be of value to anyone associated with the chemical and process industries, the breadth of coverage is aimed in particular at the technical user of chemicals who is not involved in their manufacture and therefore not necessarily aware of their properties and safe handling procedures. Thus the audience includes plant operators, technicians and their line management plus occupational medical officers and their nursing staff, safety officers, chemists, hygienists, chemical engineers, safety representatives and staff with responsibility for health and safety in laboratories and pilot plant in industry and in academic institutions. It should also be of assistance to students of occupational safety and health in the broadest sense. Although the reader is introduced to the hazards and control of radioactive chemicals, the book does not cover the special risks associated with the nuclear industry.

While some reference is necessarily made to UK legislation and codes of practice, the aim throughout is to identify hazards commonly arising in chemicals' storage, transport, handling and disposal; and to summarise measures which result in safe plant, a safe working environment and a safe system of work. It represents best current practice at the time of compilation in the mid 1980s. Thus, although there will inevitably have been some recent changes in legislation, and even advances in technology, the general principles remain unchanged and are not restricted in anyway to the UK.

Brief cameo case histories of incidents are used to exemplify some problems which can arise with chemicals, engineering and system-of-work failures, etc. Presented in a typeface different to text typeface they are included to help relate theory and principles to practice and to be useful, together with other case histories given in the sources referred to in Chapter 18, in the preparation of training materials.

Where possible the examples chosen for these case histories, and to explain fundamental principles, are restricted to a relatively small selection of common, simple chemicals and processes. These are supplemented by additional examples in numerous tables to increase the reader's awareness of the scope. Even so, while the book contains a significant amount of information in tabular form this is intended to be illustrative rather than encyclopaedic. Thus the absence of data in a table, e.g. on a specific chemical, should not be interpreted necessarily as having significance and the reader must consult the literature (such as the sources listed in Ch. 18) for information relating to specific enquiries.

The main thrust is to describe techniques for the recognition of chemical hazards and to quantify and control the risks so that chemicals can be handled safely in practice. The hazards, preventive measures, legislation, and sources of information discussed are based upon current knowledge. So far as the general fire and explosion characteristics of common chemicals are concerned, except where very large vapour clouds are possible, this is reasonably well established. With regard to the toxicity of chemicals there tends to be more uncertainty, particularly when considering chronic effects or the possibility of disease following long latency periods, for example with chemical carcinogens. Thus occupational exposure limits/threshold limit values, from which 'permissible' levels of workplace exposure are derived, are constantly under review. It is a basic tenet of chemicals handling, as discussed in detail in the appropriate chapters, that occupational exposures and releases to the environment generally should in all cases be as low as is reasonably practicable.

Acknowledgements

Books, published papers and other sources of information which have been drawn upon have generally been referenced in the text. However, we also acknowledge those sources, e.g. of case histories, which have – perhaps understandably in a book of this size and scope, but quite unintentionally – escaped due reference.

We thank numerous colleagues within Unilever and Aston University for helpful discussions during the preparation stages. In particular, recognition is due to those who took time to read and offer constructive comments on different parts of the text, namely Messrs N. Burak, G. R. Cooper, G. R. Cliffe, S. Evans, A. Jarman, J. G. Lyons, and Drs D. M. Roberts, M. J. How and T. C. Pestell (all of Unilever) and R. C. Keen of Bristol Polytechnic. Gratitude is also expressed to the Management of Unilever Research Laboratory for generously making available typing and administrative facilities.

Finally we express special thanks to our respective families without whose encouragement this book could not have been written.

P. A. C.
C. J. M.

CHAPTER 10

Hazardous chemical processes

Some chemical processes carry with them an inherent hazard. Some which are potentially dangerous are listed in Table 10.1.

Some of these when operated on a large scale bring the plant into the category of a major hazard installation as discussed in Chapter 15.

When such processes go out of control the effect can be immediate and disastrous, resulting in fire, explosion, or release of toxic material. Even accidents involving small plants can be costly in terms of injuries, loss of life, plant damage, pollution of the environment, loss of production capacity, severe financial loss, plus alienation of public confidence. However, irrespective of the process to be carried out, a hazard also arises with some chemicals because of their propensity to decompose, or to react violently on contact with other substances.

Incompatible chemicals

A common example of substances in this class is those which are water-sensitive: a selection encountered in industry and the effects of water contact, which include fire, toxic releases, or highly exothermic vigorous reaction, are given in Table 10.2.[1]

Table 10.1 Hazardous processes

Materials	Subject to explosive reaction or detonation. Which react energetically with water or common contaminants. Subject to spontaneous polymerisation or heating.
Processes	Which are exothermic. Containing flammables and operated at high pressure or high temperature, or both. Containing flammables and operated under refrigeration. In which intrinsically unstable compounds are present. Operating in, or near, the explosive range of materials. Involving highly toxic materials. Subject to a dust, or mist, explosion hazard. With a large inventory of stored pressure energy.

Table 10.2 A selection of water-sensitive materials

(On contact with water these compounds give: flammable gases (F); toxic products (T); vigorous reactions (V).)

Material	F	T	V
Acetyl bromide		T	
Acetyl chloride		T	V
Acetylcholine bromide		T	
Aluminium (powder)	F		
Aluminium alkyls	F		V
Aluminium isopropoxide	F		
Aluminium lithium hydride	F		
Aluminium selenide		T	
Aluminium phosphide	F	T	
Boron tribromide		T	
Calcium (granules)	F		
Calcium carbide	F		
Calcium hydride	F		
Calcium phosphide	F	T	
Chlorosulphonic acid		T	V
Disulphur dichloride		T	V
Ethoxides, alkaline			V
Lithium (metal)	F		V
Lithium aluminium deuteride	F		
Lithium aluminium hydride	F		
Lithium borohydride	F		
Lithium hydride	F		
Lithium methoxide	F		
Magnesium (powder)	F		
Magnesium alkyls	F		
Magnesium phosphide	F	T	
Methoxides, alkaline	F		V
Nickel sulphide		T	
Phosphorus pentasulphide	F	T	
Phosphorus sesquisulphide	F	T	
Phosphorus pentachloride		T	
Phosphorus pentabromide		T	
Potassium (metal)	F		V
Potassium borohydride	F		
Potassium methoxide	F		
Silicon tetrachloride		T	V
Sodium (metal)	F		V
Sodium aluminium hydride	F		
Sodium borohydride	F	T e	
Sodium hydride	F		
Sulphur dichloride		T	V
Sulphuric acid, fuming (Oleum)		T	V
Sulphur tetrachloride		T	V
Sulphuryl chloride		T	V
Thionyl chloride		T	V
Titanium tetrachloride		T	V
Trichlorophenylsilane	T		
Trichlorosilane	F		
Zinc (powder)	F		
Zinc alkyls		T	V
Zirconium (powder)	F		

Fatal poisoning by arsine, AsH_3, occurred when a hot tin dross containing arsenic, produced during the pyrometallurgical refining of tin from cassiterite, was sprinkled with water. Similar cases have been reported when aluminium dross was sprayed with water.[2]

Because of the increasing sophistication of modern industry, undesirable concentrations of toxic gases may be generated from what may appear to be relatively innocuous operations.

Freshly machined spheroidal graphite cast iron (SG iron) may evolve phosphine due to the presence of trace quantities of hydrolysable magnesium phosphide. Therefore, when machining large surfaces of this 'ductile iron', as in roll-turning, it is advantageous to provide adequate ventilation, to draw dust and fumes away from the operator. Alternatively, machining operations, such as grinding, may be carried out wet using a solution of potassium permanganate.

Some chemicals such as potassium and calcium carbide produce flammable gas on contact with water:

$$2K + 2H_2O \rightarrow 2KOH + H_2$$
$$CaC_2 + 2H_2O \rightarrow Ca(OH)_2 + CH\vdots CH$$

On occasions the heat of hydrolysis is sufficient to ignite the gas evolved.

One pound of sodium hydride was being disposed of in a routine manner in a yard when it came into contact with water and exploded. This caused a fire which spread to the factory and resulted in 3 men requiring hospitalisation for treatment of burns.[3]

Water-sensitive chemicals pose a special hazard because of the possibility under factory conditions of exposure to process/cooling water, or condensate, or rainwater. Depending on the specific chemical, gases may be liberated which are either flammable, e.g. hydrogen and acetylene, or toxic.

An operator was engaged upon filling drums with phosphorus oxychloride; he had personally labelled these drums and inspected them internally for the presence of water or other contaminants. Six drums were filled without incident but when the seventh drum was almost full the contents erupted violently. The top of the drum was blown open and the weigh-scales destroyed; a vent line on the second floor was broken. The operator was fortunately on the opposite side of the drum to that ripped open but was exposed to phosphorus oxychloride and hydrogen chloride vapours, the latter being generated due to hydrolysis of the contents with water on the floor.

The most likely cause of this accident was reaction of the oxychloride with water. This water may have entered after inspection, e.g. due to condensate dripping from overhead water lines, or escaped detection on inspection.[4]

$$3H_2O + POCl_3 \rightarrow 3HCl + H_3PO_4$$

In some cases reaction with water may result in over-pressurisation of sealed equipment or pipework.

An operator was engaged in the transfer of thionyl chloride from a 250 litres drum via a 2.5 cm flexible, stainless steel hose and dip pipe assembly using vacuum. The hose and dip pipe assembly leaked and a replacement was obtained. When nothing drained from the replacement hose it was exchanged for the original hose and the dip pipe subsequently placed in the drum.

When the feed valve was opened the hose ruptured and the dip pipe pulled out of the drum; the operator was struck on the chest by the hose causing multiple injuries.

The cause was a violent reaction between thionyl chloride and aqueous material, present as a contaminant in the hose, generating sulphur dioxide and hydrogen chloride.[5]

$$SOCl_2 + H_2O \rightarrow 2HCl + SO_2$$

Precautions commonly followed with water-sensitive substances include:

- Storage and use such that accidental ingress of, or contact with, water is avoided, e.g.
  Covered storage, off the ground.
  Away from sprinkler systems, safety showers or overhead water lines or condensate lines.
  Away from water taps or sinks. (The roofs of storage areas should be regularly inspected and maintained to minimise leaks.)
- Storage under an inert medium (e.g. sodium is stored under paraffin). In this case stocks should be checked regularly to ensure that an adequate level of inert medium is maintained.
- Segregation of the material from other flammable materials, e.g. solvents or combustibles.
- Provision and use of appropriate eye/face protection; overalls and gloves.

There is a wide range of chemicals which are 'incompatible', in the sense that they can be brought into contact only under strictly controlled conditions since they result in the generation of highly toxic gases. Examples are given in Table 10.3(a). The obvious precautions with such chemicals are as described in Chapters 6 and 12, to label them correctly and to segregate storage to avoid accidental mixing. In addition it is, of course, necessary to take appropriate precautions to provide protection against the inherent properties of the individual substances.

A dyehouse foreman was exposed to chlorine gas when a solution of sodium hypochlorite was added, in error, to a barrel of dyestuff containing hydrochloric acid. He died in hospital 36 hours later.[6]

Three men including a bulldozer driver were affected by fumes on a waste disposal site; all suffered respiratory congestion. Drums containing sodium hypochlorite were emptied from one skip and drums containing an acid solution from another. The bulldozer crushed the drums and admixture of the chemicals resulted in chlorine evolution.[7]

There is also a wide variety of chemicals which are 'incompatible' since on mixing a violent reaction occurs leading in some circumstances to an explosion.

Table 10.3 Incompatible chemicals

(a) Toxic hazards
Substances in column A should be stored and handled so that they cannot possibly accidentally contact corresponding substances in column B because toxic materials (column C) would be produced.

Column A	Column B	Column C
Arsenical materials	Any reducing agent*	Arsine
Azides	Acids	Hydrogen azide
Cyanides	Acids	Hydrogen cyanide
Hypochlorites	Acids	Chlorine or hypochlorous acid
Nitrates	Sulphuric acid	Nitrogen dioxide
Nitric acid	Copper, brass, any heavy metals	Nitrogen dioxide (nitrous fumes)
Nitrites	Acids	Nitrous fumes
Phosphorus	Caustic alkalis or reducing agents	Phosphine
Selenides	Reducing agents	Hydrogen selenide
Sulphides	Acids	Hydrogen sulphide
Tellurides	Reducing agents	Hydrogen telluride

* Arsine has been produced by putting an arsenical alloy into a wet galvanised bucket.

(b) Reactive hazards
Substances in column A should be stored and handled so that they cannot possibly accidentally contact corresponding substances in column B under uncontrolled conditions, when violent reactions may occur.

Column A	Column B
Acetic acid	Chromic acid, nitric acid, hydroxyl-containing compounds, ethylene glycol, perchloric acid, peroxides, and permanganates
Acetone	Concentrated nitric and sulphuric acid mixtures
Acetylene	Chlorine, bromine, copper, silver, fluorine and mercury
Alkali and alkaline earth metals, such as sodium, potassium, lithium, magnesium, calcium, powdered aluminium	Carbon dioxide, carbon tetrachloride, and other chlorinated hydrocarbons. (Also prohibit water, foam, and dry chemical on fires involving these metals – dry sand should be available)
Ammonia (anhydrous)	Mercury, chlorine, calcium hypochlorite, iodine, bromine and hydrogen fluoride
Ammonium nitrate	Acids, metal powders, flammable liquids, chlorates, nitrites, sulphur, finely divided organics or combustibles
Aniline	Nitric acid, hydrogen peroxide
Bromine	Ammonia, acetylene, butadiene, butane and other petroleum gases, sodium carbide, turpentine, benzene, and finely divided metals
Calcium oxide	Water
Carbon, activated	Calcium hypochlorite

Table 10.3 (continued)

Column A	Column B
Chlorates	Ammonium salts, acids, metal powders, sulphur, finely divided organics or combustibles
Chromic acid and chromium trioxide	Acetic acid, naphthalene, camphor, glycerol, turpentine, alcohol, and other flammable liquids
Chlorine	Ammonia, acetylene, butadiene, butane and other petroleum gases, hydrogen, sodium carbide, turpentine, benzene and finely divided metals
Chlorine dioxide	Ammonia, methane, phosphine and hydrogen sulphide
Copper	Acetylene, hydrogen peroxide
Fluorine	Isolate from everything
Hydrazine	Hydrogen peroxide, nitric acid, any other oxidant
Hydrocarbons (benzene, butane, propane, gasoline, turpentine, etc.)	Fluorine, chlorine, bromine, chromic acid, peroxide
Hydrocyanic acid	Nitric acid, alkalis
Hydrofluoric acid, anhydrous (hydrogen fluoride)	Ammonia, aqueous or anhydrous
Hydrogen peroxide	Copper, chromium, iron, most metals or their salts, any flammable liquid, combustible materials, aniline, nitromethane
Hydrogen sulphide	Fuming nitric acid, oxidising gases
Iodine	Acetylene, ammonia (anhydrous or aqueous)
Mercury	Acetylene, fulminic acid*, ammonia
Nitric acid (conc.)	Acetic acid, acetone, alcohol, aniline, chromic acid, hydrocyanic acid, hydrogen sulphide, flammable liquids, flammable gases, and nitratable substances, paper, cardboard and rags
Nitroparaffins	Inorganic bases, amines
Oxalic acid	Silver, mercury
Oxygen	Oils, grease, hydrogen, flammable liquids, solids or gases
Perchloric acid	Acetic anhydride, bismuth and its alloys, alcohol, paper, wood, grease, oils
Peroxides, organic	Acids (organic or mineral), avoid friction, store cold
Phosphorus (white)	Air, oxygen
Potassium chlorate	Acids (see also chlorates)
Potassium perchlorate	Acids (see also perchloric acid)
Potassium permanganate	Glycerol, ethylene glycol, benzaldehyde, sulphuric acid
Silver	Acetylene, oxalic acid, tartaric acid, fulminic acid*, ammonium compounds

* Produced in nitric acid – ethanol mixtures.

Table 10.3 (continued)

Column A	Column B
Sodium	See alkali metals (above)
Sodium nitrite	Ammonium nitrate and other ammonium salts
Sodium peroxide	Any oxidisable substance, such as ethanol, methanol, glacial acetic acid, acetic anhydride, benzaldehyde, carbon disulphide, glycerol, ethylene glycol, ethyl acetate, methyl acetate and furfural
Sulphuric acid	Chlorates, perchlorates, permanganates

9 litres of chloroform were added to a 250-litre drum containing 45 litres of methanol and 11.3 litres of 50% sodium hydroxide; the objective was to deactivate any traces of toxic material present in the chloroform. There was an explosion in which the bottom was blown off the drum which was projected into the air and penetrated the ceiling and roof. Fortunately no one was injured. Subsequently a literature survey confirmed that a vigorous exothermic reaction results when chloroform is added to methanolic sodium hydroxide.[8]

Similar explosive reactions occurred when waste chloroform and acetone solvents were mixed in the presence of a base owing to the exothermic formation of 1,1,1-trichloro-2-hydroxy-2-methyl propane:

$$CHCl_3 + CH_3.\overset{\overset{\displaystyle O}{\|}}{C}.CH_3 \rightarrow Cl_3C\text{-}\underset{\underset{\displaystyle CH_3}{|}}{\overset{\overset{\displaystyle OH}{|}}{C}}\text{-}CH_3$$

For this reason waste halogenated solvents should not be mixed with other waste solvent.

A concentrated mixture of nitric acid and methanol is sometimes used to clean laboratory glassware contaminated by persistent organic material. This mixture, which releases nitrogen peroxide as the effective agent, is potentially explosive and should be used in very small quantities, e.g. by rinsing the glassware with methanol followed by a rinse in concentrated nitric acid so that <1 ml of the agent is present. In a modification of this technique a larger volume of the mixture was made up and stored in a 250 ml screw cap bottle. A spontaneous explosion resulted, with glass fragments being distributed over much of the laboratory and a fluorescent light above the bench on which the bottle was left being destroyed. (There was no fire so that presumably the reagent slowly produced gaseous nitrogen peroxide leading to a 'pressure' explosion.)

(In fact with the advent of modern detergent cleaning solutions the hazardous practice of using alcohol and nitric acid should be discouraged, as should the use of chromic acid solutions.)

Further examples are given in Table 10.3(b) and numerous incident reports confirm that relatively commonplace operations may result in 'incompatibility' problems.

A workshop engineer left a small synchronous motor to soak overnight in a beaker containing about 100 ml trichloroethylene in order to clean the movement. Next morning the exposed steel parts of nearby machinery were covered with a fine film of rust and the equipment in the beaker was covered in a 'tar-like' substance, clearly due to a rapid chemical reaction. The incident was explained by the presence of a single aluminium alloy gearwheel in the equipment.[9]

Some commercial grades of trichloroethylene and trichloroethane contain inhibitors to retard this type of reaction but these may be removed by distillation, e.g. in de-greasing vats.

The problem of incompatibility of chemicals cannot be dealt with in many cases merely by reference to lists. What is required is a detailed appraisal of all the chemicals which may be present, even if not intended, and how they will react under the most extreme conditions of concentration, temperature, pressure, etc.

It has been found that formaldehyde reacts with hydrogen chloride in the presence of moisture to yield appreciable amounts of bis-chloro methyl ether. The reaction is fast and occurs both in the vapour phase and in solution. A similar reaction takes place with metal chlorides (Friedel–Crafts) catalysts.[10] Bis-chloro methyl ether is carcinogenic and therefore this hazard should be emphasised whenever formaldehyde is used in industry, or in laboratories.

Pyrophoric chemicals

Materials which are so reactive that on contact with air, or its moisture, they undergo exothermic oxidation/hydrolysis at such a rate as to cause ignition were listed as pyrophoric chemicals in Table 4.5.

Typical measures taken to reduce the hazard with pyrophoric materials include:

- Handling and storing the minimum quantities necessary at any time.
- Segregation of the material from other chemicals, particulary 'fuels', i.e. solvents, paper, cloth, etc.
- Handling in dry, chemically inert atmospheres or beneath other appropriate media, e.g. dry oil.
- Handling in solution (e.g. aluminium alkyls in petroleum solvents).
- Immediate destruction and removal of spilled materials.
- Careful selection and provision of appropriate fire extinguishers in advance.
- Provision and use of appropriate eye/face protection, overalls and gloves.

Thermodynamically unstable compounds

A different range of chemicals may undergo violent self-reaction or decomposition catalysed by mechanical shock, friction, heat, light or chemical impurities. Such compounds are characterised by the presence of groups such as
—C≡C (acetylide), $—N{=}\overset{+}{N}{=}\overset{-}{N}$ (azide), $—ClO_3$ (chlorate), —N=N— (diazo), O—N=C (fulminate), =N—Cl (N-chloroamine), —O—OH (hydroperoxide), —O—X (hypohalite), $—O—NO_2$ (nitrate ester), —O—NO (nitrite), $—NO_2$ (nitro), —NO (nitroso), —C—O—C— (ozonide, the two C atoms also joined through O——O), —C(=O)—O—OH (peracid), $—ClO_4$ (perchlorate), —O—O— (peroxide).
Common examples are included in Table 10.4. Problems with peroxides are described in Chapters 4 and 11.

Acetylene, acetylides and acetylenic compounds

Acetylene and acetylenic compounds are notoriously unstable. The presence of the endothermic acetylene group confers explosive instability over a wide range of acetylene compounds. Particularly sensitive classes of these compounds are:

- Those containing halogens.
- Derivatives of metals, especially heavy metals.

Acetylides are among the few explosives containing neither oxygen nor nitrogen. They are very sensitive to shock, heat and friction and explode readily. Their explosion produces no gas but is simply an effect arising from the large amount of heat instantaneously produced. They are in a class together with the fulminates and the azides, as primary detonants.

Because acetylides are so sensitive to shock, extreme care is essential in their handling. They should be kept cool and, if they are to be stored, they should be kept wet. Metal powders such as finely divided copper and silver can form very sensitive acetylides with acetylene or other acetylides, which although not dangerous in themselves, can cause a flash sufficient to ignite explosive mixtures of gases.

Nitrogen compounds

Azides comprise a group of chemicals characterised by the formula $R(N_3)_x$. R may be a metal, halogen, complex radical, or an organic moiety. They are thermally unstable: all explode when heated and many are shock sensitive. Some metal azides and metal azide halides are extremely explosive. The presence of NO_2 groups increases the explosive tendency. Hydrogen azide is a very sensitive explosive with a heat of explosion comparable with that of nitroglycerine. Metal azides are thermally unstable in the range 100–200 °C and at higher temperatures may explode.

Solutions of ammoniacal silver nitrate, used in conjunction with a reducing agent (e.g. glucose solution) for silvering glass, tend to form silver azide on standing. This may result in an explosion after only a few hours following solution preparation. Therefore, the correct procedure must be followed in preparation of the solution and within an hour or so of preparation any unused solution should be treated with sodium chloride to precipitate any silver as silver chloride.[11]

A flask of 'silvering' solution was left overnight and deposited silver azide crystals. When a technician agitated the flask he narrowly escaped injury when the contents decomposed violently.[12]

Commercially available diazo compounds are not generally shock-sensitive but practically all will decompose exothermically, and even explosively, on heating. They are also sensitised by many other compounds. Diazomethane, made from *N*-methyl-*N*-nitrosotoluene-4-sulphonamide – a precursor commercially available – is extremely toxic and will undergo violent decomposition. If its vapour is heated above 200 °C, it may explode violently. Explosions can occur at low temperatures owing to traces of organic matter. When working with diazomethane in any form (gas, liquid or solution), all possible precautions must be taken to obviate the risk of explosion – with implications to its storage. Excess material should be destroyed daily and not stored overnight.

Fulminates are a group of compounds which are extremely sensitive to impact and friction when dry, and should be kept moist until ready for use. Fulminates are subject to deterioration under warm conditions. When they detonate, they decompose completely – with great violence. A selection is given in Table 10.4.

Concentrated nitric acid is a very powerful oxidising agent. It can explode on contact with powerful reducing agents. Steps should be taken to avoid the storage of nitric acid with reducing agents and other 'incompatible' compounds (see Table 10.3).

Nitrates are nearly all powerful oxidising agents. All inorganic nitrates act as oxygen carriers and, under favourable conditions, they can give up their oxygen to other materials which, in turn, may detonate.

Ammonium nitrate has all the properties of the other nitrates but is also able to detonate by itself, under certain conditions. It is, therefore, a high explosive – although it is relatively insensitive to impact. In the pure state, it requires a combination of an initiator and a high explosive. Ammonium nitrate in combination with other nitro compounds forms a major high explosive for military use. It should be stored in a cool, well-ventilated place, away from acute fire hazards and easily oxidisable materials. It must not be confined because in the event of fire, violent detonation can take place (see Ch. 15).

Aromatic nitro compounds (listed in Table 10.4) are flammable. Many

Table 10.4 Selection of potentially explosive compounds

(a) Peroxy compounds

(i) Organic peroxy compounds

Acetyl cyclohexane-sulphonyl peroxide (70%)
Acetyl cyclohexane-sulphonyl peroxide (28% phthalate solution)
o-Azidobenzoyl peroxide

t-Butyl mono permaleate (95% dry)
t-Butyl peracetate (70%)
t-Butyl peractoate
t-Butyl perpivalate (75% hydrocarbon solution)
t-Butyl peroxy isobutyrate
Bishexahydrobenzoyl peroxide
Bis-monofluorocarbonyl peroxide
Bis-benzenesulphonyl peroxide
Bis-hydroxymethyl peroxide
Bis (1-hydroxycyclohexyl) peroxide
2,2-Bis (t-butylperoxy) butane
2,2 -Bis-hydroperoxy diisopropylidene peroxide
Barium methyl peroxide
Benzene triozonide

Cyclohexanone peroxide (95% dry)

Diacetyl peroxide
Di-n-butyl perdicarbonate (25% hydrocarbon solution)
2:4-Dichlorobenzoyl peroxide (50% phthalate solution)
Dicaproyl peroxide
Dicyclohexyl perdicarbonate
Di-2-ethylhexyl perdicarbonate (40% hydrocarbon solution)
Dimethyl peroxide
Diethyl peroxide
Di-t-butyl-di-peroxyphthalate
Difuroyl peroxide
Dibenzoyl peroxide
Dimeric ethylidene peroxide
Dimeric acetone peroxide
Dimeric cyclohexanone peroxide
Diozonide of phorone
Dimethyl ketone peroxide

Ethyl hydroperoxide
Ethylene ozonide

Hydroxymethyl methyl peroxide
Hydroxymethyl hydroperoxide
1-Hydroxyethyl ethyl peroxide
1-Hydroperoxy-1-acetoxycyclodecan-6-one

Isopropylpercarbonate
Isopropyl hydroperoxide

Methyl ethyl ketone peroxide
Methyl hydroperoxide
Methyl ethyl peroxide
Monoperoxy succinic acid

Nonanoyl peroxide (75% hydrocarbon solution)
1-Naphthoyl peroxide

Oxalic acid ester of t-butyl hydroperoxide
Ozonide of maleic anhydride

Phenylhydrazone hydroperoxide
Polymeric butadiene peroxide
Polymeric isoprene peroxide
Polymeric dimethylbutadiene peroxide
Polymeric peroxides of methacrylic acid esters and styrene
Polymeric peroxide of asymmetrical diphenylethylene
Peroxyformic acid
Peroxyacetic acid
Peroxybenzoic acid
Peroxycaproic acid
Polymeric ethylidene peroxide

Sodium peracetate
Succinic acid peroxide (95% dry)

Trimeric acetone peroxide
Trimeric propylidene peroxide
Tetraacetate of 1,1,6,6-tetrahydroperoxycyclodecane

(ii) Inorganic peroxy compounds

Peroxides
Hydrogen peroxide (> 30%)
Mercury peroxide

Peroxyacids
Peroxydisulphuric acid
Peroxynitric acid
Peroxy ditungstic acid

Peroxyacid salts
Sodium peroxyborate (anhydrous)
Sodium triperoxychromate
Sodium peroxymolybdate
Sodium peroxynickelate
Sodium diperoxytungstate
Potassium peroxyferrate
Potassium peroxynickelate
Potassium hyperoxytungstate
Potassium peroxy pyrovanadate
Calcium diperoxysulphate
Calcium peroxychromate
Zinc tetraaminoperoxydisulphate
Ammonium peroxyborate
Ammonium peroxymanganate
Ammonium peroxychromate

Table 10.4 (continued)

Superoxides
Potassium superoxide
Ozone (liquid > 30%)
Potassium ozonide
Cesium ozonide
Ammonium ozonide

Inorganic peracids and their salts (common examples which are particularly hazardous)
Ammonium perchlorate
Ammonium persulphate
Ammonium pernitrate
Perchloric acid (>73%)
Performic acid
Silver perchlorate
Tropylium perchlorate

(b) Halo-acetylenes and acetylides

Lithium bromoacetylide
Dibromoacetylene
Lithium chloroacetylide
Sodium chloroacetylide
Dichloroacetylene
Bromoacetylene
Chloroacetylene
Fluoroacetylene
Diiodoacetylene
Silver trifluoromethylacetylide
Chlorocyanoacetylene
Lithium trifluromethylacetylide
3,3,3-Trifluoropropyne
1-Bromo-2-propyne
1-Chloro-2-propyne
1-Iodo-1,3-butadiyne
1,4-Dichloro-2-butyne
1-Iodo-3-penten-1-yne
1,6-Dichloro-2,4-hexadiyne
2,4-Hexadiynylene bischlorosulphite
Tetra (chloroethynyl) silane
2,4-Hexadiynylene bischloroformate
1-Iodo-3-Phenyl-2-propyne
1-Bromo,-1,2-cyclotridecadien-4,8,10-triyne

(c) Metal acetylides

Disilver acetylide
Silver acetylide-silver nitrate
Digold(I) acetylide
Barium acetylide
Calcium acetylide (carbide)
Dicaesium acetylide
Copper(II) acetylide
Dicopper(I) acetylide
Silver acetylide
Caesium acetylide
Potassium acetylide
Lithium acetylide
Sodium acetylide
Rubidium acetylide
Lithium acetylide-ammonia
Dipotassium acetylide
Dilithium acetylide
Disodium acetylide
Dirubidium acetylide
Strontium acetylide
Silver trifluoromethylacetylide
Sodium methoxyacetylide
Sodium ethoxyacetylide
1,3-Pentadiyn-1-ylsilver
1,3-Pentadiyn-1-ylcopper
Dimethyl-1-propynylthallium
Triethynylaluminium
Triethynylantimony
Bis(Dimethylthallium) acetylide
Tetraethynylgermanium
Tetraethynyltin
Sodium phenylacetylide
3-Buten-1-ynyldiethylaluminium
Dimethyl-phenylethynylthallium
3-Buten-1-ynyltriethyllead
3-Methyl-3-buten-1-ynyltriethyllead
3-Buten-1-ynyldiisobutylaluminium
Bis(triethyltin) acetylene

(d) Metal azides

Aluminium triazide
Barium diazide
Boron triazide
Cadmium diazide
Calcium diazide
Chromyl azide
Copper(I) azide
Copper(II) azide
Lead(II) azide
Lead(IV) azide
Lithium azide
Lithium boroazide
Mercury(I) azide
Mercury(II) azide
Potassium azide
Silicon tetraazide
Silver azide

(e) Metal azide halides

Chromyl	Azide	Chloride
Molybdenum	Azide	Pentachloride
Molybdenum	Diazide	Tetrachloride

Table 10.4 (continued)

Silver	Azide	Chloride
Tin	Azide	Trichloride
Titanium	Azide	Trichloride
Tungsten	Azide	Pentabromide
Tungsten	Azide	Pentachloride
Uranium	Azide	Pentachloride
Vanadium	Azide	Dichloride
Vanadyl	Azide	Tetrachloride

(f) Diazo compounds

1,1 Benzoylphenyldiazomethane
2-Butan-1-yl diazoacetate
t-Butyl diazoacetate
t-Butyl 2-diazoacetoacetate
Diazoacetonitrile
2-Diazocyclohexanone
Diazocyclopentadiene
1 Diazoindine
*Diazomethane
Diazomethyllithium
Diazomethylsodium
Dicyanodiazomethane
Dinitrodiazomethane
Isodiazomethane
Methyl diazoacetate
* The precursor of this compound (N-Methyl-N-nitroso-toluene-4-sulphonamide) is available commercially.

(g) Metal fulminates

Cadmium fulminate
Copper fulminate
Dimethylthallium fulminate
Diphenylthallium fulminate
Mercury(II) methylnitrolate
Mercury(II) formhydroxamate
Mercury(II) fulminate
Silver fulminate
Sodium fulminate
Thallium fulminate

(h) Nitro compounds

(i) C-Nitro compounds

4-Chloro-2,6-dinitroaniline
2-Chloro-3,5-dinitropyridine
Chloronitromethane
1-Chloro-2,4,6-trinitrobenzene (picryl chloride)
Dinitroacetonitrile
2,4-Dinitroaniline
Dinitroazomethane
1,2-Dinitrobenzene
1,3-Dinitrobenzene
1,4-Dinitrobenzene
2,4-Dinitrobenzenesulphenyl chloride
3,5-Dinitrobenzoic acid
3,5-Dinitrobenzoyl chloride
2,6-Dinitrobenzyl bromide
1,1-Dinitro-3-butene
2,3-Dinitro-2-butene
3,5-Dinitrochlorobenzene
2,4-Dinitro-1-fluorobenzene
2,6-Dinitro-4-perchlorylphenol
2,5-Dinitrophenol
2,4-Dinitrophenylacetyl chloride
2,4-Dinitrophenylhydrazine
2,4-Dinitrophenylhydrazinium perchlorate
2,7-Dinitro-9-phenylpheanthridine
2,4-Dinitrotoluene
1-Fluoro-2,4-Dinitrobenzene
4-Hydroxy-3,5-Dinitrobenzene arsonic acid
1-Nitrobutane
2-Nitrobutane
1-Nitro-3-butene
Nitrocellulose
1-Nitro-3 (2.4-dinitrophenyl) urea
Nitroethane
2-Nitroethanol
Nitroglycerine
Nitromethane
1-Nitropropane
2-Nitropropane
5-Nitrotetrazole
Picric acid (2,4,6-trinitrophenol)
Potassium-4,6-dinitrobenzofuroxan hydroxide complex
Potassium-3,5-dinitro-2(1-tetrazenyl) phenolate
Potassium trinitromethanide ('Nitroform' salt)
Sodium 5-dinitromethyltetrazolide
Tetranitromethane
N,2,4,6-Tetranitro-N-methylaniline (tetryl)
Trichloronitromethane (chloropicrin)
2,2,4-Trimethyldecahydroquinoline picrate
Trinitroacetonitrile
1,3,5-Trinitrobenzene
2,4,6-Trinitrobenzenesulphonic acid (picryl sulphonic acid)
Trinitrobenzoic acid
2,2,2-Trinitroethanol
Trinitromethane
2,4,6-Trinitrophenol (picric acid)
2,4,6-Trinitroresorcinol
2,4,6-Trinitrotoluene (TNT)
2,4,6-Trinitro-*m*-Xylene

Table 10.4 (continued)

(ii) N-Nitro compounds	***(i) Reactive vinyl monomers***
1-Amino-3-nitroguanidine	Acrylic acid
Azo-N-nitroformamidine	Acrylonitrile
1,2-Bis(difluoroamino)N-nitroethylamine	*n*-Butyl acrylate
N,N′ Diacetyl-N,N′-dinitro-1,2-diaminoethane	*n*-Butyl methacrylate
N,N′ Dinitro-1,2-diaminoethane	4-Chlorostyrene
N,N′ Dinitro-N-methyl-1,2-diaminoethane	Divinyl benzene
1-methyl-3-nitro-1-nitrosoguanidine	Dodecyl methacrylate
Nitric amide (nitramide)	Ethyl acrylate
1-Nitro-3(2,4-dinitrophenyl) urea	Ethylene dimethacrylate
Nitroguanidine	2-Hydroxypropyl methacrylate
N-Nitromethylamine	Methyl acrylate
Nitrourea	Methyl methacrylate
N,2,4,6-Tetranitro-N-methylaniline (tetryl)	α-Methyl styrene
1,3,5,7-Tetranitroperhydro-1,3,5,7-tetrazocine	Methyl vinyl ether
	Styrene
	Vinyl acetate
	Vinyl bromide
(iii) Ammonium nitrate	

of their di- and tri-derivatives are explosive under favourable conditions. Dinitro compounds may form shock-sensitive nitrolium salts. Contact with strong reducing agents must be avoided. Picric acid is an important member of the group of hazardous aromatic nitro compounds and has been used extensively as a high explosive. It is able to corrode metals and its metallic salts are extremely sensitive to shock and heat. The acid and its hazardous derivatives should, if possible, be replaced by substances which are innocuous or potentially less harmful. Direct contact with these materials should be avoided. Picric acid and its flammable derivatives should be stored in a cool area, away from acute fire hazards with larger stocks preferably in an isolated or detached building. Picric acid itself must be stored under water.

Aliphatic nitro compounds can constitute fire and explosion hazards when exposed to high temperatures, flames or impact. Spontaneous exothermic reactions may take place with these compounds. Cellulose nitrate presents a serious hazard in most of its industrial forms. When the nitrogen content is high, it is classed as an explosive. Guncotton has a nitrogen content of 12.2–13.8% and thus nearly corresponds to cellulose trinitrate for which the theoretical nitrogen content is 14.16%. Less highly nitrated material containing 10.5–12.0% nitrogen is known as pyroxylin and approximates to a dinitrate – theoretical content 11.13%. One commercial grade of pyroxylin has a declared nitrogen content of 11.8–12.2% and is moistened with alcohol to ensure safety in transit.

A selection of aliphatic nitro compounds is included in Table 10.4.

Peroxy compounds

These compounds are characterised by —O—O— bond grouping, e.g.

	—O—O— metal	:	metal peroxides, peroxoacid salts
	—O—O— non-metal	:	peroxoacids
and	—O—X—	:	alkyl perchlorates, perchloric acid, perchloryl compounds, etc.

Mixtures of perchloric acid with oxidisable materials can be more hazardous, with regard to sensitivity to impact and shock, than many explosives. Sodium hypochlorite solution (a domestic bleach) can on heating produce sodium chlorite by disproportionation with attendant hazards:

After accidently burning the contents of a saucepan, a housewife cleaned the pan by boiling with sodium hypochlorite. Again the pan boiled dry and as a result of reaction between the chlorate and the charred remains it exploded violently causing considerable damage.[13] While the housewife could not have been in a position to foresee the consequences of her action, this demonstrates the dangers of using products for purposes for which they were not designed or marketed.

Organic peroxides can be classified as shown in Table 10.5.

Performic acid is treacherously unstable (due to its low energy of activation). A sample at −10 °C is known to have exploded when moved. TNT is relatively stable by comparison. Perchlorates have exploded when being pulverised with a mortar, some detonating when being stirred. All perchlorates have some potential for hazard when in contact with other reactive materials.

Many peroxides are sensitive to friction or even slight mechanical shock. Although some are non-flammable, others may burn fiercely in the presence of fire and explode.

Extreme caution is required in the storage and use of chemicals which can form peroxides or hydroperoxides. Common peroxidisable compounds, their hazard categories and a scheme for their identification by labelling are given in Table 11.20. Alkali metals, their alkoxides and amides may also form peroxides.

Some of these peroxides may explode even without concentration. The volatile compounds create a serious hazard since vapour near vessel outlets, or the screwed cap of a bottle or drum, may be in contact with air; the threads of a cap are particularly vulnerable since friction may initiate an explosion.

1,3-benzodithiolylium perchlorate was in preparation using a method described in the literature. A severe explosion occurred during drying in a desiccator and a post-graduate student was badly injured. The explosion

Table 10.5 Classification of organic peroxides (See also Table 10.4)

Peroxide class	General structures or characteristic group
Hydroperoxides	ROOH $R_mQ(OOH)_n$ (Q = metal or metalloid)
α-Oxy- and α-peroxy-hydroperoxides and peroxides	contain the grouping: $>C(OO^-)(O^-)$
Peroxides	ROOR′ $R_mQ(OOR)_n$ R_mQOOQR_m
Peroxyacids	$R(CO_3H)_n$ RSO_2OOH
Diacyl peroxides	RC(=O)OOC(=O)R′ ROC(=O)OOC(=O)OR RS(=O)(=O)OOC(=O)R RS(=O)(=O)OOS(=O)(=O)R RC(=O)OOC(=O)OR′
Peroxyesters	$R(CO_3R')_n$ $R'(O_3CR)_2$ ROC(=O)OOR′ ROOC(=O)OOR >NC(=O)OOR RS(=O)(=O)OOR′

was initiated when the desiccator lid was being removed because a small amount of the compound was deposited on the ground-glass flange.[14]

About 2 litres of tetrahydrofuran had been used in an instrument and were being distilled for recovery when an explosion occurred. This was followed by fire causing extensive damage. The cause was identified as an organic peroxide derived from tetrahydrofuran which had been exposed to air. Commercial materials were treated with inhibitors to prevent peroxidation; it was presumed that either the original use, or subsequent exposure to air, resulted in removal or deactivation of the inhibitor.

Organic peroxides are unstable. Rates of decomposition can be accelerated by trace impurities, reducing agents (including rags, paper, etc.) and strong acids and alkalis, in addition to heat. Often, the intense heat generated during rapid decomposition can damage the container, releasing flammable vapours. These compounds are used as oxidising agents, initiators in polymerisation, cross-linking reactions, etc. Peroxy compounds can cause severe injuries to the eyes. They require special attention to be given to their storage temperatures in order to reduce decomposition rates drastically. Ideally, the main stocks of peroxides should be stored under stringent conditions in suitably designed outbuildings with appropriately lurid signs.

Special precautions are also necessary in the transport and distribution of such chemicals.

A lorry driver whose load of methyl ethyl ketone peroxide was giving off smoke drove off the main road in West Bromwich, UK, and on to waste ground. He telephoned the fire brigade from a public house nearby. As he was returning he warned pedestrians to keep clear. The whole load exploded causing damage estimated to be equivalent to 500 pounds of TNT; 42 houses were badly damaged, 160 houses damaged and 28 adults and 4 children injured.

The labelling, segregation and transport of chemicals is discussed in Chapter 13.

Vinyl monomers

Many vinyl monomers can undergo spontaneous polymerisation. Divinyl benzene and acryloyl chloride have been known to polymerise spontaneously and explosively on storage, accompanied by the destruction of their containers.

This class of compound requires strict control with respect to storage and handling and requires the presence of stabilisers to render them 'safe'. It is essential that certain vinyl monomers are not stored under chilled conditions otherwise the pure monomer may separate from its diluent or stabiliser. However, many vinyl monomers are safely polymerised, to produce a range of useful polymers, without recourse to unusual control measures. Typical examples of reactive monomers are given in Table 10.4.

Precautions which may be adopted to reduce the hazards inherent in handling potentially explosive chemicals include:

- Working in specially designed facilities remote from other buildings, accessways or populated areas. Examples are:
  - For large-scale operations, remote handling – using remote control and possibly closed circuit TV monitoring – in concrete outbuildings or bunkers.
  - For laboratory-scale operations, handling of materials in armour plated fume-cupboards behind robust, polycarbonate plastic blast screens. Use of appropriate personal protection, e.g. face shield and thick leather gloves. Pickling of glass apparatus in nitric acid and thorough rinsing after use.
- Storage of minimum practicable quantities of material in specially designed containers in no-smoking areas. Several small containers are preferable to one larger container. Storage should be segregated from all other chemicals.
- Storage under chilled conditions, except where this will cause separation of pure material (e.g. acrylic acid).
- Where practicable, keeping samples dilute, or damp, and avoiding the formation of large crystals.
- Addition of stabilisers where possible, e.g. to vinyl monomers (see page 527).

Some of the oxygen-rich compounds listed in Table 10.4 pose a fire hazard because of their inherent oxygen content and, while not spontaneously combustible, they can facilitate the ignition of reducing agents (e.g. perchlorates in contact with wood – see Ch. 11, page 552) or support combustion in otherwise oxygen-deficient atmospheres. Precautions generally include:

- Handling and storing the minimum quantities practicable for the process or experiments in progress.
- Segregation of the materials from other chemicals particularly reducing agents, paper, straw, cloth or materials of low flashpoint.
- Handling in the dilutest possible form in clearly designated areas, away from potential ignition sources.
- Provision and use of appropriate eye/face protection, overalls and gloves.

It must be emphasised that the lists given in Tables 10.2 to 10.5 are of the commoner chemicals in the respective classes. However, it must *not* be assumed that because a specific chemical is not included in any list that it is non-hazardous; in any case of doubt reference should be made to the supplier and to the literature, for which details are given in Chapter 18. References 1 and 15 at the end of the present chapter are particularly useful for reactive hazards.

Characteristics of chemical reactions

Reaction characteristics may involve:

- Reaction in gas, liquid or solid phase (either neat or in solution, suspension or emulsion).
- Catalytic or non-catalytic.
- Exothermic or endothermic (or negligible heat loss/gain).
- Reversible or irreversible.
- First- or second-order or complex kinetics.

In practice a variety of reactors are used, viz.:
Batch, e.g. for synthetic resins, fermenters, drugs.
Continuous
– Tubular } with/without a packed bed
– Tubes in shell }
– Continuous stirred tank
– Fluidised bed.

The continuous designs tend to predominate in large-scale production and have the advantage of low inventory and reduced variation of operating variables, i.e. batch operation is an unsteady-state procedure.

> Several advantages accrue from carrying out nitration processes continuously rather than batch-wise.[16] The smaller reaction vessels allow more uniform and efficient agitation than in the much larger batch vessels and the steady-state continuous process can be much better controlled and regulated. The evolution of heat is uniform, and the rate is much lower than the peak values in the batch process, which eases removal. Moreover the inventory of potentially hazardous reaction material present at any time is considerably reduced.

Recycle of reactants, or products, or of diluent is common in continuous reactors; this may be in conjunction with heat removal in an external exchanger but, in any event, is an important means of controlling the progress of the reaction.

Numerous collections of case histories have been published which highlight the need for comprehensive physical, chemical and thermal data, data on the effect of corrosion products and impurities, and data on thermal stability in order to minimise hazards associated with chemical reactors.[17]

Some causes of runaway in reactors or storage tanks are shown in Fig. 10.1.[18] However, experience has shown that a dangerous runaway reaction is most likely to occur in those processes in which all the reactants are initially mixed together with any catalyst in a batch reactor and

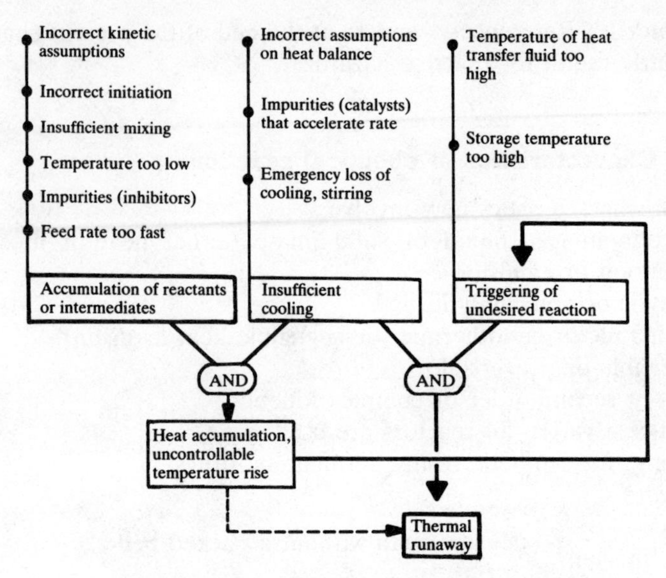

Fig. 10.1 Some causes of runaway in reactors or storage tanks[18]

where heat is supplied to start the reaction. Thus the condensation reaction between phenol and formaldehyde, using either acid- or alkali-type catalysts to produce synthetic resins, has frequently resulted in explosions.

Many chemical processes involve a high degree of technology and require equipment designed to rigid specifications coupled with sophisticated automatic control and safety devices. With some reactions which are difficult to control it is particularly important to provide protection against failure of cooling media, agitation, control or safety instrumentation, etc., as discussed in Chapter 12. Moreover, the reactor itself must be adequately designed for the operating conditions of, e.g. pressure, temperature, corrosive environment.

In July 1976 a bursting disc ruptured on a reactor at a chemical works near Milan.[19, 20] The reactor was in use to manufacture trichlorophenol at a temperature of 170 °C to 185 °C; it was heated by steam at 190 °C. It had been assumed that the reactants could not attain the runaway temperature of 230 °C, at which the hyper-poison TCDD (tetrachlorodibenzo dioxin) would be produced as a by-product.[21] The reactor was listed as working at atmospheric pressure and the bursting disc, rated at 3.5 bar, was to protect it from over-pressure during transfer of the contents using air pressure[22]; it vented directly to atmosphere just above the roof of the single-storey, reactor building.[22]

At about 6 am on the day of the accident, a Saturday, the reactor was shut down before the aciduration stage which released the trichlorophenol product. All external power was shut off including the temperature recorder which indicated 160 °C. The exact cause of an ensuing exothermic reaction is unproven[22]; the reactor contents are estimated to have reached 300 °C before the disc ruptured. Approximately half the reactor contents escaped in 20 minutes. Estimates of the amount of TCDD released vary from 0.25 to 3.0 kg.[22]

In this case safeguards against over-heating included provision for cooling using the steam coils, for dumping 3,000 litres of cold water into the reactor, and for using the reflux condenser. These all required manual actuation but even if they had been automatic they would have remained inoperative with the power off.[19]

In some cases conditions may be extreme, e.g. temperatures exceeding 500 °C or approaching absolute zero, so that construction requires special steels or other metals/alloys. High pressure or high vacuum operation may also be a factor.

The reaction rate–temperature characteristics commonly encountered with chemical reactions are illustrated by the curves in Fig. 10.2 namely[23]:

(a) Rapid increase in rate with increasing temperature. Normal characteristic.
(b) Slow increase in rate with increasing temperature. Characteristic of some heterogeneous reactions controlled by interphase transfer by diffusion.
(c) Very rapid increase at one point, i.e. the ignition point in an explosion.

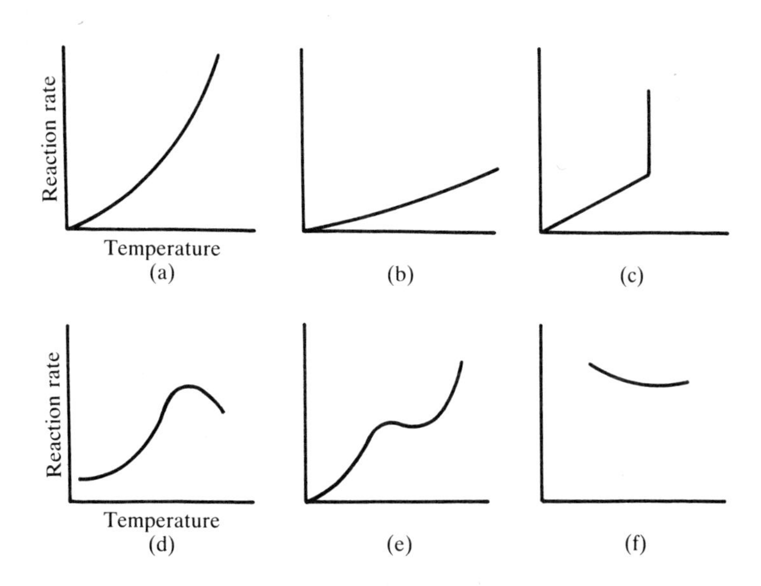

Fig. 10.2 Types of reaction rate versus temperature curves.

(d) Decrease in rate at higher temperatures. Characteristic of catalytic reactions.
(e) Reaction rate complicated by side reactions as temperature is increased.
(f) Rate decrease with temperature (e.g. reaction between nitric oxide and oxygen).

Clearly to avoid runaway conditions, i.e. as in (a), or an explosion, as in (c), it is essential to control carefully the rate of addition of reactants and temperature and eliminate any possible confusion in process control.

Confusion of the quantities of raw materials fed into a 3.4 m^3 reactor resulted in side reactions taking place. One of these involved polymerisation of sodium dichlorophenate which would be highly exothermic. Increased pressure caused the reactor to explode; some of the asbestos roofing and cladding was ripped from the building but no one was injured. Extra precautions including the provision of an adequate cooling system and a rupture disc were recommended.[24]

It is also necessary that the kinetics and thermodynamics of the reaction, and possible side reactions, are fully understood.

An explosion with subsequent fire occurred in a 4,000-litre glass-lined steel vessel. It contained 450 kg of anhydrous aluminium chloride in 1,000 kg of nitrobenzene for use in a Friedel–Craft synthesis in a different vessel. There was a breakdown in the steam supply to the jacket; the inlet valve was left open and when the supply was restored the contents were raised to some 'unknown' temperature. Subsequent investigation of the nitrobenzene–$AlCl_3$ system, and determination of the mechanism of decomposition and kinetic constants of the various reactions involved, led to the conclusion that this system is thermally unstable at any concentration. Safe handling for a relatively short period of time requires provision of adequate cooling facilities and accurate temperature control.[25]

An explosion in a conical drier blender in 1976 killed one man and caused an estimated £1.75 m. damage to plant. At the time of the explosion, about 1,300 kg of a poultry food additive had remained in the closed drier for a period of 27 hours after the drying process was completed, at a temperature maintained between 120 °C and 130 °C by the drier insulation.[26]

The explosion was caused by thermal decomposition although tests using differential thermal analysis (DTA) had shown that the product exhibited exotherms at 274 °C and 284 °C, i.e. temperatures >130 °C above those attained during drying. Other tests including shock sensitivity, flammability and thermodynamic computations had also failed to identify a potential severe hazard with the material. Afterwards testing with the new technique of accelerating rate calorimetry (ARC) showed that a typical batch could self-heat to destruction if held under adiabatic conditions at 120 °C to 125 °C for 24 hours.[26]

Hence manufacturers and users of substances liable to exothermic reaction should appraise their explosive potential in near adiabatic conditions and be aware of new techniques for measurement of explosion potential.[27]

Scale-up of reactions from a laboratory scale to a pilot-plant scale, and thence to a commercial scale, may often introduce heat transfer problems as discussed in Chapter 3. Modifications in reactor geometry, agitation arrangements, charging procedures, or control must also be introduced with care.

A new product was produced on a laboratory scale by a rapid, strongly exothermic isomerisation reaction in the presence of a catalyst. A glass reactor of the type shown in Fig. 10.3 was adapted to repeat this on a pilot scale; feed was to be introduced in small increments after the initial charge had reacted.[28]

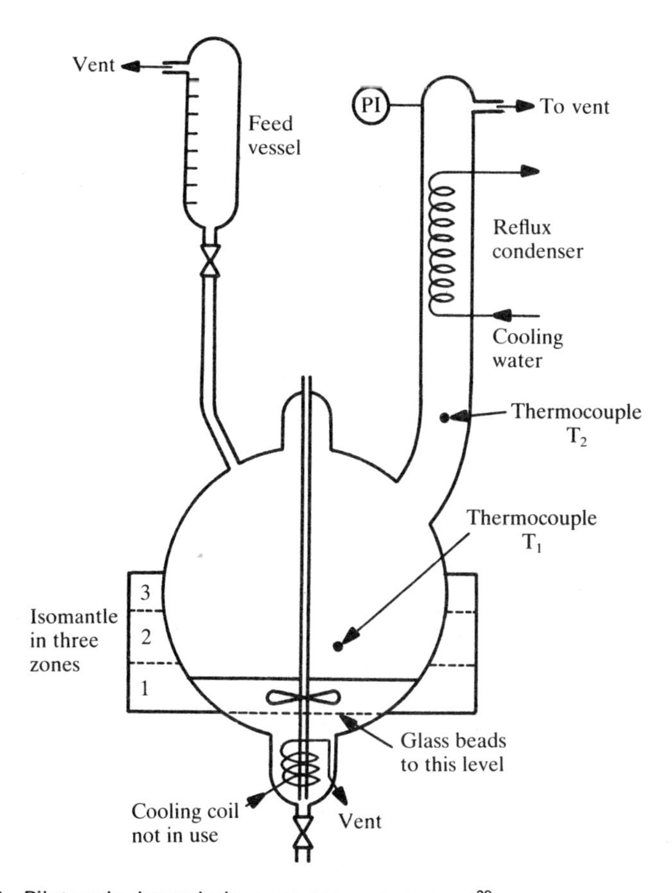

Fig. 10.3 Pilot-scale, heated glass reactor arrangement[28]

The arrangement had an approximately 5 litre volume beneath the isomantle, from which a pumping stirrer drew liquid into the bulk. However, the 10 litres required to provide a reasonable level was considered too large. Therefore the base of the vessel was filled with glass beads which immobilised the stirrer.

A 5 litre charge of reactant, to a level just above the glass beads, was warmed up with catalyst by switching on the lowest zone of the isomantle. When the vapour space temperature, indicated by the thermocouple T_1, reached 60 °C there was an internal vapour explosion which pushed apart the reactor base flange.

The reactant had a flash point of 28 °C, and an auto-ignition temperature of 250 °C, and the incident was caused by auto-ignition of vapour by the hot glass surface above the liquid level. Subsequent modifications involved use of a heel to ensure that the liquid level completely covered the isomantle zones in use, use of the stirrer, monitoring of the liquid temperature; these proved successful.[28]

(Such isomantles may reach surface temperatures of 300 °C and it is, therefore, important that glass surfaces, e.g. in laboratory stills, always remain submerged.)

Hazardous reactions

Certain types of chemical reaction included in Table 10.6 are particularly likely to involve hazards. These include oxidations, polymerisations, halogenations, hydrogenations, nitrations and alkylations. (The special case of combustion reactions is discussed in Ch. 4.) Other processes involve operation at high pressures. Some facets of these processes are summarised below.

Reaction hazards

It has been proposed that exothermic reactions may be grouped roughly in order of increasing hazard (after ref. 29):

1. (a) Hydrogenation – addition of hydrogen atoms to both sides of a double or triple bond.
   (b) Hydrolysis – reaction of a compound with water (e.g. manufacture of sulphuric or phosphoric acids from oxides).
   (c) Isomerisation – rearrangement of the atoms in an organic molecule such as a change from a straight chain to a branched molecule.
   (d) Sulphonation – introduction of an SO_3H radical into an organic molecule by reaction with H_2SO_4.
   (e) Neutralisation – reaction between an acid and a base yielding a salt and water.

2. (a) Alkylation – addition of an alkyl group to a compound to form various organic compounds.
   (b) Esterification – reaction between an acid and an alcohol or unsaturated hydrocarbon. (The hazard is increased when the acid is a strong reacting material.)
   (c) Oxidation – combination of oxygen with some substances in which reaction is controlled and does not go to CO_2 and H_2O. Where vigorous oxidising agents such as chlorates, nitric acid, hypochlorous acids and salts are used the hazard is greater.
   (d) Polymerisation – joining together of molecules to form chains or other linkages.
   (e) Condensation – joining together of two or more organic molecules with splitting off of H_2O, HCl or other compounds.
3. Halogenation – introduction of halogen atoms (fluorine, chlorine, bromine or iodine) into an organic molecule.
4. Nitration – the replacement of a hydrogen atom in a compound with a nitro group.

In general, endothermic reactions of the following types are inherently less hazardous (albeit less so when the energy source necessary to sustain the reaction involves the combustion of solid, liquid or gaseous fuel).

- Calcination – heating a material to remove moisture or other volatile material.
- Electrolysis – separation of ions by means of electric current.
- Pyrolysis or cracking – thermal decomposition of large molecules by high temperatures, pressures and a catalyst.

In addition to these generalisations specific chemical reactions are 'rated' in Table 10.6.[30]

Hazards of particular reactions

Hydroprocesses

Since hydrogen is chemically stable and relatively unreactive at ordinary temperatures most processes utilising hydrogen require a catalyst. Above 500 °C, however, it reacts spontaneously with oxygen and confined flammable mixtures explode violently if ignited: flammability limits are wide (4–75%) and minimum ignition energy is low (*c.* 0.017 mJ). Hydrogen exhibits a reverse Joule Thompson effect so that leaking gas heats up and may ignite. Furthermore, hydrogen flames tend to be non-luminous and may escape immediate detection. Because of its very low density, hydrogen disperses readily but has been involved in unconfined vapour cloud (UCVC) explosions (see Ch. 15). The largest use of hydrogen is in synthesis gas for the production of ammonia which requires high pressures. Other important processes include hydrocracking, hydrorefining,

Table 10.6 Hazard rating of chemical reactions[30]

Reaction	Degree of hazard
Reductions	
1. Clemmensen	D
2. Sodium-amalgam	D
3. Zinc–acetic acid	E
4. Zinc–hydrochloric acid	E
5. Zinc–sodium hydroxide	E
6. Ferrous ammonium sulphate	E
7. Lead tetraacetate	E
8. Meerwein–Pondorff	D
9. Lithium aluminum hydride	B
10. Dialkyl aluminum hydride	B
11. Rosenmund	A
12. Catalytic high pressure	A
13. Catalytic low pressure	B
Oxidations	
1. Hydrogen peroxide – dilute aqueous	E
2. Air or I_2 (mercaptan to disulphide)	D
3. Oppenauer	D
4. Selenium dioxide	D
5. Aqueous solution nitric acid, permanganate, manganic dioxide, chromic acid, dichromate	E
6. Electrolytic	B
7. Chromyl chloride	C
8. Ozonolysis	A
9. Nitrous acids	A
10. Peracids – low molecular weight or two or more positive groups	A
11. Peracids – high molecular weight	B
12. *t*-Butyl hypochlorite	C
13. Chlorine	C
Alkylations	
Carbon–carbon	
1. Jarousse	E
2. Alkali metal	C
3. Alkali metal alcoholate	D
4. Alkali metal amides and hydrides	C
5. Reformatsky	E
6. Michael	E
7. Grignard	B
8. Organo metallics, such as dialkyl zinc or cadmium–alkyl or aryl lithium	B
9. Alkali acetylides	A
10. Diels–Alder	D
11. Arndt–Eistert	A
12. Diazoalkane and aldehyde	A
13. Aldehydes or ketones and hydrogen cyanide	C
Carbon–oxygen	
1. Williamson	D
2. Formaldehyde – hydrochloric acid	E
3. Ethylene oxide	C
4. Dialkyl sulfate	D
5. Diazoalkane	A
Carbon–nitrogen	
1. Cyanomethylation	C
2. Chloromethylation	D
3. Ethylenimine	C
4. Ethylene oxide	C
5. Quaternization	D
Condensations	
1. Erlenmeyer	D
2. Perkin	D
3. Acetoacetic ester	D
4. Aldol	D
5. Claisen	D
6. Knoevenagel	D
7. Condensations using catalysts such as phosphoric acid; $AlCl_3$; $KHSO_4$; $SnCl_4$; H_2SO_4; $ZnCl_2$ $NaHSO_2$; $POCl_2$; HCl; $FeCl_2$	E
8. Acyloin	C
9. Diketones with hydrogen sulphide	C
10. Diketones with diamines → quinazolines	D
11. Diketones with NH_2OH → isoxazolines	D
12. Diketones with NH_2NH_2 → pyrazoles	C
13. Diketones with semicarbazide → pyrazoles	D
14. Diketones with ammonia → pyrazoles	D
15. Carbon disulphide with aminoacetamide → thiazolone	A
16. Nitriles and ethylene diamines → imidazolines	D

Table 10.6 (continued)

Reaction	Degree of hazard	Reaction	Degree of hazard
Aminations		***Hydrolysis, aqueous nitriles, esters***	E
1. Liquid ammonia	B		
2. Aqueous ammonia	E	***Simple metathetical replacements***	D
3. Alkali amides	C		
		Preparation and reaction of peroxides and peracids	
Esterifications		1. Concentrated	A
1. Inorganic	E	2. Dilute	D
2. Alkoxy magnesium halides	B		
3. Organic:		***Pyrolysis***	
Alcohol and acids or acid chloride or acid anhydride	D	1. Atmospheric pressure	D
		2. Pressure	B
Alkyl halide and silver salts of acids	E	***Schmidt reaction***	B
Alkyl sulphate and alkali metal salt of acid	D	***Mannich reaction***	D
Alkyl chlorosulphates and alkali salts of carboxylic acid	D	***Halogenations***	
		SO_2X_2, SOX_2, SX, POX_2, PX_5	D
		HX	D
Ester-exchange	D	Cl_2, Br_2,	C
Carboxylic acid and diazomethane	A	***Nitrations***	
Acetylene and carboxylic acid-vinyl ester	A	1. Dilute	D
		2. Concentrated	B

Key:

Hazardous
(A) Highly flammable, develop high pressure instantly, highly toxic.

Special
(B) Flammable, perhaps explosive, mixtures form.

(C) Flammable or generate toxic substances.

Conventional
(D) Slightly flammable, generate or use mildly toxic substances.

(E) Non-flammable, do not use or generate toxic substances.

hydroisomerisation, hydroalkylation and hydrogenation; the first two of these are operated at atmospheric pressure but the others are performed under elevated pressure. Methanol synthesis from hydrogen also requires elevated pressures. The main hazards of the above processes are therefore fire and explosion, coupled with metallurgical problems arising from hydrogen attack.

Alkylation

Alkylation processes are particularly hazardous because of the toxicity of some of the chemicals used as alkylating agents, e.g. dimethyl sulphate (a suspected human carcinogen) and hydrogen fluoride. Thermal alkylation processes require higher temperatures and pressures, with the associated problems, compared with catalytic processes.

High-pressure reactions

High inventories of stored pressure, e.g. in pressurised reactors or associated plant, can result in incidents due to a catastrophic failure of the pressure shell. Design to prevent this is discussed in Chapter 12.

Typical high-pressure processes in the chemical industry are given in Table 10.7[31] most of which are in excess of 100 bar. Table 10.8 lists processes likely to involve pressures equal to, or greater than, 100 bar. Not all these processes are gaseous, and only a limited number would reach commercial production. Moreover the trend is to develop more efficient catalysts for lower pressure reactions. However such pressures are clearly encountered in the laboratory and pilot plant stages of development.

Halogenations

The commercially important halogens are chlorine, bromine, fluorine and iodine. The reactions in which they are involved are highly exothermic and chain reactions, which may result in detonations, can occur. Chlorine substituted hydrocarbons can be synthesised by a variety of routes, e.g. hydrochlorination, oxychlorination and most commonly chlorination. Hazards accompanying chlorination reactions arise from the flammability of hydrocarbon raw materials and from the toxicity of chlorine. Furthermore, unless carefully controlled, many chlorinations may proceed at a rate leading to fire or explosion. Consequently upper and lower flammability limits of hydrocarbons in chlorine (analogous to those in oxygen)

Table 10.7 Typical high-pressure processes

Application	Pressure (atm)*	Application	Pressure (atm)*
Nitric acid	1–10	Methanol synthesis	50–350
Synthetic ethanol	65–70	Hydrogenation of coal	350–600
Hydrogenation of vegetable oil	20–350	Acetic acid synthesis	650–700
Hydrogenation of petroleum distillates	200–350	Ammonia synthesis	200–1000
Urea synthesis	200–400	Polyethylene	50–2000
Oxo process	250–300		

* Note: 1 bar = 750.1 mmHg = 14.55 psia, 1 atm = 14.7 psia.

Table 10.8 Potential processes involving pressure greater than 100 bar – basis US patent literature to 1969[32]

Product	Process	Pressure (psi)
Acetic acid	Liquid phase reaction from butene vapour or liquid phase conversion over nickel salt catalyst	450–1,500 10,500
Aluminium trialkyls	Mixed liquid/gas phase combination of olefin, aluminium and hydrogen catalysed by chlorine	1,500–2,250
Amino methyl benzoic acid	Nickel catalysed reaction between iso-phthalic acid, ammonia and hydrogen	1,500–2,000
Ammonia	Metal oxide catalysed combination of nitrogen and hydrogen from synthesis gas	1,400–3,200
Aniline	Reaction between chlorobenzene and ammonia over cuprous oxide	2,000
Benzene	Coal reduction	500–3,000
β-Naphthol	Liquid phase oxidation of naphthalene	2,400
Butyraldehyde	Catalytic oxidation of propylene	500–5,000
Caprolactam	Ammonolysis of caprolactone	3,000–5,000
Cobalt hydrocarbonyl	Liquid phase reaction between carbon monoxide and hydrogen over cobalt acetate	1,000–4,000
Cresols	From toluene via metachlorotoluene	3,200
1,4-Cyclohexane-dimethanol	Palladium catalysed reactions between dimethylphthalate and hydrogen	5,500
Cyclo-octa tetraene	Cyclisation of acetylene with nickel acetyl acetonate catalyst in THF	1,500
Dialkyl aluminium hydrides	Liquid phase reduction of aluminium trialkyl	500–2,500
Dibasic acids	Oxidation of paraffin wax with nitric acid	15–1,500
Dicobalt octacarbonyl	Liquid phase reaction of hydrogen, carbon monoxide and cobalt acetate	2,000–6,000
Diethyl ketone	Cobalt oleate catalysed reaction of ethylene, carbon monoxide and THF in hexane	3,000–5,000
Diglycolamine	Catalytic ammonolysis of diethylene glycol	1,500–4,500
Dimethylhydrazine	Reduction of nitrosodimethylamine over platinum on alumina	50–3,000
Dimethylphthalate	Liquid phase oxidation of p-xylene	250–1,500
Durene	Vapour phase disproportionation of pseudocumene over molybdena – alumina	600–3,000
Ethanol	Liquid phase oxidation of ethylene over aluminium hydroxide gel	1,500
Ethylene diamine	Ammonolysis of ethylene glycol over Ni – Cu on alumina	3,000–6,000
Ethyl morpholine	Dehydration of diglycolamine over mixed oxide catalyst	2,000–3,000

Table 10.8 (continued)

Product	Process	Pressure (psi)
Fatty acid amides	Ammonolysis of fatty acid methyl ester	2,000
Fatty alcohols	Catalytic hydrogenation of fat over silica – alumina	500–2,000
Hexachlorobicyclo-heptadiene	Reaction of hexachloropentadiene with acetylene	2,000–4,000
Hexamethylene diamine	Hydrogenation of adiponitrile over cobalt	9,000–9,600
Hexamethylene	Reaction of propargyl alcohol with air and hydrogen	150–3,000
Hydrazine	Platinum catalysed reaction of isobutylene with nitrogen dioxide and hydrogen	1,000–1,500
Hydroxy cyclohexane carboxylates	Catalytic reaction between methyl cyclohexane carboxylate, air and hydrogen	15–1,500
Isopropyl-N-phenyl-*p*-phenylene diamine	Combination of acetone and *p*-amino diphenyl-amine over copper chromite	2,700–3,000
Methanol	Direct combination of carbon monoxide with hydrogen over zinc oxide	5,000
Methyl pentenes	Catalytic vapour phase dimerisation of propylene	2,000–3,500
Neoheptanol	Reduction of neoheptanoic acid over molybdenum sulphide on activated charcoal	500–5,000
Neohexane	Dimerisation of propylene with subsequent hydrogenation	1,000–5,000
Nicotinic acid	Oxidation of 2-methyl-5-ethyl-pyridine with nitric acid	500–1,500
Norbornylene	Diels – Alder addition of ethylene to cyclo-pentadiene	500–10,000
Long chain olefins	Polymerisation of ethylene with aluminium alkyls	900–2,250
Oxoalcohols	Reduction of oxoaldehydes over molybdenum sulphide on carbon	2,500–4,500
p-Methoxyphenol	Condensation of methanol with hydroquinone over silica – alumina	200–1,500
Phenol	Catalytic oxidation of benzene with copper sulphate	1,000–1,500
Phenol	Catalytic hydrolysis of chlorobenzene with calcium hydroxide and copper	2,000–3,000
Phenol	Reductive decomposition of lignin over iron sulphate	10,000
Piperazine	Dehydration of monoethanolamine over Ni – Cu – chromia	1,500–6,000
Polyethylene	Catalytic polymerisation of ethylene with oxygen or peroxide	15,000–25,000
Propionitrile	Reaction between propylene and ammonia over cobalt oxide	1,000–3,000

Table 10.8 (continued)

Product	Process	Pressure (psi)
Styrenated drying oils	Addition of styrene to linseed oil	375–1,500
Succinic acid	Liquid phase addition of carbon monoxide and water to acetylene over cobalt carbonyl	2,250–7,500
Triethyl aluminium	Liquid phase combination of aluminium, ethylene and hydrogen	1,500–2,250
Triethylene diamine	Nickel catalysed reaction of ethanolamine and hydrogen	300–6,000
Trimethylacetic acid	Liquid phase addition of carbon monoxide and water to isobutylene over BF_3	750–1,500
Urea	Condensation of carbon dioxide with ammonia	3,000
Vinylidene fluoride-, hexafluoropropene copolymer	Copolymerisation of monomers over ammonium persulphate	250–1,500
Xylene diamines	Condensation of ammonia with phthalic acid	2,000–5,000

Table 10.9 Flammable properties of a selection of hydrocarbon derivatives (after 13)

Compound	Explosive limits* (mole %)			Auto-ignition temperature (°C)	
	Air	O_2	Cl_2	Air	Cl_2
Methane	4.0–16.0	5.0 –61.85	5.51–63	537	318
Methyl chloride	7.0–17.4	8.05–66.0	10.20–63.0	618	215
Dichlormethane	12.0–22	12.65–73.4	16.50–52.9	556	262
Methyl alcohol	5.5–40		13.80–73.5		
Ethane	3.0–15.4	3.0 –67	4.95–58.8	472	205
Propene				455	150
1,2-Dichloropropane				555	180
1,2-Dichloroethane	4.5–17.3	4.0 –67.5	16.40–36.8		
Ethyl alcohol	3.3–19		5.06–64.1		
Acetic acid	3.1–17	4.0 –60	15.83–56.0		

* Care is required in interpretation since discrepancies arise due to differences in experimental methods, temperature, etc.

exist.[31] A selection are presented in Table 10.9 with, for comparison, the flammability limits in air and in pure oxygen. It is also to be noted from the table that the auto-ignition temperatures of hydrocarbons in chlorine are lower than in air. Also many compounds form detonatable mixtures in chlorine.

Some halogenated compounds react with ease with other material to produce hazardous substances. Thus tetrachloroethane reacts with strong bases to produce unstable chloroacetylides capable of exploding or detonating spontaneously on contact with air:

$$2C_2H_2Cl_4 + 5NaOH \rightarrow C_2Cl_2 + C_2HCl + 4H_2O + 4NaCl + NaClO$$

Organic nitrogen compounds react with chlorine often to form nitrogen trichloride, which is very unstable and easily detonated. Many inorganic compounds also react with chlorine to produce unstable products such as nitrogen trichloride, chloro-oxides and perchlorates. Sodium hypochlorite decomposes to form oxygen or chlorine, depending upon conditions:

$$2\,NaClO \xrightarrow{heat} 2NaCl + O_2$$
$$NaClO + 2HCl \rightarrow NaCl + H_2O + Cl_2$$

Under certain circumstances sodium chlorate can also be produced. Chlorine may also react violently with ancillary materials in common service as exemplified in Table 10.10[31,32] and, as with all halogen systems, materials of construction require careful selection to avoid corrosion problems. This can result in valves that cannot be operated, in the formation of volatile metal chlorides which can block lines, orifices of flow meters, flame-traps, etc., or lead to loss of containment.

> A fill-in piece of PVC instead of steel was installed. After several weeks the plastic became brittle and fractured spontaneously. The line contained liquid chlorine at a pressure of above 7 bar and as a result 3–7 tonnes of the chemical were released.

Nitrations

Nitration is of commercial importance in the manufacture of, e.g. nitrobenzene (an intermediate in the preparation of aniline used in the dyestuffs industry) and of chemical explosives. Examples of the latter include TNT made by the stepwise nitration of toluene; picric acid by the

Table 10.10 Common in service materials reacting violently with chlorine

Material	Common use
Polypropylene	Filter elements for filtering chlorine
Silicone oil	A 'stable' fluid employed in instrument diaphragms, e.g. pressure gauges
Dibutyl phthalate	Pressure transmitters
Polychlorinated biphenyls	Heat-transfer fluids
Hydrocarbon oils	Diaphragm and other type pumps for chlorine service
Glycerine	Instruments in chlorine service
Drawing wax	To slip loose liners inside steel tanks or pipes

nitration and hydrolysis of chlorobenzene; and nitroglycerine and nitrocellulose by the nitration of glycerine and cellulose, respectively.

Nitroglycerine is a misnomer since the compound is an ester and not a nitro derivative. It was first made commercially in 1865 by Alfred Nobel and is still one of the most powerful explosives. The basic method of manufacture has changed little and involves nitration with mixed nitric and sulphuric acids:

$$\begin{array}{l} CH_2OH \\ | \\ CHOH + 3\,HNO_3 \\ | \\ CH_2OH \\ \text{(Glycerol)} \end{array} \xrightarrow{H_2SO_4} \begin{array}{l} CH_2{\cdot}O{\cdot}NO_2 \\ | \\ CH{\cdot}O{\cdot}NO_2 \\ | \\ CH_2{\cdot}O{\cdot}NO_2 \\ \text{Glyceryl trinitrate ('nitroglycerine')} \end{array}$$

Aromatic nitrations are also normally effected with a mixture of the acids at temperatures between 0 and 120 °C for example:

$$HNO_3 + H^+ \rightleftharpoons H_2O + NO_2^+$$

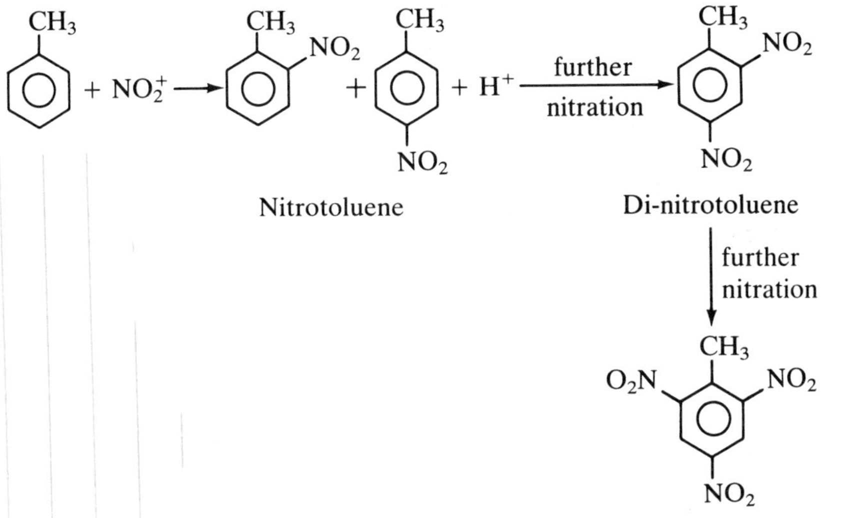

$$\text{Toluene} + NO_2^+ \longrightarrow \text{Nitrotoluene (o- and p-)} + H^+ \xrightarrow[\text{nitration}]{\text{further}} \text{Di-nitrotoluene} \xrightarrow{\text{further nitration}} \text{Tri-nitrotoluene (TNT)}$$

Nitrotoluene

Di-nitrotoluene

Tri-nitrotoluene (TNT)

of saturated hydrocarbons (e.g. methane) is usually achieved phase nitration.

rds with nitrations stem from the strong oxidising nature of g mixture, the toxicity of many nitro compounds by inhalation sorption, the highly flammable character of most nitro derivathe explosive nature of many products. The explosive tendase with increased degrees of nitration.

On 17 March, 1870 at Paterson, New Jersey, a man dropped a can of nitroglycerine. The resulting explosion killed the man and caused an explosion in the store killing several others. Almost 100 years later similar explosions were being reported.[33]

Clearly the handling of certain nitrated chemicals warrants respect and nitration processes demand precise control of reaction kinetics and temperature: at elevated temperatures of reaction the degree of exothermicity may exceed the ability to maintain adequate control and a 'runaway' reaction develops with excessive side reactions and the possibility of explosion. Accordingly, batch nitrations are normally conducted in segregated buildings in stages with careful control of the rate of addition of reactants, vigorous agitation of the two-phase system, and provision of efficient heat transfer surfaces (a jacket, internal coils, cooled baffles, or an external heat exchanger). Drowning tubs are often provided under reactors into which the batch can be dumped and the reaction quenched in the event of unacceptable rises in temperature. Nitric acid fumes are extracted for recovery.

Figure 10.4 depicts a process operator controlling the temperature of manufacture of nitroglycerine in the early days. He sat on a one-legged

Fig. 10.4 Process operator on nitroglycerine plant (Courtesy of Imperial Chemical Industries plc)

stool so that if he became sleepy he fell off and woke up. Nowadays such primitive forms of detection and alarm are replaced by more elaborate automatic safety devices to monitor and control the process remotely. Also developments in continuous flow processes have led to safer and easier control in the production of certain nitrated substances.

Polymerisations

Polymerisations are exothermic processes which, unless carefully controlled, can 'run away' and create a thermal explosion. As an illustration, styrene, with a heat of polymerisation of 17 kcal mol^{-1}, can vaporise at high polymerisation rates, rupture the reactor and produce a vapour–air mixture within the explosive limits.[31] At high temperatures, such as in a fire, styrene and its polymers undergo exothermic degradation. Other commercial polymerisations requiring strict control are those of vinyl chloride, vinyl acetate, ethylene oxide and propylene oxide.

In addition, certain processes, e.g. the manufacture of polyethylene and polypropylene, require polymerisation of feedstock at high pressures with the associated hazards discussed later. Many vinyl monomers, in particular vinyl chloride and acrylonitrile, pose a chronic toxicity hazard. However, a more detailed understanding of the chemistry of the system is crucial in order to minimise hazards.

Polymerisations can proceed via ionic or free-radical mechanisms. The polymerisation of ethylene oxide, for example, is initiated by heat, acids, bases or metal ions.

$$H\bar{O} + CH_2\overset{O}{—}CH_2 \longrightarrow HO—CH_2—CH_2—\bar{O} \xrightarrow[2)\ H_2O]{1)\ CH_2\overset{O}{—}CH_2}$$

$$HOCH_2CH_2O\ (CH_2CH_2O)_xH + \bar{O}H$$

Polyethylene glycol

Control is required to prevent an exothermic run-away reaction. Furthermore, ethylene oxide is toxic and vapour will explode in the presence of a source of ignition (e.g. open flames, glowing surfaces, static electricity). Thus, while pure ethylene oxide may undergo polymerisation only very slowly on storage, the presence of heat or impurities can initiate a rapid reaction with explosive consequences.

The exothermic polymerisation of ethylene oxide resulted in the major disaster at Doe Run (refer to Ch. 16): 30 tons of ethylene oxide in a pressure tank began to polymerise due to impurities in the material. The resulting thermal explosion caused failure of the vessel and the release of a vapour cloud which then exploded.

Many vinyl monomers on the other hand polymerise by free radical processes which can be initiated by heat, light or redox reactions, thus

$$\mathrm{ROOR} \xrightarrow[\Delta]{h\nu} 2\mathrm{RO}\cdot$$

$$\mathrm{ROOR} + \mathrm{M^{I}} \longrightarrow \mathrm{RO}\cdot + \mathrm{RO^{-}} + \mathrm{M^{II}}$$

Where M^{I} is a transition metal ion in one valency state and M^{II} is the next higher oxidation valency state.

Initiation

$$\mathrm{R}\cdot + \mathrm{CH_2 = CHX} \longrightarrow \mathrm{R{-}CH_2{-}\underset{X}{\overset{H}{C}}\cdot}$$

Polymerisation

$$\mathrm{R{-}CH_2{-}\underset{X}{\overset{H}{C}}\cdot + CH_2{=}CHX \longrightarrow R{-}CH_2{-}\underset{X}{CH}{-}CH_2{-}\underset{X}{\overset{H}{C}}\cdot}$$

$$\mathrm{R{-}CH_2{-}\underset{X}{CH}{-}CH_2{-}\underset{X}{\overset{H}{C}}\cdot + CH_2{=}CHX \longrightarrow R{-}CH_2{-}\underset{X}{CH}\,(CH_2{-}CHX)_{\mathit{x}}CH_2{-}\underset{X}{\overset{H}{C}}\cdot}$$

Termination

$$\mathrm{RH + RCH_2{-}\underset{X}{CH}\,(CH_2{-}CHX)_{\mathit{x}}\,CH_2{-}\underset{X}{\overset{H}{C}}\cdot \longrightarrow RCH_2{-}\underset{X}{CH}\,(CH_2{-}CHX)_{\mathit{x}}\,CH_2CH_2X + R\cdot}$$

Again, control of reaction and storage conditions is crucial in avoiding runaway conditions. Stabilisation is achieved by cooling monomers and checking the adequacy of inhibitors. Monomers such as styrene can polymerise explosively if the temperature is allowed to rise and for this reason the monomer should be kept below 30 °C.

A two-compartment tank was used for making blends of styrene. One compartment was full but the other was empty and was being steamed out prior to entry. Steaming had been in progress for two days when there was a jet of styrene liquid and gas with polymer ejected from the dip hatch of the other compartment. The material ignited.

Other precautions include carrying out reactions in solution, in suspension or in aqueous emulsion to assist in heat dissipation and maintenance of viscosity (which increases as polymer is produced). Efficient agitation, cooling and constant monitoring of temperature are important, with alarms where necessary to indicate failure of control features.

Since many monomers are highly flammable at room or process temperatures, inert atmospheres are also often employed. However, in some circumstances this can increase the hazard since the presence of oxygen can play a crucial role in stabilisation of the monomer. For example, many vinyl monomers undergo facile oxidation to form peroxy radicals:

$$R\cdot + O_2 \longrightarrow R{-}O{-}O\cdot$$

$$R{-}CH_2{-}\underset{X}{\overset{H}{C}}\cdot + O_2 \longrightarrow R{-}CH_2{-}\underset{X}{\overset{H}{C}}{-}O{-}O\cdot$$

As a consequence additives are added to inhibit polymerisation by impurities or to act as an anti-oxidant. Examples include hydroquinone and tertiary butyl catechol. They function by 'mopping up' free radicals:

$$2R{-}O{-}O\cdot + HO{-}C_6H_4{-}OH \longrightarrow 2ROOH + O{=}C_6H_4{=}O$$

Hydoquinone

$$2R{-}CH_2{-}\underset{X}{\overset{H}{C}}{-}O{-}O\cdot + HO{-}C_6H_4{-}OH \longrightarrow 2R{-}CH{-}\underset{X}{CH}{-}OOH + O{=}C_6H_4{=}O$$

In the absence of oxygen in the liquid phase only the retardation reaction occurs and the resultant phenoxy radical can re-initiate polymerisation.

$$R{-}CH_2{-}\overset{H}{\underset{X}{C}}\cdot \;+\; HO{-}C_6H_4{-}OH \longrightarrow R{-}CH_2{-}CH_2X \;+\; \dot{O}{-}C_6H_4{-}OH$$

a phenoxy radical

The resulting phenoxy radical re-initiates polymerisation. Clearly in order for these stabilisers to function and scavenge unwanted radicals the presence of oxygen dissolved in the liquid phase is essential. Blanketing of monomer with inert gas will deplete the oxygen content in the bulk and in certain instances (e.g. styrene) can increase the hazard.

Loss of inhibitor also poses a danger. Explosions have occurred as monomers have been distilled due to polymerisation of the pure unstabilised distillate, which has also been known to block vents. (For this reason rupture discs may be installed before, or instead of pressure relief valves on polymerisation reactors.)

> A runaway polymerisation occurred in a distillation column causing the death of one person and two others fighting the subsequent fire. The explosion damaged a storage tank and two distillation columns and moved a vessel 50 cm.

Similarly monomers can become separated from inhibitor during crystallisation. The unstabilised liquid layer formed on thawing of 'pure' monomer can undergo low temperature polymerisation and 'run away'. A prime example of such a material is acrylic acid with a melting point of 14 °C.

> On 3 January 1976 a road tanker containing 13,620 kg of glacial acrylic acid burst open with explosive violence. Upon rupturing, the gelatinous polymer, which was above its auto-ignition temperature, was thrown blazing over a 152 metres diameter semi-circle. The tanker explosion was believed to have been initiated by the thawing of a layer of crystallised acrylic acid by warm water which was being used to maintain the tank temperature above its freezing point of 14 °C. The crystallisation of the acrylic acid occurred during the loading operation and would take place without its attendant inhibitor. On thawing in the tanker, a liquid layer formed without the polymerisation inhibitor. Polymerisation started in this layer and consumed the inhibitor in the bulk leading to the subsequent rupture of the tank.

Inhibitors also become spent as a consequence of prolonged storage and monomer stocks require checking periodically to ensure sufficient stabiliser is present.

Oxidations

Feedstocks for oxidation processes are generally hydrocarbons, so that there is a hazard of fire or explosion arising from the contact of flammable material with oxygen. The reactions are highly exothermic and equilibrium favours complete reaction. Air is commonly used under controlled conditions in vapour phase oxidations.

Liquid phase reactions are potentially more hazardous since the mass of flammable material present is greater. The common oxidising agents used, all of which require appropriate handling precautions and control measures, include salts of permanganic acid, hypochlorous acid and salts, sodium chlorate and chlorine dioxide, all chlorates, all peroxides, nitric acid and nitrogen tetroxide, and ozone. Careful attention must be paid to efficient heat removal, the potential accumulation of peroxides or hydroperoxides either as products, intermediates or by-products and the elimination of ignition sources inside or outside the reactors. Some liquid-phase oxidation processes are listed in Table 10.11.

A review of fires at hydrocarbon oxidation units[34] indicated that more explosions are caused by the decomposition of oxidation products than as a result of formation of a flammable hydrocarbon–air (or oxygen) mixture. For example several explosions have been caused by the decomposition of ethylene oxide (see page 525) due to small nearby fires, overheated pumps, polymerisation in a distillation column, and the presence of contaminants. Similarly, explosions have occurred at phenol plants due to the decomposition of cumene hydroperoxide:

$$C_6H_5CH(CH_3)_2 \xrightarrow{[O]} C_6H_5C(CH_3)_2OOH \longrightarrow O{=}C(CH_3)_2 + C_6H_5OH$$

cumene — cumene hydroperoxide — acetone — phenol

Also, of those incidents involving ignition of flammable mixtures many were not attributable to the process as such. For example, at the time of the Flixborough explosion the plant was under nitrogen pressure and the flow of air had not been started.

Nevertheless, explosions at oxidation plants have occurred as a result of the process due to a variety of reasons, including:

- combustion of hydrocarbon in the air inlet line or near the sparge pipe;
- a choke in the liquid overflow pipe from an oxidiser which boiled dry while air feed continued;
- failure or disarming of control instruments.

Table 10.11 Examples of liquid-phase oxidation processes

Product	Feedstock
Acetic acid and acetic anhydride	Butane, hexane and naphtha
Benzoic acid and phenol	Toluene
Phenol	Cumene
Cyclohexanol and cyclohexanone	Cyclohexane
Acrylonitrile	Propylene
Methyl isobutyl ketone	Pentene
Vinyl acetate, acetaldehyde and acetone	Ethylene
Methyl ethyl ketone and acetic acid	Butane
Ethylene and propylene oxides	Respective olefins
Terephthalic acid	*p*-Xylene

Several 'near-miss' incidents have resulted in formation of flammable atmospheres which fortunately failed to ignite.

A 'near-miss' incident occurred on a cumene oxidation plant equipped with a trip. In the event of an unacceptable rise in reactor temperature, this shut off the air automatically and opened a valve to dump the reaction mixture into a vat of water [see Fig. 10.5]. A spurious high temperature trip closed the air valve and opened the dump valve. The trip rectified itself and the dump valve stayed open but the air valve re-opened because the solenoid switching the air supply to the control valve would not stay in the tripped position. Thus air was allowed to enter the almost empty reactor and form an explosive mixture. Because no ignition source was present the reactor did not explode.

The solenoid in question has been replaced by a latching solenoid which will not reset of its own accord once a trip has been initiated.[34]

Precautions are generally dependent upon a combination of temperature control and limiting the concentrations of the oxidant and oxidising

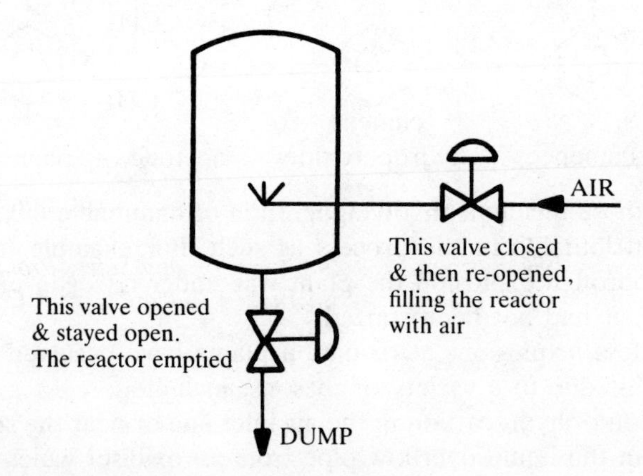

Fig. 10.5 Valve arrangement on cumene reactor

agents to control both the rate of reaction and to operate as far as possible from the explosive limits. Inventories of reactants and unstable intermediates should be kept as low as practicable. Careful design and installation of control features are essential and instruments should be tested regularly and not be allowed to be over-ridden. Guidance is given in reference 35 on oxygen analysers for hydrocarbon oxidation plants covering safety of design, sample systems, installation, calibration and maintenance.

References

(*All places of publication are London unless otherwise stated*)

1. Bretherick, L. 'Reactive hazards', in G. D. Muir (ed.) *Hazards in the Chemical Laboratory*. Chemical Society 1977.
2. Habashi, F. & Ismail, M. I., *Canadian Mining & Metallurgical Bulletin*, **68**, 1976, 99.
3. Anon., *RoSPA Bull.*, 1985 (Mar.), 2
4. *Manufacturing Chemists Assoc., Case History 2286.*
5. *Manufacturing Chemists Assoc., Case History 1808.*
6. *Annual Report of H.M. Chief Inspector of Factories, 1967*, 73. HMSO.
7. Health and Safety Executive, *Health and Safety: Industry & Services*, 1975 21. HMSO.
8. *Manufacturing Chemists Assoc., Case History 1913.*
9. Universities Safety Association, *Safety News*, No. 16 (Mar. 1982), 5.
10. Rohn & Hass Co., press release, 27 Dec. 1972; CIB 9, 9, 424.
11. *Safety in Chemical Laboratories and in the Use of Chemicals* (3rd edn). Imperial College of Science and Technology. Dec. 1971, 5.
12. Gaston, P. J., *The Care, Handling and Disposal of Dangerous Chemicals*. Institute of Science Technology 1979, 29.
13. Dokter, T., J., *Hazardous Materials*, 1985, **10**, 73
14. Universities Safety Association, *Safety News*, No. 10 (Nov. 1978), 11.
15. Bretherick, L., *Handbook of Reactive Chemical Hazards* (2nd edn). Butterworths 1979.
16. Evans, F. W., Meyer, P. & Oppliger, W., 'Loss Prevention and Safety Promotion in the Process Industries', *Proceedings of 2nd International Symposium*, Dechema, 1978, IV, 191–7.
17. *Loss Prevention Bulletins 003, 005, 006, 012, 013, 028.* Institution of Chemical Engineers.
18. *4th International Symposium on Loss Prevention and Safety Promotion in the Process Industries*; Vol. 3 – *Chemical Process Hazards. Instn Chem. Engrs, Symp. Series*, No. 82, 1983, A12–A21.
19. 'Seveso: lessons from an escape', *The Economist*, 17 June 1978, 101.
20. Theofanos, T. G., 'A physicochemical mechanism for the ignition of the Seveso accident', *Nature*, 1981, **291**, 640.
21. Fawcett, H. E., 'Dioxin (TCDD), dibenzofurans, and related compounds' in *Hazardous and Toxic Materials*. Wiley 1984, 8, 219–248.

22. Marshall, V. C., 'Seveso: an analysis of the official report', *Chemical Engineer*, July 1980, 499.
23. Walas, S. M., *Reaction Kinetics for Chemical Engineers*. McGraw-Hill 1959.
24. Health and Safety Executive, *Health and Safety: Manufacturing & Services*, 1976, 28.
25. *Chem. Eng. Tech.*, 1976, **4**, 8.
26. Health and Safety Executive *The Explosion at the Dow Chemicals Factory, Kings Lynn, 27th June 1976*, 1977. HMSO.
27. Rowbottom, D. W., Laird, J. H. & Beveridge, G. S. E., *Chem. Eng.*, 1980, **354**, 155–60.
28. Anon., *Loss Prevention Bulletin 047*. Institution of Chemical Engineers, 1981, 21–8.
29. *Fire & Explosion Index Hazard Classification Guide* (*The Dow Index*) (5th edn). American Instn of Chemical Engineers 1981, 16.
30. Shabica, A. C., *Chem. Eng. Prog.*, 1963, **59**, (9), 57.
31. Carson, P. A. & Mumford, C. J., *J. Occupational Accidents*, 1979 **2**, 85.
32. Sittig, M., *Organic Chemical Process Encyclopaedia*. Noyes Development Corporation: New Jersey 1969.
33. Bond, J. *Loss Prevention Bulletin*. Institution of Chemical Engineers 1985 (065), 21
34. Kletz, T. A., *Loss Prevention*, 1979, **12**, 96
35. Institution of Chemical Engineers, *Guide Notes On Safe Application Of Oxygen Analysers To Hydrocarbon Oxidation Reactions In Chemical Process Plant*. Instn Chem. Engr. 1983

CHAPTER 11

Safety in laboratories and pilot plants

Introduction

Chemical laboratories are encountered in most manufacturing or process industries, research establishments, schools, colleges, hospitals and government scientific departments. The handling of chemicals on this scale is at the opposite end of the spectrum to major hazards and it is the wide variety of substances used, rather than their bulk, and the proximity of the workers which present the main problems.

Sometimes, work on semi-technical and pilot-plant scales is considered to be laboratory development because of the novelty of the processes and the relatively small-scale of usage compared to production facilities. This chapter is structured into four main sections;

1. A discussion of the general risks associated with laboratory work based on accident statistics, and as a corollary
2. A mention of the main relevant administrative controls
3. A discussion of the safety aspects of common laboratory procedures, and
4. A description of the hazards ard precautions associated specifically with use of chemicals on a laboratory-bench scale and at the pilot-plant stage of product development.

Hazards of working in laboratories

The main features of work in laboratories which tend to distinguish them from many other workplaces include:

- The diverse background of employees, viz. scientists (chemists, physicists, biologists, etc.), engineers, technicians of varying skills and experience, administrative staff, students, tradesmen, process operators, and unskilled ancillary staff such as storemen, cleaners, and wash-up assistants. (Particular consideration needs to be given to the system of work to protect transient workers such as cleaners, maintenance staff, contractors and security staff.)

- The wide variety of potential hazards encountered which stem from the wide range of technologies and the vast selection of chemicals used.
- The nature of the work – often non-routine and sometimes at the frontiers of knowledge.
- The large number and variety of 'samples' sometimes received on a routine basis but varying widely in toxicity (e.g. in quality control laboratories) or pathogenicity (e.g. in hospital pathology laboratories).

Official statistics on laboratory hazards are sparse. Where data are reported, [1–5, 7–9] accident classification schemes differ between authors, making quantitative conclusions difficult.

A comparison of the accidents for research and development (Table 11.1) with those of industry in general (Table 11.2) tends to indicate a close similarity of type for the majority of accidents in both cases.[1] This observation was also noted in a recent report on safety in Research.[2] It acknowledged that, although laboratories were potentially dangerous places, their accident rates were lower than in many other workplaces and the type of accidents experienced paralleled those in most other indoor environments.

One report[3] of school accidents in the USA suggests 12% are caused by 'classroom activities'. Of these 6% occur in general science, 2% in chem-

Table 11.1 Accidents in research and development services in 1981

Type	% of Total
Over-exertion	25
Falls	31
Struck by objects	17
Mechanical	11
Striking objects	9
Chemical	6

Table 11.2 Causation of accidents in all industries in 1981

Type	% of Total
Machinery	26
Materials handling (lifting)	24
Falls	16
Other	14.5
Muscular	11
Striking against	6
Chemicals	2
Fires/explosions	0.5

istry and 2% in biology. Table 11.3 summarises the nature of accidents in these three categories. And again, though potentially hazardous it is suggested that serious student injuries are prevented by close supervision.

The types of accidents occurring in medical laboratories are summarised in Table 11.4.[4]

Data from a 10-year study of clinical laboratory accidents reveal that parts of the body most affected were fingers, hands and arms, with the main causes listed in Table 11.5.[5]

Minor accidents recorded for a 2-year period in a large government chemical laboratory employing 100 scientists plus 10 other staff are given in Table 11.6.

The importance of eye protection was emphasised by several authors: many serious injuries could have been prevented by use of eye protection while many 'near misses' would have been serious accidents if protection had not been used.

One analysis of laboratory accidents claims that 65% were due to human error and 20% to equipment problems. It suggests that the 20–29

Table 11.3 Accidents in school science laboratories in USA

Nature of accidents in science	%
Glass	21
Chemicals	18
Animals	13
Another person involved	10
Laboratory utensils	6
Metal item	4
Thermometer	4

Table 11.4 Nature of accidents – medical laboratories

Accident	Frequency (% of total)
Cuts (glassware)	40.7
Cuts (knives, needles)	15.8
Cuts (other)	16.3
Burns	7.2
Splashes, spills, leaks	5.7
Falls	4.1
Assaults	3.0
Eyes	1.9
Mouth pipetting	1.4
Animal bites	1.0
Gassings	0.8
Explosions	0.6
Concussions and sprains	1.5

Table 11.5 Causation of accidents – medical laboratories

Cause	Frequency (average of % quoted for university and private hospitals)
Glass	30
Microtome knives	7
Chemicals	3
Burns	16
Blades	15
Serum	5
Others	10
Needles	14

Table 11.6 Type of minor accident – chemical laboratories

Type of minor accident	Number
Cuts, bruises (hand tools)	8
Cuts (glassware)	7
Cuts, bruises (lifting and carrying)	7
Cuts (sharp objects other than glass)	5
Bumps, falls	5
Chemical splashes (hands and face)	4
Burns on hot surfaces	2
Cuts, bruises (power tools)	1
Chemical splashes (eyes)	1
Explosion in glove box	1
Foreign body in eye	1
Inhalation of toxic vapour	1

years age group are most vulnerable and that women were involved in significantly fewer laboratory accidents than men.[6]

In addition to chemists others at risk from laboratory work are technicians, cleaners and maintenance staff. Indeed it was deduced that these groups (after adjusting accident figures for the population in each category) are at more risk in university laboratories than undergraduates.[7]

One survey[8] of aspects of laboratory safety and working practices revealed that in laboratory accidents a significant number resulted in:

- death (mainly from gassings, fires and explosions, and infections);
- partial or total loss of sight (resulting from explosions, and reagents or glassware entering the eyes); and
- dermatitis or allergy (stemming in the main from use of formalin, organic solvents, inorganic chemicals and polymer catalysts and monomers). This was the major reason for laboratory workers changing or ceasing employment. (See also Table 5.16.)

Interestingly, in a separate survey[9] 11% of respondents also reported allergic reactions and it is common experience that many chemists suffer from some form of dermatitis at some period. Table 11.7 summarises potentially serious incidents occurring in laboratories in which questionnaire respondents worked.[8]

Over recent years the average number of gassing accidents reported in general in the UK are shown in Table 11.8 for solvents and in Table 11.9 for non-solvents.[1]

Apropos long-term hazards, several epidemiological studies on chemists have been undertaken. However, results with regard to expected risk from cancer are conflicting. Further work is required and the Royal

Table 11.7 Potentially hazardous laboratory incidents

Incident	% Reporting occurrence at least once in the year
Fires needing the use of extinguishers but not the fire brigade	21.6
Explosions or implosions	12.9
Accidental spillage or release of pathogens into the laboratory atmosphere	9.5
Accidental spillage or release of radioactive isotopes	9.3
Fires requiring the use of the fire brigade to extinguish them	2.0

Table 11.8 Gassing accidents (solvents)

Solvent	Average number of accidents notified each year
Trichloroethylene	10
Dichloroethylene	5
Formaldehyde	5
Other chlorinated hydrocarbons	6

Table 11.9 Gassing accidents (non-solvents)

Chemical	Average number of accidents notified each year
Chlorine	43
Carbon monoxide	16
Ammonia	10
Sulphur oxides	15
Hydrogen sulphide	9
Hydrogen chloride	7
Nitrous fumes	5
Phosgene	4

Society of Chemistry is currently undertaking a survey of the effects of chemicals on its members.

It is clear that chemical-type accidents represent a relatively small proportion of laboratory accidents; even half of the chemically related accidents reported in Table 11.1 stemmed from extremes of temperature while the remainder resulted from the harmful nature of the substance. Nevertheless, chemical accidents can be particularly severe causing serious injury and even death, and can result in major material damage. As shown by Table 1.1 a relatively high percentage of accidents in laboratories requiring first aid involve chemicals.

Few figures are available with regard to other hazards. However, a pilot study of conditions for use of radiation in universities concluded that actual practice often failed to meet the required minimum standards, even though the dose/response relationship of radiation and safety procedures to avoid adverse effects are well understood.[10] A report from the Health and Safety Executive also indicated that insufficient care was devoted to use of ultraviolent radiation in some academic institutions.[10] Problems identified in Occupational Safety and Health Administration (OSHA) type inspections of academic laboratories in the USA included improper electric wiring, unguarded machinery, improper storage of chemicals, inadequate ventilation and fume-cupboard face velocities, lack of eyewash and safety shower facilities plus a host of lesser issues associated with housekeeping, waste disposal, blocked-emergency escape routes, etc.[11]

While firm conclusions are difficult to draw from the foregoing statistics it is clear that to minimise accidents with chemicals in the laboratory due consideration must be given to the people (employees, students and visitors), the equipment and processes, and the inherent properties of the chemicals. These are discussed separately.

Precautions

Personnel considerations/administrative controls

As with other activities, responsibility rests with management to devise a safety policy and to make provisions for implementation of the policy's aims and objectives both in terms of hardware and manpower. The safety organisation must be described, areas or responsibility clearly defined, and a means of monitoring the effectiveness of the policy established as discussed in Chapter 17. At all stages, the background of the various groups must be taken into account, and arrangements made to ensure workers with no knowledge of chemicals are not inadvertently exposed to hazardous materials (e.g. wash-up assistants, cleaners, maintenance staff).

Safe systems of work must be established such as 'permit to work' systems for any maintenance work in high risk areas. This should eliminate the possibility of tradesmen being inadvertently exposed (e.g. to emissions when working on fume-cupboard stacks).

In general the policy should be based on a 'safe place' strategy, with detailed Codes of Practice established. Where work is non-routine a more flexible approach may be necessary and general guidelines provided (e.g. advising on how to handle certain classes of chemical rather than detailed procedures for specific substances). It is to be appreciated that for success the latter approach relies more heavily on employee ability, experience and co-operation.

In addition to these general requirements for a safety policy more specific aspects of laboratory safety should be identified in a set of 'laboratory rules' drafted to reflect individual company needs. No single person can cope with the complexity of day-to-day hazards in laboratories, especially large research establishments. As a consequence management must provide for a team approach to accident prevention and to ensure compliance with the rules. Regular inspections should be carried out by members of the management, and the safety committee. A comprehensive check-list is a useful *aide-mémoire*.[12] The team should report with formal suggestions for improvements. Key issues to consider in drafting rules on laboratory safety include the following topics.

Education

All employees should know the laboratory rules and understand the fundamental principles upon which they are based. They should be trained to understand all hazards they may encounter and how to utilise any special facilities provided to minimise risk. The nature of common laboratory accidents underscores the need for training, ranging from general actions to follow in the event of an emergency (including a sound grasp of first aid) to the development of good individual manipulative skills.

Supervision/working alone

First-line supervisors, including teachers in schools and colleges, must themselves be adequately trained to provide instruction and supervision of more junior members of their team or their students. They should:

- set a good example by observing all rules, using protective equipment, and demonstrating an enthusiastic attitude towards safety;
- maintain discipline and enforce rules;
- know the preventative and remedial action in emergency situations;
- check the adequacy of safety provisions in areas under their control;
- assume responsibility for students, visitors or outside contractors.

> A student was injured when a container of methanol ignited in a classroom. A mathematics classroom was being used for a physical science class because of crowding in the school. No storage, ventilation or other laboratory equipment was present. Open-flame alcohol lamps were used since gas outlets were unavailable. Methanol was kept in a plastic jug on a counter. Alcohol spilled and ignited. As a student attempted to extinguish the lamp there was an explosion. Her clothing was ignited and she received, severe second- and third-degree burns. The court held the teacher's conduct was justifiably called into question.[13]

Generally, it is advisable not to work alone in laboratories. However, on occasions lone working during 'silent' hours may be necessary. The level of precaution will be dictated by the degree of risk involved – which is often underestimated. This may range from periodic checks by security patrolmen in the case of clerical or domestic tasks, to arranging for the patrolmen to act as 'second man' during a stage where slight risk may be involved (e.g. topping-up liquid nitrogen flasks during the weekend). An alternative may be to rely on the presence of another laboratory worker within calling distance, e.g. in a communicating laboratory with door open. During the normal work hours, supervisors may need to arrange cover with colleagues whenever staff are depleted owing to meal breaks, or attendance at meetings, lectures, etc.

No work of a potentially hazardous nature should be undertaken without the presence of an experienced qualified assistant. Whenever persons are on the premises out of normal hours security should be informed of their presence together with the nature and purpose of their work. As discussed later, arrangements are also necessary for out-of-hours operation of unattended apparatus, and the restriction of entry to authorised personnel.

Housekeeping/behaviour

Generally the behaviour of employees and the standard of housekeeping reflect the attitude towards safety: poor housekeeping can lead to accidents (cf. Heinrick triangle, Ch. 1). General guidelines are:

- Work should be carried out systematically and in a tidy manner. This is aided by preplanning of experimentation.
- Running and horseplay should be prohibited in the laboratory.
- The use of laboratory coats is always advisable since they offer a first line of defence against accidental contamination and can be quickly removed and discarded.
- No unauthorised experiments should be permitted.
- Work areas, passageways, stairs, etc., should be kept clear and free from obstruction. Access to exits, emergency equipment, controls, etc., should never be blocked.
- If it is necessary to reach up to equipment, use should be made of a proper step-stool or step-ladder.

- Special facilities should be provided for waste chemicals, glassware and burnable rubbish (e.g. paper).
- Spillages and breakages should be dealt with immediately. This is aided by the provision of drip trays and absorbents.
- Rigs which may use liquids, including water, should be constructed with a drip tray of suitable size. This must be sufficient to contain all the liquid in the apparatus if it is to be left loaded and unattended at any time. The area of the tray should cover all parts of the apparatus where leaks are possible. (As with bunds for elevated storage tanks, the area of the tray should allow for the trajectory of possible leaks.) Any offtakes from these trays may be permanently piped to drains where appropriate.
- All containers should be adequately labelled.
- Floors and benches should be in good condition and cleaned regularly. Walls and windows should be kept clean. Work areas should be uncluttered.
- Equipment should be cleaned and stored when not in use. Chemical-stock containers should be wiped clean, stoppered and returned to their allocated storage location as soon as the required sample has been taken.
- Flexible tubing should be inspected regularly and replaced if it appears damaged or perished.
- Flexible cables, tubing, etc., should be as short as practicable and arranged neatly so as not to create hazards. Personal protection should be cleaned and in good order. It should be housed in specified locations when not in use.
- Lockers, cupboards, sinks and wash-up areas should be kept clean and tidy.
- Oily rags and cleaning materials should be kept in metal containers with close-fitting lids (to reduce the potential fire hazard).
- Warning notices, overnight running permits, etc., must be kept up to date and removed when they expire.

Accident reporting

There should be provisions for reporting and investigation of all accidents and near misses with a view to identifying the cause and preventing a recurrence, as opposed to apportioning blame.

Eye protection

Suitable eye protection must be worn whenever hazardous chemicals are handled and stored, and provisions must be made for visitors. Eye protection must be worn in other locations as instructed (e.g. in laser areas).

Eating and drinking

Eating and drinking should be prohibited in laboratory areas, particularly where chemicals are present. However, provision should be made for breaks (e.g. office areas, restrooms, etc.). It follows that the making of beverages and drinking from laboratory glassware is not permitted.

Two technologists developed cyanosis shortly after eating breakfast together. Though this suggested food poisoning, two others ate the same food and suffered no ill-effect. It was found that the two affected had used the salt cellar in the laboratory which had allegedly been filled with 'reagent grade NaCl'. Further investigation revealed that $NaNO_3$ from an adjacent jar had accidentally been used.[14]

Smoking

The prohibition of smoking in all laboratory areas may be relevant even where flammable chemicals are absent (since it eliminates one potential source of ignition, prevents ingestion of contaminants from the fingers, and prevents inhalation of any toxic thermal degradation products formed on contact of vapours with the hot cigarette). Convenient smoking areas should be identified.

Emergencies

Management must ensure provision for dealing with all foreseeable emergencies (fire, explosion, toxic gas release, floods, people trapped in lifts or with machinery, bomb threats, major catastrophies, etc., as relevant). Appropriate staff should be adequately trained and procedures rehearsed. Audibility tests of alarms and evacuation exercises should be carried out on a regular basis.

A number of standard fire precautions should be adopted in addition to special requirements to reduce the probability of a fire starting when handling highly flammable or explosive chemicals.

The laboratory normally forms part of a building with built-in fire-fighting provisions. A standard procedure should be established for action in the event of a local fire; a notice detailing this procedure identifying escape routes and listing key telephone numbers should be prominently displayed at strategic points in the laboratory. A typical sequence is:

- Raise the alarm, e.g. by breaking the nearest alarm-point glass.
- Inform the security and emergency service by telephone.
- If it is safe to do so, attempt to extinguish the fire. (This may include moving other materials which could be involved in an escalation.)
- If the fire is uncontrollable with the available extinguishers, close the laboratory doors and leave the building.

The fire stop doors, provided to prevent the spread of smoke in the event of fire, must never be propped open.

Whereas, as discussed in Chapter 4, a water hose may be used to fight fires of ordinary combustible materials, e.g. cardboard, paper, wool or rags, it should not be used on electrical fires or on fires involving liquids immiscible with water, e.g. organic solvents. Where there is a risk of electrical or chemical fires carbon dioxide or dry powder extinguishers are more usual. These are usually located near the exits, so that an escape route is kept clear while fire fighting; additional extinguishers may be provided near to apparatus containing flammable liquids to provide some opportunity to prevent a *small* fire escalating.

Unattended operations

Unattended operations are a major source of fire, explosion and flood.

> After the laboratory had closed for the weekend, fire broke out in a 23-litre insulated pot containing a synthetic flammable liquid. The liquid was undergoing a test at elevated temperature maintained by a hot plate equipped with a temperature control. Failure of this control was believed to be responsible for overheating and ignition of the liquid. Fortunately the fire was discovered in time by a watchman and eventually extinguished.[15]

It should be the exception rather than the rule to leave equipment running unattended. Depending upon the risk, fail-safe features may need to be incorporated into equipment such as heaters and condensers. Monitoring would need to be both technical and administrative. (For example, with the exception of distilled water stills, other distillations are hazardous because of the equipment and/or the chemicals involved.) Unattended equipment should be labelled to indicate the chemicals involved, how to isolate the equipment and any emergency procedures. Aspects to consider when leaving laboratory operations unattended are given in Table 11.10.

Unauthorised experiments

All experimental work must be approved by line management and all unauthorised practical work should be prohibited. Chemicals must not be obtained from outside suppliers, nor, must any chemical be dispatched from the site, without the supervisor's specific knowledge.

Communal areas

The responsibility for common areas such as sub-stores, cold rooms, satellite libraries, etc., should be vested in identified personnel.

Table 11.10 Features to consider for 'unattended operations'

Technical considerations

- Equipment should be left in a safe condition and electrically isolated at the end of any work period. Water supplies should be isolated or disconnected when equipment is not in use to avoid 'flooding' incidents.
- Fluctuation in water pressure, or complete failure, can have serious consequences if water-cooled condensers are an integral part of the apparatus. Flow sensors linked to automatic shut-down may be installed if manpower is unavailable. Examples include automatic termination based on time, temperature, weight or volume.
- Surges and complete failure of electrical power can occur. Although complete failure rarely presents a hazard, e.g. distillations normally stop, bumping could occur once the power supply resumes. Constant voltage transformers may be incorporated to rectify voltage fluctuation and relay drop-out devices installed to prevent power once failed from returning to the equipment.
- Devices should be used to stop automatically the distillation at a safe stage if manpower is unavailable. Again examples include automatic termination based on time, temperature, weight or volume.
- The chance of breakage of unattended apparatus should be minimised. Equipment should be assembled without undue strain, and only properly annealed and undamaged glassware used.
- Distillation receivers should be large enough to contain entire distillates in case the distillation rate is faster than anticipated.

Administrative precautions

Arrangements with security patrols for regular inspection of the equipment should take account that their technical knowledge may be limited. This is aided by the following steps:

- Obtaining permission from the supervisor for any experiment left unattended.
- Completion of a log book at the control centre, reception or gate-house. This should indicate the name of the experimenter, the nature of the experiment and the hazards.
- Display of an 'overnight running' card near the experiment indicating the above details plus what action to take in the event of failure of any of the services. The home address and telephone number of the experimentor and his supervisor/deputy should be included.
- Surveillance by close-circuit television. Similarly, administrative and where necessary, special technical precautions (e.g. close-circuit TV) may be taken for unattended endo- or exothermic reactions.

Equipment/techniques considerations

Since chemical laboratories vary in purpose so do the operations performed in them. It is not possible to cover the hazards and precautions for all laboratory equipment and techniques. Emphasis is placed here on the more commonly encountered situations which also serve to illustrate the general approach. For large laboratories elements of Chapter 2 are also relevant.

Laboratory design

Properly designed laboratories are a prerequisite to the safe handling of chemicals. The hazardous properties of the substances and the risks with associated equipment, must be considered at the design stage. Detailed discussion of laboratory design is beyond the scope of this book and the reader is referred to references 16 and 17. The salient features are summarised. Clearly the design must conform to statutory and local authority regulations. There should be full consultation about design with, among others, fire authorities and insurers.

At least two escape routes, suitable for use by anyone in the building day or night, should lead from work areas and high hazard areas. Large constructions may require compartmenting. Design should prevent smoke/fumes entering corridors, stairways and foyers but in the event of smoke-spread it should allow for ready ventilation. Emergency exit doors should be easily opened (outwards) and the route must be free of obstruction. Multi-storey buildings will generally need to be constructed so that walls, floors, ceilings and supports are of at least 2-hour fire resistant specification, although this will depend upon the hazard in adjacent compartments. Fire stops are needed to maintain compartmenting where ducts and services pass through walls, etc. For single storey laboratories it may be adequate to construct the buildings of non-combustible material.

Potentially very hazardous materials or processes should be confined to segregated parts of the laboratory, preferably located remote from the main building. These facilities may require special features such as airlocks, changing rooms, provision for scrubbing effluent, detectors, alarms, glove-boxes, and in extreme cases may be of blast-proof construction with provision for remote operation.

The design of the general laboratory should be simple and give cognisance to safety, ease of cleaning and maintenance, and to provide a bright and comfortable working environment. Walls and ceilings should ideally be smooth and impervious, vermin proof and have slow flame-spread characteristics. Extra thickening may be required in speciality structures (e.g. to provide shielding in radioactive laboratories). Ledges should be avoided where practicable. The manner in which service pipes pass through the wall needs careful attention, e.g. to seal any special laboratory and ensure integrity of fire compartmentalisation is maintained. Floors should preferably be of a concrete base finished with a covering (preferably without joints) to contain spillage. Tables 11.11 and 11.12 give guidance on laboratory floor coverings with respect to their resistance to chemical attack and ease of decontamination respectively.[17] Floors should be unpolished to avoid a slippery surface.

Table 11.11 Summary of resistance to chemical attack of floor coverings (24 hr exposure)

Attacking agent	Linoleum	Crestaline 67% PVC	Lefco fully vitreous ceramic tile
Acetone	2	3	1
Animal fats	1	1	1
Beer*	1	1	1
Butyl alcohol	2	2	1
Carbon tetrachloride	2	3	1
Chloroform	2	3	1
Chromic acid 3N	2	1	1
Diethyl ether	2	2	1
Ethyl alcohol	1	1	1
Glacial acetic acid	2	1	1
Hydrochloric acid	2	1	1
Hydrogen peroxide (10%)	1	1	1
Mineral oils	1	1	1
Nitric acid 7N (36%)	2	1	1
Paraffin	2	1	1
Petrol	2	1	1
Potassium hydroxide	3	1	2
Sodium hydroxide	3	1	2
Sulphuric acid 10N (38%)	2	1	1
Trichloroethylene	2	3	1
Vegetable oils	1	1	1
Water	1	1	1
Xylene	2	3	1

* Drinking should not permitted in laboratories.
1 = satisfactory; 2 = slight attack; 3 = attacked.

Table 11.12 Ease of decontamination of the surface of various materials

Material	Ease*
Stainless steel	0.01
Industrial polyvinylchloride, grey	0.01
Industrial polyvinylchloride, white	0.05
Polypropylene on glass fibre base	0.2
Plastics laminate	1.5
Plastics laminate with abraded surface	1.4
Plastics laminate treated with hypochlorite abrasive cleaner	5.7
Plastics laminate, aged	4.5
Polyurethane-varnished hardwood	33.2
Vinyl flooring	0.03
Vinyl flooring plus asbestos filler	4.4
Vinyl flooring plus rubber filler	10.0
Linoleum	6.9

* Percentage of contaminant ($^{60}Co/^{134}Cs$) remaining after decontamination by British Standard procedure.

Bench surfaces should have good resistance to mechanical, thermal and chemical abuse and should be easily decontaminated. Formica melamine resin plastic, polyurethane, or pyroceram-coated hardwood are generally superior for all-purpose use to plastics such as PVC or polypropylene. Laboratory furniture and fittings should be designed and installed to facilitate decontamination and such as not to hamper escape in the event of emergency.

It is advisable to locate main services requiring frequent attention by maintenance engineers external to the building. Services should be grouped, colour coded, and valves/switches identified.

Electrical services should be designed, installed and maintained to good practice by qualified electricians. Wiring or modifications of electrical equipment by laboratory workers should be discouraged (see Ch. 2). A labelled, mains-isolation switch should be conveniently located near the exit to the laboratory. UK regulations require all conductors and equipment exposed to weather, water, corrosion, flammable or explosive atmospheres, or used for any special purpose, be so constructed or protected as not to present a danger from such exposure or use. Sufficient sockets should be provided to prevent the necessity for long cables. The use of multi-plug adaptors should be kept to a minimum to reduce the hazard from trailing cables. Sockets should be of the separate switch type and should not be located immediately below a water tap or outlet. It is essential to avoid flexible cables coming into contact with chemicals, or being pinched between clamps or structures, or having step-ladders or containers placed on them. Fluorescent lighting is usually provided. Explosion-proof lights will be required in special locations such as solvent stores. Emergency lighting should be provided and may, in some circumstances, be a legal requirement. Electrical hazards are discussed further on page 584.

Water is used for wash-up and as a heat transfer medium in condensers and hot-water baths. Tempered potable water should be used for emergency showers and eye-wash fountains. Sewers are comparable to ventilation ducting as the means of transporting effluent from the laboratory; since sewers combine with other foul-water drains it is important that undiluted chemicals, or flammable solvents, should not be disposed of via this route. (The nature and level of effluent should be in accordance with legislation and local authority requirements – see Ch. 14.) They should be capable of carrying the maximum water flow via a trap. Whenever the trap is emptied, e.g. during maintenance, it should be flooded to prevent noxious and offensive odours entering the laboratory from the drains. Nowadays drains are usually plastic; these can be affected by organic solvents. Lead drains can be attacked by acids. Joints in old tiled gullies should be checked periodically for deterioration. If systems are inadequately designed, or if sinks, etc., are abused, chemists and wash-up and maintenance ancillaries can be put at risk.

Steam is used particularly in development laboratories and/or pilot plants as a heat transfer medium, to heat small jacketed vessels, to facilitate steam distillation, etc. Unlagged steam pipes can cause burns and heat build-up (causing discomfort and activation of heat detectors giving 'false alarms' on the fire detection system). In addition, surface temperatures can exceed auto-ignition temperatures of a few flammable vapours, e.g. carbon disulphide. Steam pipes and vessel jackets should be lagged with asbestos-free material, and metal flanges should be covered. Where solvent or oil spillages are possible the exterior surfaces of the insulation should be sealed or metal clad. Slight defects can develop in inner linings of jacketed vessels which can cause reactors to overheat. Pressure-release valves should be vented externally or in extraction ducts.

Gases should preferably be piped from outside[18] with non-return valves and pressure-relief systems. For hazardous gases automatic detection and alarm arrangements may need to be incorporated where practicable. For fuel gas (e.g. natural gas) an easily accessible control valve should be fitted outside the work area on the main supply pipe to enable isolation in the event of emergencies. Though Bunsen burners are still common, they are no longer the favoured means of heating equipment and their use has declined. The numbers of gas points should be minimised. Where compressed air is supplied, consideration must be given to the air quality. Compressor input supply should be clean and not contaminated with fumes; hence inlets should be situated away from equipment vents or fume extraction exhausts. High-pressure systems should have suitable (Schrader type) fittings. High-pressure hoses should be secure. General-purpose vacuum lines should be restricted to low vacuum only. For higher vacuums, individual pumps should be employed.

General ventilation

In the UK the Health and Safety at Work Act places the duty on employers to provide and maintain a safe working environment for employees and to ensure their activities do not pose a risk to others, such as the neighbouring populace. Laboratory ventilation is clearly crucial to ensure compliance with requirements, to provide wholesome air, to remove contaminants and to maintain the comfort of employees. The ventilation system can be conveniently divided into general and local ventilation (the requirements for factories are covered separately by s.4 and s.63).

For general ventilation reliance on natural air movement is insufficient and a mechanically aided arrangement is required. It is difficult to be specific over requirements but, generally, the system should provide 10–12 air changes per hour in the laboratory, account being taken of air extracted via hoods, fume-cupboards, etc. Air movement should be from offices and corridors into the laboratory and air is generally exhausted directly to atmosphere with no recirculation. This results in laboratories being under negative pressure with respect to other parts of the building. Air intakes should be located so as to prevent contamination (e.g. from fume-cupboard stacks) from re-entering the building via the ventilation system.

Contamination is best prevented from entering the working atmosphere by containment. Failing this, localised ventilation at source is preferred. This can be achieved, e.g. in pilot plants by positioning trunking over a leaking gland or over a temporarily opened filling port. It is also used in workshops, e.g. during welding. Depending upon the nature of the hazard the air can be recycled into the workplace environment via a filter, electrostatic precipitator, etc. This is not acceptable where toxic gases or vapours arise. Alternatively, emissions may be exhausted externally.

Fume-cupboards

If hazardous chemicals cannot be substituted by safer materials then all work with toxic, flammable or objectionable substances should ideally be contained. When this is impracticable fume-cupboards are a crucial asset for controlling the working milieu and affording protection for both workers and observers.

Basically, a fume-cupboard is a partial enclosure ventilated by an induced airflow through an adjustable aperture (see Fig. 11.1). Fume-cupboards are designed to prevent corrosive, toxic, flammable or socially unpleasant gases, vapours, fumes, etc., escaping from small-scale experiments into the working atmosphere. They also offer some protection from spillages, splashes and eruptions or minor explosions. The system comprises five main components, viz. the cabinet or chamber, ducting (including dampers), fan, and stack plus attendant services. The airborne contaminant is exhausted remotely from the working environment, though some speciality designs rely on recirculation of laboratory air via a filter bed. Air removed from the laboratory by the fume-cupboard system must be replaced, and the fume-cupboard requirements and performance specification must be considered along with the general laboratory ventilation. The location of air intakes must be chosen to prevent contamination of the make-up air. Also the location of diffusers in laboratories must prevent eddy formation in the fume-cupboard. In theory air can be supplied from corridors via grilles though the fire protection must not be weakened.

When working with fume-cupboards consideration must be for the user and others present in the laboratory, maintenance staff and the local community. Factors which influence the degree of protection afforded

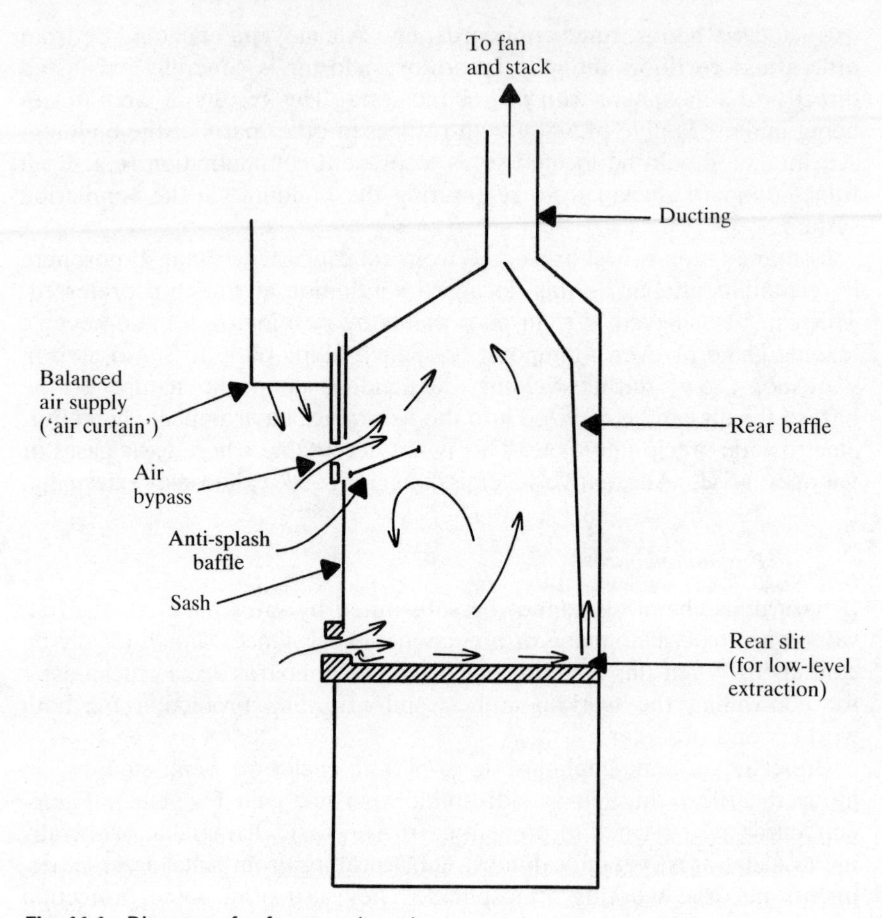

Fig. 11.1 Diagram of a fume-cupboard

include design specification (including performance rating and materials of construction), correct installation, siting within the laboratory, maintenance and good working practice.

A major factor influencing fume-cupboard contaminant performance is the velocity of air entering the face of the working aperture. Guidance on air velocity varies in the literature[19] but 100 ft/min (0.5 m/second) with the sash set at its maximum working-opening seems to be generally acknowledged as a minimum requirement, and indeed in the UK is a statutory obligation for work involving radioactive chemicals.[20]

For very volatile and highly toxic chemicals more efficient systems are required. Where necessary, physical stops should prevent the sash from being raised beyond its normal height if this would lead to face velocities below 100 ft/min. At the very least, signs should be displayed on the facia prominently marking this critical height. A minimum working sash height of 500 mm has been recommended.[21]

Each fume-cupboard should be marked with the face velocity, the date tested and the date of scheduled retest. Air velocity measurements are made using an anemometer and advice is given in the literature.[19,22] Reliance on face velocity as the sole criterion for fume-cupboard performance can be misleading: high degrees of containment have been achieved in fume-cupboards with face velocities of 80 ft/min[23] while others of poor aerodynamic design perform unsatisfactorily even with velocities in excess of 100 ft/min due to eddy formation within the chamber. At face velocities as low as 80 ft/min the degree of containment is greatly influenced by factors such as draughts from open windows, doors or even caused by adverse meteorological conditions, or as people pass the open sash of the fume-cupboard.

Thus the face velocity must be adequate to overcome eddies created by external influences yet not so high as to cause turbulence within the chamber resulting in leakage, or adversely to affect apparatus, or generate unacceptable noise levels, or consume expensive energy unnecessarily.

Design – Subleties of style in fume-cupboard design influence the degree of containment. Thus eddies in the cupboard are reduced by use of rear baffles. These provide for both high- and low-level extraction to smooth the flow of air in the chamber and to enable vapours denser than air to be scavenged by low-level extraction. Turbulence is minimised by air bypass arrangements. These compensate for changes in air velocity across the fume-cupboard face as the sash height is adjusted: air is bled into the top of the fume-cupboard as the sash is lowered. Design must guard against chemical spray through the top opening in the event of eruption in the fume-cupboard. Smooth air-flow is also assisted by use of fish-tail connections between ducting and fume-cupboard ceiling and by picture frame facias comprising 45° fairings. Even the shape of the sash grip can be important. Noise levels generally should not exceed 60 dB(A).

All components (e.g. ducting, dampers, fans, motors, stacks) should be adequately labelled for easy identification. Choice of material of construction will be dictated by degree of resistance to chemical attack, ease of cleaning and decontamination, resistance to absorption of hazardous chemicals, mechanical strength (e.g. abrasion) and thermal stability (including flammability), ease of construction and working, and cost. The inner walls of the cabinet can be of 316- or 317-type stainless steel, vitreous tile or fire-resistant boards (not asbestos based) on a wooden framework. Neither aluminium nor plastics (in general) are suitable: the former because of its low melting temperature, and the latter because of its low softening temperature. Fume-cupboard bases should be of impervious materials of adequate durability: a selection of materials is given in Table 11.13.[19,22] Some fume-cupboards require special attention; thus those to be used for radioactive chemicals should be strong enough where necessary to support lead or concrete shields.

Wood should not be used in the construction of fume-cupboards used for perchloric acid work since it may eventually become impregnated with the acid/perchlorates. The wood is then liable to ignite if the surface temperature reaches 100 °C due to heat from common sources, e.g. Bunsen burner or hot plates.

Sashes should be fitted with toughened glass or polycarbonate (see Table 11.14).[19,22] Sash cords should be preferably of stainless steel and arranged so that failure of a counter-weight does not result in free fall of the sash. The counter-weights should be guarded so that on failure they are prevented from falling on operators or on equipment.

The purpose of the extract ducting and fans is to remove contaminant safely without leakage and to discharge it remotely at an acceptable concentration. Ideally each fume-cupboard should be equipped with its

Table 11.13 Materials of construction: fume-cupboard base

Materials	Comment
Asbestos-based materials	Not generally acceptable.
Timber, etc.	Only acceptable if sealed with suitable plastic or stainless steel.
316- or 317- type stainless steel	Attacked by acids but easily decontaminated. Welds most susceptible to chemical attack.
Glazed ceramic slab	Porous beneath thin glazing but otherwise of good chemical and heat resistance.
Ceramic tiles	As for ceramic slab but adsorption can occur in between tiles. Acid resistant epoxy jointing compounds and furfuryl resins are preferred.
Glass reinforced plastics	Materials of high glass-fibre content possess chemical resistances and some heat resistance. Good mechanical properties but difficult to control quality.
Plastics	Good chemical resistance but some have low softening temperatures and pose a hazard. Rigid PVC has some limited useful outlets.
Melamines and phenolics	Offer good range of chemical resistance characteristics and do not easily burn. Thicknesses in excess of 6 mm are required when supported and 10 mm when unsupported. Bonding of, e.g. Formicas difficult to guarantee.
Lithium aluminium silicate glass ceramic	Resistant to abrasion and chemical attack but very difficult to work.
Glass	Generally chemically resistant though attacked by hydrofluoric acid. Not flammable but brittle to mechanical or thermal shock.

Table 11.14 Materials of construction: fume-cupboard sash

Materials	Comment
Ordinary window glass	Very weak and very dangerous in the event of explosions within the cabinet or on shock and should never be used. Requires edge protection.
Toughened glass	Withstands considerable impact but susceptive to point loads causing crystallisation and cannot be drilled.
Laminated glass	Holds together after fracture but has lower resistance to thermal shock than toughened glass. Certain organic solvents may attack plastic interlayers.
Wired glass	This is intended only for protection against spread of fire. Minor explosions within the chamber can produce shrapnel by spalling. It should never be used for sashes.
Polycarbonate	Both its resistance to temperature and its surface hardness are less than those of glass but it possesses higher resistance to impact and to HF. It is affected by ammonia, amines and amides.

own extract system. Alternatively, each system should be fitted with a damper to adjust the air flow through each fume-cupboard. Duct work should be designed to good engineering practice to ensure smooth flow, minimise noise and energy consumption, and to prevent settling of droplets or particles and build-up of condensation, etc.

Transport velocities should be around 7.5 m/second as a compromise between noise and duct size for most applications involving gases and vapours. Velocities 2 or 3 times higher would be required to transport particulate matter. Sharp bends in the ducting should be avoided and for sensitive analytical work it may be prudent to incorporate a single bend in the ducting near the top of the fume-cupboard to prevent condensation or deposits falling back into the cupboard. Horizontal sections of ducting should be avoided otherwise condensation can collect and lead to corrosion; a sloped section with drain tap is preferred. Supports for ducting should not transmit unnecessary stress to joints. Inspection covers prove useful. Ducts should be of circular rather than square cross-section and preferably of short lengths with bolted flanges and gaskets, though socket and spigot joins are also used. Ideally ducts should be fire resistant and of high softening temperature, be non-absorbent and resistant to chemical attack. Rigid PVC and polypropylene are used for most purposes with service temperatures up to 60 °C and 90 °C respectively. Joints in plastic ducting should be welded to prevent leakage. Fire dampers are generally not recommended. Where ducts pass through adjoining compartments of a building they should be encased in material having the same fire resistance as that compartment. Generally, it is advisable to seek advice from the local fire authorities when planning new or modified laboratories.

Centrifugal fans and mixed flow fans are widely used in fume-cupboard systems. They should be of galvanised mild steel, polypropylene or PVC. Plastic fans or induction venturi systems are advised for especially hostile environments (e.g. prolonged use of perchloric acid). The fan rating should be adequate to overcome the resistance of the ducting and to draw the required air through the extract system with capacities of 110% of the operating requirements (speed range 720–1,440 rpm). They should be sited and mounted so that noise and vibration are minimal. They should be located as near as possible to the discharge end of the duct work: they must not be put directly above the fume-cupboard otherwise positive pressure, noise and corrosion may develop. Connections of fan inlet and outlet should be flexible. Fan motors (either direct or belt driven) should be outside the air stream and, if in a potentially explosive area, of flameproof construction. Condensate drains should be fitted at the lowest point of the fan casing. Access platforms should be provided where necessary to afford ample and safe working areas around fans. Where fans are located in plant rooms the latter should be adequately ventilated of fumes and leaks from ducts.

Emissions must be exhausted safely from the laboratories so as not to constitute hazards or nuisance to employees (either from re-entry into buildings via windows or air-intakes, or if working on the roof), or to the neighbouring community. Factors which determine the risk include the nature and concentration of exhaust contaminant, effluent velocity, stack height, building geometry, terrain, prevailing climate, etc. Mathematical dispersion and dilution models exist to take account of these parameters though none is completely reliable. In the main, good practice ensures the level of contaminant extracted by fume-cupboards is small and heavily diluted on discharge. Experience has shown that discharge should be via high velocity jet cowls with a minimum discharge velocity of 7–10 m/second. Occasionally, fume scrubbers or dust collection units are justified. Unacceptable terminations include Chinese hats, covered boxes, T- or low-blow arrangements. The siting of stacks *vis-à-vis* air intakes needs consideration. Ideally stack heights above roof level should range from 3 m for single-storey buildings to 8 m for 10-storey constructions: in reality compromise is often required because of restrictions due to planning permission.

Installation and commissioning – Fume-cupboard efficiency can be reduced by draughts caused by people passing by, poorly sited air-supply grilles, open doors and windows, and nearby bulky fixtures. For this reason fume-cupboards should not be located near to busy thoroughfares, doors, other non-related ventilation units, etc. The siting of fume-cupboards should not hamper emergency escape routes, and indeed, in certain cases additional escape routes are required (e.g. under-bench tunnels). The design and construction should ensure structural integrity in the event of an explosion. Figure 11.2 depicts the complete collapse of a fume-cupboard as a result of a relatively minor explosion.

Laboratory engineers should maintain a watchful eye on the installation of units by contractors. Commissioning checks include visual inspection and performance tests such as face velocity measurements, smoke tests and containment tests. If the system is satisfactory, static pressure after

Fig. 11.2 Collapse of fume-cupboard as a result of explosion

each hood and after each junction, air volume/velocity in each duct, coefficient of entry for each hood, static pressure differences across the fan (and filter where installed), and noise levels should be measured where relevant. The position of rear slit setting, vanes on air curtains and the dampers should be recorded.

Maintenance – In order to prolong the life of the fume-cupboard system and to ensure performance continuously matches design specification, regular maintenance of the system is essential. The frequency of maintenance will be dictated by circumstances. One suggestion for a maintenance schedule for fume-cupboards is given in Table 11.15 though experience will indicate whether a different frequency is more appropriate.

When ducting is removed, all open pipework should be capped to contain any hazardous deposits therein. Before ducting is re-used, stored or disposed of, it should be taken outside the building and hosed thoroughly with water, inside and out.

Work may be required on the fans due to blockages, failure of bearings, loss of efficiency, etc. The motor must be electrically isolated before

Table 11.15 Specimen maintenance schedule for fume-cupboards

Six-monthly intervals:

(a) Visual inspection (e.g. general cleanliness, position of rear slit or air vanes, presence of anti-splash baffles, jambs running smoothly, etc.).
(b) Face-velocity measurements.

Twelve-monthly intervals:

(c) Remove the baffles and clean both the baffles and the rear of the chamber.
(d) Wash the entire interior surface of the chamber with dilute detergent solution.
(e) Inspect the sash mechanism for corrosion and damage.
(f) Inspect the fans and their motors, drives and bearings for correct running. Drain contamination from fan-casing.
(g) Check the position of the balancing damper.
(h) Dismantle the sash mechanism and inspect it for corrosion and damage.
(i) Check the condition of the services to the cupboard, including their controls.
(j) Inspect the fire damper and the release mechanism.
(k) Check that the motors of the extract-air and supply-air fans do not overheat and test for worn bearings. Check for excessive noise and the state of the flexible couplings, and the anti-vibration mountings are free, and that the pulleys are tight and correctly aligned and that diffusers and louvres are not obstructed. Measure the fan and motor speeds and the electric current flowing through the motor.
(l) Inspect the fan impellers for wear.
(m) Check the stability and condition of the discharge stack.
(n) Inspect the condition of the extract ducting, particularly the joints. Check the need for cleaning the interior of the ducting through the inspection ports, particularly at bends where substances may be impacted.
(o) Check the make-up air balance and its temperature.
(p) Check alarm mechanisms (flowrate indicator if attached).

attempting work on the fan. Any materials accumulating inside the casing should be considered potentially hazardous, checked with litmus (neutralised, if necessary) and washed out with water if miscible; alternatively, materials should be collected for disposal, including any solvents used during cleaning.

The sequence of dismantling operations for fume-cupboards, ducting, etc., should be from upper to lower components.

Work on the fume-cupboard outlets at roof level should be undertaken only when it is known that no other fume-cupboards in the vicinity are discharging hazardous fumes. Work on the roofs is generally carried out during 'out-of-normal' hours. Consideration must be given to the necessary precautions for work during normal hours.

A 'permit to work' should be completed by the appropriate laboratory manager or his representative before any work is undertaken on fume-cupboards in radiotracer laboratories or where explosion-proof electrics are involved. Special arrangements may be required according to existing levels of contamination and other conditions. Before the equipment is returned to the laboratory after maintenance, the engineers should carry out a performance check and report the results to the laboratory managers concerned.

When the nature of the work in a fume-cupboard changes (e.g. from radiotracer to non-radiotracer work), before the permit-to-work system is deferred the laboratory manager responsible must ensure that any previous hazards associated with the fume-cupboard system no longer exist. This will be done by monitoring and cleaning procedures, etc., where necessary.

Containment tests must be performed on fume-cupboards following major changes in the laboratory, such as installation of additional fume-cupboards, additional walls provided, walls removed, etc.

Before cleaning or maintenance work is attempted, approved protective clothing should be worn, e.g. overalls, safety footwear, cap, dust mask, gloves and eye protection together with any additional equipment required because of the condition and previous usage history of the fume-cupboard. Scientific staff should remove all apparatus, tubing, broken glass, residues, debris, chemicals, etc. (apart from 'permanent' fittings) from the fume-cupboard and immediate adjoining areas must be completely cleared when dismantling cupboards. The laboratory manager where necessary should arrange for monitoring of any known hazards caused by recent splashes or spillages of chemicals. Monitoring for mercury (see Ch. 7) in particular should be considered because of its extensive use in most laboratories, ease of entrapment, volatility and toxicity. Any dust contamination should be removed with a special vacuum cleaner which incorporates a high-efficiency disposable filter bag; attention should be given to the inside surfaces of the cupboard and associated duct work. Splashes should be tested with litmus and if necessary neutralised. When the foregoing procedures have been implemented the cupboards should be washed down. The engineering personnel should check with the manager immediately prior to commencing any work to confirm that conditions are safe, and that their activities will not create a hazard to others working in the laboratory. Before painting commences, the fan should be switched on in order to remove any paint fumes.

By definition, residual volatiles are unlikely to present a hazard to maintenance staff since these chemicals will have been extracted, but there may be problems associated with dusts (e.g. from powdered substances settling in the duct or caused by duct corrosion) and viscous materials resulting from collection and condensation in the ducts with possible chemical reaction.

The fan should be isolated and the dampers closed before work is commenced, to minimise 'chimney effects'. The manipulation of large items of ducting requires the attendance of at least two tradesmen.

Use of fume-cupboards – Unless fume-cupboards are used properly even well-designed and installed facilities will not provide optimum protection. It is line management's responsibility to ensure experiments are undertaken in a safe manner in the appropriate location and with adequate facilities. Hence users of fume-cupboards should be trained in their correct use; the training needs of other relevant personnel (e.g. maintenance staff) must also be considered.

Fume-cupboard settings (e.g. position of vanes on air curtains, rear slit) should not be interfered with or features (e.g. anti-splash panels) removed unless authorised. Any defects should be reported immediately.

All work involving toxic gases, vapours or liquids should be undertaken in fume-cupboards of adequate performance (see later). Fume-cupboards are inappropriate for handling highly toxic, dusty materials which should be handled in glove boxes or specially designed ventilated enclosures. Walk-in fume-cupboards should not be used for volatile toxic materials unless their adequacy is demonstrated in each instance by environmental monitoring. Arrangements should be made to prevent highly flammable substances forming levels in the fume-cupboard atmosphere above 20% of their lower explosive limit. Suitable fire extinguishers should be at hand and nearby workers informed of the hazards and precautions.

Potentially explosive materials or substances which are both highly volatile and highly toxic should be substituted for safer alternatives if possible. Where toxic materials are essential they must be handled in aerodynamically designed fume-cupboards of above average face velocity (e.g. two-speed fans). These special fume-cupboards should be equipped with an airflow sensor and fan-failure alarm.

Before a fume-cupboard is used it should be empty and shown to be working. Only one experiment should be performed at a time in the fume-cupboard. The experiment should be designed so as to minimise quantities of materials in use and the rate of reaction, and to contain vapours/gases wherever possible. Vapours from highly toxic mixtures should be scrubbed or absorbed. Deliberate flooding of the chamber, e.g. by boiling off solvent (distil under controlled conditions, instead) should be avoided.

All equipment required for the experiment should be placed in the fume-cupboard before starting the work. Apparatus should be assembled, preferably over a spill-tray, to the rear of the chamber such that the sash can be fully lowered. The spill-tray, bulky apparatus, etc., must not block bottom, rear extract slots. Arm movements in and out of cupboard and rapid movements which can cause eddies must be avoided. The operator should avoid leaning inside the chamber when handling chemicals or once the experiment is set up.

Fume-cupboards should not be used as a store and should not be cluttered with superfluous apparatus: stocks of chemicals should be removed from the cupboard once the equipment has been set up.

If the materials are particularly hazardous it is advisable not to sit in front of the fume-cupboard, and the operation should always be with the sash pulled down as low as practicable (preferably fully lowered):

> Eleven children and a teacher were treated for acid burns after an explosion during a school chemistry demonstration. The 12 were sprayed with sulphuric acid when hydrogen exploded inside a fume-cupboard. They received burns to their faces, hands and eyes. Although the fume-cupboard took the brunt of the explosion the sash was partly open which allowed the acid to spray into the laboratory.[24]

The fan should be kept running until the experiment is complete and the chemicals have been removed from the chamber. Cupboards should be cleaned at the end of each experiment and the sash fully lowered. For work in fume-cupboards left overnight the sash should be fully lowered and the fume-cupboard switched on: some systems are arranged whereby the fume-cupboard switches on/off automatically with the general area ventilation. The reader is also referred to Table 11.10. After any spillage in the fume-cupboard the affected area should be decontaminated. If there is a major spillage of hazardous material the sash should be closed and the laboratory evacuated.

For handling explosive chemicals (e.g. organic peroxides), conventional fume-cupboards (as described herein) are suitable for quantities up to about 0.5 g, so long as hand and face protection (e.g. polycarbonate or 'Woolwich' triplex screens in strong frames) is used. The provision of a safety screen does not negate the need for appropriate eye protection, since adjustments to apparatus, or sampling or observations may involve working close to the apparatus. (Hand protection will also be necessary.) For work involving 0.5 to 5 g of these materials the same arrangements apply but the work is best undertaken in a 'Dangerous Chemicals' laboratory which is specially equipped and located remote from the main areas of activity. For work involving greater quantities of explosive materials special armoured fume-cupboards may be required.

When using perchloric acids, in addition to precautions necessary for handling these materials, use should be restricted to a specific fume-cupboard and all vapours trapped/scrubbed to prevent release into the fume-cupboard or ducting. Special care is needed when stripping these fume-cupboards.

Containment tests – Because of the difficulty of relying solely on face velocity as the sole criterion for assessing the efficiency of fume-cupboards, environmental monitoring is the only true method for checking on containment. This should be considered when either highly toxic, volatile chemicals are being used or if there is real doubt about fume-cupboard performance. Clearly, this approach is impracticable for every experiment. For this reason containment tests have been devised to give an indication of likely leakages from fume-cupboards.

Smoke tests – Release of smoke (e.g. titanium tetrachloride, amine hydrochlorides) from tubes or bombs into the chamber highlights the pattern of air movement both inside the hood and on discharge from the stack, the presence of vortices, and effects of cross-drafts, leaks from ducting, etc.

Containment index[19,22] – Gas is released into the fume-cupboard at a known rate and any leakage is measured outside the fume-cupboard. This enables a containment index, L, to be calculated thus

$$L = \log \frac{r}{C_0} = hwv \text{ which approximates to } \log \frac{C_i}{C_0}$$

where:
- r = rate of release of gas inside the fume-cupboard
- C_0 = concentration of gas escaping from the plane of the sash
- h = maximum, working sash-opening
- w = fume-cupboard aperture width
- C_i = concentration of gas inside the fume-cupboard
- v = average face velocity

Gas should be chosen which is of relatively low toxicity and easy to detect. Examples include Freons (which can be monitored by a portable IR spectrometer, as mentioned in Ch. 7), or sulphur hexafluoride (which can be detected by electron-capture cells).

Spill tests – These aim to illustrate the effectiveness of fume-cupboards in containing vapours from spillages. Volatile solvent capable of detection at low atmospheric concentrations (e.g. methylchloroform) is poured into a spill-tray inside the switched-on fume-cupboard and allowed to evaporate. The concentration of vapour outside the plane of the sash at various sash-height settings indicates efficiency of containment in such emergency spillage situations. In such circumstances the practice should be to fully lower the sash (and evacuate the laboratory if necessary). Significant leakage with the sash down indicates inadequate containment.

Glassware

Most laboratory ware used for handling and storing chemicals is of glass because of its transparency, chemical resistance and hardness. Glass is a major source of injury in laboratory accidents as mentioned earlier; however, most of these accidents can be prevented by training, supervision and personal diligence.

Glass is a supercooled liquid. There are two main types of glass for laboratory equipment, soda glass and borosilicate (e.g. Pyrex) glass. Because of its superior thermal and chemical resistance properties the latter is most common. Whereas borosilicate glass requires a gas–oxygen flame for working, soda glass with its lower softening point can be worked in a gas–air flame. However, unless skilfully annealed, soda glass is more prone to fracture than borosilicate glass. Soda glass is reserved for reagent bottles, measuring equipment, policemen, tubing, etc., whereas borosilicate glass is essential for apparatus to be heated or subjected to vacuum.

All forms of glass tend to be fragile and should not be subjected to severe mechanical or thermal shock. This factor must be considered in the storage, choice, assembly and general manipulation, cleaning and disposal of glassware.

Storage – Overcrowding glass items in drawers, cupboards, on shelves, etc., should be avoided. Drawers should slide smoothly and not be too shallow. Small, round items may be stored in drawers so long as they are cushioned (e.g. with cotton wool). Large items should be stored on shelves with heavy, cumbersome articles at low levels and round-bottom vessels seated on ring supports. Tall items should be stored at the rear and smaller ones at the front. Stop-cocks should be removed and ground glass joints dismantled before storing such items. Rods and tubing should be stored horizontally without protrusion of long items.

Choice of equipment – Apparatus should be inspected before use and dirty or damaged glassware (e.g. with jagged edges, scratches, hairline or star cracks) discarded. This is particularly important when choosing equipment for work involving pressure or vacuum. Equipment with defects should be disposed of in a specially designated glass bin, or sent for repair by a competent glassblower with access to facilities for adequate annealing. In addition to normal, visual inspection, strong and polarised light is useful for identifying strains in glassware. High-frequency Telsa coils are useful for highlighting defects in equipment under vacuum.

Before sending glassware for repair it should be decontaminated and thoroughly cleaned (see later) and be free of flammable chemicals, mercury or other substances hazardous to a glassblower. In general no markings should be scratched on to glassware since this results in a significant reduction in its strength. All glass tubing, rods, etc., should be fire-polished before use in order to remove razor sharp edges capable of lacerating skin.

During assembly of a high-vacuum apparatus a one-litre flask was being connected to the main rig by the usual glass-blowing techniques. The flask had been washed with water, then acetone, and was eventually evacuated and brought up to atmospheric pressure several times to remove the acetone. However, the flask exploded when the flame of the torch was applied. The flying fragments broke several fluorescent light tubes, smashed part of the vacuum rig, and cut the glassblower about the face and hands.[25] (As a consequence of a similar incident one glassblower rejects acetone-rinsed apparatus, or washes it in water and dries the glassware for 30 min at 150–175 °C in a vented drying oven.)[25]

The correct size vessels should be chosen to accommodate the chemicals at the required temperatures and leave a minimum of 20% free space. For glassware heated in isomantles it is extremely important that the correct size, round-bottomed flask is used, otherwise overheating of the elements may occur (see Ch. 10).

Equipment for use under pressure should ideally be round bottomed or specially constructed of thick-walled glass (see later). Laboratory containers must not be used for beverages or food.

Glassware handling

Assembling apparatus – Apparatus should be assembled well on to the bench, or well inside the fume-cupboard to prevent equipment from being knocked by passers-by or by the fume-cupboard sash. A spill tray is advisable to contain liquid contents in the event of glass breakage, particularly when hazardous chemicals (e.g. acids, alkalis, solvents) are involved.

Equipment such as flasks should be assembled vertically starting from the bottom. As few clamps as necessary are advisable to secure equipment firmly so that there is little leeway for misalignment. Only the correct clamps should be used and reliance should not be on clamps at the neck of flasks alone. Heavy vessels should also be supported with rings. Apparatus supported by rings and clamps should be positioned so that the centre of gravity is over the base of the retort stand and not to one side. With the exception of the bottom clamp, all clamps should initially be fastened loosely, tightening only once alignment has been checked. Joints must not be forced. (Ball-and-socket joints may be more suitable than cone and sockets.)

Space is required below the equipment to lower flasks in an emergency. Ancillary equipment such as rubber tubing, electric cables, etc., should be arranged tidily so that it cannot be accidentally caught. Rubber tubing supplying cooling water, vacuum, etc., should be secured at both ends; it should be sound (i.e. with no visual signs of perishing).

Handling carboys and Winchesters – Extra care is needed when handling any old-style glass carboys or Winchester bottles. The wall thickness may be less than 1 mm and they should not therefore be knocked or lifted by the neck. The containers (even empties) should be transported in specially designed carriers. Some supply houses use Winchesters covered in plastic coating to render them unbreakable if dropped.

Full carboys should be vented. Storage of any containers at heights should be avoided, particularly if the contents are hazardous.

Cutting/breaking glass tubing or rods – The correct procedures for cutting glass tubing/rods is to rotate a cloth-covered (for hand protection) tube on a bench against a lightly held file or glass knife, wetting the scratch and pulling the tubing apart without bending it.

Inserting/removing tubing from bungs – The correct procedure for inserting glass tubing into bungs, in order is (i) fire polish the end of the tubing; (ii) ensure the hole in the bung is the same size as the tubing; (iii) lubricate the tube with glycerol; (iv) hold the tube with a cloth to protect the hands and allow no space between the bung and the fingers when inserting the tube. Pressure and torque should be minimised. (A similar technique should be used for removing tubing taking care not to twist the tubing in the bung. If the tube is difficult to extract cut away the bung from the tube, again taking precautions to protect the hands.)

Removing 'frozen' glass-to-glass joints – It is usually easier to prevent joints freezing than to free joints once they jam. For example, alkali reagents should never be stored in ground-glass stoppered bottles or burettes. Where problems have arisen careful techniques are required to remove frozen burette taps, glass bottle tops, etc., such as those described below, when hand and eye protection should be worn.

1. Tap the joint gently with a wooden implement (e.g. spatula handle) working directly over a cushioned bench top.
2. Heat the joint in hot water or gently rotate in the smoking flame of a Bunsen burner. (Ensure first that the materials are not incompatible with water in the former instance and not flammable in the latter.)
3. Soak the joint in detergent solution or solvent, again after consideration of any adverse reaction with the contents.
4. A frozen desiccator lid can be loosened using a fine metal object as a wedge. Ensure atmospheric pressure has first been restored in vacuum desiccators.

Vacuum work – Glassware under vacuum can implode resulting in glass fragments flying with considerable velocity, chemicals splattering and possibly fire. Whenever glassware is under vacuum the area should be designated an 'eye protection area' and approved protective spectacles or face shields worn as appropriate. Any glassware under vacuum should be considered as having potential to implode either spontaneously or if accidentally knocked. For this reason damaged, or even repaired, glassware (unless skilfully annealed) should not be used. Equipment under moderate vacuum is as likely to implode as that under high vacuum, since failure results from inability of the equipment to withstand the pressure differential across the walls. This is essentially similar in both cases. Thus the pressure differential across the walls of a flask under moderate vacuum such as that achieved with a water aspirator (typically 10 mmHg) is $760 - 10 = 750$ mmHg, while that obtained with a high-vacuum pump (typically 10^{-5} mmHg) is $760 - 10^{-5} \simeq 760$ mm. Only purpose-built glassware should be subjected to vacuums. Generally, round-bottomed or pear-shaped vessels are required. Vacuum should never be applied to flat-bottomed vessels unless constructed of thick-walled glass (e.g. Buchner flasks).

During a vacuum filtration a laboratory worker suffered severe burns to the arms, face and eyes (even though safety spectacles were worn) when the flask imploded. Investigation revealed the modified equipment, including a 3-litre flat bottomed flask, was unsuitable for vacuum work.[26]

Glassware can be covered with friction tape or close-fitting plastic mesh. Large items (e.g. in excess of 1 litre) should be assembled in fume-cupboards or behind blast screens. Desiccators should only be used under vacuum when absolutely essential. Even then plastic versions (e.g. polycarbonate) are generally preferred. Glass desiccators subjected to vacuum should be of Pyrex glass and always taped, or guarded under a metal cage when under reduced pressure. Dewar flasks should be of borosilicate glass and be protected with tape or an outer metal casing to contain flying glass on failure. Domestic thermos flasks with thin walls are not adequate substitutes for thick-walled Dewar flasks; explosions can result from pressure build-up as cryogenic material vaporises on warming (see Ch. 3).

> When a domestic thermos flask was used for holding a blood sample with dry-ice refrigerant a serious explosion occurred as the top was being removed. Two men received multiple cuts to the face and one received injury to both eyes. The accident occurred as a result of build-up of pressure within the Dewar which had no vent.[27]

Traps should be inserted in line between the pump and equipment under vacuum to prevent 'suck-back'.

High-pressure work – High-pressure chemical reactions should be undertaken in purpose-built metal vessels (autoclaves) in specially-designed, cubicles. This is discussed in Chapter 12. Occasionally, glass 'Carius' tubes can be used for pressures of a few atmospheres (Fig. 11.3). These

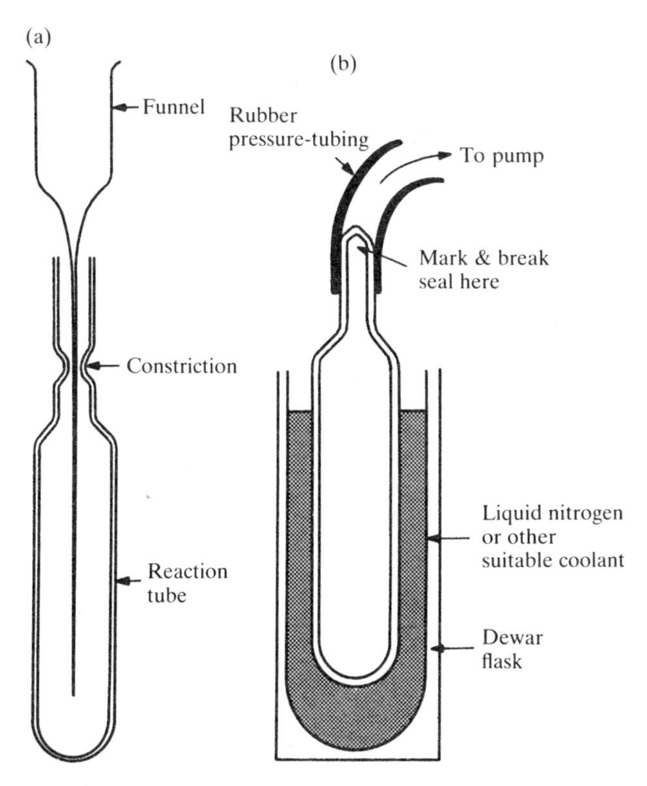

Fig. 11.3 (a) Charging Carius tube; (b) opening Carius tube

are specially constructed, thick-walled, Pyrex tubes (unless reagents are to be exposed to UV radiation when special silica versions should be employed). Tubes can be sealed in the conventional manner or fitted with PTFE taps. The precautions necessary with Carius tubes are summarised in Table 11.16. Precautions are also required when handling commercially available ampoules of reagents.

> Two accidents occurred as a result of opening sealed soda glass ampoules of dimethylamine. The amine boils at 7 °C at atmospheric pressure and at 20 °C the vapour pressure is 1.7 atm. As a research worker chilled the ampoule in acetone/cardice mixture at −21 °C in preparation for opening it there was an explosion causing the operator to lose the sight of one eye. As a result more stringent precautions were introduced but on a further occasion a similar incident occurred. Fortunately the operator was better protected and suffered no injury. It was discovered that the ampoule glass was of variable thickness and in parts perilously thin.[28]

These ampoules should be opened in a fume-cupboard with the sash as low as practical. Gloves and a face shield should be worn. The container should be cooled in ice/salt bath before chilling further if this is deemed necessary.

Table 11.16 Precautions for using Carius tubes

A. General

Inspection
All tubes should first be inspected and damaged tubes (hair-line or star cracks, etc.) should be rejected.

Pressure limits
The tubes are designed to withstand 7 atm pressure. Anticipated pressure should carefully be calculated *before* a reaction is commenced to ensure that this maximum pressure will not be exceeded.

B. Preparing the tube

Cleaning
Solvents such as acetone, alcohol, etc., should be avoided as cleaning agents because of dangers of ignition during sealing (see D). If their use is essential, however, the solvent wash should be followed by an aqueous rinse (or an alcohol wash then an aqueous rinse if the solvent is water-immiscible) and finally the tube should be carefully 'flamed out'.

'Necking'
To facilitate sealing of the tube when the reagents have been introduced, a constriction should first be made in the neck of the tube. Care should be taken to ensure that the wall of the constriction is of the same thickness as that of the tube walls (see Fig. 11.3(a)). Occasionally, 'necking' may have to be done after the reagents have been introduced.

Flaming
Certain experiments demand a completely anhydrous environment in which case the tube is put under high vacuum and heated over its entire length with a hair-drier or a cold Bunsen burner flame. If solvent has been used and flaming is simply to remove surplus water from the aqueous rinse it is advisable to flame the tube out before necking off, in

Table 11.16 (continued)

case any residual solvent is present. In general, however, flaming-out solvent-rinsed glassware is not recommended.

C. Introducing reagents

Volatiles
Gases and volatile liquids are best introduced under vacuum (i.e. the tube is connected to a gas rig, evacuated, cooled in liquid nitrogen and then opened to the volatile reservoir until the required volume of reagent has condensed into the tube).

Solids and high-boiling liquids
These materials are best introduced via a funnel to avoid contamination of tube neck at the point of sealing. This prevents problems of compound ignition or vitrifaction on heating (see Fig. 11.3(a)). However, because of particle size of certain reagents it may be necessary occasionally to introduce solids before the tube is 'necked'. Again, however, a funnel should be employed to prevent solid from adhering to the tube neck. The funnel is best made from a 'pulled out' boiling tube.

D. Sealing

PTFE taps
PTFE taps are available for glass-blowing on to the neck of reaction tubes. These are convenient to use and will withstand the maximum pressure recommended for the Carius tube.

Glass seals

(a) Under vacuum
It is advisable to freeze reaction tubes in liquid nitrogen under vacuum before sealing. The thickness of the seal should equal the thickness of the tube walls: a thin seal will not withstand high pressures and a thick seal can cause strain at that point and thus be as hazardous as a thin seal.

(b) Under air
Liquid air soon collects in tubes open to the atmosphere if refrigerated in liquid nitrogen. When such tubes are sealed and allowed to warm explosions can occur. Thus, if tubes are to be sealed while open to the atmosphere, a warmer refrigerant than liquid nitrogen should be employed.

E. Warming tubes
All refrigerated tubes after sealing should be placed in metal guard-tubes in the fume-cupboard to warm to ambient temperatures. If further heating is required furnaces are available to take guard tubes and their contents: this operation should be done in a special dangerous chemicals laboratory. Under no circumstances must tubes be heated without using guard tubes.

F. Opening tubes

With PTFE seals
The tube should first be cooled (e.g. in liquid nitrogen), in case excess volatile reactants/products are present, before opening the tap to condensation traps under vacuum. If the tube is to be opened to the atmosphere care must be taken that, on cooling, liquid air is not allowed to condense in the tube.

With glass seals
Using a glass-knife a scratch is first put in the neck of the tube just below the seal. Rubber pressure-tubing is carefully forced over the tube neck, the tube cooled and vacuum applied. Manipulation of the rubber tubing allows the seal to be broken.

Cleaning glassware – As soon as apparatus is finished with it should be cleaned. Apparatus should always be 'decontaminated' before sending for 'wash-up'. Detergent solutions should be used in preference to more aggresive chemicals for cleaning wherever practical. A check should be made to ensure cleaning brushes contain no exposed sharp metal points to scratch glassware and excessive pressure on the brush should be avoided. Rinsed glassware may be dried in ovens, except graduated equipment which should be air-dried at ambient temperature. Inflexible, thick wire should not be used to unblock burette tips or taps.

Where chemical cleaning is essential the safest available agents should be used. (Aqua regia and chromic acid are both corrosive and oxidising agents.) Mixtures of ethanol and nitric acid can form explosive products and should be avoided. Ultrasonic cleaning is a safer alternative. If acetone is used to aid drying, and small amounts are involved, then residues can be discarded to drain flushing with copious quantities of *cold* water. Larger volumes of acetone must be collected as 'waste solvent'. Solvent-rinsed apparatus should not be 'flamed'. Precautions are necessary to avoid contact of the skin with cleaning, oxidising or reducing agents (see Ch. 5).

Plastic laboratory equipment

Occasionally plastic apparatus (desiccators, beakers, etc.) can be substituted for glassware to avoid the hazards associated with use of the latter. However, its stability and integrity can be affected by chemicals and its compatibility with the contents must be considered.

Table 11.17 gives a general guide to the chemical resistance of a selection of plastics to various chemicals; the original reference contains more comprehensive listings.[29] However, since other factors (e.g. temperature) affect resistance, equipment should be tested under the proposed conditions of use. Strong oxidising agents should not be used in contact with plastic except Teflon FEP. Plastic equipment should be kept away from heat sources. Most plastic can be cleaned as glassware with detergent solution. Solvents should be avoided unless they are known not to affect the plastic. Ultrasonic cleaning can also be used though strong oxidisers must be avoided.

Low-temperature work

This section deals with the hazards and precautions associated with 'refrigeration' (generally understood to cover temperatures down to *c.* −73 °C). 'Cryogenics' (temperatures between −73 and 272 °C) is covered on pages 644 to 646.

Table 11.17 Chemical resistance of plastics*[27]

Chemical	Resins †						
	CPO	LPE	PP	PMP	FEP/ ETFE	PC	PVC
Acetaldehyde	GN	GF	GN	GN	EE	FN	GN
Acetamide, sat.	EE	EE	EE	EE	EE	NN	NN
Acetic acid, 5%	EE	EE	EE	EE	EE	EG	EE
Acetic acid, 50%	EE	EE	EE	EE	EE	EG	EG
Acetone	EE	EE	EE	EE	EE	NN	FN
Adipic acid	EGφ	EE	EE	EE	EE	EE	EG
Alanine	EE	EE	EE	EE	EE	NN	NN
Allyl alcohol	EE	EE	EE	EG	EE	EG	GF
Aluminium hydroxide	EG	EE	EG	EG	EE	FN	EG
Aluminium salts	EE	EE	EE	EE	EE	EG	EE
Amino acids	EE	EE	EE	EE	EE	EE	EE
Ammonia	EE	EE	EE	EE	EE	NN	EG
Ammonium acetate, sat.	EE	EE	EE	EE	EE	EE	EE
Ammonium glycolate	EG	EE	EG	EG	EE	GF	EE
Ammonium hydroxide, 5%	EE	EE	EE	EE	EE	FN	EE
Ammonium hydroxide,	EG	EE	EG	EG	EE	NN	EG
Ammonium oxalate	EG	EE	EG	EG	EE	EE	EE
Ammonium salts	EE	EE	EE	EE	EE	EG	EG
n-Amyl acetate	GF	EG	GF	GF	EE	NN	FN
Amyl chloride	NN	FN	NN	NN	EE	NN	NN
Aniline	EG	EG	GF	GF	EE	FN	NN
Antimony salts	EE	EE	EE	EE	EE	EE	EE
Arsenic salts	EE	EE	EE	EE	EE	EE	EE
Barium salts	EE	EE	EE	EE	EE	EE	EG
Benzaldehyde	EG	EE	EG	EG	EE	FN	NN
Benzene	FN	GG	GF	GF	EE	NN	NN
Benzoic acid, sat.	EE	EE	EG	EG	EE	EG	EG
Benzyl acetate	EG	EE	EG	EG	EE	FN	FN
Benzyl alcohol	NN	FN	NN	NN	EE	GF	GF
Bismuth salts	EE	EE	EE	EE	EE	EE	EE
Boric acid	EE	EE	EE	EE	EE	EE	EE
Boron salts	EE	EE	EE	EE	EE	EE	EE
Brine	EE	EE	EE	EE	EE	EE	EE
Bromine	NN	FN	NN	NN	EE	FN	GN
Bromobenzene	NN	FN	NN	NN	EE	NN	NN
Bromoform	NN	NN	NN	NN	EE	NN	NN
Butadiene	NN	FN	NN	NN	EE	NN	FN
n-Butyl acetate	GF	EG	GF	GF	EE	NN	NN
n-Butyl alcohol	EE	EE	EE	EG	EE	GF	GF
sec-Butyl alcohol	EG	EE	EG	EG	EE	GF	GG
tert-Butyl alcohol	EG	EE	EG	EG	EE	GF	EG
Butyric acid	NN	FN	NN	NN	EE	FN	GN
Cadmium salts	EE	EE	EE	EE	EE	EE	EE
Calcium hydroxide, conc.	EE	EE	EE	EE	EE	NN	EE
Calcium hypochlorite, sat.	EE	EE	EE	EG	EE	FN	GF
Carbazole	EE	EE	EE	EE	EE	NN	NN
Carbon bisulphide	NN	NN	EG	FN	EE	NN	NN

Table 11.17 (continued)

Chemical	Resins †						
	CPO	LPE	PP	PMP	FEP/ ETFE	PC	PVC
Castor oil	EE	EE	EE	EE	EE	EE	EE
Cedarwood oil	NN	FN	NN	NN	EE	GF	FN
Cellosolve acetate	EG	EE	EG	EG	EE	FN	FN
Caesium salts	EE	EE	EE	EE	EE	EE	EE
Chlorine, 10% in air	GN	EF	GN	GN	EE	EG	EE
Chlorine, 10% (moist)	GN	GF	GN	GN	EE	GF	EG
Chloroacetic acid	EE	EE	EG	EG	EE	FN	FN
p-Chloroacetophenone	EE	EE	EE	EE	EE	NN	NN
Chloroform	FN	GF	GF	FN	EE	NN	NN
Chromic acid, 10%	EE	EE	EE	EE	EE	EG	EG
Chromic acid, 50%	EE	EE	EG	EG	EE	EG	EF
Cinnamon oil	NN	FN	NN	NN	EE	GF	NN
Citric acid, 10%	EE	EE	EE	EE	EE	EG	GG
Citric acid, crystals	EE	EE	EE	EE	EE	EE	EG
Coconut oil	EE	EE	EE	EG	EE	EE	GF
Cresol	NN	FN	EG	NN	EE	NN	NN
Cyclohexane	GF	EG	GF	NN	EE	EG	GF
Decalin	GF	EG	GF	FN	EE	NN	EG
o-Dichlorobenzene	FN	FF	FN	FN	EE	NN	GN
p-Dichlorobenzene	FN	GF	EF	GF	EE	NN	NN
Diethyl benzene	NN	FN	NN	NN	EE	FN	NN
Diethyl ether	NN	FN	NN	NN	EE	NN	FN
Diethyl ketone	GF	GG	GG	GF	EE	NN	NN
Diethyl malonate	EE	EE	EE	EG	EE	FN	GN
Diethylene glycol	EE	EE	EE	EE	EE	GF	FN
Diethylene glycol ethyl ether	EE	EE	EE	EE	EE	FN	FN
Dimethyl formamide	EE	EE	EE	EE	EE	NN	FN
Dimethylsulphoxide	EE	EE	EE	EE	EE	NN	NN
1,4-Dioxane	GF	GG	GF	GF	EE	GF	FN
Dipropylene glycol	EE	EE	EE	EE	EE	GF	GF
Ether	NN	FN	NN	NN	EE	NN	FN
Ethyl acetate	EE	EE	EE	EG	EE	NN	FN
Ethyl alcohol	EG	EE	EG	EG	EE	EG	EG
Ethyl alcohol, 40%	EG	EE	EG	EG	EE	EG	EE
Ethyl benzene	FN	GF	FN	FN	EE	NN	NN
Ethyl benzoate	FF	GG	GF	GF	EE	NN	NN
Ethyl butyrate	GN	GF	GN	FN	EE	NN	NN
Ethyl chloride, liquid	FN	FF	FN	FN	EE	NN	NN
Ethyl cyanoacetate	EE	EE	EE	EE	EE	FN	FN
Ethyl lactate	EE	EE	EE	EE	EE	FN	FN
Ethylene chloride	GN	GF	FN	NN	EE	NN	NN
Ethylene glycol	EE	EE	EE	EE	EE	GF	EE
Ethylene glycol methyl ether	EE	EE	EE	EE	EE	FN	FN
Ethylene oxide	FF	GF	FF	FN	EE	FN	FN
Fluorides	EE	EE	EE	EE	EE	EE	EE
Fluorine	FN	GN	FN	FN	EG	GF	EG
Formaldehyde, 10%	EE	EE	EE	EG	EE	EG	GF

Table 11.17 (continued)

Chemical	Resins †						
	CPO	LPE	PP	PMP	FEP/ ETFE	PC	PVC
Formaldehyde, 40%	EG	EE	EG	EG	EE	EG	GF
Formic acid, 3%	EG	EE	EG	EG	EE	EG	GF
Formic acid, 50%	EG	EE	EG	EG	EE	EG	GF
Formic acid, 98–100%	EG	EE	EG	EF	EE	EF	FN
Fuel oil	FN	GF	EG	GF	EE	EG	EE
Gasoline	FN	GG	GF	GF	EE	FF	GN
Glacial acetic acid	EG	EE	EG	EG	EE	GF	EG
Glycerine	EE	EE	EE	EE	EE	EE	EE
n-Heptane	FN	GF	FF	FF	EE	EG	FN
Hexane	NN	GF	EF	FN	EE	FN	GN
Hydrochloric acid, 1–5%	EE	EE	EE	EG	EE	EE	EE
Hydrochloric acid, 20%	EE	EE	EE	EG	EE	EG	EG
Hydrochloric acid, 35%	EE	EE	EG	EG	EE	GF	GF
Hydrofluoric acid, 4%	EG	EE	EG	EG	EE	GF	GF
Hydrofluoric acid, 48%	EE	EE	EE	EE	EE	NN	GF
Hydrogen	EE	EE	EE	EE	EE	EE	EE
Hydrogen peroxide, 3%	EE	EE	EE	EE	EE	EE	EE
Hydrogen peroxide, 30%	EG	EE	EG	EG	EE	EE	EE
Hydrogen peroxide, 90%	EG	EE	EG	EG	EE	EE	EG
Isobutyl alcohol	EE	EE	EE	EG	EE	EG	EG
Isopropyl acetate	GF	EG	GF	GF	EE	NN	NN
Isopropyl alcohol	EE	EE	EE	EE	EE	EE	EG
Isopropyl benzene	FN	GF	FN	NN	EE	NN	NN
Kerosene	FN	GG	GF	GF	EE	GF	EE
Lactic acid, 3%	EG	EE	EG	EG	EE	EG	GF
Lactic acid, 85%	EE	EE	EG	EG	EE	EG	GF
Lead salts	EE	EE	EE	EE	EE	EE	EE
Lithium salts	EE	EE	EE	EE	EE	GF	EE
Magnesium salts	EE	EE	EE	EE	EE	EG	EE
Mercuric salts	EE	EE	EE	EE	EE	EE	EE
Mercurous salts	EE	EE	EE	EE	EE	EE	EE
Methoxyethyl oleate	EG	EE	EG	EG	EE	FN	NN
Methyl alcohol	EE	EE	EE	EE	EE	FN	EF
Methyl ethyl ketone	EG	EE	EG	EF	EE	NN	NN
Methyl isobutyl ketone	GF	EG	GF	FF	EE	NN	NN
Methyl propyl ketone	GF	EG	GF	FF	EE	NN	NN
Methylene chloride	FN	GF	FN	FN	EE	NN	NN
Mineral oil	GN	EE	EE	EG	EE	EG	EG
Nickel salts	EE	EE	EE	EE	EE	EE	EE
Nitric acid, 1–10%	EE	EE	EE	EE	EE	EG	EG
Nitric acid, 50%	EG	GN	GN	GN	EE	GF	GF
Nitric acid, 70%	EN	GN	GN	GN	EE	FN	FN
Nitrobenzene	NN	FN	NN	NN	EE	NN	NN
n-Octane	EE	EE	EE	EE	EE	GF	FN
Orange oil	FN	GF	GF	FF	EE	FF	FN
Ozone	EG	EE	EG	EE	EE	EG	EG
Perchloric acid	GN	GN	GN	GN	GF	NN	GN

Table 11.17 (continued)

Chemical	Resins †						
	CPO	LPE	PP	PMP	FEP/ ETFE	PC	PVC
Perchloroethylene	NN	NN	NN	NN	EE	NN	NN
Phenol, crystals	GN	GF	GN	FG	EE	EN	FN
Phosphoric acid, 1–5%	EE	EE	EE	EE	EE	EE	EE
Phosphoric acid, 85%	EE	EE	EG	EG	EE	EG	EG
Phosphorus salts	EE	EE	EE	EE	EE	EE	EE
Pine oil	GN	EG	EG	GF	EE	GF	FN
Potassium hydroxide, 1%	EE	EE	EE	EE	EE	FN	EE
Potassium hydroxide, conc.	EE	EE	EE	EE	EE	NN	EG
Propane gas	NN	FN	NN	NN	EE	FN	EG
Propylene glycol	EE	EE	EE	EE	EE	GF	FN
Propylene oxide	EG	EE	EG	EG	EE	GF	FN
Resorcinol, sat.	EE	EE	EE	EE	EE	GF	FN
Resorcinol, 5%	EE	EE	EE	EE	EE	GF	GN
Salicylaldehyde	EG	EE	EG	EG	EE	GF	FN
Salicylic acid, powder	EE	EE	EE	EG	EE	EG	GF
Salicylic acid, sat.	EE	EE	EE	EE	EE	EG	GF
Salt solutions	EE	EE	EE	EE	EE	EE	EE
Silver acetate	EE	EE	EE	EE	EE	EG	GG
Silver salts	EG	EE	EG	EE	EE	EE	EG
Sodium acetate, sat.	EE	EE	EE	EE	EE	EG	GF
Sodium benzoate, sat.	EE	EE	EE	EE	EE	EE	EE
Sodium hydroxide, 1%	EE	EE	EE	EE	EE	FN	EE
Sodium hydroxide, 50% to sat.	EE	EE	EE	EE	EE	NN	EG
Sodium hypochlorite, 15%	EE	EE	EE	EE	EE	GF	EE
Stearic acid, crystals	EE	EE	EE	EE	EE	EG	EG
Sulphuric acid, 1–6%	EE	EE	EE	EE	EE	EE	EG
Sulphuric acid, 20%	EE	EE	EG	EG	EE	EG	EG
Sulphuric acid, 60%	EG	EE	EG	EG	EE	GF	EG
Sulphuric acid, 98%	EG	EE	EE	EE	EE	NN	NN
Sulphur dioxide, Liq., 46 psi	NN	FN	NN	NN	EE	GN	FN
Sulphur dioxide, Wet or Dry	EE	EE	EE	EE	EE	EG	EG
Sulphur salts	FN	GF	FN	FN	EE	FN	NN
Tartaric acid	EE	EE	EE	EE	EE	EG	EG
Tetrachloromethane	FN	GF	GF	NN	EE	NN	GF
Tetrahydrofuran	FN	GF	GF	FF	EE	NN	NN
Thionyl chloride	NN	NN	NN	NN	EE	NN	NN
Titanium salts	EE	EE	EE	EE	EE	EE	EE
Toluene	FN	GG	GF	FF	EE	FN	FN
Tributyl citrate	GF	EG	GF	GF	EE	NN	FN
Trichloroethane	NN	FN	NN	NN	EE	NN	NN
Trichloroethylene	NN	FN	NN	NN	EE	NN	NN
Triethylene glycol	EE	EE	EE	EE	EE	EG	GF
Tripropylene glycol	EE	EE	EE	EE	EE	EG	GF
Turkey red oil	EE	EE	EE	EE	EE	EG	EG
Turpentine	FN	GG	GF	FF	EE	FN	GF
Undecyl alcohol	EF	EG	EG	EG	EE	GF	EF
Urea	EE	EE	EE	EG	EE	NN	GN

Table 11.17 (continued)

Chemical	Resins †						
	CPO	LPE	PP	PMP	FEP/ ETFE	PC	PVC
Vinylidene chloride	NN	FN	NN	NN	EE	NN	NN
Xylene	GN	GF	FN	FN	EE	NN	NN
Zinc salts	EE	EE	EE	EE	EE	EE	EE
Zinc stearate	EE	EE	EE	EE	EE	EE	EG

* Key to classification code:
E – 30 days of constant exposure cause no damage. Plastic may even tolerate exposure for years.
G – Little or no damage after 30 days of constant exposure to the reagent.
F – Some signs of attack after 7 days of constant exposure to the reagent.
N – Not recommended; noticeable signs of attack occur within minutes to hours after exposure. (However, actual failure might take years.)

† Resins (code):
CPE: Conventional (low-density) polyethylene
LPE: Linear (high-density) polyethylene
PP: Polypropylene
PMP: Polymethylpentene
FEP: Teflon FEP (fluorinated ethylene propylene). Teflon is a Du Pont registered trademark
ETFE: Tefzel ethylene-tetrafluoroethylene copolymer. (For chemical resistance, see FEP ratings.) Tefzel is a Du Pont registered trademark
PC: Polycarbonate
PVC: Rigid polyvinyl chloride

ϕ 1st letter: at room temp → EG ← 2nd letter: at 52°C.

Ice crystals are commonplace in laboratories and are used to induce condensation, crystallisation/solidification and as ice-baths to control the rate of chemical reactions. As with oil-baths, ice-baths should not be left unattended if the consequences of warm up would be hazardous, such as a 'runaway' reaction. Similarly, refrigeration and freezers are used to chill mixtures for crystallisation, and for storing thermally unstable chemicals (e.g. certain organic peroxides). Many laboratory explosions have resulted when ordinary household refrigerators have been used to store volatile, highly flammable solvents. The amount of energy required to ignite air–vapour mixtures in the flammable range is only 200 microjoules. Hence the electrical contacts of a thermostat or light inside domestic refrigerators possess sufficient energy to act as a means of ignition. Many volatile solvents have an appreciable vapour pressure at low temperatures causing vapour concentrations to accumulate to explosive limits within the confined space of the cabinet. Thus 5 ml of benzene can produce an explosive mixture in a 0.1 m^3 refrigerator. (See also Ch. 4 page 143).

The whole top floor of a main laboratory was badly damaged in a fire involving 8 Winchesters of ether left on the bench. The ignition source was thought to be a thermostatically controlled oil-bath, control switches in one of the two refrigerators in the room, or a spark from the electric motor in the thermostatically operated air extraction unit.[27]

A beaker containing an organic solute dissolved in hexane was placed in a domestic refrigerator to cool in order to induce crystallisation. A severe explosion resulted during the 'quiet hours' and fortunately no one was injured. However, the extent of the damage to the refrigerator and to the laboratory can be gauged from Figs 11.4(a) and 11.4(b).

It is unacceptable to rely on prominent signs warning against the use of ordinary refrigerators for storing flammable solvents. Rather, explosion-proof versions should be purchased, or the domestic versions modified so that thermostat contacts are moved outside the cabinet and the interior light and circuitry removed. Such modified refrigerators are unsuitable for positioning in locations where flammable atmospheres may exist. Laboratory refrigerators should be placed against fire-resistant walls.

All materials placed in the refrigerator should be capped and clearly marked identifying the contents, the name and location of the 'owner', and the date when storage commenced. Such labels must be durable.

Fig. 11.4(a) Result of explosion inside domestic refrigerator

Fig. 11.4(b) Laboratory damage caused by explosion in refrigerator

Samples should be placed in secondary plastic containers, or in suitable trays to contain the specimen in the event of container failure. Consideration must be given to the 'compatibility' of samples to be stored with materials already in the refrigerator. Potentially explosive chemicals should not be stored in the general laboratory refrigerator. Special freezers should be used, located in remote parts of the building and equipped with an alarm in the event of power failure. Food and drinks must never be stored with chemicals.

A regular defrosting programme should be maintained and the inventory checked. Unwanted samples should be discarded.

In cold stores there must be adequate means of escape in the event of emergency, and a mechanism to prevent anyone becoming accidentally locked inside the store or, alternatively, an audible alarm button provided inside near the door. The general fire alarm must be audible inside the cold store. For very low temperatures protective clothing will be required and the duration of exposure limited as described in Chapter 3. This may also need to be a two-man operation. An emergency lighting system may need to be provided in the store.

Work at elevated temperatures

Elevated temperatures are often employed to change certain physical properties (e.g. viscosity, phase) or to increase the rate of certain processes (e.g. vaporisation, chemical reactivity). Such work above ambient temperatures requires a heat source and controller, a heat transfer medium and a monitor of temperature.

Clearly, hot apparatus can cause burns and the risk is compounded by dropping items such as vessels of hot chemicals. Hot glassware should not be placed on cold or wet surfaces. Since it may not always be apparent to colleagues when equipment is hot (particularly immediately following shut-down), warning notices should be prominently displayed adjacent to the equipment. A process may need to be monitored either manually or instrumentally when changes in volume or state can occur. Thus a vessel may crack or explode if heated to 'dryness', and solidification in a condenser can cause hazardous back-pressure (e.g. during steam distillation).

Whatever source of heat is used, generally the temperature should be raised gradually and evenly. No means of ignition is permissible when handling highly flammable solvent or gas. Hazardous operations should never be left unattended unless equipped with high temperature cut-out devices. Similarly, safety devices should be incorporated to switch off the heat source in the event of coolant failure. Where temperatures of <100 °C are adequate hot water/steam-baths are the favoured method of heating apparatus since they present few hazards, though ventilation may be required to remove steam.

For higher temperatures oil-baths are commonly used, particularly for heating irregular-shaped vessels. The bath should be of metal or thick-walled porcelain and not glass which is less robust. Ordinary mineral oil can be used safely below 200 °C; above this temperature it may flash. As oil ages (and darkens) however, its flashpoint is lowered due to oxidation; therefore it is imperative that oil is regularly replaced with fresh material. Dow Corning 550 silicone oil is suitable for most heating requirements up to 300 °C. Since ventilation is required it is normal to assemble such equipment in fume-cupboards. Caution is required to prevent water, distillate, etc., from leaking into hot oil. Ideally hot oil should be left to cool before moving the bath, and in any event should be arranged on a laboratory jack-stand so that it can be lowered quickly and safely from around the flasks in an emergency. Suitable impervious heat-resistant gloves should be worn during such an operation. Heat to any oil-bath is usually provided electrically from a hotplate or immersion heater which poses potential electrical and thermal hazards. Portable immersion heaters should be secured firmly to the bath. As with any electrical equipment regular inspection and maintenance is essential. With hotplates, checks should be made to ensure old and corroded bimetallic thermostats have not fused so as to deliver full current continuously to the hotplate.

Electric isomantles are commonly used for higher working temperatures. They should always be used in conjunction with a temperature controller and never plugged directly into the mains. Only the correct size mantles should be used and they should be protected from chemical spillage and corrosive atmospheres whenever practical. Liquid levels in flasks should not be allowed to fall below the mantle surface since liquid splashing on the overheated glass may decompose or crack the vessel. In addition to normal periodic inspection of the electrical equipment, the glass-fibre coating should be checked for damage, wear or chemical contamination.

Air-baths, though less common, are sometimes used. Because of their poor heat capacity characteristics the bath must commonly be heated to 100 °C above the desired working temperature. The heating element must be completely encased. Hot-air guns (e.g. portable hair driers) are sometimes employed to dry apparatus, heat the upper parts of distillation rigs, etc. The element becomes red hot and cannot be encased. This together with the switches, which are not spark-free, represent an ignition source and these devices should be prohibited wherever flammable atmospheres may exist.

Ovens, unless specially designed, are generally unsuitable for heating highly flammable solvents. They should always be equipped with thermostats, and internal fans should be completely guarded, or interlocked with the door to prevent operation when the door is open. Efficient traps should be inserted in the line between pumps and vacuum ovens, and thick-walled inspection windows should be taped and inspected periodically for defects. The contents of any oven should be inspected before the temperature is raised.

Special precautions are required with molten salt baths to avoid accidental ingress of moisture. Shields are necessary to contain splashes. Oxidisable materials must never be allowed near nitrate salt baths.

Muffle furnaces are used to give very high temperatures (often *c*. 800–1,000 °C). Since external surfaces can sometimes be hot enough to cause burns when the system is operational, a warning light should be incorporated to indicate when the furnace is operating and highly flammable solvents, etc., prohibited from the area. Furnaces are often used to melt metal or to decarbonise organic matter and, as a consequence, the vent from the furnace should exhaust via an extract ventilation duct. Furnace doors should be counterbalanced to facilitate opening and closing. Gloves and specially designed long-reach tongs should be used to insert and remove specimens from the hot furnace. Hot specimens should be rested on a heat-resistant surface to cool in a marked, designated area. Where very high temperatures are involved goggles may be required to protect the eyes from glare (see Ch. 2).

Furnaces, Bunsen burners, hotplates and ovens should stand on heat-resistant surfaces. Equipment should have an air gap between it and the bench top to prevent charring.

For any hot work the temperature must be monitored. Many devices are available ranging from pyrometers in furnaces to mercury in glass thermometers. The correct device should be selected otherwise the consequences could be hazardous. Glass thermometers are, of course, fragile, particularly the bulb; mercury is both toxic if spilled and problematic to clean up. (See page 627.)

Atomic absorption flame spectrometers

Atomic absorption flame spectrometers, used for the determination of chemical elements, pose a combination of hazards from the use of compressed gases, flames, corrosive liquids, flammable solvents, toxic products, UV radiation and high temperatures. The equipment can be used safely so long as correct precautions are instituted, such as the fitting of sensors and interlocks to prevent flashback and associated hazards by inhibiting ignition or shutting off the flame automatically in the event of operational abnormalities. It is helpful to study the different stages in performing atomic absorption spectrometric (AAS) analysis and the different components of the instrument.

Wet oxidation – Prior to elemental analysis many organic compounds must be oxidised, usually with a mixture of concentrated sulphuric and nitric acids. The operation should only be undertaken by experienced staff wearing laboratory coats, face shields and gloves. It should be carried out in a well-ventilated fume-cupboard and a wash bottle should be at hand together with a beaker of 10% aqueous sodium bicarbonate solution. A trial oxidation should be performed on 0.1 g scale, and then scaled up to 1 g for analysis only if there is no obvious danger. The procedure is to warm the mixture of 1 g of compound and 2 ml concentrated sulphuric acid in a long-necked flask for a few minutes until charring occurs. Cool before adding concentrated nitric acid dropwise until no further reaction occurs. Dilute to the required volume with water. At all times the neck of the flask should be directed towards the rear of the fume-cupboard.

Solvents – Highly flammable or chlorinated solvents should be avoided whenever possible: the latter produce toxic degradation products. Where organic solvents are necessary the least volatile and highest flashpoint solvent is recommended. Ensure the minimum quantity of solvent is near the flame and a portable aluminium screen may be inserted between the sample and the mixing chamber/burner.

Hazardous samples – Solutions which could react in an undesirable manner with acetylene (see later) should not be aspirated, e.g. hypochlorites release chlorine which could react violently with acetylene. When working with silver, copper or mercury compounds the burner, mixing chamber and waste-drain must be washed immediately on completion of analysis. Such compounds can react with acetylene to produce unstable acetylides which may explode, especially when dry (see Ch. 10).

Exhaust gases – Many exhaust gases may be toxic and should be vented to an extract ventilation hood. Duct casing should be flameproof and free of solder joints. The fan should be installed as near to the outlet as possible. All joints on the discharge side should be airtight. Backdraft dampers should be fitted; roof-stack heights must ensure that effluent cannot re-enter the building. The ventilation system can be linked either to give an audible/visible warning on failure, or such that the gas supply is cut off automatically.

Compressed gases – Compressed gases are dealt with on page 627 which should be consulted. In AAS, gases are used as fuel and as oxidants. Common mixtures are air–acetylene and nitrous oxide–acetylene. In addition to the considerations given in the section on compressed gases, the following are relevant. The ventilation system should be on before any gas cylinder valves are opened. When the equipment is not operational the gases must be switched off at the cylinder and surplus gas bled from the lines before the ventilation system is switched off. Pure oxygen must not be used as oxidant. Gas lines, valves and other fittings which could come into contact with nitrous oxide should be oil- and grease-free, which could otherwise undergo spontaneous combustion. An approved heater may be installed on the cylinder head of nitrous oxide cylinders to prevent 'freeze up' during gas discharge. Acetylene must never be used at pressures above its critical explosive pressure. Metal tubing should be chosen to avoid chemical reaction with acetylene (e.g. avoiding copper). Checks should be made periodically using detectors for acetylene leaks in the laboratory, particularly near ceilings. Preferably, continuous background detectors should be installed to give audible alarms or automatic shut-down on detection. All joints should be tested periodically for leaks using soap solution, as should any new connection.

Hoses – Hoses should be arranged so that they cannot be damaged or accidentally knocked.

Burner head – The correct burner head should be selected for the gas mixture in use and should be seated and secured prior to lighting flames. Once the flame is alight the burner compartment shield should be closed during the operation of the equipment. The flame must never be left unattended. Special eye protection must be worn if the nitrous oxide/acetylene flame is to be viewed because of the emission of UV radiation (see Ch. 2). Deposits must never be removed from the burner head while the flame is burning otherwise flashbacks may result. Dislodged deposits may be blown upwards from the burner slot during cleaning and care must be taken to ensure these do not enter the eyes.

Burner box controls – Before turning on fuel gas at the cylinder, mains power to the burner box control should be OFF and the 'GASES' control in the 'SHUTDOWN' position. When adjusting 'FUEL' flow for air–acetylene mixtures the 'GASES' control should be 'ON' to allow both gases to flow rather than pure acetylene. Adjustments should be rapid to prevent build-up, to minimise any opportunity for flammable atmospheres to develop.

Drain tube – When a burner is connected to a waste vessel (which must be unbreakable), a liquid channel between the burner and the vessel must be provided, and the head of liquid in the tube should exceed the burner pressure with a minimum drop of 5 cm of water. A typical arrangement is shown in Fig. 11.5. Failure to make this provision may result in explosive mixtures of fuel and oxidant gases venting into the laboratory.

Lamps – Hands must be dry and fingers clear of terminal pins when inserting the hollow cathode lamp plug into the high voltage socket. A live cathode lamp must never be viewed directly because of the emission of UV radiation.

Chromatographic equipment

Chromatography is a common analytical technique for the separation of mixtures, based on the partition of components between two immiscible and generally heterogeneous phases. One phase is stationary with a large surface area and the second is mobile moving over the stationary phase. Variations, depending upon the nature of the two phases, include liquid-column chromatography, gas–liquid chromatography (GLC), high-performance liquid chromatography (HPLC) and thin-layer chromatography (TLC).

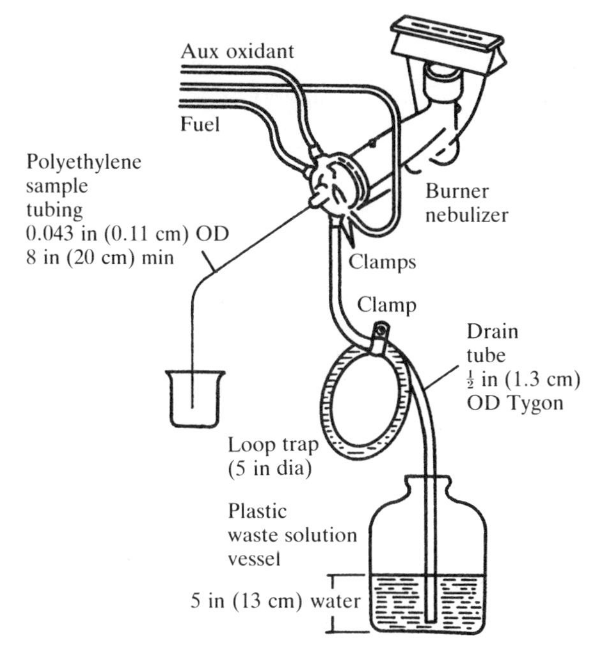

Fig. 11.5 Typical arrangement for drain tube from atomic absorption spectrometer

Many systems involve the use of organic solvents and all the necessary precautions for handling these flammable and/or toxic, volatile materials should be followed, as indicated elsewhere. The specification of ventilation systems needs to take account of the fact that excess air velocities may cause temperature differentials on columns with adverse effects on the separation.

Solid supports can present dust hazards when filling columns (e.g. in GLC) or coating plates (e.g. in TLC). Adequate precautions, including the use of ventilation and respiratory protection, may be required in the packing of columns, etc., where purchase of prepacked columns is not possible or non-dusty techniques (e.g. using 'wet' solid support) cannot be adopted. Similarly, scraping TLC plates has reportedly produced aerosols in the operator's breathing zone even when undertaken in a fume-hood.[30] Where hazardous components are involved, this operation should be undertaken in a glove box to prevent the necessity to rely on personal protection and extensive decontamination of the fume-hood. TLC also often relies on spraying plates with hazardous reagents to render components 'visible'. Ventilation booths are required to contain aerosols, and respiratory, face and hand protection should be worn as appropriate. Hazards are also encountered as a result of the sharp corners and clipped edges of glass TLC plates. Defective plates should not be used. Heat-resistant gloves should be worn when removing hotplates from ovens.

A hazard in HPLC stems from the considerable pressures sometimes used; columns should be made of metal, or housed in sealed, metal jackets. Eye and hand protection should be worn.

To unpack a silica column a chemist removed the metal frit, connected a pump and pumped iso-octane through the columns at a high flow-rate. Nothing happened so he scratched some silica out with a wire. This triggered a rapid expulsion of silica and solvent as a jet on to his thumbs and the side of his hand. The injury appeared slight but became increasingly painful throughout the night preventing him from sleeping. The injury worsened and led to 2-weeks' hospitalisation including surgery.[31]

Clearly, had a more toxic solvent been used, or had the jet entered the eye, the outcome would have been more serious.

Pumps should be fitted with low-pressure detectors coupled to an audible alarm or automatic shut-down mechanism to prevent solvent being continuously pumped into the laboratory in the event of line-fracture.

The main hazards associated with GLC include use of compressed gases (including hydrogen), high temperatures, cryogenic materials and syringes. These hazards are dealt with elsewhere.

Centrifuges

Centrifugation is a common technique for separation of heterogeneous solid/liquid, liquid/liquid phases. Accidents involving centrifuges due to misuse or equipment failure can be serious, particularly on production-scale units. Even smaller, laboratory-scale versions produce high *G* forces and are often located in a general laboratory thereby exposing several workers to risk. (The periphery of a rotor of 10 cm radius when spinning at 5,000 rpm, is travelling at 160 km/h.)

Detailed advice on centrifuges is given by among others the National Safety Council,[32] the Institution of Chemical Engineers,[33] and in a new British Standard.[34] The latter also deals with the requirements for use with flammable materials and accessories for use with potentially hazardous biological or chemical materials.

Hazards associated with centrifuges include:

- Physical contact with the rotating head.
- Failure of the rotor head or bucket due to overloading or corrosion (buckets can shear off and the energy released in the case of an ultra-centrifuge rotor is equivalent to a head-on collision of a car travelling at 160 km/h).
- Failure of bearings due to lack of maintenance.
- Severe vibration owing to unbalanced rotor with centrifuge possibly falling from bench.

- Formation of hazardous aerosols when microbiological specimens or toxic chemicals are centrifuged.
- A means of ignition, igniting vapour–air mixtures when highly flammable solvents are used.
- Cuts to the fingers when removing broken glass from bottles within buckets (these cuts may result in serious ingress of toxic or biologically active materials).

Precautions are therefore required when using centrifuges. Thus the manufacturer's operating instructions should be strictly followed, especially with regards to balancing tolerances and operating speeds for different loads and rotors. Operators of high-speed centrifuges must be trained and supervised. Records should be kept for certain rotors so that they can be de-rated as instructed, and centrifuges and ancillary components should be inspected and maintained regularly by competent engineers.

Bench-top centrifuges should be anchored, or at least sited a safe distance from the edge of the bench and so that vibration cannot result in bottles falling from shelves. The centrifuge should be housed in a robust case and should be correctly earthed. Lids should be interlocked to prevent operation with the lid open. A less satisfactory arrangement relies on a warning notice, e.g. 'Close the lid before start-up and do not open until rotors stop.' 'Never stop the rotors by friction such as using a finger or bung.' A notice should be attached to indicate the minimum delay period between switch-off and opening the cover. The load must be distributed evenly by matching buckets, trunions, etc. They must be seated properly before start-up and it is necessary to ensure that centres of gravity of paired tubes are equidistant from the axis of rotation. The correct tubes (not test tubes) must be used, complete with rubber cushions in the buckets. Only tubes of the precise diameter and length are permissible.

The speed should be increased gradually and the manufacturer's recommended maximum speed never exceeded. Should abnormalities develop the centrifuge should be stopped immediately. The centrifuge must be stopped by turning the speed control to zero, never at the mains. If the rotor head is removed the correct handles are required and it must be stored safely. Gloves should be worn when removing tubes from within buckets, since breakage of a tube may result in contaminated glass fragments and hence a serious puncture wound. Only explosion-proof versions are permissible for use with flammable liquids. Centrifuges for cold work must be specifically designed for locating in cold stores. Free-standing, large high-speed ultracentrifuges should ideally be located in a centrifuge room and the mains power switch located outside the room. Operating instructions should be displayed near the equipment and a permit to work, or similar restriction, instituted for maintenance of high-speed machines. Machines used with biologically active materials should be properly sterilised before storage.

Ultraviolet light

Laboratory uses of UV light include sterilisation, spectroscopy, fluorescence in thin-layer chromatography, and in photochemistry. The hazards of UV radiation and the precautions are discussed in Chapter 2. For laboratory work, the source can (and should) be encased. With high-pressure mercury lamps used in photochemistry, measures should prevent exposure of personnel to scattered or reflected light. (The explosion hazard associated with these lamps also needs attention.) Appropriate goggles should be worn when aligning the lamp or whenever exposure may result (e.g. from the sides).

Lasers

The hazards of lasers and precautions for their use are described in Chapter 2. Basically measures should entail engineering control and preventing anyone looking down the laser beam.

Electrical hazards

The hazards of electricity and general precautions are discussed in Chapter 2 and on page 28. Although statistics do not indicate electricity to be a major cause of laboratory accidents, it has been responsible for many laboratory fires, and serious injury and death of laboratory workers from electric shock are not unknown.

An employee was found unconscious on the floor of the laboratory after receiving an electric shock from exposed electric contacts on a magnetic stirrer as he placed his hand under the bottom edge of the stirrer to lift it.[34]

A graduate student was fatally electrocuted while using a high-voltage electrophoresis apparatus in a university biology laboratory. There was a short circuit between the high-voltage cable and the connector shell which attached it to the electrophoresis tank. The student touched the energised connector shell and a grounded portion of the apparatus simultaneously after he turned it up to 5,000 volts.[35]

Furthermore, during recent years there has been a significant increase in the use of electrical equipment in laboratories (i.e. for stirrers, heaters, refrigerators, pumps and instruments). Consequently vigilance is essential in the correct choice, installation, use and maintenance of such apparatus to prevent accidents.[36]

A recent review[37] examined the causes of electrical-equipment fires in hospital laboratories: the lessons are equally applicable to other types of laboratory. It concluded that many of the fires could have been prevented by use of thermal safety cut-out devices though thermostats and cut-outs have themselves finite lives. Only equipment designed to recognised standards should be purchased. A programme of regular inspection and a policy of periodic replacement of old equipment are also crucial. Equipment should be left switched on only when essential with special precautions for unattended operations as discussed earlier. The last person leaving the laboratory in the evening should be responsible for checking that all electrical equipment is switched off unless labelled to the contrary. The hazards of domestic refrigerators are described elsewhere. Figure 11.6 summarises a common sequence of events, highlights fault conditions and enables preventative factors to be identified.

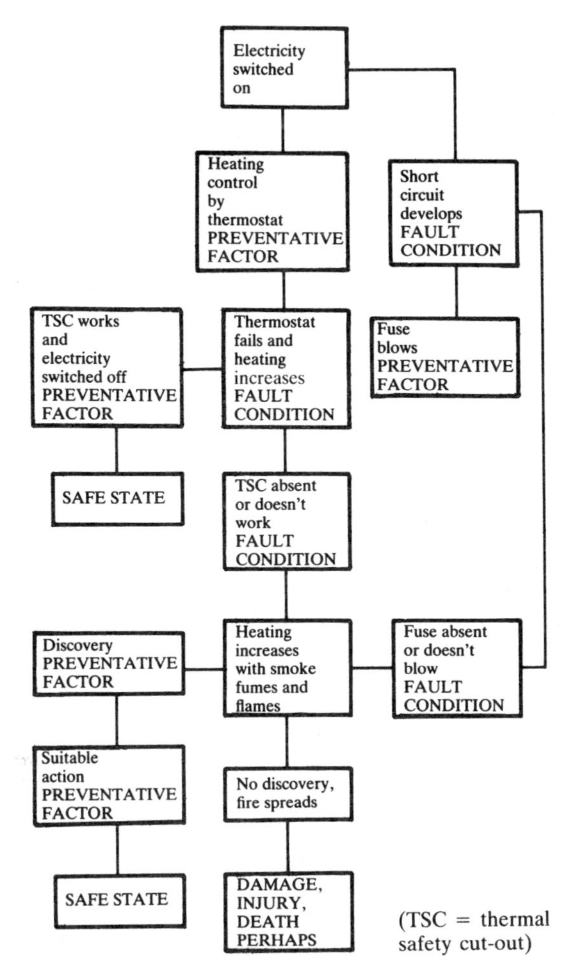

Fig. 11.6 Event tree for electrical equipment fires in laboratories[37]

Chemical hazards and precautions

General

Fundamentally, chemicals may be hazardous in the laboratory because they have flammable, explosive, toxic, sensitising, corrosive, radioactive, etc., potential. These hazardous groups are dealt with individually. However, all chemicals can create hazardous conditions if the properties are not understood and the materials improperly handled. Consideration has to be given to a chemical from the time it arrives in the laboratory, during storage, transportation throughout the laboratory, use and disposal. Management must consider foreseeable contingent dangers such as emergencies and spillages; advise on first aid in the event of personal contamination must be available. The following general guidelines should be followed whenever chemicals are encountered in the laboratory.

Purchase

Storage requirements (along with precautions for use, disposal, first aid, etc.) require consideration before ordering the material. Free samples, or other samples for analysis or use, require equal attention as purchased materials. Chemicals must not be imported into the laboratory other than by authorised routes. The minimum quantity practicable should be ordered and if data sheets are not available they should be requested at the same time as ordering the material.

Storage

All stored containers must be adequately labelled with durable labels to identify contents and hazards.

> A verdict of accidental death was recorded on a man who died after drinking paraquat solution. He had taken a drink from a bottle in his garden shed where he kept home brewed beer in lemonade bottles. At that time there were said to have been 30 deaths throughout the world as a result of people drinking paraquat solution from unlabelled containers.[38]

Appropriate warning signs are necessary where hazardous materials are stored. Eye protection and emergency equipment should be provided.

Stores should be maintained in a clean and tidy condition. Materials may be arranged systematically noting that storage in alphabetical order could result in chemicals standing above, below or adjacent to chemicals with which they are incompatible. Large and heavy containers should be kept at convenient heights (e.g. below benches) for easy access. Only small bottles should be above waist-height – unless special racks and facilities are provided. Shelves must not be overloaded, bottles should not be stored on the open floor, or one on top of another. Larger vessels should not be stacked free-standing, more than two tiers high. Containers should be stored in cool, well-ventilated and vibration-free areas, away from direct sunlight.

A fire destroyed the chemistry stockroom, the ceilings of two classrooms and records of several teachers. Combustible acoustical fibreboard ceiling tiles and an open dumb-waiter shaft contributed to the rapid spread of fire. The cause was concentration of sun's rays by bottles, etc.[39]

Liquids with a freezing-point around 20 °C must be maintained above this temperature to prevent failure of glass vessels as the substance solidifies. Examples include glacial acetic acid, acrylic acid and methacrylic acid. Furthermore, the acrylic acids should not be allowed to freeze since separation of pure material from its inhibitor may lead to explosion on subsequent melting as discussed in Chapter 10.

Some chemicals may require refrigerated conditions. As discussed earlier spark-producing components must be avoided in all refrigerators used for storing chemicals. Numerous fires and explosions have been reported due to ignition of a flammable vapour–air mixture within a refrigerator when the internal electrical circuits have operated; the following is an example in somewhat unusual circumstances.

Several mice were killed with ether prior to being dissected. They were placed in a domestic refrigerator in a laboratory overnight. The reduction in temperature caused their bodies to contract and ether was expelled from their lungs. This mixed with air within the refrigerator to form a mixture within the flammable range which is believed to have been ignited by a spark from the internal thermostat. The explosion destroyed the refrigerator and the ensuing fire caused extensive damage to the laboratory.[40]

Suitable carriers must be available when chemicals are withdrawn from stores – particularly for highly toxic materials, flammable solvents and corrosive materials.

Two students were carrying four Winchesters of concentrated sulphuric acid down the corridor. The bottles were being held by the necks. Two bottles clanked together and broke. Both men slipped on the acid and fell into the acid and broken glass. Another person coming to their aid also fell. Severe acid burns and lacerations were suffered by all three of the men.[41] (This acid – 'oil of vitriol' – results in very slippery conditions on spillage; similar accidents have occurred in factories, e.g. on metal plating lines.)

A technician collected a Winchester of nitrobenzene from the stores in the basement; because he was in a hurry the usual practice of using an approved Winchester carrier was not followed. The goods lift was occupied and, contrary to normal practice, he went to use a passenger lift. The Winchester was placed on the floor while waiting but was accidentally kicked over when the lift doors opened. The odour of nitrobenzene was detectable near the lift shaft within minutes on all 15 floors. Smoke extract fans were operated but the odour persisted in the basement for several weeks.[40]

Chemical stocks should be issued in chronological order – first in, first out. They should be used before a quoted expiry date for usage or otherwise arrangements made for their disposal.

Chemicals no longer required should be disposed of safely (see page 590). Containers (including drums) used for waste chemicals and solvents should be clearly marked with their contents. Old labels and markings should be removed or obliterated. Empty drums should be kept in the drum compound. Regular inspections are desirable for all equipment, containers and labels (in particular where corrosive materials are stored).

Adequate lighting should be provided, flameproof where necessary, in all storage areas: smoking, naked lights and all other sources of heat and means of ignition should be prohibited in all chemical storerooms. Appropriate fire extinguishers should be available near to the entrance to the store, in addition to fire buckets, etc., kept inside. Emergency exits must not be blocked and, externally, there must be clear access for fire engines and other services – not encumbered by drums, scrap, etc. More detailed advice on the storage of chemicals is given in a recent book devoted to this subject.[42]

Use

When using chemicals attention must be given to the label, e.g. by reading it twice and studying the warnings. Unlabelled material must not be used pending analysis. Hazards require checking on data sheets or from literature sources. If in doubt further enquiries are necessary. Unless indicated to the contrary it is prudent to assume all chemicals or samples to be hazardous, particularly for new, research specimens.

All containers of liquids should be opened carefully; the ambient temperature and the temperature of the container may result in some internal pressure. Opening should be performed slowly, and any pressure released while the cap or bung is still secure. Eye protection, gloves, laboratory coat, etc., should be worn as appropriate when handling chemicals and smoking, eating and drinking should be prohibited. The cap on the stock container should be replaced as soon as the sample has been withdrawn and the bottle should be returned to the storage location. The smallest amount necessary should be drawn from storage for use in each experiment/analysis: reagents should be used in small quantities, adding concentrated reagents to water, not water to concentrated solution, particularly if reactions are exothermic (e.g. dilution of concentrated sulphuric acid).

Experiments must never be rushed or reagents added faster than instructed (particularly when reactions are exothermic or involve a phase change).

During the synthesis of hydrocarbons of the cyclopentene and cyclopentane series by the Dieckmann reaction an explosion occurred that seriously injured two men and caused damage to the building and laboratory equipment. The first stage of the synthesis, addition of a dispersion of sodium in toluene to adipic ester dissolved in toluene, had been completed and the second stage, addition of methyl iodide to the reaction flask was in progress when the explosion happened. Apparently the methyl iodide was added too quickly, causing the reaction to boil so rapidly that the capacity of the reflux condenser was exceeded. Toluene vapours surged from the condenser and two stoppers in the 3-neck flask came loose allowing some of the reaction mixture, including pieces of sodium, to boil over. On being exposed to the air a piece of sodium on the floor ignited and almost immediately thereafter the toluene–air mixture in the room exploded.[43]

Personal contact with chemicals should be avoided. Chemicals must not be tasted to identify them and special care is required if attempting to identify a substance by smell. Hands should always be washed after handling chemicals and a high standard of personal hygiene encouraged. Volatile materials should be handled in a fume-cupboard.

Nowadays, pipette bulbs or aspirators should be used to provide vacuum, never mouth suction, especially for toxic or corrosive chemicals. A clean bung or stopper, never the finger, should be used when shaking reagents in a test-tube. The top of the test-tube should be pointed away from both the user and other workers. Similarly, liquids should not be poured towards nearby workers.

Samples (bottles, flasks, bags, etc.) should be carried in baskets or other suitable carriers; never by the neck (see page 563). Vessels containing hot material should not be stoppered while cooling (see Ch. 3).

Two 2-litre Erlenmeyer flasks containing hot propanol were stoppered and left to cool. One hour later owing to the formation of a reduced pressure in the flask, one of the vessels imploded spraying a technician who was nearby which resulted in her needing to use the eye-wash facilities.[44]

Appropriate warning notices should be displayed where hazards may arise. To avoid confusion and complacency, these notices should be removed promptly when the hazard is passed, e.g. on completion of an experiment.

Spillages/disposal

Procedures for dealing with a chemical spillage should be agreed prior to using the substance. Spillages on the outside of bottles must be wiped off immediately, and before the container is returned to store. Indeed any spillage must be dealt with immediately. For non-hazardous solids these can be dampened down and brushed up – collecting the material for disposal. Liquids can be wiped up with tissues, cloths, etc., which should then be disposed of in a labelled receptacle. The actions for more hazardous substances will be dictated by circumstances. (In general the experimenter should warn colleagues, evacuate the area, get help, and ensure the situation is safe and that they are adequately protected before dealing with the incident.) Only small quantities of inert, water-soluble waste should be discarded to drain, flushing with copious quantities of cold water. Bulk or more hazardous materials should be collected separately for disposal in suitable, well-labelled containers.

Solvent waste for disposal may be collected under the following categories:

1. Halogenated solvent (e.g. Genklene, carbon tetrachloride, chloroform).
2. Hydrocarbons (e.g. toluene, petrol, cyclohexane).
3. Other water-immiscible solvents (e.g. ethers) – possibly contaminated with water-miscible solvents or hydrocarbons.

Water (layer or emulsions) must be removed, particularly where a can is used for collection. Reactive solutes such as acids must be neutralised before discarding the solvent. This is to avoid possible corrosion of containers or dangerous reactions when wastes are mixed.

Methods for the disposal of chemicals are given in the literature[45,46] and should be included on suppliers' data sheets. Some suppliers provide an indication of suitable disposal methods in their catalogue. Thus Table 11.18 illustrates a guide given by one supplier for the disposal of small quantities of chemicals; a few suitable examples from the catalogue have been included.[47] Clearly, attention must be given to the chemical and physical properties of the substance, together with local laws and regulations (see Ch. 14). Chemical waste-disposal operations must be undertaken with the same precautions adopted in synthetic chemistry. Thus consideration must be given to stoichiometry, order and rate of addition, heat of reaction, evolution of gases, pH, efficient agitation, atmospheric sensitivity, etc. The disposal reactions should be done in an efficient fume-cupboard in appropriate apparatus and the operators should use relevant personal protection and other appropriate protective equipment. Reactions should be tested on a small scale with dilute solutions. The reactions may need to be left for several hours to one day after mixing of the reagents in order to ensure the process runs to 'completion' prior to disposal of 'decontaminated' waste. (In fact complete reaction

Table 11.18 Methods of disposal of small quantities of chemicals (with examples)

Method	Examples
A Dissolve or mix the material with a combustible solvent and burn in a chemical incinerator equipped with an afterburner and scrubber.	2,6-Dichlorobenzonitrile, oxalic acid
B The material should be ignited in the presence of sodium carbonate and slaked lime (calcium hydroxide). The substance should be mixed with vermiculite and then with dry caustics, wrapped in paper and burned in a chemical incinerator equipped with an afterburner and scrubber.	Ethanesulphonyl chloride, 2-fluorobenzoyl chloride.
C This combustible material should be burned in a chemical incinerator equipped with an afterburner and scrubber.	2,4-dimethyl styrene, isobutyric acid.
D Burn in a chemical incinerator equipped with an afterburner and scrubber but exert extra care in igniting as this material is highly flammable.	(Dichloromethyl) trimethyl silane, 1-heptene
E To a solution of the product in water, add an excess of dilute sulphuric acid. Let stand overnight. Remove any insolubles and bury in a chemical landfill.	Strontium chloride hexahydrate
F Cautiously dissolve the material in water. Neutralise immediately with sodium carbonate. (First add a little hydrochloric acid followed by sodium carbonate if the material does not completely dissolve.) Add calcium chloride in excess of the amount needed to precipitate the fluoride and/or carbonate. Separate the insolubles for disposal in a chemical landfill.	Aluminium fluoride trihydrate, Copper(II) fluoride
G Under an inert atmosphere, cautiously add the material to dry butanol in an appropriate solvent. The chemical reaction may be vigorous and/or exothermic. Provisions must be made to safely vent large volumes of highly flammable hydrogen and/or hydrocarbon gases. Neutralise the solution with aqueous acid. Filter off any solid residues for disposal as hazardous waste. Burn the liquid portion in a chemical incinerator equipped with an afterburner and scrubber.	Ethylaluminium dichloride
H Neutralise the solution and add a filtering agent (10 g per 100 ml). Evaporate liquid and bury residual solid in a chemical landfill.	Europium atomic absorption standard solution
I Dissolve the solid in (or dilute the solution with) a large volume of water. Carefully add a dilute solution of acetic acid or acetone to the mixture in a well ventillated area. Provisions should be made to safely vent the hydrogen gas given off during the decomposition. Check the acidity of the solution and adjust to pH 1 if necessary. Let stand overnight. Neutralise the solution (to pH 7). Evaporate the solution and bury the residue in a chemical landfill.	Lithium borohydride, sodium borohydride

Table 11.18 (continued)

Method	Examples
J Cautiously acidify a 3% solution or a suspension of the material to pH 2 with sulphuric acid. Gradually add a 50% excess of aqueous sodium bisulphite with stirring at room temperature. An increase in temperature indicates that a reaction is taking place. If no reaction is observed on the addition of about 10% of the sodium bisulphite solution initiate it by cautiously adding more acid. If manganese, chromium or molybdenum is present, adjust the pH of the solution to 7 and treat with sulphide to precipitate for burial in a chemical landfill. Destroy excess sulphide, neutralise and flush the solution down the drain with plenty of water.	Calcium hypochlorite, chromium(VI) oxide, sodium persulphate
K Please contact the Technical Services Department for disposal method. Be sure to mention the name, catalogue number and quantity of material.	Calcium carbide, sodium azide
L The material should be dissolved in (1) water, (2) acid solution or (3) oxidised to a water-soluble state. Precipitate the material as the sulphide, adjusting the pH of the solution to 7 to complete precipitation. Filter the insolubles and bury in a chemical landfill. Destroy any excess sulphide with sodium hypochlorite, neutralise the solution and flush down the drain with plenty of water.	Chromium(III) chloride, lead(II) acetate trihydrate
M A slurry of the arenediazonium salt in water is added gradually to a stirred solution of 5–10% excess 2-naphthol in 3% aqueous sodium hydroxide at 0–20°C. After 12 hours, the resulting azo dye is filtered and incinerated or buried in a chemical landfill. Neutralise the remaining solution before disposal.	Benzene diazonium hexafluorophosphate
N For small quantities: cautiously add to a large stirred excess of water. Adjust the pH to neutral, separate any insoluble solids or liquids and package them for hazardous-waste disposal. Flush the aqueous solution down the drain with plenty of water. The hydrolysis and neutralisation reactions may generate heat and fumes which can be controlled by the rate of addition.	Aluminium bromide, aluminium trisec-butoxide
O Bury in a landfill site approved for the disposal of chemical and hazardous waste.	Bismuth(III) iodide, lead(II) carbonate
P Material in the elemental state should be recovered for reuse or recycling.	Devarda's alloy, Wood's metal
Q Cautiously make a 5% solution of the material in water or dilute acid. There may be a vigorous, exothermic reaction and fumes may be generated due to hydrolysis of the material. Control any reaction by cooling and by the rate of addition.	Hafnium sulphate, germanium tetrachloride

Table 11.18 (continued)

Method	Examples
Gradually add dilute ammonium hydroxide to pH 10. Filter off any precipitate for disposal in a chemical landfill. If no precipitation occurs, gradually adjust the pH from 10 to 6, stopping when precipitation occurs.	
R Catalysts and expensive metals should be recovered for reuse or recycling.	Platinum on activated carbon, palladium
S Treat a dilute solution (pH 10–11) of the material with a 50% excess of commercial laundry bleach. Control the temperature by the addition rate of bleach and adjust the pH if necessary. Let stand overnight. Cautiously adjust pH to 7. Vigorous evolution of gas may occur. Filter any solids for burial in a chemical landfill. Precipitate any heavy metals by addition of sulphide and isolate for burial. Additional equivalents of hypochlorite may be needed if the metal can be oxidised to a higher valence state. For metal carbonyls, the reaction should be carried out under nitrogen.	Lithium cyanide, tungsten hexacarbonyl
T Cautiously make a 5% solution of the product in water, vent because of possible vigorous evolution of flammable hydrogen gas. Acidify the solution to pH 1 by adding 1 M sulphuric acid dropwise. Acidification will cause vigorous evolution of hydrogen gas. Allow the solution to stand overnight. Evaporate to dryness and bury the residue in a chemical landfill.	Borane-dimethylamine complex
U Take the material (or its solution) and make a 5% solution in tetrahydrofuran. Cautiously add the solution dropwise to an ice-cooled, stirred basic solution of commercial bleach. Oxidation may release flammable hydrocarbon gases which must be vented safely. Let stand overnight. Adjust the pH to 7 and destroy excess hypochlorite with sodium bisulphite before disposal of solution.	Triphenylboron-sodium hydroxide adduct
V Under an inert atmosphere, cautiously add dry butanol or a mixture of dry butanol in an appropriate solvent, to a solution of the material in tetrahydrofuran. The chemical reaction may be vigorous and/or exothermic. Provision must be made for safely venting a large volume of flammable hydrogen gas. When gas evolution ceases, cautiously add a basic hypochlorite solution dropwise to the reaction solution. Let stand overnight. Neutralise the solution and treat with sodium bisulphite to destroy any excess hypochlorite. Filter any solid for burial in a chemical landfill. Burn liquid portion in a chemical incinerator equipped with an afterburner and scrubber.	Sodium triethylborohydride

is often not achievable, i.e. some unreacted chemicals may remain, and appropriate precautions should be taken accordingly.)

Flammable chemicals

Any material which is combustible is 'flammable'. The general considerations for assessing the risk and handling such chemicals are discussed in Chapter 4. In the laboratory materials which pose a particular fire risk are certain compressed gases (covered later), highly flammable organic solvents, and pyrophoric chemicals including some water-sensitive materials.

Highly flammable solvents

Organic solvents find wide laboratory use; some are toxic or highly flammable or both. The main fire hazards stem from their volatility, low flash point, the vigour with which they burn, and the speed of spread of flame. For example, the more volatile solvents liberate heat up to 10 times faster than does wood. The vapours from even small amounts of solvents can also explode in confined spaces.

For example 100 litres of air (small cupboard or glove box) can be consumed by 10 g of diethyl ether and the immediate pressure in the container could attain 10 atm.

The legal definition of a highly flammable liquid varies with country (see Ch. 4). However Table 11.19 lists some common examples of laboratory solvents which are generally considered as being highly flammable.

Aspects of the compound to consider in assessing the risk with highly flammable liquids are the flash point, auto-ignition temperature, explosive range and vapour pressure (and the toxicity). Problems associated with reliance on flash point are given in Chapter 4. The determination of these properties is described in references 48 and 49.

In the laboratory it is also necessary to consider the process (temperature, pressure, the presence of co-reagents, and means of ignition, etc.). Only 200 millijoules are required to ignite flammable vapours in the explosive range. Important examples of ignition sources include open flames, cigarettes, matches, hot surfaces, sparking electrical contacts and static electricity. Non-polar solvents (e.g. hydrocarbons) can easily accumulate static charges as they possess high insulating values and do not allow the charge to leak away. Dispersion of polar liquids in non-polar solvents can create static electricity even more rapidly than non-polar solvents alone. Even the action of crystallisation can produce static.

Table 11.19 Fire and hygiene data for a selection of highly flammable organic solvents

Compound	Flash point (°C)	Auto-ignition temp (°C)	Flammability range (%)	Recommended OEL* (ppm)
Acetaldehyde	−38	185	4–57	100
Acetone	−18	538	3–13	1000
Acetonitrile	6	524	4–16	40
Acrolein	−26	278	3–31	0.1
Acrylonitrile	0	481	3–17	2†
Allyl acetate	21	374	—	—
Allyl alcohol	21	378	3–18	2
Allyl amine	−29	374	2–22	—
Allyl bromide	− 1	295	4–7	—
Allyl chloride	−32	392	3–11	1
Allyl chloroformate	31	—	—	—
n-Amyl acetate	25	379	1–7.5	100
iso Amyl acetate	23	380	1–7.5	—
sec-Amyl acetate	32	—	1–7.5	125
Amylamine	7	—	—	—
Amylene	2	273	1.5–9	—
iso-Amyl formate	26	—	—	—
Amyl mercaptan	18	—	—	—
Benzene	−11	562	1.4–8	10
n-Butyl acetate	27	399	1.4–7.6	150
iso-Butyl acetate	18	423	1.3–7.5	200
n-Butyl alcohol	29	365	1.4–11	50
iso-Butyl alcohol	28	427	1.7–10.9	—
sec-Butyl alcohol	24	406	1.7–9.8	150
tert-Butyl alcohol	10	478	2.4–8	100
Butyl mercaptan	2	—	—	—
iso-Butyl methyl ketone	17	460	1.2–8	—
Butyl vinyl ether	− 9	—	—	—
Butyraldehyde	− 6.7	230	2.5–	—
n-Butyronitrile	26	—	—	—
Carbon disulphide	−30	100	1–44	10†
Chlorobenzene	29	638	1.3–7.1	75
Chloroprene	−20	—	4–20	1
Cyclohexane	−20	260	1.3–8.4	300
Cyclopentane	− 7	—	—	—
Cyclopentanone	26	—	—	—
Di-*n*-butyl ether	25	194	1.5–7.6	—
1,3-Dichloro-2-butene	27	—	—	—
1,1-Dichloroethane	− 6	458	5.6–11.4	200
1,2-Dichloroethylene	2–4	—	9.7–12.8	200
1,2-Dichloropropane	16	557	3.4–14.5	75
1,1-Diethoxyethane	−20	230	1.7–10.4	—
Diethylamine	<−26	312	1.8–10.1	25
Diethylcarbonate	25	—	—	—
Diethylketone	13	452	—	200
3,4-Dihydro-2H-pyran	−18	—	—	—
Di-Isopropylamine	− 1	—	−84	5
Di-isopropylether	−28	443	1.7–7.9	250
Dimethoxymethane	−18	237	—	1000
Dimethoxypropane	− 7	—	—	—

Table 11.19 (continued)

Compound	Flash point (°C)	Auto-ignition temp (°C)	Flammability range (%)	Recommended OEL* (ppm)
Dimethylcarbonate	19	—	—	—
Dimethylsulphide	<−18	206	2.2–19.7	—
1,4-Dioxane	12	180	2–22.2	50
Ethanethiol	< 27	299	2.8–18	0.5
Ethanol	12	423	3.3–19	1000
Ethyl acetate	− 4.4	427	2–9	400
Ethyl benzene	15	432	1–7	100
Ethyl butyrate	26	463	—	—
Ethylene dichloride	13	413	6–16	10
Ethyl formate	−20	455	2.7–13.5	100
N-Ethylmorpholine	32	—	—	20
Ethyl vinyl ether	<−46	202	1.7–28	—
Furan	<0	—	—	—
Gasoline	−43	280–456	1.4–7.6	—
Heptane	− 4	223	1.2–6.7	400
Hexane	−22	261	1.1–7.5	100
Hexene	<− 7	—	1.2–6.9	—
Mesityl oxide	31	344	—	25
Methyl acetate	− 9	502	3.1–16	200
Methyl acrylate	− 3	—	2.8–25	10
Methyl alcohol	12	464	6–37	200
2-Methyl-1-butene	<− 7	—	—	—
2-Methyl-2-butene	<− 7	—	—	—
N-Methylbutylamine	13	—	—	—
Methyl butyrate	14	—	—	—
Methyl chloroformate	12	504	—	—
Methyl cyclohexane	− 4	285	1.2–	—
4-Methyl cyclohexene	− 1	—	—	—
Methyl ether	−14	350	3.4–18	—
Methyl ethyl ether	−37	190	2–10.1	—
Methyl ethyl ketone	− 7	515	2–10	200
Methyl formate	−19	456	5.9–20	100
2-Methyl furan	−30	—	—	—
Methyl isobutyl ketone	23	460	1.4–7.5	100
Methyl isobutyrate	13	482	—	—
Methyl methacrylate	10	421	2.1–12.5	100
Methyl *n*-propyl ketone	7	505	1.6–8.2	200
1-Methyl pyrrole	16	—	—	—
Octane	13	220	1.0–4.66	300
Piperidine	16	—	—	—
iso-Propenyl acetate	16	—	1.9–	—
Propionaldehyde	− 9.7	207	2.9–17	—
Propionitrile	2	—	3.1–	—
iso-Propyl acetate	4	460	1.8–7.8	—
n-Propyl acetate	14	450	2–8	200
Propyl alcohol	12	433	2.1–13.5	200
Propylamine	−37	318	2.0–10.4	—
Propyl benzene	30	450	0.8–6	—
iso-Propyl ether	−28	443	1.4–7.9	—
Pyridine	20	482	1.8–12.4	5
Pyrrolidone	3	—	—	—

Table 11.19 (continued)

Compound	Flash point (°C)	Auto-ignition temp (°C)	Flammability range (%)	Recommended OEL* (ppm)
Styrene	31	490	1.1–6.1	100
Tetrahydrofuran	−14	321	2–11.8	200
Thiophene	− 1	—	—	—
Toluene	4.4	536	1.4–6.7	100
2,2,4-Trimethyl pentane	−12	418	1.1–6.0	—
2,4,4-Trimethyl-2-pentane	2	—	—	—
Valeraldehyde	12	—	—	—
Vinyl acetate	− 8	427	2.6–13.4	10
Vinyl ether	<−30	360	1.7–27	—
Xylenes	29–32	460–528	1–7	100

* 8 hr TWA Occupational Exposure (Recommended) Limits–HSE Guidance Note EH 40/1984.
† Control Limits–HSE Guidance Note EH 40/1984.

Laboratory coats and clothing of certain synthetic fibres (e.g. nylon) are prone to static generation.

Many organic solvents are immiscible with water and possess a specific gravity of less than 1. This means they will float on water and if the vapours are ignited the flame can spread down drains, gulleys, etc., back to the source. The solvents also have a high vapour density causing vapours to fall and travel considerable distances to ignition sources across bench tops, floors, down service ducts, drains, lift shafts, stairwells, etc. (see Ch. 3). Once ignited the flame can flash back to the source.

Assuming the solvent cannot be substituted by a safer alternative then the general precautions should be directed towards ensuring *at least* one side of the fire triangle is eliminated, as described in Chapter 4. Generation of flammable air–vapour mixtures may be prevented by minimising quantities of solvent, containing the vapours (e.g. avoiding liquid in open vessels), providing ventilation or inerting with, e.g. nitrogen. To exemplify, one laboratory has instituted the following restrictions on quantities of solvent permitted in the general laboratories.

- Not more than 2.5 l may be used in any one experiment.
- Not more than 10 l may be used in any one work area.
  (Work on a large scale must be undertaken in specially designed semi-technical laboratories or pilot plants.)
- The total daytime holdings of any work area shall not exceed 40 l.
- The total overnight storage limit is 2 l with not more than 500 cm^3 in any one container.

Furthermore, in the work areas, during the daytime, bottles are kept in metal solvent-trolleys fitted with antistatic wheels. Each evening these trolleys (of the type shown in Fig. 11.7) are transferred from the laboratory to purpose designed solvent sub-stores of fire-resistant structure, to reduce the fire load of the main building overnight. (Clearly unless such specialist transport is used the hazards associated with solvent movement could exceed those of leaving an excess in the laboratory.)

Another source[46] suggests the maximum quantity of flammable liquid permitted in unprotected glass or plastic should be 0.5 l per m^2 of laboratory with containers no larger than 1 litre, and a maximum volume in ordinary metal cans of 2.5 l per m^2. In some cases mild-steel cabinets fitted with overlapping hinged lids are used to house small stocks of laboratory solvents in bottles; the permitted stock is limited to the capacity of the cabinet.

Sources of ignition can be reduced by prohibition of smoking or naked

Fig. 11.7 Trolley for solvents

lights in the vicinity. Vessels containing highly flammable liquids should be positively identified with durable labels. Solvents should never be heated with an open flame. Electrical equipment may be explosion proof or inerted, or non-electrical counterparts (e.g. steam baths for heating, air-driven motors for stirrers) may be used.

On the bench scale (other than in exceptional circumstances) it is often impracticable or impossible to use explosion-proof electrical equipment. In such circumstances the volume of solvent must be limited, the equipment ventilated (e.g. in fume-cupboards), and any sparking electrics (temperature controllers for isomantles) moved outside of the fume-cupboard.

An insidious ignition source stems from the use of sodium metal (usually as wire) to dry certain solvents. Inadvertent contact of this with water (e.g. by spillage near sink) can cause the solvent to ignite. Such water-reactive materials are dealt with later.

Storage – Highly flammable liquids have to be stored in accordance with legal and local authority requirements. Stocks should be kept cool and out of direct sunlight: flasks can act as a magnifying glass if left in sunlight.

Quantities should be limited as appropriate. Ideally, the store should be a single-storey, purpose-built, room segregated from the main laboratory and constructed from non-combustible materials, with a bunded floor. The door should open outwards and be locked to prevent unauthorised entry. The store should be labelled to indicate the nature of the contents, and that smoking and the presence of naked flames are prohibited. The store should be adequately ventilated and indeed forced extract ventilation may be required if highly toxic, flammable chemicals are stored or dispensed. Drums should be earthed. Electric lighting should be of explosion-proof design and, if necessary, a gas fire-suppression system installed, capable of automatic discharge with facilities to switch to manual operation when the store is manned to prevent the system from discharging automatically when operators are present. Certain solvents, such as alcohol, should be stored under bond, and all other chemicals (particularly oxidising agents) should be prohibited as should wood, paper and other combustible materials.

Dispensing of highly flammable liquids (e.g. from drums into Winchesters) must be undertaken only by trained personnel using air-operated hand pumps and wearing appropriate personal protection. In some circumstances it is advisable for drums to be earthed and the possibility of static discharge eliminated by attention to materials of equipment (e.g. non-plastic funnels) and personal protection (avoid nylon overalls and use conducting footwear), and where relevant earthing of the receptor. Local extract ventilation is likely to be required (with attention to electrics).

Wastes – Waste solvent should be collected for disposal in labelled, metal containers allocated for this purpose. Special safety cans are available which are of welded construction with a spring loaded cap and flame arrestor to prevent flashback. Acetone and chloroform must not be mixed since the mixture can explode under certain circumstances as described in Chapter 10.

As with other hazardous chemicals segregation of wastes is an important consideration.

A student disposed of a quantity of nitric acid by pouring it into a Winchester bottle reserved for organic waste. He replaced the cap carefully but two other people were injured in an explosion which followed.[50]

Spillages must be dealt with immediately by shutting down equipment, removing all means of ignition, stopping the source of spillage, evacuating the area, alerting the fire authorities, providing ventilation, etc. Depending upon the scale of the spillage, self-contained breathing apparatus should be worn (in addition to usual personal protection) when dealing with the emergency unless environmental monitoring has shown atmospheric levels to be safe.

Peroxides – The fire and explosion risk is increased by the presence of thermally unstable materials. Thus, many ethers (e.g. ethyl ether, isopropyl ether, dioxane and tetrahydrofuran) absorb oxygen from the air and form unstable peroxides.[51–55] These may detonate violently if concentrated by evaporation or distillation.

A chemist inherited a pint bottle of isopropyl ether. He grasped the bottle in one hand, pressed it to his stomach, and twisted the tightly stuck cap with the other hand. Just as the cap broke loose the bottle exploded, practically disembowelling the man and tearing off several fingers. He died within 2 hours.[51] [See Table 11.20.]

Peroxides do not necessarily take long to form in freshly distilled and unstabilised ethers. Peroxide formation in tetrahydrofuran and ethyl ether reportedly begins within 3 days and 8 days respectively. Formation is accelerated in opened, partially-emptied containers. While dry ether oxidises faster than wet material, the presence of moisture does not prevent peroxide formation. Some commercially-available ethers contain inhibitors (such as hydroquinone, diphenylamine, l-naphthol, stannous chloride, Dowex-l) to extend the induction period for oxidation.

Precautions with ethers include purchasing the smallest practicable volume. All containers of ethers should be labelled to include details of:

- date of receipt or opening;
- date for discarding;
- initials of user (if single user).

Table 11.20 Examples of peroxidisable compounds

List A	
Peroxide hazard on storage; discard after three months	
Isopropyl ether	
Divinyl acetylene	
Vinylidene chloride	
Potassium metal	
Sodium amide	
List B	
Peroxide hazard on concentration; discard after one year	
Diethyl ether	Dicyclopentadiene
Tetrahydrofuran	Diacetylene
Dioxane	Methyl acetylene
Acetal	Cumene
Methyl isobutyl ketone	Tetrahydronaphthalene (Tetralin)
Ethylene glycol dimethyl ether (glyme)	Cyclohexene
Vinyl ethers	Methylcyclopentane
List C	
Hazardous due to peroxide initiation of polymerisation; discard after one year*	
Methyl methacrylate	Chlorotrifluorethylene
Styrene	Vinyl acetylene
Acrylic acid	Vinyl acetate
Acrylonitrile	Vinyl chloride
Butadiene	Vinyl pyridine
Tetrafluoroethylene	Chloroprene

* Under conditions of storage in the liquid state the peroxide-forming potential increases and certain of these monomers (especially butadiene, chloroprene, and tetrafluoroethylene) should then be considered as List A compounds

Ethers in glass should be kept tightly closed when not in use and shielded from light and heat. Time limits should be established for keeping opened containers of the various ethers (e.g. see Table 11.20).[53]

A system of work is required for the regular safe disposal of outdated samples of ether. All work with ethers (particularly distillations) should be undertaken in fume-cupboards fitted with toughened glass sashes, or behind blast screens.

Ethers should be tested for peroxides prior to distillation. Some methods for testing for the presence of peroxides in ether are given below.[53,55] The tests should be undertaken in a fume-cupboard and all ignition sources eliminated. An approved face-shield and gloves should be worn.

Ethyl ether

Shake ether (10 ml) for 1 minute with a freshly prepared 10% solution (1ml) of potassium iodide in a glass-stoppered cylinder (25 ml) of colourless glass, protected from light. The mixture should then be viewed transversely against a white background. The absence of colour is favourable while the appearance of a yellow colour indicates the presence of a more than 0.005% peroxide. In the latter case the ether should be discarded.

Tetrohydrofuran or dioxane
To the ether (50 ml) add glacial acetic acid (6 vol.), plus chloroform (4 vol.) and potassium iodide (1 g). On titration with thiosulphate solution (0.1 N) the peroxide content can be calculated thus,

$$\frac{\text{titration} \times \text{normality} \times 1.7}{\text{weight of ether}}$$

It is important to understand the chemistry when applying such techniques in order to ensure the appropriateness of the method.

Solvent containing residues of peroxypropionic acid was treated with a calculated excess of aqueous sodium sulphite. A negative peroxide test was obtained. The solvent was water washed and vacuum distilled. Towards the end of the distillation the still contents exploded, fortunately causing only minor injuries. On investigation the presence of diacyl peroxide was detected in the treated solvent. This peroxide was not reduced by the sulphite – an excess of which invalidated the iodide test for detection of peroxide.[56]

An explosion occurred during the distillation of the products of reaction of *N*-methyl acetamide with aqueous sodium hypochlorite. The reaction initially produced *N*-chloro-*N*-methyl acetamide. Iodine titrations confirmed that this *N*-chloro compound had been destroyed during the latter stages of the reaction. Similarly the iodine titrations indicated that 99.7% of the hypochlorite had been consumed in the process. The most likely cause of the explosion was sodium chlorate which, it was subsequently shown, was not detectable by the iodine test.[57]

Besides ethers other classes of compound capable of undergoing peroxidation include acetals, allylic compounds, dienes, vinylacetylenes, vinylmonomers, amides, various olefins including alicyclic versions such as cyclohexene, cyclo-octene, and aromatic hydrocarbons containing tertiary benzylic hydrogens such as cumene (see Table 11.4).

Pyrophoric compounds

Some materials react so exothermically with oxygen in the air (or its moisture) that ignition temperatures are reached without application of external heat sources. Ensuing fires can spread rapidly engulfing other chemicals and combustible matter. Examples of pyrophoric substances are given in Chapter 10.

A summary of the precautions for pyrophoric materials is given below:

- The quantity in store and in use should be limited and eye protection worn at all times.

- Stores should be of fire-resistant construction and ignition sources should be excluded: storage arrangement must prevent contact with water (from fire sprinklers, water-pipes, steam pipes, radiators, showers, taps, sinks, leaking roofs and seepages of any kind) with water-sensitive chemicals. Storage above ground level is preferred if accidental flooding is a possibility.
- Chemicals stored under 'inert media' must be inspected regularly (e.g. every 3 months) to ensure complete immersion by the medium (e.g. sodium must be stored under paraffin and never under halogenated hydrocarbons).
- Pyrophoric substances should be stored separately from other chemicals particularly flammable solvents, paper, cloth, etc.
- Dry powder extinguishers and sand buckets should be provided.
- Face- and dry hand-protection should be worn when handling water-sensitive materials and accidental contact with water or aqueous reagents should be prevented. Solvents, etc., should be removed from the work area.
- All sodium wire used as a 'drying agent' (e.g. for diethyl ether) should be removed/destroyed before rinsing empty vessels with water prior to returning the bottles to the store. Substitution of sodium sand or wire with molecular sieves or sodium/lead alloy is a preferred arrangement.
- Acetylides should be stored cool, and moist, and separate from metal wool or finely-divided metal powders (e.g. copper, silver).
- Hydrides should be kept dry and apart from oxygen and oxidising agents.

Explosive chemicals

Chemical explosions can be divided into four categories, viz.:

- Those resulting from build-up of pressure within an enclosed container: even 'inert' substances can be involved here.
- Ignition under confined conditions of flammable mixtures of air and gas, vapour or finely-divided combustible particulate.
- Thermal explosions arising from 'run-away' or uncontrolled exothermic reactions.
- Violent reaction or decomposition of chemicals existing in a metastable state.

The first three categories are referred to in Chapters 4, 10 and 15 and this section is restricted to the hazards and precautions for handling the last group. Detailed discussion is beyond the scope of this text and reference should be made elsewhere.[55,58]

The term 'explosive' covers a wide class of materials of which a synopsis is given in Table 11.21.[58] Countries tend to have their own laws and regulations governing definition, acquisition and utilisation of explosives. For example in the UK a substance has been classed as 'explosive'

Table 11.21 Explosive materials and their application

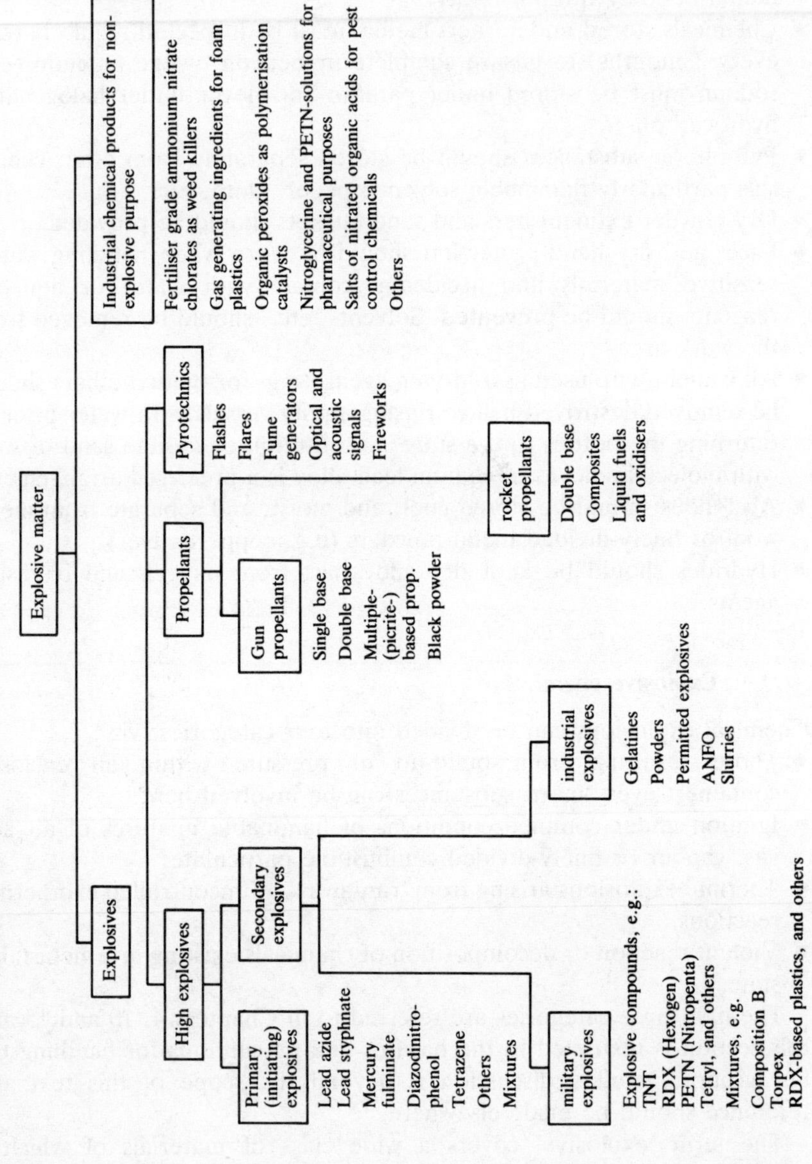

if it explodes under the effect of a flame or if it is more sensitive to shock or friction than dinitrobenzene.[48] Methods are described in the literature for determining such parameters as thermal stability; mechanical sensitivity to shock and friction; self-ignition temperature; oxidising properties of solids; flammability of solids, liquids and gases; pyrophoricity; flammability of water-sensitive materials; etc.[48,49,59,60] In general, legislation requires purchasers to demonstrate competence and maintain storage records. Detailed considerations of facilities and building design are essential as are distances between these and inhabited buildings as shown in Table 11.22.[58] Equations exist to enable safe distances to be calculated such as the expression

$$d = f \times \sqrt[3]{M}$$

where: d = distance in metres; M = maximum amount of explosive in kg present in the building at any one time; f = a variable factor ranging typically from 1.5 between two barricaded stores, 8 from non-dangerous parts of a plant and 20 to domestic dwellings in the vicinity of the factory.

Many of the above explosives are found only in specialist undertakings. However, some chemicals commonly encountered in laboratories are capable of violent and explosive self-reaction or decomposition on initiation by mechanical shock, heat, friction, light or the presence of chemical impurities. Typical examples are given in Chapter 10. Some are

Table 11.22 Distance for storage of explosives in barricaded magazines as approved by the Institute of Makers of Explosives, USA

Magazine charge up to	Distances Inhabited buildings	Distances Passenger railways	Distances Public highways	Separation of magazines
Metric tons	(m)	(m)	(m)	(m)
0.136	82.4	33.6	33.6	7.3
0.227	97.6	39.7	39.7	9
0.454	122	48.8	48.8	11
0.907	154	62.5	56.4	14
1.36	177	71.7	63.9	16
1.81	194	77.8	64.1	18
2.27	209	83.9	68.6	19
4.54	264	105	79.3	24
9.07	297	133	88.5	30
13.6	345	153	104	34
18.1	389	168	116	38
22.7	427	180	128	41
36.3	517	210	156	50
45.4	554	227	166	56
90.7	619	285	186	87
136	694	328	210	117

available as commercial entities while others are more often found as unstable intermediates formed during the course of chemical reactions, or during storage (e.g. peroxide formation on storage of ethers and other substances with electron-rich moieties). Further reference to the latter is made on page 507. Table 11.23 lists some common oxidising agents classified according to their stability.[61]

The main potentially explosive chemicals can be divided into groups according to structure, viz. acetylides and acetylenic compounds, nitrogen compounds, peroxy compounds, vinyl monomers, etc., as discussed in Chapter 10.

Table 11.23 Some common oxidising agents

Classification of oxidising agents according to their stability	
Relatively stable	Aluminium nitrate
• Increase the burning rate of combustible materials	Ammonium persulphate
	Barium nitrate/peroxide
	Calcium nitrate/peroxide
• Form highly flammable or explosive mixtures with finely divided combustible materials	Cupric nitrate
	Hydrogen peroxide solutions (8–27.5% by weight)
	Lead nitrate
	Lithium peroxide/hypochlorite
	Magnesium nitrate/perchlorate
	Nickel nitrate
	Nitric acid (concentrations up to 70%)
	Potassium dichromate/nitrate/persulphate
	Silver nitrate
	Sodium dichromate/nitrate/nitrite/perborate/persulphate/chlorite (up to 40% by weight)
	Strontium nitrate/peroxide
	Zinc peroxide
Moderately unstable/reactive	Ammonium dichromate
• Undergo vigorous decomposition on heating	Barium chlorate
	Calcium chlorate/hypochlorite
	Chromium trioxide (chromic acid)
• Explode when heated in a sealed container	Hydrogen peroxide solutions (27.5–91% by weight)
	Nitric acid (concentrations over 70%)
• Cause spontaneous heating of combustible materials.	Potassium bromide/chlorate/permanganate peroxide
	Sodium chlorate/permanganate/peroxide/chlorite (over 40% by weight)
	Strontium chlorate
Unstable	Ammonium chlorate/perchlorate/permanganate
• Explode when catalysed or exposed to heat, shock or friction	Benzoyl peroxide
	Guanidine nitrate
	Mercury chlorate
	Methyl ethyl ketone peroxide
• Liberate oxygen at room temperatures	Potassium superoxide

Dusts Any combustible solid material, in finely-divided form, can give rise to a dust explosion hazard as described in Chapter 4 if the powder is suspended in air in sufficient concentration (i.e. at or above the lower explosive limit), and subjected to a source of ignition. Pressure waves from the initial explosion can whip deposited dust into the air in front of the advancing flame with the result that the explosion may be extended far beyond the original dust cloud, in the form of a secondary explosion.

For most practical purposes the lower explosive limit for the majority of flammable dusts is of the order of 10–20 g m^{-3} At this concentration a dust cloud would appear as a dense fog. The level of dust may be much lower than this under normal storage conditions but high values may be attained as a result of the primary explosion. Explosive clouds are ignited in hot enclosures at temperatures above 400 °C. However, if dust can settle on a hot surface, smouldering is liable to be encountered above 150 °C: ignition temperature of such a deposit may be more properly regarded as the highest safe temperature in the context of a dust explosion hazard. Explosive clouds are ignited by electrical discharges in excess of about 10 mJ, but contact with repeated discharges of lower energy could engender smouldering. Dust clouds can also be ignited by the effects of mechanical friction or impact, but prolonged friction or repeated impact is generally necessary to be effective. Naked flames will immediately ignite an explosive cloud.

The hazard of dust explosions is encountered with many materials of natural origin, with plastics and certain other organic chemicals, light metals (e.g. aluminium, magnesium, titanium) and with sulphur.

An excellent recent publication[60] discusses dust explosion hazards, indicates basic preventative requirements and describes methods to determine explosive properties of powders and dusts. It lists explosibility data for a range of materials a selection of which is given in Table 11.24. The original publication should be consulted for a comprehensive listing of some 300 materials. It emphasises caution in interpretation since results are affected by particle size, chemical purity, moisture content, etc., and advises determination of such criteria for a specific material.

Precautions for handling explosive materials The main hazards associated with working with explosive chemicals at the laboratory scale (i.e. up to 5 g) are in the event of explosion, projection of glass fragments (e.g. from apparatus) accompanied by a pressure pulse, and possible fire (e.g. from solvents). The key safety strategies are evaluation of explosion hazard by testing, working at small scales, dilution, and simplicity of operation. The following important precautions exemplify the general principles:

- The quantities of potentially explosive materials in store and in use should be strictly limited.

Table 11.24 Dust explosibility data for selected chemicals

Description of dust	Minimum ignition temperature (°C)	Minimum explosible concentration (g/m^3)	Minimum ignition energy (mJ)	Maximum explosion pressure (bar)	Maximum rate of pressure rise (bar/s)
*Acetamide	560	nd	nd	0	0
Acetoparaphenetidine	nd	nd	12	5.0	545
Acetyl salicylic acid	550	15	16	6.0	531
5 Acetyl salicylamide	450	10	20	5.5	216
Acrylic powder, 38 μm	420	30	25	7.8	567
Acrylonitrile-butadiene-styrene copolymer	400	nd	nd	nd	nd
Acrylonitrile-vinylidene chloride copolymer	nd	50	23	4.6	314
Adipic acid	550	35	60	6.6	276
Alkyd powder coating	360	28	22	nd	nd
Aluminium, 6 μm	nd	30	13	6.4	1331
Aluminium, 17 μm	610	40	28	7.0	621
Aluminium, 100 μm	nd	nd	nd	5.4	135
Aluminium, 1000 μm	nd	nd	nd	0	0
Aluminium stearate 20 μm	nd	nd	nd	7.0	1432
D(-)alpha amino phenylglycine	500	20	15	5.8	297
Anthraquinone	670	nd	nd	nd	nd
Anthracene	nd	nd	7	5.8	835
Atrazine, technical	530	90	28	5.0	121
Azodicarbonamide	nd	600	130	5.7	24
Benzoic acid	600	11	12	6.8	749

Benzophenone hydrazone	400	100	40	2.8	16
Benzoyl peroxide	nd	nd	31	6.2	238
Benzthiazole disulphide	nd	nd	nd	5.0	362
4-benzthizole-2-sulphenyl morpholine	nd	nd	nd	5.6	188
Bisphenol A	570	12	11	5.6	814
Bronze	440	nd	nd	nd	nd
*Cadmium sulphide	700	nd	nd	nd	nd
*Cadmium sulphoselenide	710	nd	nd	nd	nd
Cadmium yellow	390	nd	nd	nd	nd
*Calcium citrate	470	nd	nd	nd	nd
*Calcium gluconate	550	nd	nd	nd	nd
Calcium pantothenate	430	50	80	7.3	318
Calcium propionate	530	nd	nd	nd	nd
Calcium silicide	540	60	5	7.1	818
Calcium stearate, 23 μm	570	60	25	8.6	1152
Caprolactam	430	70	60	5.5	118
2 Carbamoyl oxymethyl-1-methyl-5 nitro-imidazole	440	50	36	9.2	352
Carbon, 13% volatiles	590	nd	45	3.0	28
Cellulose acetate	340	35	20	7.9	449
Chloro amino toluene sulphonic acid	650	nd	nd	nd	nd
p-Chloro-*o*-toluidine hydrochloride	650	nd	nd	nd	nd
Coal, 36% volatiles	490	nd	nd	nd	nd
Coal, Pittsburg 74 <μm	530	30	nd	nd	nd
Coal, pulverised 150 μm	550	nd	nd	5.4	41
Coal, rank 202 75 μm	nd	nd	nd	0.3	0.7

Table 11.24 (continued)

Description of dust	Minimum ignition temperature (°C)	Minimum explosible concentration (g/m^3)	Minimum ignition energy (mJ)	Maximum explosion pressure (bar)	Maximum rate of pressure rise (bar/s)
Coal, rank 502 75 μm	nd	nd	nd	5.3	135
Coal, rank 702 75 μm	nd	nd	nd	5.1	114
Coal, rank 802 75 μm	nd	nd	nd	5.0	45
Coal, rank 902 75 μm	nd	nd	nd	5.6	124
Cyclohexanone peroxide	nd	nd	21	5.8	386
N-Cyclohexylthio phthalimide	nd	nd	nd	5.7	148
Detergent, high non-ionic	410	nd	nd	nd	nd
Detergent, low non-ionic	560	nd	nd	nd	nd
Detergent, standard ABS	520	nd	nd	nd	nd
Dextrin	440	nd	nd	nd	nd
Dextrose monohydrate	350	nd	nd	nd	nd
Dibutyl tin maleate	600	nd	nd	nd	nd
Dibutyl tin oxide	530	12	7	5.9	366
Dicyclohexyl phthalate	nd	nd	nd	5.5	297
O-Diethyl phthalimidopho sphonothiate	350	30	210	4.9	80
Dihydro streptomycin sulphate	670	nd	nd	nd	nd
Dimethyl acridan	540	nd	nd	nd	nd
Dimethyl diphenyl urea	490	nd	nd	nd	nd
Dinitro aniline	470	nd	nd	nd	nd
Dinitrobenzoyl chloride	380	nd	nd	nd	nd

2,2,4 – Dinitrophenyl thiobenzthiazole	nd	nd	nd	9.5	863
Dinitro stilbene disulphonic acid	450	nd	nd	nd	nd
Dioctyl tin oxide	nd	nd	nd	5.7	238
Diomethane	nd	nd	nd	5.8	566
Diphenyl guanidine	nd	nd	nd	4.6	304
Diphenyl guanidine – 1.5% dedusting powder	540	nd	28	nd	nd
Diphenyl guanidine-meraptobenzothiazole	nd	nd	5	4.9	466
Diphenyl guanidine phthalate	nd	nd	nd	4.6	335
Diphenylol propane	nd	12	11	5.6	814
4,4 – Dithiomorpholine	nd	nd	nd	7.9	431
Epoxy, powder coating	430	10	15	8.3	615
Epoxy resin	490	12	9	5.4	902
Epoxy resin, polyester	500	10	25	5.7	383
Ferrochromium	600	2000	nd	nd	nd
Gallic acid	570	30	30	5.1	104
Heptabarbitone	nd	nd	nd	4.9	559
Hydrocarbon resin	nd	nd	nd	4.9	1078
Hydroxy ethyl cellulose	420	nd	nd	nd	nd
Hydroxy ethyl methyl cellulose	410	nd	nd	nd	nd
Iron	510	200	110	3.3	145
Iron, sponge	400	100	40	3.1	27
Isinglass	520	nd	nd	0	0
Lauryl peroxide	nd	nd	12	6.2	442
Lead stearate, dibasic	nd	nd	12	nd	nd

Table 11.24 (continued)

Description of dust	Minimum ignition temperature (°C)	Minimum explosible concentration (g/m³)	Minimum ignition energy (mJ)	Maximum explosion pressure (bar)	Maximum rate of pressure rise (bar/s)
Lead sulphate, tribasic	nd	nd	nd	0	0
Lignin	450	40	20	7.1	345
Lycopodium	420	20	12	8.8	990
Magenta	570	nd	nd	nd	nd
Magnesium stearate	580	40	25	7.6	1604
Manganese, electrolytic	nd	nd	nd	4.7	162
Manganese ethylene bis dithio-carbamate	270	70	35	nd	nd
Mefenamic acid	nd	30	nd	6.6	790
Melamine formaldehyde resin	410	20	50	6.4	125
2-Mercapto benzthiazole	nd	nd	nd	4.8	490
Methyl benzoquate	380	nd	20	4.9	418
Methyl butadiene styrene	480	50	165	2.3	14
2,2 – Methylene dis-4-ethyl 6 – tertiary butyl phenol	310	nd	nd	5.2	504
Methyl cellulose	480	nd	nd	nd	nd
Methyl methacrylate	440	20	13	7.4	607
Monochloracetic acid	620	nd	nd	nd	nd
β-Naphthol	670	nd	nd	nd	nd
Nigrosine hydrochloride	630	nd	nd	nd	nd
p-Nitro-*o*-anisidene	400	nd	nd	nd	nd
Nitrocellulose	nd	nd	30	18	1442

Nitrodiphenylamine	480	nd	nd	nd	nd
Nitrofurfural semi-carbazone	240	nd	nd	10.0	593
m-Nitro *p*-toluidine	470	nd	nd	nd	nd
p-Nitro *o*-toluidine	470	nd	nd	nd	nd
Nylon 11	480	5	32	6.5	325
Orthotolyl biguanide	nd	nd	nd	5.7	338
Oxalic acid	nd	nd	nd	0	0
Paracetamol	380	nd	10	5.5	1042
Pectin	390	75	35	9.2	552
Penicillin, N-ethyl piperidine salt	310	nd	nd	nd	nd
Phenol formaldehyde	450	15	10	5.3	621
Phenol formaldehyde, paper filled	450	20	40	5.7	69
Phenol formaldehyde resin	450	15	nd	7.4	449
Phenolic resin	420	20	5	8.2	1148
Phenolic resin, laminate	nd	nd	nd	4.2	41
Pitch	640	30	20	6.1	414
Polyester (31% inorganic) 30 μm	420	70	25	7.2	414
Polyester resin	400	nd	nd	nd	nd
Polyester – fibreglass resin	nd	nd	nd	6.4	804
Polyethylene 14 μm	470	30	5	6.9	1646
Polyethylene, commercial	nd	nd	57	4.4	31
Polyethylene, granular	nd	nd	nd	3.2	17
Polyethylene, ground	400	nd	nd	4.3	31
Polyethylene, high density	nd	10	17	5.1	659
Polyethylene glycol	320	60	150	6.6	40
Polypropylene	380	nd	43	3.1	21

Table 11.24 (continued)

Description of dust	Minimum ignition temperature (°C)	Minimum explosible concentration (g/m³)	Minimum ignition energy (mJ)	Maximum explosion pressure (bar)	Maximum rate of pressure rise (bar/s)
Polypropylene copolymer	nd	nd	nd	4.1	31
Polypropylene – glycol maleate phthalate	nd	nd	nd	5.9	238
Polystyrene, bead	nd	10	nd	4.4	42
Polystyrene foam, fire retardant grade	nd	40	nd	3.8	56
Polyurethane, crumb	510	40	80	4.6	100
Polyvinyl acetate	450	40	70	4.8	69
Polyvinyl chloride	510	nd	nd	0	0
Polyvinyl chloride, dispersion resin	550	nd	nd	0.2	0.3
Polyvinyl chloride, plasticised blend (70/30 PVC/DOP)	nd	80	170	4.9	3.7
Polyvinylidene chloride	670	nd	nd	nd	nd
Potassium phthalimide	540	250	>2000	2.1	11
Procaine penicillin	450	nd	nd	5.5	473
Propyl iodine	470	nd	nd	nd	nd
Pyrogallol	nd	nd	nd	6.6	288
Rubber	380	nd	nd	nd	nd
Rubber, crumb	440	nd	nd	5.8	469
Rubber, cyclised	nd	100	nd	5.4	185
Rubber, latex	450	nd	nd	nd	nd

Rubber, spray dried – silica detackifier	nd	nd	nd	3.9	31
Saccharin	560	nd	nd	nd	nd
Salycylamide	nd	nd	nd	5.2	252
Salicylic acid	590	25	nd	5.8	469
Sebacic acid	nd	nd	nd	5.1	28
Silicon	900	nd	nd	7.1	849
Soap	430	20	25	5.4	194
Sodium acetate	560	30	35	6.2	318
Sodium carboxy methyl cellulose	320	1100	440	3.4	28
Sodium 2,2-dichloropropionate	520	300	nd	4.4	59
Sodium formate	550	>1500	>2000	0.4	3
Sodium glucaspaldrate	600	nd	nd	nd	nd
Sodium glucoheptonate	600	nd	nd	nd	nd
Sodium monochloracetate	550	nd	nd	nd	nd
Sodium propionate	470	nd	nd	4.8	48
Sodium stearate	nd	nd	nd	5.5	32
Sodium toluene sulphonate	530	nd	nd	nd	nd
Sodium xylene sulphonate	490	nd	nd	nd	nd
Soot	nd	nd	nd	0	0
Sorbic acid	440	20	15	7.4	690
Starch	470	nd	nd	6.3	45
Starch, cold water	490	nd	nd	nd	nd
Stearic acid	nd	nd	nd	6.7	344
Steel, mild	450	nd	nd	3.1	42
Steel, stainless	nd	nd	nd	0	0

Table 11.24 (continued)

Description of dust	Minimum ignition temperature (°C)	Minimum explosible concentration (g/m³)	Minimum ignition energy (mJ)	Maximum explosion pressure (bar)	Maximum rate of pressure rise (bar/s)
Streptomycin complex	nd	nd	nd	5.2	30
Streptomycin sulphate	700	nd	nd	nd	nd
Sulphur	190	20	15	5.4	325
Tallow, hydrogenated	620	nd	nd	nd	nd
Tartaric acid	350	nd	nd	nd	nd
Terephthalic acid	nd	nd	nd	6.3	424
Tetramethyl thiuram disulphide	nd	nd	nd	5.5	141
Tinuvin 320	510	30	5	6.7	1604
Toner, electro photographic 5 μm	nd	10	20	6.9	990
Tributyl tin fluoride	nd	nd	nd	5.5	383
3,4,4 – Trichlorocarbanilide	nd	nd	nd	3.9	104
2,2,4 – Trimethyl 1,2-Dihydroquinoline	nd	nd	nd	5.0	549
Trimellitic anhydride	nd	300	nd	5.6	304
*Urea	900	nd	nd	0	0
Urea formaldehyde, moulding powder	450	40	80	6.2	249
Urea formaldehyde, moulding powder paper filled	430	70	49	6.8	41
Urea formaldehyde, moulding powder wood filled	430	25	40	6.0	83

Urea formaldehyde, resin	430	20	34	7.6	110
Vitamin B_2 (riboflavin)	nd	nd	nd	4.2	98
Vitamin B_{12}	370	nd	nd	nd	nd
Wax, paraffin	340	nd	nd	nd	nd
Wax, synthetic	nd	nd	nd	5.3	926
Wood	360	nd	nd	6.2	393
Wood, flour 30 μm	380	70	100	8.8	592
Wood pulp, dried	450	nd	nd	7.3	1371
Wood pulp, 10% H_2O	nd	nd	nd	7.4	879
Wood pulp, fluffed	470	20	nd	nd	nd
Zinc 17 μm	690	850	400	6.0	116
Zinc mecaptobenzthiazole	nd	nd	nd	4.9	311
Zinc stearate	420	20	14	7.8	1566
Zircaloy 2	380	100	10	3.5	21
Zirconium, 10 μm	300	10	5	5.6	1621
Zirconium hydride 10 μm	320	50	8	6.4	1260

*Group 'B' dust
nd – parameter not determined

- Stores should be specially designed and constructed of non-combustible material, and located away from other hazards (for example brick 'coal bunkers' are suitable for small samples, while purpose-built constructions with explosion-proof lights, etc., are required for larger quantities). They should be designated NO SMOKING areas and be well labelled.
- Stores should be used exclusively for these materials and in particular other combustible material such as fabric, paper, organic solvents should not be stored therein.
- Generally the substances in this class are unstable when heated or exposed to light and they should be stored cool and in the dark. However, for liquids with added stabiliser the cooling may cause separation of the material from the stabiliser. Similarly precipitation of a potentially explosive compound from a diluent may occur on cooling. In both cases this can represent a hazardous situation.
- Stores should be ventilated and sound, e.g. no cracks in floors, no rusty window frames, no water seapages, etc.
- Stores should be clean, tidy and locked. Contamination must be avoided and a high standard of housekeeping maintained.
- Heat sources should not be permitted nearby.
- Material should be purchased in several small containers rather than one large container and always stored in original containers. The integrity of the labels should be checked.
- Use must be restricted to experienced workers who are aware of the hazards and the necessary precautions.
- Records of usage should be kept and stock rotated. Old material should be disposed of.
- Work should be on a scale of less than 0.5 g for novel but potentially explosive material until the hazards have been fully evaluated (see later) and less than 5 g for established, commercially available, substances such as peroxide free-radical initiators.
- Skin contact, inhalation and ingestion must be avoided. Splashes in eyes or on skin should be washed away immediately with copious quantities of water. Medical attention should be sought. Similarly, if material is swallowed medical aid is required immediately.

For the above scales, eye protection should be worn and work should be undertaken in a standard fume-cupboard behind a well-anchored polycarbonate screen. It is advisable to wear a protective apron and hand protection, but whether leather gauntlets or tongs should be used will be dictated by circumstances. In general such measures are recommended but it should be ensured that they do not precipitate a hazard as a result of loss of tactile sensitivity (e.g. dropping a flask, overtightening clamps, exerting excessive pressure when assembling apparatus). The material of fabrication of gloves needs consideration. Thus PVC but *not* rubber is suitable for tert-butyl peroxide.

- Sources of ignition such as hot surfaces, naked flames, etc., must be avoided and smoking prohibited where explosives are used. Accidental application of mechanical energy should be avoided (e.g. material should not be trapped in ground-glass joints): siezed stoppers, taps, etc., must not be freed by the application of force. In minimising risk of static electricity, laboratory coats made of natural fibre rather than synthetic fabrics are preferred. However, in such circumstances it is even more important to neutralise any spillage on the coat immediately since delay could result in the impregnated garment becoming a fire hazard.

A process worker in a chemical plant manufacturing organic peroxides was provided with an impermeable plastic apron with a bib in addition to his ordinary overalls. At lunch-time he removed his apron before visiting the canteen but he left on his overalls. After the meal he and his companions lighted cigarettes. Suddenly the legs of his overalls burst into flame and the flames spread to his sleeves. Colleagues extinguished the flames but only after the operator had suffered severe burns.[62]

- To prevent glass fragments from flying in the event of an explosion, use should be made of metal gauzes to screen reaction flasks, etc., or cages, e.g. for desiccators. Awkward size or shape vessels may be covered with cling film.
- Whenever possible a stabiliser or diluent should be used and separation of the pure material should be avoided.
- Any waste material (and contaminated cloths, tissues, clothing, etc.) must be rendered safe by chemical means or by controlled incineration of dilute solution where practical prior to disposal via drains, burying, etc. (see Ch. 14).
- In the event of fire, the area should be evacuated, the alarm raised and the fire brigade summoned. Only if it is clearly safe to do so should the fire be tackled with *an appropriate* extinguisher (see Ch. 4).

For new materials scale-up should only be considered once the explosive potential behaviour of the substance has been established. The former can be calculated. Thus if the enthalpies of formation of the compound are known then the heat of reaction of specified breakdown products can be calculated providing their enthalpies are known. Similarly, a heat of combustion from reaction with oxygen can be calculated. However, such thermodynamic estimates indicate the 'explosive potential' not the 'sensitivity' of the compound, since only the amount of energy release is described and not the rate at which it is liberated. Nevertheless this itself can be useful in deciding whether or not scale-up is advisable. This 'energy hazard potential' can be calculated using a computer program known as CHETAH developed by ASTM, which uses four criteria[63]:

1. The amount of each element present in the compound and the thermodynamic data from the structural relationship are used to define those products which could be formed from the substrate and which, if formed, would release the maximum amount of energy. The energy-hazard potential is graded as shown in Table 11.25.
2. The heat of combustion of the compound in excess oxygen with the maximum heat of reaction as shown in Fig. 11.8.
3. A graduation of energy hazard potential based on the 'oxygen balance' concept as summarised in Table 11.26.
4. A gradation of energy hazard potential as shown in Table 11.27.

$$Y = 10\ \Delta H^2_{max} W/n$$

Where: ΔH^2_{max} = maximum heat of decomposition; W = weight of composition in grams; n = number of moles in the composition.

Table 11.25 Hazard potential grading based on heat of reaction

Energy hazard potential	Maximum heat of reaction* (kcal/g)
Low	More positive than −0.3
Medium	−0.3 to more positive than −0.7
High	−0.7 or more negative

* Heat of reaction = heat of decomposition.

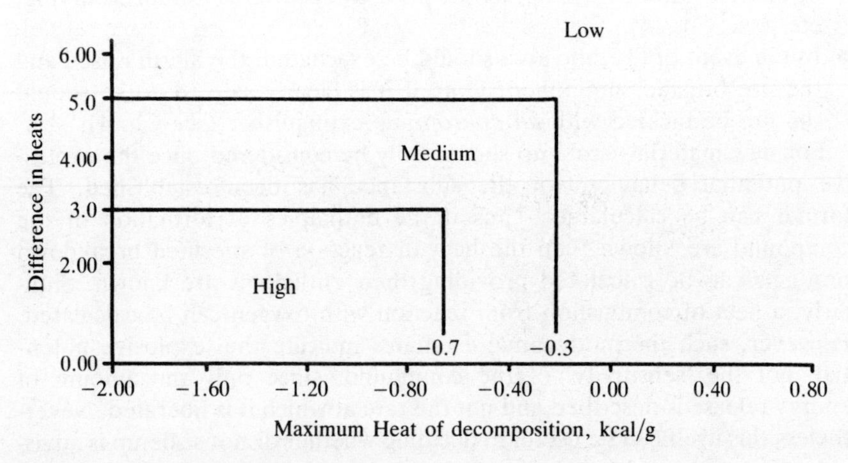

Fig. 11.8 Heat of combustion v. heat of reaction

Table 11.26 Hazard potential grading based on oxygen content

Energy hazard potential	Percentage computed oxygen balance
Low	more positive than 240.0 or more negative than −160.0
Medium	between 240.0 and 120.0 or between −160.0 and −80.0
High	between −80.0 and +120.0

Table 11.27 Grades of energy hazard potential

Energy hazard potential	*Y*
Low	below 30
Medium	30–110.0
High	above 110.0

There is an appreciable probability that a material will be sensitive to shock if its calculated CHETAH criterion (1) is *high* or the criterion (2) or (3) is high or medium.

The nature of the explosive behaviour can only be efficiently gauged from a battery of experimental tests[48,59,60] under a variety of conditions: no simple test can be considered in isolation. Tests range from simple small-scale bench operations to more sophisticated larger scale evaluation.

Examples of the former are given below. It is emphasised that, for various reasons, the results are only intended to be used to indicate whether it is safe to scale up to 5 g with caution. Too much reliance should not be placed on low or negative responses. Clearly these tests must be undertaken only by trained personnel with appropriate precautions.

Flame test – A few crystals (or drop of solution) are heated in the non-luminous flame of a Bunsen burner. Melting with quiet burning is at one end of the spectrum while cracking, flashing-off or flaring are considered as hazardous.

Impact – Impact sensitivity can be gauged by striking a few crystals of the compound on a metal last with the ball of a ball-pein hammer. Ignition, smoking, cracking or other sign of decomposition should be considered as hazardous.

Differential scanning calorimetry/differential thermal analysis/hot stage microscope – Providing the decomposition reaction follows the general rate law, the activation energy, heat of decomposition, rate constant and half-life for any given temperature can be obtained on a few milligrams using the ASTM method. Hazard indicators include heats of decomposition in excess of 0.3 kcal/g, short half-lives, low activation energies and low exotherm onset temperatures, especially if heat of decomposition is considerable.

The DTA or hot-stage microscope can be used under ignition conditions to obtain an ignition temperature. The nature of the decomposition can also be observed at a range of temperatures. Observations, such as decomposition with evolution of gases prior to ignition, should be regarded as potentially hazardous.

Bomb calorimetry – Using oxygen and an inert gas enables the heat of combustion and the heat of decomposition to be evaluated respectively.

Deflagration – A melting-point test has been described for diazo compounds. The first 1 mm of a melting-point tube filled with *c.* 10 mg of test compound is inserted in a melting-point apparatus heated at 270 °C. Once decomposition starts the tube is removed. The decomposition rapidly propagates through the entire mass for unstable diazo compounds whereas no such propagation is reported for stable versions.

Many more elaborate specialised techniques have been described for assessing the response of compounds to mechanical, thermal or detonative stimuli. These include impact and friction tests, flashpoint, ignitability, Koenen steel tube and Dutch pressure-vessel tests, time/pressure and time/temperature/pressure tests, self-heating tests, trough and open channel burning tests, ballistic mortar tests, card-gap tests, underwater test methods, etc.

Toxic chemicals

A detailed discussion of toxic hazards is given in Chapter 5 and the general measures for their control are described in Chapter 6. Because of the nature of laboratory work and the small volumes of chemicals used, the duration of exposure to a toxic chemical is less than in the process industry. However, the opportunity for exposure may be increased by the plethora of chemicals and techniques encountered in the laboratory. The level of precautions required for the safe use of toxic materials in the laboratory is dependent upon scale and frequency of use, nature of operation, physical properties, toxic nature and potency, likely routes of entry, duration of exposure, etc. Some materials pose a variety of hazards (flammability and acute toxicity).

As discussed in Chapter 5, the key modes of entry of chemicals into the body are injection, ingestion, skin absorption, and inhalation. Injection is rarely encountered in industry but can happen accidentally in laboratories as a result of careless use of syringes or via cuts by contaminated glass or metal. Similarly, swallowing is an uncommon occupational route but workers must be alerted to the dangers so that inadvertent ingestion can be prevented. A wide range of chemicals can enter the bloodstream by absorption through skin or mucous membranes. Corrosive chemicals are particularly hazardous and the eyes and skin are especially vulnerable. While some materials cause damage at the point of entry others are systemic poisons. Compounds such as dimethyl sulphoxide can pass readily through the skin and act as a vehicle for materials otherwise incapable of skin absorption. Again, materials incapable of passing through intact skin, can enter the bloodstream via skin lesions. Inhalation is by far the commonest route of entry of chemicals and precautions should be aimed at containment or removal of contaminant at source: respiratory protection should be considered as a back up facility for short-term exposures including emergencies.

The process conditions can dictate the most likely route of entry used. Thus, skin absorption may be relevant for a high-boiling liquid at room temperature, while inhalation may become feasible when using the same substance at elevated temperatures.

The guidelines below indicate the main essential precautions, though all aspects of safe laboratory practice are applicable. Local Regulations such as Carcinogenic Substances Regulations 1967 (UK) or OSHA's listing of carcinogens and accompanying regulations for usage (USA) must be adhered to as appropriate. Precautions for work with compressed gases, corrosive chemicals, and radioactive substances are covered elsewhere.

Guidelines for working with toxic chemicals

Administration – Policy should recognise staff experience. Thus junior inexperienced technicians should not be required to handle highly toxic chemicals. Such substances should not feature in school-laboratory practical work, and only under strict supervision in undergraduate classes. The supervisor must authorise all work involving toxic substances. Where these are particularly hazardous (e.g. carcinogens) formal authorisation should be recorded. He must also consider the purchase, storage, use and disposal of toxic chemicals. He is responsible for ensuring the provision of adequate protection and emergency facilities as relevant (e.g. fume-cupboards, eye protection, eye-wash fountains, emergency showers, self-contained breathing apparatus, antidotes). In the case of highly toxic materials, records should be kept indicating materials used, quantity, user and dates; scales of operation should be kept to a minimum.

Operators, researchers, students, etc., should be informed of hazards, instructed and trained in the precautions and emergency precedures, and supervised. Eating, drinking, smoking, chewing of gum, taking of medicine, application of cosmetics, biting of nails, storage of food, etc., should all be prohibited in laboratories, particularly where highly toxic substances are used or kept, and laboratory equipment should not be used for food or drinks.

Storage – All materials should be adequately labelled to indicate the contents and warn of the hazards. Appropriate warning signs should be displayed in the stores. Volatile toxic substances require ventilated storage conditions with stores maintained under negative pressure. Highly toxic materials should be kept in secondary containers (or spill dishes) to contain the substance in the event of leakage. Security should limit access to the stores and prevent unauthorised withdrawal of stock. Certain toxic substances such as those poisons particularly hazardous on ingestion, or carcinogens, should be stored under lock and key in poison cabinets and withdrawals logged. Water should be prevented from coming into contact with those chemicals with which it reacts to produce toxic products (see Ch. 11) and incompatible chemicals which produce toxic products on contact with each other should be segregated (see Ch. 10).

Use and disposal – Highly toxic gases and hyper-poisons should ideally be used only in a restricted area and preferably in a specially designed dangerous-chemical laboratory. This can be equipped with unique services and special features (e.g. extra efficient fume-cupboards fitted with effluent treatment facilities, wrist-operated taps, disposable paper towels, air locks, changing rooms, etc.). Toxic gases and volatile liquids should be handled in an efficient fume-cupboard. Experiments should be designed to contain the substance; thus, waste gases and vapours should be scrubbed, condensed or filtered through an HEPA system. Techniques should prevent aerosol generation (e.g. liquids should not be dropped from a height, or blown out of pipettes). Hyper-poisons and toxic, dusty solids should be manipulated in a glove-box. The draft in a fume-cupboard may be sufficient to encourage fine powders to become airborne and decontamination of the fume-cupboard may become necessary since transport velocities are likely to be inadequate to prevent particulate from settling in the ventilation system. Glove-boxes under negative pressure should be ventilated at the rate of at least two complete air changes per hour with an internal pressure of at least 1.3 cm of water below the ambient atmosphere. Effluent should be scrubbed or passed through an

HEPA filter. Positive pressure glove-boxes (often employed to provide inert atmospheres) should be inspected for leaks and located in a fume-cupboard. Experiments should be undertaken in spill trays or the fume-cupboard base lined to contain spillage.

No chemical should be tasted. If it is essential to check the odour of a chemical, it should be done with caution. For example the proper method for 'testing' a sample in a beaker or test-tube for smell is to hold it away from the face and to waft the vapour from above it gently towards the nose. 'Unknown' materials should be assumed to be toxic and consideration should be given to the need for medical screening before handling sensitisers with the possibility of periodic health examinations. Mouth pipetting should be prohibited.

For operations involving highly toxic chemicals which are volatile or dusty, consideration should be given to the need for environmental analysis, particularly on preliminary runs until procedures have been established as adequate for routine adoption. If the process cannot be contained the need for respiratory protection should continue until safety is demonstrated quantitatively. Even then when very toxic materials are in use the operator may need to use respiratory protection to guard against the eventuality of an accident. Laboratory coats, gloves and face protection should be worn when handling skin sensitisers or materials capable of absorption through the skin. (Attention to choice of glove material is essential: Ch. 6 lists a range of fabrics for a variety of chemicals.) Depending on the relative risk, impervious aprons and disposable overalls may be more appropriate than laboratory coats. Laboratory coats should not be worn when visiting the canteen, library, etc. On completion of work protective clothing should be removed (washing plastic or rubber protection before doing so) and the hands and face (at least) washed thoroughly before leaving the restricted area or after handling the toxic substance. Contaminated personal protection should be sealed in labelled plastic bags to await treatment of disposal. Equipment should be decontaminated to remove toxic residues prior to submission for wash-up.

Laboratory vacuum pumps should be protected by a trap or filter. Decontamination of pumps should be undertaken in fume-cupboards or similar ventilated booths. Wastes should be collected, labelled and segregated, and treated or disposed of via a specialist contractor. Transportation from restricted area to collection point must be supervised.

All accidents or incidents involving toxic materials must be reported to line management and the safety department. Prompt first aid may be crucial. All complaints should be fully investigated as should lost time attributed to occupational exposure.

While many toxic chemicals are encountered in laboratories, the following comments on mercury serve as an illustration of chronic and acute hazards and the range of handling precautions required. These include substitution, containment, careful choice of 'process' conditions, personal protection, atmospheric analysis and emergency measures.

Mercury is particularly common in laboratories in, e.g. barometers, U-tube manometers, portable vacustats, thermometers, polarographs, diffusion pumps, fluorescent tubes, ozonisers, discharge lamps. Mercury vapour is a strong cumulative poison and the symptoms include tremor, tension, agitation, headaches, insomnia, blueing of the gums, loosening of the teeth, vomiting and diarrhoea. The liquid attacks many metals including copper, brass, aluminium and gold. Splashes break up into very small mobile droplets which spread rapidly and get into cracks in benches or the floor. Hence cleaning up a spillage is difficult. Precautions for work with mercury are given in Table 11.28.

Corrosive chemicals

The hazards of corrosive chemicals are discussed in Chapter 5 but basically the main problems surround their effects on tissue (e.g. skin, eyes, mouth and respiratory system) and materials of construction.

Good laboratory techniques should be followed to minimise the possibility of exposure, e.g. using hand, face (or eye) protection, fume-cupboards, etc. Where larger quantities are involved (e.g. pilot plants) use more substantial protective clothing and ensure self-contained breathing apparatus is available. Dust control procedures should be adopted when handling dusty material (e.g. local extract ventilation).

Barrier creams of an appropriate grade may be helpful but operators need training in their use. They are of no value in treating skin complaints caused by exposure to irritants; medical advice should be sought for all such complaints. A high degree of personal hygiene is required and hand to face contact should be minimised. Safety pipettes, should be used; the use of mouth pipetting should be discouraged. Good standards of housekeeping should be maintained and suitable emergency showers and eye-wash fountains should be located close to the sources of hazard; these must be maintained.

All containers containing corrosives should be made of compatible materials and suitably labelled. (NB Acids can react with mild steel to generate hydrogen, causing pressure build-up within the drum and a potentially explosive atmosphere within the head space.) Storage above head height should be avoided. Bottles of acid or alkali require support on resistant spill trays. Strong oxidising agents should be stored in glass or other inert container: corks and rubber bungs should be avoided. They need to be segregated from reducing agents. Any heating requires use of an isomantle and *not* an oil bath.

Spillages should be neutralised and mopped up or flushed to drain with copious quantities of water, depending upon volume.

Table 11.28 Precautions for working with mercury in the laboratory

- The use of mercury should be avoided if possible. Dial gauges and oil manometers are normally acceptable substitutes for a mercury manometer.
- Mercury should be stored in air-tight containers or under water and the strength of the container in relation to the weight of mercury stored must be considered.
- Containers of mercury should be sealed, labelled and stored under cool, ventilated conditions.
- Mercury should only be handled over a suitable tray in a fume-cupboard where practicable. Because of the tendency to splash, the tray should be near to the apparatus. All rigs containing mercury should be fitted with catchpots.
- Rings should not be worn when dealing with mercury. Gloves are recommended. In any case the hands (and gloves) should be washed with soap and water after dealing with mercury.
- Mercury should not be left exposed to the atmosphere. A layer of water/oil is suitable for reducing the vapour, and also keeps the mercury clean.
- Mercury is heavy and moves with considerable force, apparatus should be made strong enough to allow for this. Stout polythene or PVC is to be preferred to glass.
- No one should use hot mercury without first seeking expert advice.
- In the event of a spillage the following action is recommended:
  - Inform people in the vicinity to avoid the area of spillage.
  - Switch off heating equipment close to the spillage if this can be done safely.
  - If possible, inform line management and the safety office.
  - Wait for a suitably experienced person if one is available, and mark off the area of spillage (with chalk, angle-iron, wood, etc.) plus a notice.
  - If no one is available carefully sweep together obvious globules with a piece of plastic, cardboard, etc., collect them into a suitable container; small droplets can be collected using a glass capillary tube coupled to a hand or water pump and Buchner flask. This should be labelled 'dirty mercury' and sealed (or a layer of water added).
  - Collect small globules as amalgam by sprinkling zinc powder on the contaminated area; collect the powder carefully without excessive brushing.
  - Neutralise invisible drops by sprinkling sulphur over the area (and beyond) brushing it well into the cracks, and if possible leaving overnight before sweeping up. A note should be left for the cleaner.
- Following a spillage of mercury it may be advisable to have a check made on the atmospheric concentration of mercury vapour (see Ch. 7).
- Contaminated absorbent, broken glassware, rubber gloves should be placed in a labelled sealable plastic bag or bottle and sent for disposal.
- Contaminated equipment needs careful attention (e.g. vacuum pumps may require an oil change and to be run in a ventilated area and the exhaust gases monitored; glassware should be cleaned with nitric acid, rinsed and dried prior to sending to the glass blowers; contaminated drains may require attention by the plumbers and the U-bend water traps inspected/cleaned).

Compressed gases

Compressed gases are used in laboratories as reactants, as fuels, for inerting and for carrying samples into analytical apparatus (e.g. gas–liquid chromatography). Typical definitions of 'compressed gases' include in the UK 'Gases with a gauge pressure exceeding 1.5 kg/cm^2'

and 'Liquids with a vapour pressure exceeding 3.0 kg/cm^2'. Liquefied flammable gas in the UK is defined as any substance which at a temperature of 20 °C and a pressure of 760 mm Hg would be a flammable gas but which is in liquid form as a result of the application of pressure or refrigeration or both. Liquefied petroleum gas is commercial butane, propane and mixtures thereof. 'Commercial butane' means a hydrocarbon mixture consisting predominantly of butane, butylene or any mixture thereof. 'Commercial propane' means a hydrocarbon mixture consisting predominantly of propane, propylene or any mixture thereof.

The US Department of Transport definition is 'any material or mixture having in the container either an absolute pressure greater than 276 kPa (40 lb f/in^2) at 21 °C or an absolute pressure greater than 717 kPa (104 lb f/in^2) at 54 °C, or both, or any liquid flammable material having a Reid vapour pressure greater than 276 kPa (40 lb f/in^2) at 38 °C'.

All compressed gas cylinders are so constructed that they are safe for the purpose for which they were intended when first put into service and they should be hydrostatically pressure tested periodically by the supplier. However, serious accidents may result from their misuse, abuse or handling. The greatest care should be exercised in transportation, handling, storage and the disposal of such cylinders.

Compressed gases can often be more dangerous than chemicals in the liquid or solid form because of the potential source of high energy, low boiling point of the contents, ease of diffusion of escaping gas, low flashpoint of highly flammable materials, absence of visual and/or odour detection of leaking materials. Also the containers tend to be heavy and bulky.

Compressed gases, therefore, present a unique hazard from their potential physical and chemical dangers. Unless cylinders are secured they may topple over, cause injury to operators, become damaged themselves and cause contents to leak and, if the regulator shears off, the cylinder may rocket like a projectile or 'torpedo' dangerously around the laboratory. Other physical hazards stemming from the high pressure of a cylinder's contents include accidental application of a compressed gas/air hose or jet on to an open cut or wound, whereby the gas can enter the tissue or bloodstream and is particularly dangerous. A further hazard exists when compressed-air jets are used to clean machine components in work places: flying particles have caused injury and blindness.

Low boiling-point materials can cause frostbite on contact with living tissue. Although this is obvious with cryogenics, such as liquid nitrogen or oxygen (see page 644), it is not always appreciated that cylinders of other liquefied gases become extremely cold and covered in 'frost' as the contents are discharged.

Precautions also have to be instituted to protect against the inherent properties of the cylinder contents, e.g. toxic, corrosive, flammable properties (summarised in Table 11.29) as illustrated below.

Table 11.29 Compressed gases* – Hazards and construction materials for services.

Gas	Hazard†	Materials of construction for ancillary services‡	
		Compatible	Incompatible
Acetylene	F	Stainless steel, aluminium, wrought iron	Unalloyed copper, alloys containing more than 70% copper, silver, mercury, and cast iron
Air	O	Any common metal or plastic	
Allene	F	Mild steel, aluminium, brass, copper or stainless steel	Copper, silver and their alloys, PVC and neoprene
Ammonia	C F T	Iron and steel	Copper, zinc, tin and their alloys (e.g. brass), and mercury
Argon		Any common metal	
Arsine	F T	Stainless steel and iron	
Boron trichloride	C T	Any common metal for *dry* gas Copper, monel, Hastelloy B, PVC polythene and PTFE if *moist* gas is used	Any metal incompatible with hydrochloric acid when moist gas is used
Boron trifluoride	C T	Stainless steel, copper, nickel, monel, brass, aluminium for *dry* gas up to 200 °C. Borosilicate glass for low pressures. For *moist* gas, copper and polyvinylidene chloride plastics	Rubber, nylon, phenolic resins, cellulose and commercial PVC
Bromine pentafluoride	C T O	Monel and nickel	
Bromine trifluoride	C T O	Monel and nickel	
Bromotrifluoroethylene	F T	Most common metals so long as gas is dry	Magnesium alloys and aluminium containing more than 2% magnesium
Bromotrifluoromethane		Most common metals	
1,3-Butadiene	F T	Mild steel, aluminium, brass, copper or stainless steel	PVC and Neoprene plastic

Table 11.29 (continued)

Gas	Hazard[†]	Materials of construction for ancillary services[‡]	
		Compatible	Incompatible
Butane	F	Any common metal	
1,Butene	F	Any common metal	
Carbon dioxide	T	Iron, steel, copper, brass, plastic for dry gas. For moist gas use stainless steel or certain plastics	For moist gas avoid materials attacked by acids
Carbon monoxide	F T	Copper-lined metals for pressures less than 34 bar. Certain highly alloyed chrome steels	Iron, nickel and certain other metals at high pressures
Carbon tetrafluoride		Any common metal	
Carbonyl fluoride	C F T	Steel, stainless steel, copper or brass for dry gas. Monel, copper or nickel for moist gas	
Carbonyl sulphide	F T	Aluminium and stainless steel	
Chlorine	C T O	Extra heavy black iron or steel for dry gas. Drop forged steel, PTFE tape. Moist gas requires glass, stoneware (for low pressures) and nobel metals. High silica, iron, Monel and Hastelloy show some resistance	Rubber (e.g. gaskets)
Chlorine trifluoride	C T O	Monel and nickel, PTFE and Kel-F, soft copper, 2S aluminium and lead are suitable for gaskets	
Chlorodifluoromethane		Steel, cast iron, brass, copper, tin, lead, aluminium at normal conditions Neoprene or boprene rubber and pressed fabrics are suitable for gaskets	Silver, brass, aluminium, steel, copper, nickel can cause decomposition at elevated temperatures. Magnesium alloys and aluminium coating more than 2% magnesium. Natural rubber

Gas	C	F	T	Suitable materials	Unsuitable materials
Chloropentafluoroethane				Neoprene or boprene rubber and pressed fabrics are suitable for gaskets	Silver, brass, aluminium, steel, copper, nickel can cause decomposition at elevated temperatures. Magnesium alloys and aluminium coating more than 2% magnesium. Natural rubber
Chlorotrifluoroethane		F	T	Most common metals	
Chlorotrifluoromethane				As for chlorodifluoromethane	
Cyanogen		F	T	Stainless steel, Monel and Inconel up to 65 °C. Glass lined equipment. Iron and steel at ordinary temperatures	
Cyanogen chloride	C		T	Common metals for dry gas. Monel, tantalum, glass for moist gas	
Cyclobutane		F		Most common metals	
Cyclopropane		F		Most common metals	
Deuterium		F		Most common metals	
Diborane		F	T	Most common metals. Polyvinylidene chloride, polyethylene, Kel-F PTFE graphite and silicone vacuum grease	Rubber and certain hydrocarbon lubricants
Dibromodifluoromethane				Copper or stainless steel	Aluminium for wet gas
1,2-Dibromotetrafluoroethane	C			Most common metals for dry gas. Stainless steel, titanium and nickel for moist gas	Zinc
Dichlorodifluoromethane				As for chlorodifluoromethane	
Dichlorofluoromethane				As for chlorodifluoromethane	
Dichlorosilane	C	F	T	Nickel and nickel steels and PTFE	Stainless steel for moist gas
1,2-Dichlorotetrafluoroethane				As for chlorodifluoromethane	
1,1-Difluoro-1-chloroethane		F		Most common metals under normal conditions	Hot metals can cause degradation to toxic corrosive products
1,1-Difluoroethane		F		Most common metals under normal conditions	Hot metals can cause degradation to toxic corrosive products

Table 11.29 (continued)

Gas	Hazard[†]	Materials of construction for ancillary services[‡]	
		Compatible	Incompatible
1,1-Difluoroethylene	F	Most common metals	
Dimethylamine	C F T	Iron and steel	Copper, tin, zinc, and their alloys
Dimethyl ether	F T	Most common metals	
2,2-Dimethyl propane	F	Most common metals	
Ethane	F	Most common metals	
Ethyl acetylene	F	Steel and stainless steel	Copper other metals capable of forming explosive acetylides
Ethyl chloride	F T	Most materials for dry gases	
Ethylene	F	Any common metal	
Ethylamine	C F T	Iron and steel. Reinforced neoprene hose	Copper, tin, zinc and their alloys
Ethylene oxide	F T	Properly grounded steel	Copper, silver, magnesium and their alloys
Fluorine	C T O	Brass, iron, aluminium, magnesium and copper at normal temperatures. Nickel and Monel at higher temperatures	
Fluoroform		Any common metal	
Germane	F T	Iron and steel	
Helium		Any common metal	
Hexafluoroacetone	C T	For dry gas monel, nickel, Inconel, stainless steel, copper and glass – Hastelloy C-line equipment	
Hexafluoroethane		Any common metal for normal temperatures. Copper, stainless steel and aluminium up to 150 °C	

Hexafluoropropylene				Any common metal for dry gas	
Hydrogen		F		Most common metals for normal use	At elevated temperature and pressure hydrogen embrittlement can result
Hydrogen bromide	C		T	Most common metals when dry. Silver, platinum and tantalum for moist gas. Heavy black iron for high-pressure work. High-pressure steel, Monel or aluminium pipe.	Most metals when gas is moist. Galvanised pipe or brass or bronze fittings
Hydrogen chloride	C		T	Stainless steel, mild steel for normal conditions of temperature and pressure. When moist use silver, platinum or tantalum. Moist or dry gas use backed carbon, graphite. High pressure work in heavy black iron pipework. High pressure Monel or aluminium iron bronze valves	Galvanised pipes or brass or bronze fittings
Hydrogen cyanide		F	T	Low-carbon steel at normal temp. and stainless steel for higher temperatures	
Hydrogen fluoride	C		T	Steel in the absence of sulphur dioxide contaminants in the gas and at temperatures below 150 °F. Monel, Inconel, nickel and copper for liquid or gas at elevated temperature	Cast iron or malleable fittings
Hydrogen iodide	C		T	Stainless steel, mild steel under normal temperature and pressure	Moist gas corrodes most metals
				Silver, platinum and tantalum, carbon, graphite for wet gas. At higher pressures use extra heavy black iron pipe. High-pressure steel, Monel or aluminium-iron-bronze values	Galvanised pipe or brass or bronze fittings
Hydogen selenide		F	T	Aluminium and stainless steel are preferred but iron, steel or brass are acceptable.	

Table 11.29 (continued)

Gas	Hazard[†]	Materials of construction for ancillary services[‡]	
		Compatible	Incompatible
Hydrogen sulphide	F T	Aluminium preferred. Iron and steel are satisfactory. Brass, though tarnished, is acceptable	Many metals in the presence of moist gas
Isobutane	F	Most common metals	
Isobutylene	F	Most common metals	
Krypton		Most common metals	
Methane	F	Most common metals	
Methyl acetylene		Most common metals	Copper, silver, mercury and their alloys
Methylamine	C F T	Iron and steel	Copper, tin, zinc and their alloys, Avoid mercury
Methyl bromide	C F T	Most common metals when dry	Aluminium and its alloys
3-Methyl-1-butane	F	Most common metals	
Methyl chloride	F T	Most common metals when dry	Zinc, magnesium rubber and neoprene particularly when moist aluminium is forbidden
Methyl fluoride	F T	Most common metals	
Methyl mercaptan	F T	Stainless steel and copper-free steel alloys and aluminium. Iron and steel for dry gas	
Methyl vinyl ether	F	Most common metals	Copper and its alloys
Neon		Most common metals	
Nickel carbonyl	F T	Most common metals for pure gas. Copper or glass-lined equipment for carbonyl in the presence of carbon monoxide	

Gas	C	F	T	O	Suitable materials	Materials to avoid
Nitric oxide			T	O	Most common metals for dry gas. For moist gas use 18:8 stainless steel, PTFE	
Nitrogen					Any common metal	
Nitrogen dioxide	C		T	O	Most common metals for dry gas. For moist gas use 18:8 stainless steel	
Nitrogen trifluoride			T	O	Nickel and Monel are preferred. Steel, copper and glass are acceptable at ordinary temperatures	Plastics
Nitrogen trioxide	C		T	O	Steel for *dry* gas otherwise use 18:8 stainless steel	
Nitrosyl chloride	C		T	O	Nickel, Monel and Inconel. For moist gas tantalum is suitable	
Nitrous oxide				O	Most common metals	
Octofluorocyclobutane			T		Cast iron and stainless steel below 120 °C, steel up to 175 °C, Inconel, nickel and platinum up to 400 °C	Avoid the metals opposite above 500 °C
Oxygen				O	Most common metals	On grease or combustible materials
Oxygen difluoride			T	O	Glass, stainless steel, copper, monel or nickel up to 200 °C. At higher temperatures only nickel and Monel are recommended	
Ozone		F	T	O	Glass, stainless steel, Teflon hypalon, aluminium tygon PVC and polythene	Copper and its alloys, rubber or any composition thereof, oil, grease or readily combustible material
Perchloryl fluoride			T		Most metals and glass for dry gas at ordinary temperatures	Many gasket materials are embrittled At higher temperatures many organic materials and some metals can be ignited Some metals such as titanium show deflagration in contact with the gas under severe shock
Perfluorobutane					Most common materials	

Table 11.29 (continued)

Gas	Hazard[†]	Materials of construction for ancillary services[‡]	
		Compatible	Incompatible
Perfluorobutene	T	Most common materials whey dry	
Perfluoropropane		Most common metals	
Phosgene	C T	Common metals for dry gas. Monel, tantalum or glass lined equipment for moist gas	
Phosphine	F T	Iron or steel	
Phosphorus pentafluoride	F T	Steel, nickel, Monel and Pyrex for dry gas. For moist gas hard rubber and paraffin wax	
Phosphorus trifluoride		Steel, nickel, Monel and the more nobel metals and Pyrex for dry gas	
Propane	F	Most common metals	
Propylene	F	Most common metals	
Propylene oxide	F T	Steel or stainless steel preferred though copper and brass are suitable for acetylane-free gas PTFE gaskets	Rubber
Silane	F T	Iron, steel, copper, brass	
Silicone tetrafluoride	C T	Most common metals for the dry gas. Steel, Monel and copper for moist gas	
Sulphur dioxide	C T O	Most common metals for dry gas. Lead, carbon, aluminium and stainless steel for moist gas	Zinc

Gas	C	F	T	O	Suitable materials	Unsuitable materials
Sulphur hexafluoride					Most common metals. Copper, stainless steel and aluminium are resistant to the decomposition products at 150 °C	
Sulphur tetrafluoride	C		T		Stainless steel or 'Hastelloy C' lined containers. Glass suitable for short exposures if dry. 'Tygon' for low-pressure connections	Glass for moist gas
Sulphuryl fluoride			T		Any common metal at normal temperatures and pressures	Some metals at elevated temperatures
Tetrafluoroethylene		F			Most common metals	
Tetrafluorohydrazine			T	O	Glass, stainless steel, copper or nickel to temperatures of 200 °C. For higher temperatures use nickel and monel	
Trichlorofluoromethane			T		Steel, cast iron, brass, copper, tin, lead, aluminium under normal, dry conditions	Some of the opposite at high temperatures magnesium alloys and aluminium coating more than 2% magnesium. Natural rubber
1,1,2-Trichloro-1,2,2-trifluoroethane					As above	As above
Trimethylamine	C	F	T		Iron, steel, stainless steel and monel. Rigid steel piping	Copper, tin, zinc and most of their alloys
Vinyl bromide		F	T		Steel	Copper and its alloys
Vinyl chloride	C	F	T		Steel	Copper and its alloys
Vinyl fluoride		F			Steel	Copper and its alloys
Xenon					Most common materials	

* This is a guide only and no substitute for consulting suppliers' literature.
† Even non-toxic gases are potentially hazardous owing to asphyxiation.
‡ Irrespective of material, all equipment must be adequately designed to withstand process pressures.
Key: C = Corrosive; F = Flammable; T = Toxic; O = Oxidising

Flammable explosive hazards

As discussed in Chapter 4 there is a wide variation in the flammability range for different gases. For example, ethylene oxide has explosive limits of 3–100% V/V in air while butane has a relatively narrow range of 1.9–8.5% V/V. Also the flashpoint of a highly flammable gas under pressure is always lower than room temperature. This is compounded by high rates of diffusion, and explosive mixtures can soon develop with air. Spontaneous ignition of high-pressure leakages has been reported.

Few commercially available gases are lighter than air under normal conditions. Examples include acetylene, ammonia, ethylene, carbon monoxide, helium, hydrogen and methane. However, even these on escaping may be cooler than ambient air and as a consequence fall initially. Leaks of gases lighter than air tend to be self-venting but in buildings can be trapped at higher levels than the leak source. Hydrogen can present an explosive hazard in this way, and similarly, ammonia a toxic one.

> Hydrogen leaked from a piped-in supply to a laboratory bench turret and was ignited by a spark from a flush-mounted electric socket. Fortunately damage was limited. The potential for leaks should have been reduced by redesign of the turret and regular pressure testing introduced to monitor for leaks (e.g. soap solution). The safety of the electric equipment should have been studied and gas lines and valves all labelled.[64]

Where hazardous gases are used on a continuous basis with provision for unattended overnight running (e.g. in a large gas–liquid chromatography laboratory) consideration should be given to the installation of a detection and alarm system with facilities to shut off the supply at source remotely from the laboratory in an emergency. The arrangement for a hydrogen system at one laboratory is shown in Fig. 11.9. Features include air-operated valves (nowadays explosion-proof electric switches are available to replace the pneumatic system), solenoid valves, hydrogen detectors, audible alarm and warning lights. Detector heads must be positioned with care above the instrument but taking into account air-flow patterns in the room which can be determined with smoke tubes. The air-operated valves are normally open thus allowing hydrogen to flow but they will close automatically if hydrogen is detected in the room (e.g. at 20% LEL), or when the emergency shut-off button in the room is operated. In either case the 'hydrogen detected' sign will illuminate and the audible alarm will sound. The system is of fail-safe design, the hydrogen supply being shut off in the event of failure of electricity or air supply, or failure of a detector or amplifier. For very short pipe runs from cylinder to laboratory the necessity for the duplicate solenoid and air-operated switches in the room is questionable. In reality a bank of two or more

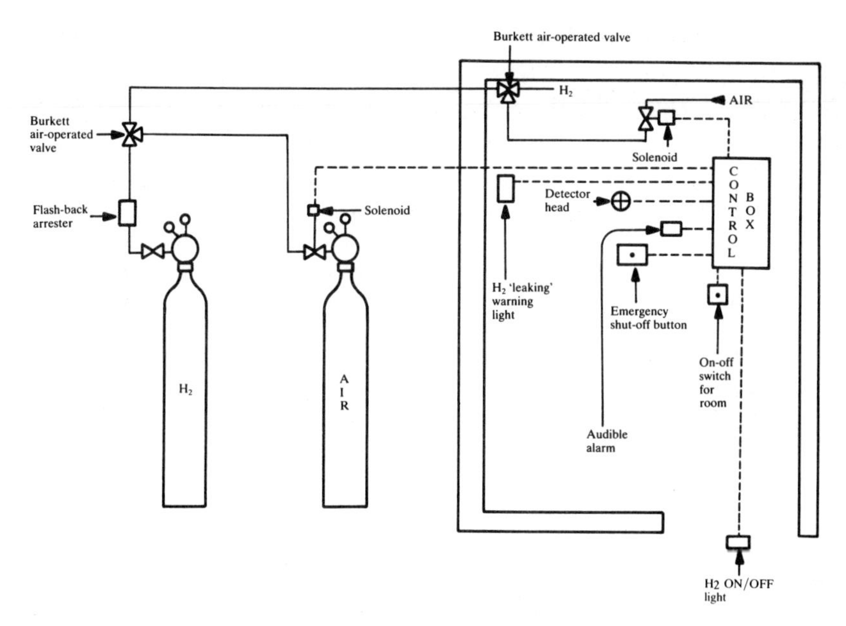

Fig. 11.9 Hydrogen-supply arrangement. Detector heads must be positioned with care above instrument but taking into account air-flow patterns in the room which can be determined with smoke tubes

hydrogen cylinders may be used so as to provide continuity of supply when a cylinder is exhausted.

Gases heavier than air can travel considerable distances and accumulate at a lower level than the source of the leak – thus producing an explosion or other hazard at a remote point (see Ch. 3).

Some gases, particularly acetylene, can decompose spontaneously when stored under pressure. Above pressures of 9 psig undissolved (free) acetylene will begin to dissociate and revert to its basic elements. This is an exothermic process which can result in explosions of great violence. For this reason, acetylene is dissolved in acetone contained in a porous material inside the cylinder. Voids in the porous material can result from settling (e.g. if the cylinder is stored horizontally) or through damage to the cylinder in the form of denting. If a void should occur in the porous mass then the acetylene may decompose. This decomposition may be initiated by some mechanical shock such as dropping the cylinder.

It should be noted from Table 11.29 that alloys containing >70% of copper must on no account be used in contact with acetylene due to the risk of formation of copper acetylide which may explode.

Even small increases in the oxygen concentration greatly increase fire hazards, and oxygen-enriched atmospheres will often permit the burning of substances that are not readily combustible in ordinary air. Thus care is essential when handling oxygen cylinders or cylinders of oxidising agents. Spontaneous and vigorous fires can occur when chlorine, fluorine or chlorine trifluoride react with certain substances. Some gases are pyrophoric.

Silane gas at 350 psi leaked from the high-pressure side of the regulator on a cylinder in a university department and ignited spontaneously. The fire was extinguished by a technician wearing breathing apparatus using a CO_2 extinguisher. However, the gas reignited and the premises were evacuated. Firemen removed 15 cylinders from the premises. At one point three explosions occurred when an ammonia cylinder and a compressed air cylinder became involved in the fire, and the valve of a diborane cylinder sheared. Silane is pyrophoric even when substantially diluted with nitrogen. The laboratories and systems have been redesigned and the cylinders are now sited outside and the gas piped in.[65]

Toxic hazards

The effects of toxic materials on the body are described in Chapter 5. Because of the high pressure of compressed gases, leaking gas can soon form toxic concentrations within a laboratory and inert gases can quickly deplete the oxygen content of the air to dangerously low levels.

Corrosive hazards

Corrosive hazards are discussed in Chapter 5. Gases in this category are toxic and also sometimes explosive. Most are lachrymatory and will irritate the skin: some can be absorbed through the skin.

Corrosive gases should be stored for the shortest possible periods before use in dry areas, and a water supply is essential to help deal with leaks. They should not be stored near instruments or other devices sensitive to corrosion.

General precautions

All cylinders should be handled as potential sources of high energy. Cylinders should only be filled by specialists. Gas must never be transferred from one cylinder to another. No attempt should be made to modify or repair cylinders.

Particular attention is necessary for cylinder storage:

- Stores should be a purpose-built compound, ideally outside the laboratory. It should be fire resistant, well ventilated, located away from ignition or heat sources, dry and equipped with proper drainage, and protected from tampering by unauthorised personnel.

- Smoking and the use of naked flames should be prohibited from the compound area.
- 'Empty' and full cylinders should be segregated. Oxidents and fuel gases (see Table 11.29) should be separated by a minimum of 6 m.
- An adequate supply of water should be provided and maintained for fire fighting, first aid, and dilution of corrosive spillages.
- Stocks should be kept to a minimum.
- Cylinders in stock should be inspected periodically for corrosion.

All operators and technicians should know and understand the properties, hazards and precautions associated with each gas before using it. Suitable eye protection and stout gloves should be worn when handling cylinders. A cylinder must never be lifted by the cap and cylinder caps should be left in place to protect the valve stem from damage until it has been secured in position ready for use. Purpose-designed trolleys are advisable for transportation around the laboratory and stores. Care should be taken that cylinders are not dropped, or knocked against one another or a wall, etc., or used as rollers. They should not be exposed to temperatures in excess of *c.* 50 °C (see Ch. 4). Turnover of stock should be reasonable (3 months for corrosive gases).

Before connecting a cylinder the complete gas system should be checked for suitability (e.g. pressure rating, materials of construction). The appropriate regulator and tools are required on every cylinder – adaptors and home-made modifications should be prohibited. Every gas cylinder has a matching regulator connection – the connection must not be forced, it may be the wrong type. Regulators are *not* flow regulators, these must also be used when downstream pressure may fluctuate. Provision is required for traps to prevent contaminant siphoning back into a cylinder, or check-valves to prevent gaseous contamination from flowing back. In-line flame arresters should be installed with flammable gases (e.g. hydrogen, acetylene).

The smallest cylinder practicable should be used – 'lecture' cylinders are particularly useful where small volumes of hazardous gas are required. For non-corrosive gases these are equipped with packless, leak-free valves with convenient hand-wheel control. However, since the outlets are often the same regardless of the gas, extra care is needed when changing lecture bottle equipment from one gas service to another. Cylinders that cannot be positively identified should not be used. Cylinders should be situated in well-ventilated areas preferably equipped with both high- and low-level ventilation. In laboratories hazardous gases should be used in fume-cupboards and cylinders of any size require support in a vertical position with chains, straps or stands to prevent them falling. Where practicable cylinders should be stored outside and gas piped into the building (especially for hazardous gases). This minimises the otherwise considerable hazard in the event of fire in the laboratory.

It also provides ventilation in the event of a leakage and helps prevent gassing accidents. It also prevents gangways and thoroughfares from becoming blocked. Where it is impractical to house toxic gases outside they should ideally be removed from the laboratory each night and returned to stores.

Compressed gases should not be used to blow away dust or dirt and gases must never be directed at individuals (e.g. skylarking). Rapid release of compressed gas will cause an unsecured gas hose to whip dangerously and may also build up static charges which can ignite highly combustible gas. A flame involving a flammable gas should not be extinguished until the source gas has been shut off: it may otherwise re-ignite and explode. Empty cylinders should be marked in some way (e.g. as 'MT'), the cylinder valve closed, regulator removed immediately and the cap replaced. Cylinders should never be emptied completely, i.e. a slight positive pressure is required to prevent entry of contamination. The main cylinder valve, as well as the regulator valve, should be closed when not in use.

Excessive pressure in glass apparatus must be avoided, e.g. by the provision of liquid loop vents. Safety relief devices are required when using liquefied gases in all-liquid transfer lines to relieve sudden changes in hydrostatic or vapour pressure build-up. Pressure-relief valves should be vented to safe places, i.e. avoiding access ways.

Valves and regulators with compressed oxygen must be clean and oil free otherwise explosions may result and, depending upon application, oil and water droplets should be removed from compressed air using an in-line filter.

On start-up, the cylinder valve should be opened slowly until full cylinder pressure is indicated on the gauge, then adjustment of flow made with the regulator valve. On shut-down, the cylinder valve should be closed before shutting off the regulator valve to permit gas to bleed from the regulator. Cylinder valves must not be forced. Regulators require checking periodically. For most materials (air, nitrogen, argon) quick monitoring checks are all that are required. For corrosive gases, weekly inspections are recommended. Checks include:

- Zero reading on pressure gauge when gas is discharged from the 'system'.
- High pressure gauge reading is the cylinder pressure, and the reading does not increase when the regulator outlet is closed.
- 'Crawl' due to wear on the regulator valve and seat assembly.
- Leakages between cylinder and regulator.
- Regulators should be overhauled completely on a 3–6 monthly basis for corrosive gases and annually for others. Maintenance should only be undertaken by competent persons.

When not in use regulators for corrosive gases should be stored dry at room temperature after flushing with dry nitrogen. The valve on an unregulated cylinder should not be 'cracked' open. For particularly hazardous gases (e.g. hydrogen, acetylene) in confined spaces gas leak detectors are required coupled to audible/visual alarms. These can detect leaks too small to cause bubbles when using soap solution for detection of leaks. Lines of hazardous gases must be checked for leaks periodically using a leak detector, explosimeter, etc. Only short lengths of plastic piping should be used which must be inspected for perishing: soft rubber tubing should not be used for piping gases from cylinders.

If a cylinder of acetylene is stored in a horizontal position and is not damaged it should be placed in the upright position for half an hour before use. The supplier's literature should be consulted and/or reference[66] for advice on materials of construction of ancillary equipment. Thus copper, tin, zinc and their alloys should be avoided for ammonia and, in particular brass regulators must not be used. Instruments containing mercury that may be exposed to ammonia should not be used. Table 11.29 summarises these aspects for the more commonly encountered gases in laboratories.

Cylinders must not be placed where they can become part of an electrical circuit. For highly flammable gases, sources of ignition must be avoided (see Ch. 4), and cylinders should be segregated from incompatible materials and not be located near to a heat source (e.g. radiator, oven, steam pipe).

A cylinder should not be dragged; it should be rolled on the edge of its base (e.g. to transfer from trolley to position of use).

Cylinders should be colour-coded and bear a label to identify the contents and hazards. Colour codes alone are unreliable since they differ between suppliers and particularly from different countries, since no international standard exists. Furthermore, colour-coded caps may be interchangeable between cylinders of different gases. All gas lines from cylinders should be clearly labelled and colour-coded so as to identify the gas in service.

Cylinders should be positioned so that the cylinder valve is accessible at all times. Only the cylinder keys and other special tools provided should be used; pliers, adjustable wrenches, etc., must not be used. Lines of hazardous gases not in use should be flushed and fitted with positive blanks.

When handling gas cylinders personal protective equipment is required to protect the face, hands and feet. Additionally, ventilation and respiratory protection would be required for highly toxic materials.

Even when changing regulators on cylinders housed out of doors, suitable respiratory protection will be required along with normal personal protection for highly toxic gases such as chlorine, phosgene, hydrogen cyanide.

Oxygen must be kept away from *all* reducing matter – even wood saturated with oxygen has exploded when subjected to shock.

Spring-loaded pressure-relief devices should be attached to containers of liquid gases (e.g. liquid hydrogen, helium, nitrogen and oxygen).

Cryogenic materials

Liquefied gases find use as a source of inert medium (e.g. nitrogen), a reagent (e.g. fluorine), or to achieve lower temperatures than are practicable using normal refrigeration techniques (e.g. below −73 °C), i.e. cryogens. The most common cryogens are given in Table 11.30.

For speciality work, such as purification of gases by fractional condensation, a range of 'slush baths' is often used, prepared by cooling organic liquids (commonly 'solvents') to their melting point by adding liquid nitrogen. From the examples (given in Table 11.31) of substances used, some of the solvents clearly pose serious toxic and flammable risks and should be substituted by safer alternatives unless very strict handling precautions are instituted.

All cryogenic materials can act as asphyxiants and the air–oxygen concentration can soon be reduced below a safe level by evaporation of a small volume of liquid cryogen in a confined space (see Table 11.30).

The conveyance of solid carbon dioxide in lifts is, therefore, inadvisable in case of prolonged breakdown. Prolonged breathing of pure oxygen may also have an adverse physiological effect while other cryogenic substances such as fluorine and carbon dioxide are toxic and the majority possess no warning odour. Direct contact of the substance

Table 11.30 Properties of selected cryogens

Gas	Boiling point (°C)	Volume of gas (l) produced on evaporation of 1 l of liquid
Helium	−269	757
Hydrogen	−253	851
Neon	−246	1438
Nitrogen	−196	696
Fluorine	−187	888
Argon	−186	847
Oxygen	−183	860
Methane	−161	578
Krypton	−151	700
Xenon	−109	573
Chlorotrifluoromethane	− 81	—
Carbon dioxide	− 78.5	553

Table 11.31 'Slush baths' and working temperatures

Bath liquid	Temperature (°C)
Carbon tetrachloride	− 23
Chlorobenzene	− 45
Solid carbon dioxide	− 63
in acetone or methylated spirits*	− 78
Toluene	− 95
Carbon disulphide	−112
Diethyl ether	−120
Petroleum ether	−140

* Liquid nitrogen is omitted from this mixture and the solvent is used to improve the heat transfer characteristics of cardice.

with the skin can result in burns akin to those produced by high-temperature contact, and the eyes are particularly vulnerable.

Some materials such as hydrogen and methane are highly flammable while oxygen can greatly increase the flammability of ordinary combustible matter. Any material with a lower boiling point than oxygen (e.g. hydrogen, neon, nitrogen) can cause oxygen to condense from air, again giving rise to a fire hazard particularly if hydrogen or organic matter such as grease, wood, etc., is present.

Because of the large expansion ratio from liquid to gas, evaporation can result in the build-up of high pressures. Since cryogens are normally stored just below their boiling point, gas is ever present and must be allowed for in container design and capacity.

Choice of service materials and joints for cryogenic work must account for the reduced ductility at low temperatures: some materials become extremely brittle at cryogenic temperatures. Materials which are used (depending upon the application) include stainless steel, copper, brass, bronze, monel, aluminium, Teflon and nylon. Consideration of adjacent plant is also important.

A 1,000 ton liquid-nitrogen storage tank was ruptured causing cold liquid to fall on to other pipework which fractured due to embrittlement. The released material from the pipework drifted in a cloud into another works where, following ignition, three men were very badly burnt and died.[67]

A carbon-steel line carrying hydrogen at 325 psig was suddenly subjected to liquid-nitrogen temperatures causing it to rupture violently.[68]

Precautions for handling cryogenic substances include a consideration of materials of construction and services which must be scrupulously clean, free of oil and grease, and chosen to resist brittle fracture. Work areas should be well ventilated. Contamination of fuel and oxidant gases/liquids must be avoided.

When flammable gases are used, additional high- and low-level ventilation may be required together with background detectors linked to automatic shut-down procedures and audible alarm in the event of atmospheric concentrations reaching hazardous levels (e.g. 40% of lower explosive limit). All ignition sources must be eliminated.

When toxic variants are used, depending upon circumstances, additional localised ventilation, monitors and automatic shut-down procedures and alarms, or self-contained breathing apparatus may be required.

Where personal contact with the cryogen is possible, face shields and impervious gloves (large enough to be removed quickly) should be worn. Clothing must be arranged so as to prevent trapping pockets of cryogenic liquids next to the skin. Watches, rings, bracelets, etc., should be removed (e.g. when dispensing cryogenic liquid).

Chemicals at the pilot-plant scale

Pilot plants are conventionally used to study process design, and to obtain process and product safety data to aid progression of successful developments from laboratory bench to the market place. Hence they are intermediate in scale between laboratory and production units.

Though pilot plants are small compared with production facilities, accidents even at this experimental stage can be costly in terms of injury and material loss.

> An acetaldehyde oxidation pilot plant was purged with nitrogen. Oxidation was achieved by the controlled flow of oxygen into a column. Unfortunately, at high pressure nothing prevented seepage of oxygen into the nitrogen lines in the event of leaking valves or operator error. On one such occasion, oxygen apparently leaked into the purging system and thence into the acetaldehyde drum. An exothermic reaction occurred accompanied by a large increase in pressure and detonation. The explosion killed four people.[69]

The hazards and precautions for pilot plant parallel those in laboratories and production units in many ways. However, there are certain special problems associated with the nature of work and scale of operation on pilot plants which warrant mention.[70]

Transfer of information from chemist to pilot-plant manager is crucial for safe scale-up. Causes of problems include:

- Inadequate testing at the bench. Physical properties of all ingredients should be established together with an assessment of their fire and explosion characteristics and preliminary acute toxicity. The course and kinetics of the process should have been investigated in the laboratory under a variety of variable process conditions. The effect of likely impurities should be known. Contamination can arise from change in materials of construction (from laboratory glassware to metal plant equipment), raw material specification (from laboratory reagents to bulk feedstocks), from side reactions or from leaks.

- Failure to report all observations including minor adverse effects in laboratory experiments. Slight abnormalities such as 'bumping' in laboratory flasks could have dire consequences in a 20,000-litre jacketed reactor.
- Failure to set up a mechanism formalising exchange of data and discussion between relevant parties.

In contrast to many production units, pilot plants like laboratories are housed indoors thereby increasing the opportunity for build-up of dangerous concentrations of hazardous vapours, gases, dusts, fumes, mists, etc. While standard laboratory equipment (e.g. fume-cupboards, glove boxes, blast shields) enable hazardous chemicals to be handled with ease at the bench, special arrangements may be required to ensure the same degree of containment and safety when handling the same materials on the pilot plant. Once hazards are fully evaluated and business opportunities assessed at the production scale then either special control measures will be deemed unnecessary or easily justified. However, more persuasive skills and enlightened management are helpful when proposing capital expenditure for safety at the intermediate and more speculative pilot-plant stage of a project.

Multi-purpose pilot plants often consist of temporary construction. They require careful control to ensure safe design and installation with attention to details of ergonomic considerations. Safe systems of work must prevent trial mock-ups and unauthorised modification and maintenance by 'enthusiastic' scientists.

The scale of operation in laboratories rarely presents problems in physical handling of chemicals. Similarly, in production, the quantity of materials involved usually necessitates mechanical conveyance and dispensing. The intermediate scale of pilot plants, however, often relies on man-handling of drums, kegs, sacks, etc., with the attendant personal hazards. About 40,000 accidents a year are reported to the Factory Inspectorate as being caused by the handling, lifting and carrying of goods. Furthermore, manual materials transfer and manual control (compared with automation on the production scale) tend to bring operators in closer proximity to chemicals.

Many of the requirements for handling chemicals at the 'bench scale' are relevant to their use on pilot plants. Additional precautions are highlighted in the following guidelines.

Guidelines for pilot plants

The pilot-plant manager should be clearly responsible for safety on the pilot plant. Pre-start-up safety check meetings should be held involving chemists, pilot-plant manager, project and design engineers, safety and hygiene specialists, operators and maintenance personnel. The team should discuss properties of materials (feedstocks, intermediates, by-products and products), the process flow-sheet, disposal, emergencies, and spillages, with a view to identifying and solving problems and drafting written instructions.

A check-list is given in Table 11.32. Experimental design should ensure generation of adequate good quality data to aid transfer to full-scale production. For novel processes, or hazardous ingredients, environmental monitoring may be relevant for assessing adequacy of precautions and working practices, developing in-house hygiene standards and in dealing with employee complaints allegedly stemming from exposure to process materials. Monitoring programmes at this stage should also be considered as a valuable aid in designing for scale-up to full production.

Suitable storage is required for all pilot-plant materials to prevent stock piling on the plant. Quarantine storage of raw materials in pilot plants (and production) may be necessary in certain circumstances (for pharmaceuticals, food processing, etc). All materials should be carefully labelled to identify contents, owner and hazards. Storage tanks for liquids should be closed to avoid contamination or evaporation into the workplace. Closed tanks may need to be vented. If contents are flammable vents should be fitted with flame traps, containers earthed and the area designated 'NO SMOKING, NO NAKED LIGHTS'. Drums (normally 200 l) must not be pressurised. They should be stored, bung uppermost, away from heat sources. They must be chocked if laid on their side to prevent them rolling. 'Full' and 'Empty' drums should be segregated. Empty drums should be rolled using both hands (not kicked). Full drums and (depending upon size and weight) sacks, etc., should be handled with the aid of mechanical devices by trained operators. Carboys should not be pressurised, rolled, exposed to heat, or carried by the neck. All plant personnel including supervisors should be trained to appreciate their physical limitations and in human kinetics to ensure the correct procedures are used for man-handling goods.

Only properly secured access equipment should be used and climbing on pipework, plant fittings, etc., must be prohibited. Head protection should be worn whenever the possibility exists to walk underneath an operator or fitter working on a platform or ladder and where regular overhead work takes place. Hoses should be coiled and replaced (e.g. on wall) immediately after use for washing down. They should never be left trailing across walkways and standing water should always be removed from plant floor.

Table 11.32 Check-list for pilot-plant safety meeting

Materials	Trade and chemical names of feedstocks, products, by-products. Their structure, composition, supplier, molecular weight, physical properties, tentative specification, stability characteristics, toxicity (acute and chronic effects), flammable/explosive properties, hazardous reactions. Labelling requirements. Shelf-life. Sensitivity to contamination.
Process	Flowsheet, materials, service and manpower requirements, equipment and instrumentation. Reaction equations and conditions. Specification of, and systems for, monitoring where relevant of extract ventilation, temperature, light, humidity, pH, vibration, noise, radiation, catalyst, airborne chemical contaminants, micro-organisms, electrical continuity, etc. Effluent treatment and restrictions, side reactions. Process kinetics and thermochemistry.
Operating procedures	Written instructions for start-up, normal operation, shut-down, emergency shut-down, clearance of blockages, reprocessing of materials, temporary modifications. Personal protection overall (special requirements), suits, spats, armlets, headgear, gloves, footwear, ear defenders, eye protection, respirators, requirements for visitors. Medical screening, biological monitoring. Personal hygiene for normal operation (washing, skin cleansing, double locker system) and emergency (showers, eye washes, first aid), fire-fighting facilities. Procedures in the event of fire/explosion, toxic release or serious accident.
Analysis	Analysis of reactants and products. Physical and chemical control of process.
Storage	Segregation, conditions, housekeeping, stock levels.
Disposal	Collection of combustibles, labelling of waste, segregation, disposal route.
Ergonomics	Man/machine interface. Difficulties caused by cramped conditions. Access, means of lifting and transportation, provision of machine guards.
Maintenance	Issue, control, testing and repair of lifting gear and access aids. Maintenance schedules. Permits to work (entry into confined spaces, flame-cutting, line breaking, electrical work, equipment removal, roofwork, machine guards, open vessels, use of spark-proof tools).
	Systems for formal approval for plant or process modifications, hazard assessment prior to implementation and in use. Up-dating of all instructions, notices, procedures. Removal of obsolete plant and lines.
Inspections	Safety inspections on ventilation, fire protection, alarms, plant, tools, hoses, stores, pressure safety valves, explosion reliefs, monitoring and control hardware, protective clothing and equipment, drains, bund walls, electrical equipment.
Communication	Log books, batch/recipe sheets, research reports, minutes of safety check meetings (with actions). Identification of lines, vessels, valves, drums. Intershift records, warnings for specific and temporary hazards. Permits to work, Codes of Practice. System for reporting plant defects, process deviations, hazardous occurrences, accidents.
Training	Management (refreshers, HAZOPS). Operators (general, specific operating instructions, emergency procedures, personal protection, monitoring, first aid, fire-fighting).

References

(*All places of publication are London unless otherwise stated*)

1. Luxon, S. G., 'Accident and dangerous occurrence statistics in the United Kingdom', in *Health and Safety in the Chemicals Laboratory – Where Do We Go From Here?* Royal Society of Chemistry, Special Publication No. 51, 1984.
2. *Safety in Research – A Study Group Report*. Royal Society, 1981.
3. Gerlovich, J. A. & Downs, G. E., *Better Science Through Safety*. Iowa State University Press 1981.
4. Harrington, J. M., MD thesis. London University, 1976.
5. Ederer, G. M., Tucker, B. & Vikmanis, A., *Safety in the Chemical Laboratory*, ed. N.V. Steere. Reprints from the *J. Chem. Educ.*, 1967, (reprinted 1974), **3**, 44.
6. Phillips, G. B., *Causal Factors in Microbial Laboratory Accidents and Infections*. US Army Biol. Labs No. 2: Fort Detrick, Maryland, 1965.
7. James, J. C. paper presented at Conference on Safety in Laboratories, University College of Wales, Aberystwyth, 25 Sept. 1979.
8. Dewhurst, F., *International Environment and Safety*, 1981 (Dec.), 18.
9. Royal Society of Chemistry, RECAP 1st Report, 1980 (June), 10.
10. Buttolph, M. A., *Health and Safety at Work*, 1979 (Mar.), 26.
11. Schmidt, R. L., *J. Chem. Educ.*, 1977, **54**, A145.
12. Weston, R., *Laboratory Safety Audits and Inspections*. Northern Publishers (Aberdeen), 1982.
13. Anon., *J. Chem. Educ.*, 1981, **58** (11), A329.
14. Chemical Industry Safety and Health Council of the Chemical Industries Association, *Chemical Safety Summary*, 1981, **52**, 219.
15. National Fire Protection Association, *Occupancy Fire Record: Laboratories*. NFPA: Boston 1958, 6.
16. Lees, R. & Smith, A. F. (eds), *Design, Construction and Refurbishment of Laboratories*. Ellis Horwood, 1984.
17. Everett, K. & Hughes, D., *A Guide to Laboratory Design*. Butterworths, 1975.
18. Maple, P. E. M., *Chem. in Brit.*, 1983, **19** (10), 819.
19. Hughes, D., *A Literature Survey and Design Study of Fumecupboards and Fumecupboard Dispersal Systems*, Occupational Hygiene Monograph No. 4. Science Reviews, 1980.
20. The Ionising Radiations (Unsealed Radioactive Substances) Regulations, 1968.
21. Everett, K., *A Guide to the Design and Installation of Laboratory Fume Cupboards*, Hygiene Technology Guide Series No. 1. British Occupational Hygiene Soc., 1975.
22. British Standards Institution, *Laboratory Fumecupboards*, Draft for Development DD80 (3 parts), 1982.
23. Roach, S., *Annals Occup. Hyg.*, 1981, **24** (1), 105.
24. *Harrowgate Adviser* (North Yorks.), Oct. 23 1981.
25. Steere, N. V. (ed.), *Safety in the Chemical Laboratory*. Division of Chemical Education of the American Chemical Society, 1974, **1**, 118.

26. *Manufacturing Chemists Assoc., Case History 1177.*
27. British Chemical Industry Safety Council, *Quarterly Summary*, 1969, **40**, (157), 6.
28. Ward, A. F. H., *J. Roy. Inst. Chem.*, 1960, 84, 451.
29. Shugar, G. J., Shugar, R. A. & Bauman, L., *Chemical Technicians Ready Reference Handbook* (2nd edn). McGraw-Hill, 1973.
30. Sansone, E. B. & Losikoff, A. M., *Amer. Ind. Hyg. Assoc. J.*, 1979 (40), 543.
31. Guiochon, G., *J. Chromatography*, 1980 (189), 108.
32. National Safety Council, 'Centrifuges', Data Sheet 591, USA.
33. A Working Party of the I. Chem. E., *User Guide for the Safe Operation of Centrifuges with Particular Reference to Hazards and Hazardous Atmospheres*. Institution of Chemical Engineers, 1976.
34. British Standards Institution, *BS 4402: Specification for the Safety Requirements for Laboratory Centrifuges*, 1982.
35. British Chemical Industry Safety Council, *Quarterly Summary*, 1968, **39** (154), 16.
36. Spencer, E. W., Ingram, V. M. & Levinthal, C., *Science*, 1966, **152**, 1722.
37. Kennedy, D. A., *Laboratory Practice*, 1985, 34 (3), 90.
38. Anon., *Brit. Med. J.*, 1969, **4**, 817.
39. National Fire Protection Association, *Occupancy Fire Record: Laboratories*. NFPA: Boston 1958, 4.
40. Universities Safety Association, *Safety News*, 1972 (May), No. 2, 3.
41. Anon., 'Safety in the chemical laboratory', in N.V. Steere (ed.). Reprints from *J. Chem. Education, 1967* (reprinted 1974), **1**, 116.
42. Pipetone, D. A. (ed.), *Safe Storage of Laboratory Chemicals*. Wiley 1984.
43. National Fire Protection Association, *Occupancy Fire Record; Laboratories*. NFPA, Boston 1958.
44. British Chemical Industry Safety Council, *Quarterly Safety Summary*, 1967, **38** (152), 46.
45. Bretherick, L., (ed.) *Hazards in the Chemical Laboratory* (3rd edn). Royal Society of Chemistry 1981.
46. Manufacturing Chemists Assoc., *Guides for Safety in the Chemical Laboratory* (2nd edn). Van Nostrand Reinhold 1972.
47. *Aldrich Catalogue Handbook of Fine Chemicals, Aldrich Chem. Co. 1985–1986.*
48. Health and Safety Commission, *Code of Practice No. 9. Methods for the Determination of Physico-Chemical Properties*, 1982. HMSO.
49. British Standards Institution, *BS 2000: Methods of Test for Petroleum and its Products* (various parts).
50. Buttolf, M. A., Symposium on Safety in University Chemical Engineering Departments. Institution of Chemical Engineers, Dec. 18, 1979.
51. Steere, N. V., (ed.) 'Safety in the chemical laboratory'. Reprints from *J. Chem. Education, 1967* (reprinted 1974), **1**, 68.
52. Steere, N. V., (ed.) ibid, 1967 (reprinted 1974), **1**, 71.
53. Jackson, H. L., McCormack, W. B. M., Rondestvedt, C. S., Smeltz, K. C., & Viele, I. E., ibid., 1967 (reprinted 1974), **3**, 114.
54. Varjavandi, J. & Mageli, O. L., ibid., 1967 (reprinted 1974), **3**, 118.

55. Steere, N. V. (ed.) *Handbook of Laboratory Safety*. The Chemical Rubber Co., 1971.
56. Letter from Manly, T. D., *Chem. in Brit.*, 1982, **18** (5), 341.
57. Private Communication.
58. Meyer, R., 'Explosives', *Verlag Chemie* (2nd edn), West Germany, 1981.
59. Connor, J., 'Loss Prev. Safe. Promot. Process Ind. Prepr.', in *1st Int. Loss Prev. Symp.*, C. H. Buschmann (ed.), Elsevier: Amsterdam 1974.
60. Field, P., *Explosibility assessment of industrial powders and dusts*. Department of the Environment: HMSO 1983.
61. Anon., *Fire Prevention*, 1978 (127), 30.
62. British Chemical Industry Safety Council, *Quarterly Safety Summary*, 1972, **43** (169), 6.
63. Treweek, D., *Ohio J. Sciences*, 1978, **78** (5), 245.
64. Chemical Industry Safety and Health Council of the Chemical Industries Association, *Chemical Safety Summary*, 1981, **52**, 220.
65. Anon., *Fire Prevention*, 1983 (165), 45; Universities Safety Association, *Safety Newsletter*, 1983 (6), 1.
66. Braker, W. & Mossman, A. L., *Gas Data Book* (6th edn). Matheson Gas Products, New York, 1981.
67. Anon., *Chemical Week*, 8 Feb. 1978, 18.
68. American Institution of Chemical Engineers, 'Nitrogen wash incident', *Ammonia Plant Safety*, 3, 1961.
69. National Fire Protection Association, *Occupancy Fire Record: Laboratories*. NFPA: Boston 1958, 8.
70. Manufacturing Chemists Association, 'Recommended safe practices and procedures: safety in the scale-up and transfer of chemical processes', *Safety Guide No. 14*.

CHAPTER 12

Safety in chemical-process plant design and operation

A survey of 317 case histories of major fire and explosion losses in the chemical and allied industries in the USA is summarised in Table 12.1 (after ref. 1).

Although analysis is difficult, these data suggest that a surprising number of incidents were attributable to 'under-design' in its widest sense.

These were major incidents but, as summarised in Chapter 1, there is some correlation between such losses and the more frequent minor incidents and near-misses. Certainly inadequacies in design are often a contributory factor in accidents involving chemicals, and in not mitigating the effects as far as reasonably practicable, or in permitting prolonged exposure to atmospheric contaminants. In this context designers have to consider that when a dangerous process or operation is continually repeated there will be mistakes or accidental slips due to wavering attention or the urgency of completing the work, and 'have in mind not only the careful man, but also the man who is inattentive to such a degree as can normally be expected'.[2] They must also consider the potential for

Table 12.1 Hazard factors in major fires and explosions

Hazard factor	% of total
Plant site problems	3.5
Inadequate plant layout and spacing	2.0
Structures not in conformity with use requirements	3.0
Chemical process problems	10.6
Material movement problems	4.4
Operational failures	17.2
Equipment failures	31.1
Inadequate material evaluation	20.2
Ineffective loss prevention programme	8.0
	100.0

equipment, or an instrument or protective device, to fail in service at some finite rate – which is a function of the quality of the item and its suitability for the particular duty, the use of it within design limits, and the frequency and quality of the inspection and maintenance procedures.

Detailed analysis often shows that minor changes in design, location of equipment, or substitution of materials, can produce a safer design and an economical solution. The danger of omitting features essential for the safe operation of the plant is increased if such considerations are left to a late stage of design, since pressures of cost and time are then considerable. A reasoned assessment should commence at the development stage.

In this chapter safety considerations in the design of chemical and process plants are summarised. Safety aspects in equipment design are reviewed but clearly for detailed design procedures and information reference should be made to standard texts.[3–5] Reference should also be made to other literature for related hazard quantification calculations.[6,7] Safe operating procedures are also discussed, excluding those considered under 'management' in Chapter 17.

Process design

Flowsheeting

The initial stages of design of any chemical/process plant involve the preparation of flowsheets or line diagrams. These diagrams are refined and modified and ultimately form a basis for piping layout drawings, plot plans, instrumentation drawings, etc. However, since decisions taken during the early stages significantly affect subsequent stages, it is important for safety considerations to receive full attention throughout 'flowsheeting'.[8]

In the design procedure it is usual to draw on data from the literature (see Ch. 18), from laboratory experiments and from pilot-plant trials. In scaling-up designs and operating procedures to a full-size plant the process designers must take account of many of the phenomena and operating parameters already discussed. For example:

Use of commercial feedstocks and less pure chemicals.

Scale-up effects in mass transfer (see page 114) and in heat transfer (page 93) and in methods of transferring materials generally.

Effects of differing residence times.

Effects of storing large inventories of raw materials, intermediates and products (page 893).

Effects of 'continuous' operation on accumulation of residues.

Effects of any differences in materials of construction.

Difference in the degree, and quality, of operating supervision; use of a greater degree of automatic control.

If the mode of operation changes from jobbing, or batch-wise, to fully continuous then additional detailed considerations arise.

Selection of process materials

The process materials should be assessed with regard to both their physical properties and any hazards to which they may contribute. Clearly any which are flammable, toxic, potentially explosive, oxidising, compressed, corrosive, at extremely high or low temperatures (or, in special cases, radioactive) are hazardous. Typical information requirements are listed in Table 12.2. A hazardous materials inventory needs to be compiled and the relevant properties noted over the complete range of process conditions.

Overall consideration has to be given to all the possible in-process materials and materials used as 'processing aids', i.e.

In-process materials	*Processing aids*
Feeds, catalysts	Heat transfer fluids
Intermediates	Recycles
Products, by-products	Refrigerants (e.g. ammonia, Freon liquefied hydrocarbons)
Solvents	
Additives	Fire-fighting chemicals

Selection of process route

At the earliest stage the safest reasonably practicable process route should be selected.[9] In many cases, however, the process route will be pre-determined. Processes which are potentially hazardous are listed in Table 10.1 and some considerations in reaction process selection and design are summarised in Table 12.3.

Certain operations within any particular process are also inherently hazardous. Examples are given in Table 12.4.

In assessing the safety of a process route, consideration should be given to whether materials added to the process for specific purposes, e.g. to assist flow, separation/settling, heat transfer (diluents), catalysis or mixing, significantly increase the hazards.

An explosion in a 3.1 tonne kettle resulted in it being projected some 300 m. A polyether–alcohol intermediate at approximately 100 °C had been pumped into the kettle at 30 hr and 14 hr prior to the incident; it remained there unagitated. Exposure occurred to air via an open vent. Subsequent thermal stability investigations indicated that this could cause an oxidising reaction, perhaps sufficient to cause a rise in temperature to 300 °C. At >300 °C a rapid exothermic decomposition would have commenced generating large quantities of gas; this was believed to have been self-sustaining until the kettle ruptured.[10] This reinforces the need to simulate plant conditions when designing plant and operating exothermic reactions.

Table 12.2 Information required for hazardous chemicals

Name of chemical; other names	
Uses	
General description of hazards	
General description of precautions	
Fire-fighting methods	
Regulations	
Sources of advice on precautions	
Characteristics: evaluate as appropriate under all process conditions	
Formula (chemical structure)	
Purity (identity of any contaminants), physical state, appearance, other relevant information	
Concentrations, odour, detectable concentration, taste	
Physical characteristics:	
Molecular weight	Particle size. Size distribution
Vapour density	Foaming/emulsification characteristics
Specific gravity	Critical temperature/pressure
Melting point	Expansion coefficient
Boiling point	Surface tension
Solubility/miscibility with water; in general	Joule–Thompson effect
Viscosity	Caking properties
Corrosivity	
Contamination factors (incompatibility), oxidising or reducing agent, dangerous reactions	
Flammability information:	
Flash point	Vapour pressure
Fire point	Dielectric constant
Flammable limits (LEL, UEL)	Electrical resistivity
Ignition temperature	Electrical group
Spontaneous heating	Explosion properties of dust
Toxic thermal degradation products in a fire	
Reactivity (instability) information:	
Acceleration rate calorimetry	Drop weight test
Differential thermal analysis (DTA)	Thermal decomposition test
	Influence test
Impact test	Self-acceleration temperature
Thermal stability	Card gap test (under confinement)
Lead block test	JANAF
Explosion propagation with detonation	Critical diameter
	Pyrophoricity
Toxicity information:	
Toxic hazard rating	
Hygiene Standard (e.g. O.E.L., T.L.V.)	
Maximum allowable concentration (MAC)	
Lethal concentration (LC_{50})	
Lethal dose (LD_{50})	
Biological properties	
Exposure effects:	
Inhalation (general)	
Respiratory irritation	
Ingestion	
Skin/eye irritation	
Skin and respiratory sensitisation	
Radiation information:	
Radiation survey	
Alpha/beta/gamma/neutron exposure and contamination	

Table 12.3 Considerations in reaction process selection and design

Investigate potentially unstable reactions and side reactions such as spontaneous combustion or polymerisation.
Consider the risk that poor mixing or inefficient distribution of reactants and heat sources may, by malfunction or due to design error, give rise to undesirable side reactions, hot spots, reactor runaway, fouling, etc.
Consider whether the reaction can be made less hazardous by changing the relative concentration of reactants or other reactor operating conditions.
Assess whether side reactions produce poisonous or explosive material, or cause dangerous fouling.
Investigate whether the materials absorb moisture from the air and then swell, adhere to surfaces, form toxic or corrosive liquid or gas, etc.
Determine the effect of all impurities on chemical reactions and characteristics of process mixtures.
Ensure that the materials of construction are compatible with each other and the chemical process materials.
Investigate if hazardous materials can build up in the process (e.g. traces of combustible and non-condensible materials or toxic intermediates or by-products).
Allow for all aspects of catalyst behaviour, e.g. ageing, poisoning, disintegration, activation and regeneration.

Table 12.4 Examples of potentially hazardous operations

- Vaporisation and diffusion of flammable/toxic liquids or gases.
- Dusting and dispersion of combustible/toxic solids.
- Spraying, misting or fogging of flammable/combustible materials or strong oxidising agents.
- The mixing of flammable materials/combustible solids with strong oxidising agents (see Ch. 9).
- The separation of hazardous chemicals from inerts or diluents.
- The increase in temperature and/or pressure of unstable liquids.

Selection may be made between batch and continuous processing, but for large-scale operations the latter predominates because of economic considerations. However the choice between single or multiple streams has a significant effect upon process reliability, the inventory in individual items of equipment, and upon business interruption potential. Some points for comparison between batch and continuous processing are listed in Table 12.5.

Batch reactors tend to require more frequent cleaning than continuous reactors, or cleaning between each consecutive series of batches if used for more than one product, e.g. dyestuffs or synthetic resins. This may introduce additional hazards due to inadequate preparation for

Table 12.5 Relative advantages of batch v. continuous processing

Batch processes
- Easier to isolate.
- More labour intensive, therefore may provide more opportunity for operator intervention in an emergency.
- Easier control of product purity, and identification of materials through the process and after sale.
- With detailed instructions and operating procedures the chance of operating error, or damage to equipment, may be reduced.
- Longer response time.

Continuous processes
- Less material held up in the process.
- More amenable to automatic control.
- Unsteady state or cyclic fluctuations (e.g. on start-up and shut-down) are less frequent.
- Less requirement for cleaning/entering vessels or equipment.
- Any potentially hazardous intermediates can be processed without storage.

cleaning,[11,12] an inadequate cleaning procedure,[12] or incomplete removal of the liquids used for cleaning purposes.[12]

The type of continuous reactor may also be designed to minimise inventory; for example for fast reactions, such as sulphonations, film reactors may be used. (These also offer process advantages with regard to temperature control.)

Documentation

The primary drawing for representation of a process is the process flow diagram. This characteristically shows the major equipment together with the principal flow routes and controls. The key temperatures and pressures anticipated during normal operation are also illustrated. Material flows and compositions and the design duties, and sizes, of major equipment items may be included. Hence a comprehensive representation of the process is provided in a readily observable/alterable form. A sample flow diagram is shown in Fig. 12.1.[13]

The piping and instrument line diagrams (P&I diagrams) or engineering line diagrams (ELD) are the basic working documents during the design and construction phases. They will normally include:
- All process equipment required for start-up, shut-down, emergency and normal operation of the plant, including valves, blinds and removable spools.
- An identification number, an identifier of the material of construction, diameter and insulation requirements for each line.

- Direction of flow.
- Identification of main process and start-up lines.
- All instrumentation, control and interlock facilities with indication of action on instrument air failure.
- Key dimensions/duties of all equipment.
- Operating and design pressures and temperatures for vessels and reactors.
- Equipment elevations.
- Set pressures for relief valves. rupture discs, etc.
- Drainage requirements.
- Special notes on piping configuration as necessary (e.g. 'no pockets', 'gravity drainage', etc.).

A range of standard symbols are used on these drawings.[14]

Flowsheet preparation

Flowsheeting starts with the basic process calculations. Each stage of the process must then be designed to meet the safety requirements. Potential hazards must be identified, evaluated, and either removed or have provision made to restrict them. However, processes are highly integrated; hence each process step affects the operation of other steps.

Therefore the process is divided into sections; each section is analysed for internal safe operation and for its effect on the safety of other sections.

Convenient sub-divisions are[9]:

1. Reaction (which sections determine the dynamics of the whole system).
2. Separation, e.g. distillation, absorption, adsorption, liquid extraction, filtration, drying, size separation. (See Ch. 9.)
3. Storage sections (solids, liquids and gases).

Safety check-lists

Basic check-lists which exemplify the questions which must be answered during basic process consideration are given in Table 12.6 (after ref. 8).

Apparently minor differences in mechanical design, process or layout can cause unexpected primary process changes, e.g. if a process which has been run successfully at one plant is transferred to another.[15]

A process for the batch solution polymerisation of vinyl acetate in toluene in a 9,000 litre glass-lined reactor under atmospheric reflux was transferred to a new location. When the initial exothermic reaction went faster than anticipated foaming occurred; this partially plugged the reflux return line and the reactor's heat removal capacity was exceeded. Runaway occurred as described in Chapters 3 and 10. Toluene and vinyl acetate were released

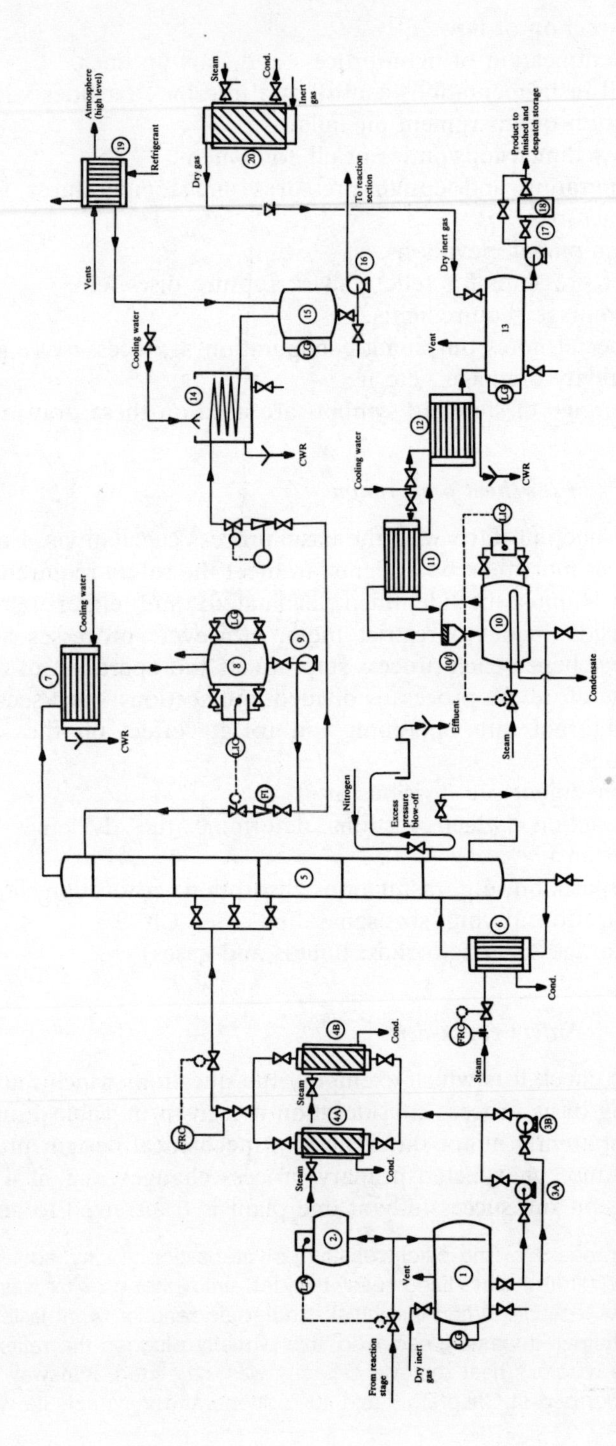

Fig. 12.1 Process flowsheet – drying and separation of organic chemicals[13] (pre-project engineering stage)

Figure 12.1 *continued*

Size/capacity	55 m^3	9 m^3	1.6 m^3/h 30 m HD	1.6 m^3/h (each)	0.6 m dia. 48 trays	5.6 m^2	1.8 m^2	0.25 m^3	1.4 m^3/h 10 m HD	5.6 m^2	9.2 m^2	2.3 m^2
Description	Feed tank	Measuring tank	Column feed pumps	Dryers (alumina)	Plate column	Reboiler-thermosyphon	Condenser	Reflux drum	Reflux pump	Re-run kettle boiler	Re-run condenser	Sub-cooler
Item number	1	2	3 A/B	4 A/B	5	6	7	8	9	10	11	12

Size/capacity	30 m^3	0.65 m^2	9 m^3	42 m^3/h 20 m HD	56 m^3/h 30 m HD	0.05 m^3	7 candle	4 m^2	0.15 m dia. × 0.6 m
Description	Low XYZ product receiver	Product cooler	High XYZ product receiver	High XYZ product pump	Low XYZ product pump	Spray arrestor	Filter	Vent condenser	Inert gas dryer
Item number	13	14	15	16	17	10/1	18	19	20

Table 12.6 Check-list – basic process considerations

1. Materials and reaction

Identify all hazardous process materials, products and by-products. Produce material information sheets for each process material.

Query the toxicity of process materials: identify short- and long-term effects for various modes of entry into the body and different exposure tolerance.

Identify the relationship between odour and toxicity for the process materials, and whether the odour of the materials is obnoxious.

Determine the means to be used for industrial hygiene recognition, evaluation and control (see Ch. 7).

Determine the relevant physical properties of process materials under all process conditions. Query source and reliability of data.

Determine the quantities and physical states of material in all stages of production, handling and storage. Relate these to the hazard and the second degree hazards.

Identify any danger the product might present to hauliers, warehousemen, railway workers, and public while in transit between plant and customer or around the factory.

Consult the supplier of any process material about its properties and characteristics and about safety in storage, handling and use (see Ch. 13).

Identify all possible chemical reactions. Consider both planned and unplanned reactions and 'excursions'.

Determine the mutual dependence of reaction rate and variables, and establish the limiting values to prevent undesirable reactions, excessive heat development, etc.

Handle unstable chemicals so as to minimise their exposure to heat, pressure, shock and friction (see Ch. 10).

Investigate and document potentially unstable reactions and side reactions, e.g. spontaneous combustion or polymerisation.

Query the risk that poor mixing or inefficient distribution of reactants and heat sources, either by malfunction or design error, can give rise to undesirable side reactions, hot spots, reactor runaway, fouling, etc.

Can the reaction be made less hazardous by changing the relative concentration of reactants or other reactor operating conditions (e.g. dilution, lower temperature, lower pressure)?

Can side reactions produce poisonous or explosive material, or cause dangerous fouling (e.g. of heat transfer surfaces)?

Can the material absorb moisture from the air and then swell, adhere to surfaces, form toxic or corrosive liquid or gas?

Determine the effect of all impurities on chemical reactions and hazard characteristics of process mixtures.

Are the materials of construction compatible with each other and the chemical process materials under all foreseeable conditions?

Can hazardous materials build up in the process (e.g. traces of combustible or toxic or unstable materials)?

Allow for all aspects of catalyst behaviour, particularly with respect to ageing, poisoning, disintegration, activation and regeneration?

Table 12.6 (continued)

2. General process specification
Is the scale, type and integration of the process correct (bearing in mind the potential safety and health hazards, and the effect on the business should the process be severely damaged)?

Identify the major safety hazards of the process and eliminate where possible. Locate critical areas on the flow diagrams and layout drawings. Is the selection of a particular process route or other design option more appropriate on safety grounds?

Can the process sequence be changed to improve the safety of the process? Are all process materials necessary or could less hazardous materials be selected?

Are emissions of material necessary? If so, are they discharged safely and in accordance with conventional good practice and legislation.

Can any unit or item be eliminated and does this improve safety (e.g. by reducing inventory or improving reliability)?

Check that the process design is correct. Are normal conditions described adequately and are all parameters of interest controlled?

Are the operations and heat transfer facilities properly designed, instrumented and contolled to minimise the occurrence of hazardous events?

Has the scale-up of the process been carried out correctly?

Does the process fail-safe in respect of heat, pressure, fire and explosion?

Has second-chance design been used?

into the pilot-plant building when a head gasket failed in the reflux condenser. The rupture disc on the reactor burst at 3.5 bar and the liquid and vapour vented above the building. Ignition occurred in the mixture inside the building and the ensuing explosion resulted in one fatality, injuries to six employees and severe building damage.

Subsequently three major changes were found to have occurred with the transfer.

1. Vinyl acetate was stored under inert gas instead of dry air. Lack of oxygen rendered the inhibitor ineffective and resulted in the fast initial reaction (see Chapter 10).
2. The gasket was unsuitable for exposure to toluene/vinyl acetate.
3. The 3.5 bar rating on the rupture disc allowed an excessive over-pressure which contributed to the gasket failure.[16]

Similarly even apparently trivial changes in process materials must be monitored.

In a process involving the reaction of an aromatic amine with a chloronitro compound the formation of ferric chloride, which was known to catalyse exothermic side reactions, was prevented by the use of synthetic soda ash as an acid acceptor. On one occasion, after 20 years' successful use, the synthetic soda ash was substituted by natural soda ash.

> In the non-aqueous medium of the reaction, the difference in crystallinity of natural soda ash rendered it less effective as an acid acceptor. This allowed acid build-up during the principal reaction resulting in ferric chloride formation from reaction with the mild steel vessel. This catalysed exothermic side reactions which caused vessel over-pressurisation leading to a release of flammables and an explosion in which one man died and five others were injured.[15]

(The importance of controlling modifications of *all* types is discussed in Ch. 17.)

The plant and individual items of equipment in it will be designed for normal operation at set throughputs within a range of operating parameters. However, different conditions will pertain during start-up, stand-by operation or any type of shut-down. A check-list of the major possible deviations from 'normal' operation is given in Table 12.7; all of these need to be accounted for by the designers and in the preparation of operating procedures and operator training manuals.

Table 12.7 Check-list – major deviations from normal operation (after ref. 17)

Start-up and shut-down

What else apart from normal operation can happen? Is suitable provision made for such events?

Can the start-up and shut-down of plant, or the placing of plant on hot stand-by, be expedited easily and safely?

In a major emergency can the plant pressure or the inventory of process materials, or both, be reduced effectively and safely?

Are the limits of operating parameters outside which remedial action must be taken known and measured (e.g. temperature, pressure, flow, concentration)?

To what extent should the plant be shut down for any deviation beyond the operating limits? Does this require the installation of alarm or trip or both? (To what extent is manual intervention expected?)

Does material change phase during the start-up and shut-down of plant from its state in normal operation? Is this acceptable? (Does it involve expansion/contraction, solidification, etc.?)

Can effluent and relief systems cope with large or abnormal discharge during start-up, shut-down, hot stand-by, commissioning and fire-fighting?

Are adequate supplies of utilities and 'miscellaneous' chemicals available for all plant activities (e.g. absorbents for spillage control)?

Is inert gas immediately available in all locations where this may be required urgently? Is there a stand-by supply?

During start-up and shut-down is any material added which can create a hazard on contact with process or plant materials?

Is the means of lighting flames of burners and flares safe on all occasions?

Location and layout

Chemical plants are generally located on sites determined by consideration of many factors other than safety, e.g. availability of, and easy access to, raw materials; ease of product distribution; amenability to pollution control requirements; availability of existing services and infrastructure; availability of suitable labour, etc.[16] However, with new plants hazard assessment and control, and the agreement of the local community to the proposed siting, are becoming increasingly important factors.[18] Open discussion with the community of potential risks and of the provisions made to minimise them, is important[19]: scale models are useful for this purpose.

In a study of safety factors influencing the siting of major hazard installations (see Ch. 15) consideration of each possible explosion hazard on a site (i.e. UVCE, BLEVE, thermal detonation, CVCE, dust explosion or explosion due to a rapid release of pressure) in terms of TNT equivalence led to a prediction that the potential hazard of a UVCE would generally be predominant. Siting distances based on a *cordon sanitaire*, and assuming criteria for acceptable damage, were considered realistic. This also allows for minimisation of environmental problems due to noise and air pollution.[20]

However, consideration of the possible effects of toxic releases involving either bulk toxics or an extremely limited number of hypertoxic materials led to the conclusion that siting policy based on 'containment' of the effects by dispersion was incompatible with environmental and financial considerations; put simply, the distances involved would be too great.[21,22] Therefore, hazard analysis and high integrity plant and procedures are needed.

Plant layout

The way a plant is laid out has a considerable effect upon safety and loss prevention. The positions of the storage and process areas, laboratories, control room, switch-house, cooling tower bank, flare stacks, effluent treatment plants and dispatch areas and the distances between them must be decided at an early stage in design. The topography, the direction of any prevailing wind, the tendency for flooding and the likelihood and extent of future development on, or alongside, the site should be taken into account. Appropriate consideration should be given in both layout and design to precautions against vandalism and terrorism.[23]

Different parts of a manufacturing plant may be placed adjacent to each other, to minimise inter-unit transfer distances, but layout of sites in the chemical and process industries tends to be such that units are segregated. This segregation also takes into account housekeeping and the requirements for access for construction and maintenance activities, e.g. crane access.

All considerations in process plant siting and layout are dealt with by Mecklenburgh[24]; only those relevant to safety and loss prevention are summarised here. Segregation is intended to reduce the risk of an accident in one plant, e.g. a fire or explosion, giving rise to 'domino', i.e. knock-on, effects to other equipment or plants.

Following the explosion at Doe Run (see page 881), fortunately the piping around chlorine storage tanks was not seriously damaged so that it was possible to isolate 1,000 tonnes of inventory by valving. This was accomplished despite the hazardous conditions prevailing during the first quarter hour after the incident.

The effects considered are:

- Damage from blast or missiles from an explosion, as discussed in Chapter 15.
- Direct spread of fire or explosion or radiant heat, as discussed in Chapter 4.
- Leakage or spillage of hazardous liquids or gases from damaged plant.
- Structural collapse, falling masonary, etc.

Initially the 'critical' items of a plant are identified and their interrelationships evaluated.[24] Items may be critical for a variety of reasons including:

- Business considerations, e.g. for continuity of production or of high financial value.
- Required for safe operation/shut-down, e.g. control rooms.
- Highly hazardous, e.g. bulk toxic storage.
- Particularly vulnerable to damage (see Fig. 12.2).
- Source of utilities, e.g. cooling-water facilities, inert-gas systems, 'fire-fighting' water.
- Potential ignition source, e.g. fired heaters.

Each part of a site can be assessed for degree and type of risk, i.e.:

Initiator	e.g. potential ignition source or containing highly reactive or corrosive or toxic chemicals.
Transmitter	e.g. equipment or storage areas containing flammable materials.
Receivers	e.g. populated buildings (control rooms, offices), emergency services, utilities.

Plants or units are provisionally grouped into segments containing risks of similar and compatible type. Grouping may be to reduce the spread of an area containing toxic or corrosive chemicals, for ease of operation, or to facilitate fire protection.

In laying out segments, access must be left for normal traffic and for emergency services. Hazardous units should, whenever possible, be located away from the high-occupancy buildings, e.g. offices, laboratories, control rooms and from public roads or housing. The special cases of major hazards are discussed in Chapter 15. Generally loading and off-loading facilities involving regular traffic should be located on the periphery of the process area. Plots are generally limited to 100 m by 200 m.

The hazards involved in each segment/plot are identified; these are considered with regard to both the nature of the hazard, the inventory of materials and the operating conditions. Groupings can be assessed by use of quantitative procedures, i.e. the Dow[25] or Mond[26] indices, and adjusted to obtain the best overall safety and loss prevention potential. A safe distance is calculated between plants or plant items to determine the plant layout. Flammable materials must be separated by a sufficient distance from potential ignition sources, for example for licensing the keeping of petroleum spirit in the UK there is a prohibition on naked lights within 50 ft (15.2 m).[27] Separation distances between blocks may be 15 m or 30 m, or greater, depending on the hazard potential when mitigated by the various precautions (e.g. explosion relief, steam or water curtains, bunds and dikes) and the potential loss, or risk to personnel, in the event of an incident. Separation distances to restrict radiant heat transfer between plants in the event of a fire,[28] or to minimise blast damage following an explosion,[29,30] can be calculated by empirical methods[5,24].

Recommended minimum separation distances for plants involving a major fire potential are also given in various codes.[31–33] However, where a potential vapour cloud explosion hazard exists (see page 140) consideration has to be given to the effects of explosion over-pressure. To supplement Fig. 15.6, Fig. 12.2 shows the effect of blast over-pressure on vulnerable plant items.[34] Calculation of the energy release in a possible explosion scenario enables the over-pressure–*v*.–distance relationships to be plotted.[30,35] Distances to account for toxic release of a given duration, e.g. instantaneous or prolonged, and size can also be calculated but tend to be large so that, in addition to providing high-integrity plant, alternative strategies, e.g. steam or water curtains become favourable; refuges are also necessary.

Power lines, which may require looping to key items, should be located to be accessible to authorised personnel but out of danger. If possible pipe tracks should be laid out to avoid areas surrounded by bunds and roads.

A road tanker hit a pipe which ran over the top of a bund wall. This was one of several pipelines between a storage area and loading area. The pipe

Fig. 12.2 Effect of explosion over-pressure on plant items[34]

Equipment	Over-pressure psi																									
	0.5	1.0	1.5	2.0	2.5	3.0	3.5	4.0	4.5	5.0	5.5	6.0	6.5	7.0	7.5	8.0	8.5	9.0	9.5	10.0	12.0	14.0	16.0	18.0	20.0	20.0
Control house steel roof	a	c	d				n																			
Control house concrete roof	a	e	p	d		n																				
Cooling tower	b			f		o																				
Tank: cone roof		d				k							u													
Instrument cubicle			a			l m						t														
Fired heater				g	l					t																
Reactor: chemical				a				i					p					t								
Filter				h					f										v		t					
Regenerator						i				i p					t											
Tank: floating roof						k							u												d	
Reactor: cracking							i							i						t						
Pipe supports							p					s o														
Utilities: gas meter									q																	

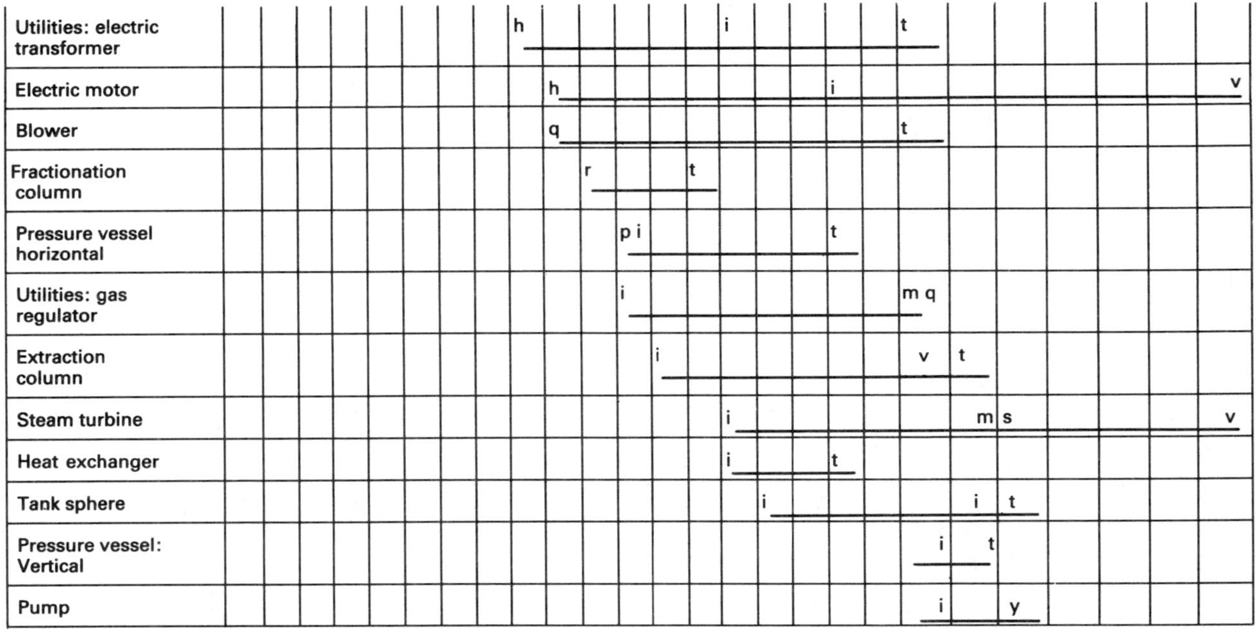

Utilities: electric transformer									h						i					t						
Electric motor										h								i								v
Blower										q										t						
Fractionation column											r			t												
Pressure vessel horizontal												p i						t								
Utilities: gas regulator												i								m q						
Extraction column													i							v	t					
Steam turbine															i						m	s				v
Heat exchanger															i			t								
Tank sphere																i					i	t				
Pressure vessel: Vertical																				i	t					
Pump																				i		y				

Code: Lines indicate range from 1% damage to 99% damage to equipment listed in left-hand column. Other damage is shown by letters with the following meanings.

a. Windows and gauges break
b. Louvers fall at 0.3–0.5 psi
c. Switchgear is damaged from roof collapse
d. Root collapses
e. Instruments are damaged
f. Inner parts are damaged
g. Brick cracks
h. Debris-missile damage occurs

i. Unit moves and pipes break
j. Bracing fails
k. Unit uplifts (half-filled)
l. Power lines are severed
m. Controls are damaged
n. Block walls fail
o. Frame collapses

p. Frame deforms
q. Case is damaged
r. Frame cracks
s. Piping breaks
t. Unit overturns or is destroyed
u. Unit uplifts (0.9 filled)
v. Unit moves on foundation

(1 psi = 6.9 KN/m^2)

broke off inside the bund and a bund fire was started when the tanker's engine ignited the spillage.[36]

While the control room should preferably be in a 'safe' area to avoid the need for pressurisation, reinforcement, etc., in practice, this cannot always be achieved.

Equipment layout[24]

Equipment is normally laid out in the sequence of process flows. However, tanks and pumps, or distillation columns provided with common access platforms, or hazardous operations may be grouped in specific areas. Equipment requiring crane or trolley beam access may also be grouped together.

All equipment should preferably be at ground level. Elevation is, however, unavoidable for process reasons (e.g. with distillation or absorption columns, spray driers or vertical reactors) or to provide for gravity flow, or pump suction head, requirements. Installation of heavy equipment at high level should be avoided wherever practicable, i.e. it should preferably be located at grade or on adequate supports.

A special vent gas silencer and tail pipe, with a total installed weight >30 tons, was installed to reduce noise levels when venting process gas during start-up and shut-down of an ammonia plant. The silencer was on special steel supports off the side of a steam drum structure with its base at approximately 23 m elevation.

During a shut-down a tube ruptured in one of the primary waste heat boilers whilst process gas from the reformer section was being vented through the silencer. Large volumes of steam and water discharged into the vent silencer resulted in vibration of such magnitude that its supporting steel framework collapsed and the silencer fell into the ammonia plant. A major section of the pipe rack, fortunately with depressurised flammable gas lines, was destroyed.[37]

Space is needed for materials handling, e.g. solids storage (drums, bins or sacks) and manual transportation (e.g. fork-lift truck access). Equipment positioning is also determined by maintenance and cleaning needs, e.g. area for removal of internals such as heat exchanger tube bundles, or head-room for removal of agitators.

Access

Valves should be easily accessible and instruments easily readable without climbing over equipment, etc. Any emergency valves or stop buttons should be safely located, e.g. remote, outside bunds, or behind protective barriers.

In process plants a recommended minimum requirement for access is that emergency equipment should be able to approach an operating area from at least two sides if the area exceeds 925 m^2.[25]

A block system layout for a petrochemical plant, operating a main process with several stages and a number of subsidiary processes, is shown in Fig. 12.3(a)[38]; 6-m roads give access for fire appliances into the process area from several different directions with no dead ends. The

(a)

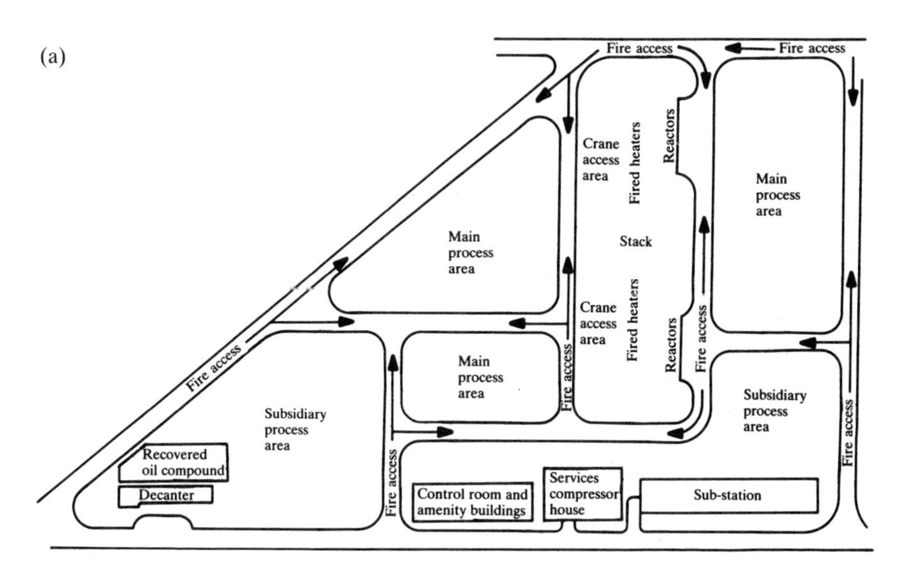

(b)

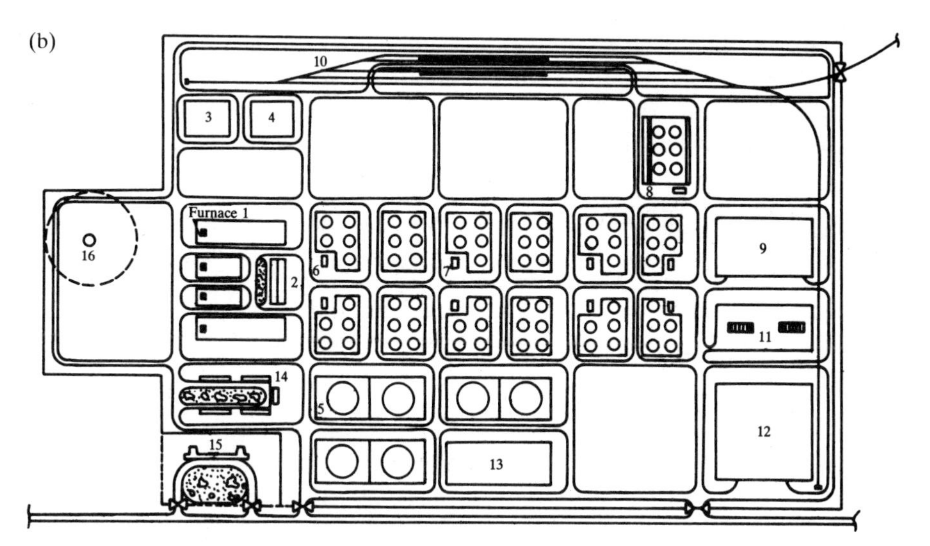

Fig. 12.3 Block layouts. (a) Process area of a petrochemical plant with 6-m roads; (b) 'Typical' petroleum refinery[39]

effective separation distance between blocks generally reaches 15 m with larger distances around the heaters, provided by the crane access areas, and the reactors, due to lay-bys. A 'typical' petroleum refinery layout is shown in Fig. 12.3(b).[39]

Clear routes are desirable for operators, avoiding kerbs and level changes as far as is practicable. In general, access ways should be 0.7 m wide for a single person or 1.2 m wide for two persons. Two or more escape exits are required from each working area. Stairways are preferred to ladders for main access; ladders and stairways should be designed to published standards.[24,40] Fixed ladders are acceptable for escape from structures. Any doors should be clearly marked and open in the direction of escape.

Lighting

An adequate standard of lighting is required, to minimise tripping/bumping accidents. Additional lighting may be needed where physical or chemical hazards arise and over instruments or escape routes. Emergency lighting should be provided in key areas including escape routes.

Control rooms

On modern chemical plants the control room constitutes the centre for safe operation, and if necessary shut-down. Therefore, its integrity, and that of supplies to it, must be maintained.

Control room design has been modified in recent years on those plants where there is a potential for a major explosion to occur, albeit as a remote possibility (see Ch. 15). For example blastproofing is now recommended on ethylene plants, the building being of reinforced concrete with no windows.[41]

> At Flixborough many of the men who were outside survived while all those in the control room were killed. If they were not killed by blast they would have been killed by the building collapsing on them.[42] Design to withstand a peak reflected pressure of 10 psig for 0.02 seconds is suggested. The installation of heavy equipment above the control room should not be allowed.

Apart from the difficulty arising from possible explosion overpressures, windows pose a problem because radiant heat-transfer from a large fire close to the room could ignite all combustible materials within it. Combustion would raise the temperature in the room and vitiate the atmosphere. Unless a positive pressure of clear air is maintained during a local fire, toxic, combustion gases may enter via broken windows.

The absence of windows provides some inconvenience to operators who can only see the plant on television monitors. Therefore flammable gas detectors are recommended in critical areas of the plant to provide warning of any hydrocarbon leaks; these detectors activate alarms in the control room and can also activate steam curtains in the critical area to assist dispersal of vapour.[43]

The design 'rules' adopted by one major company include[44]:

Total window area not more than 7% of wall area.

Window frame sizes not more than 0.25 m^2.

Laminated glass; two 3 mm panes with 1.9 mm polyvinyl butyral interlayer.

Glass frame fixed in a strong flexible way to the frame, with 30 mm rebates.

Catch bar inside window to retain frame if blown in.

The building design philosophy is to determine the development of 'plastic hinges'; the building will collapse when a sufficient number of 'plastic hinges' develop under load.

Drainage

Adequate design of drainage is essential, since poor drainage of flammable liquids – sometimes with fire water – has been a contributory factor in a large number of losses involving flammable liquid spills.[25]

A preferred design is where any spill of flammable liquid is directed away from the process unit to an impounding basin: It is also desirable for the inside grade in diked, i.e. bunded, areas to slope to an impounding basin located a minimum of one tank-diameter from the tank. A minimum of 2% slope leading to an adequate sized drainage trench is necessary to provide for the removal of a large spill from a process unit.[25]

Detailed design

Equipment, instruments and ancillaries are designed to perform a variety of unit operations (Ch. 9) and/or chemical processes (Ch. 10) and linked together to provide a complete manufacturing unit. (Such units may also be linked with the product from one providing the feedstocks to others.) Many of the decisions taken during the detailed design have a critical effect upon plant safety.

Two concepts widely applied at this stage are 'fail-safe' and 'second-chance' design. Fail-safe requires that in the event of a failure of services, or of an instrument or process control loop, the system reverts to the least hazardous position. The simplest example is of a control valve which may either fail 'open' or 'shut'. 'Second-chance' design refers to the availability of a back-up facility to prevent a hazard arising, or to mitigate its effects, if an error is made in operation or equipment fails. Typical examples include:

- Provision of pressure relief on vessels.
- Provision of explosion relief on equipment or buildings.
- Provision of bund walls around storage tanks.
- Emergency shut-down systems.
- Provision of trips activated by high or low levels of process parameters.
- Provision of alarms, e.g. for flame failure, high or low level of process parameters.

The chance of errors in design, fabrication and maintenance of the equipment – which may have hazardous consequences – may be reduced by several simple strategies, e.g.

- Standardisation of suppliers and types of pumps, valves and instruments.
- Avoiding, where practicable, complex equipment which requires technically advanced maintenance (unless this is to be contracted out to specialists).
- Using a single grade of steel throughout a unit, or avoiding the use of alloy steels.
- Eliminating redundant block valves in interconnecting pipework between equipment.

Services

In addition to electricity the operation of a chemical plant generally requires the following services (utilities):

- Steam – for high- and medium-temperature heating and, at lower pressure, for space heating: also for process uses, e.g. purging inerting or snuffing.
- Process water and boiler feed water.
- Water for fire fighting.
- Cooling water – from a recirculatory system (see Ch. 9, page 464) or a river or the sea.
  (Cooling facilities may also involve fin-fan coolers.)
- Compressed air – for instruments, B.A. supply, etc.
- Inert gas – nitrogen, carbon dioxide or treated flue gas (see Ch. 4, page 132) for inerting, pressurisation, purging or process uses.
- Provisions are also required for disposal of treated liquid effluents, e.g. to a foul water drainage system, and of surface water.

The major consideration with these supplies is that they are adequately sized to cope with the maximum forseeable demands. Consideration also has to be given to their reliability and, where appropriate, the provision of either a stand-by facility, e.g. a diesel-electric set for essential electrical lighting and equipment, or an emergency reservoir to cope with shutdown.

Freedom of services from contamination is also important, e.g. if steam is used for fire-fighting generally to dilute leaks – the supply should be separate from the normal steam supply to eliminate any possibility of ingress of hydrocarbon leaks; process contamination of compressed air can be avoided by physical disconnection. Segregation of surface water drains and foul-water drains (i.e. for effluents) is an important feature.

Experience has shown that any evaluation of the effectiveness of auxiliary power supplies should take account of the possibility of common mode failure.[46] Furthermore, it is not considered good practice to use cooling water reservoirs for fire water, since in the event of a fire, both may be required.

The reliability of the inert gas supply, and its freedom from contamination, is vital. Whatever type is used, e.g. nitrogen, carbon dioxide or inert gas from a generator, the capacity must be adequate for routine operations and for emergencies.[47]

The interior of a control unit was purged with nitrogen to prevent any flammable vapour present, e.g. due to a leak or spillage, diffusing in. A pressure switch served to isolate the power supply if the nitrogen pressure fell below a pre-set value.[48] One day an explosion blew out the panels on the front of the unit and injured a person standing there. The nitrogen was found to have become contaminated with flammable vapour from elsewhere on the plant; the pressure switch had been disarmed by setting it to zero and when a failure occurred in the nitrogen supply the fall in pressure allowed air to diffuse into the unit and create a flammable mixture.

Several large tanks with a total volume of 1000 m^3 were blanketed with nitrogen. To conserve nitrogen when the tanks were being filled or emptied, they were connected to a gasholder. However, the system leaked and, in the absence of regular checks, the oxygen content of the nitrogen reached 15%.[49] When an explosion occurred in one tank it would have been transmitted along the vent to the others but for the fact it had been disconnected from the vent system.

Ventilation – general

Open construction is preferable whenever practicable for areas processing flammable liquids or gases. This provides both general ventilation to assist in the dispersion of gas or vapour leaks and the maximum possible explosion-venting area. It also facilitates fire fighting.

A leak of ethylene gas occurred from a joint in a high pressure pipeline in a poorly ventilated ground floor area beneath a compressor house. The leak ignited after a few minutes and the explosion caused serious damage, such that half the building required re-building, and resulted in four fatalities.[45]

Dust collection equipment and filters should also be located in an open area free from other processes.

A minimum recommended ventilation requirement for any enclosed area in which flammable liquids or gases are handled is six air changes per hour, or 1 m^3 or air per minute per m^2 of floor area.[25]

As an example of the level of precautions required in cases where loss of containment may result in a serious loss, the recommended practice for compressor houses handling flammable gases now includes[45]:

- Location of compressors in an open-sided building.
- Installation of gas detectors.
- Installation of remotely operated valves to enable leaking compressors to be isolated and blown down from a safe distance.
- Surrounding compressors and ancillaries by a water curtain to prevent escaping vapour reaching a source of ignition.

Ventilation – local

With local ventilation, the mass movement of air is used:

- To remove a contaminant near its source of escape into the workplace atmosphere.
- To dilute the level of contaminant in ambient air to acceptable concentrations.
- To control the temperature and humidity of the workplace atmosphere.
- To replace exhausted air.

Within factories in the UK where in connection with any process, there is given off dust, fume or other impurity of such a character and to such extent as to be likely to be injurious (or any substantial quantity of dust of any kind) there is a statutory duty – where the process makes it practicable – to provide and maintain exhaust appliances as near as possible to the point of origin of the dust or fume or other impurity so as to prevent it entering the air of any workroom.[50] Reference was made to some examples in Chapter 6. Flammable gases, vapours, or mists may also be reduced below hazardous concentrations by local exhaust ventilation.

Good design and maintenance is of paramount importance for the system to be effective. The following is restricted to a discussion of some general principles of exhaust ventilation and specialist references should be consulted[51,52] for more detailed considerations. The system will normally consist of a hood, the entry point of which may vary from a purpose-built enclosure and canopy to a slotted piece of ducting; ducting which conveys the exhaust air and contaminents; a fan; equipment for contaminants collection/removal, and a stack for dispersion of decontaminated air.

The driving force for air flow is a pressure difference, the total pressure (TP) being the sum of static pressure (SP) and velocity pressure (VP). Energy input considerations must be sufficient to overcome resistances and energy losses in the system and induce air to flow at the design rate. Sources of pressure losses in exhaust systems include hood entry, fittings, duct friction, and air cleaning devices. Thus turbulance created as air is accelerated from rest to enter a duct or opening results in a pressure drop. This entry loss (h_e) plus the acceleration energy needed to move air at a given velocity comprise the hood's static pressure, SP_h i.e.

$$SP_h = h_e + VP$$

For most designs

$$h_e = F_h VP$$

where F_h is the hood-entry loss factor

For simple hoods, therefore

$$SP_h = F_h VP + VP = (1 + F_h)\, VP$$

For slot, plenum or complex hood arrangements there are two losses, viz. through the slot and into the duct. In such cases

$$SP_h = h_{e(slot)} + VP + h_{e(duct)} = [F_{(slot)} \times VP_{(slot)}] + VP + [F_{(duct)} \times VP_{(duct)}]$$

The coefficient of entry, C_e, is the efficiency of hood entry in converting static pressure to velocity pressure. It is defined as the ratio of flow by hood static pressure to the theoretical flow if the hood static pressure were converted to velocity pressure.

$$C_e = \frac{VP}{SP}$$

Ideally $C_e = 1$

When the hood-entry loss is a simple expression C_e is related to F_h by the following equation

$$C_e = \left[\frac{1}{1 + F_h}\right]^{\frac{1}{2}}$$

Listings of C_e and h_e in terms of VP are published.[53]

Exhaust ducts require a minimum air velocity to move the contaminant through the duct without settling. This is the transport velocity, examples of which are given in Tables 12.8 and 12.9.

Table 12.8 Typical recommended transport velocities

Material, operation, or industry	Minimum transport velocity (ft/min)*	Material, operation, or industry	Minimum transport velocity (ft/min)*
Abrasive blasting	3,500–4,000	Lead dust	4,000
Aluminium dust, coarse	4,000	With small chips	5,000
Asbestos carding	3,000	Leather dust	3,500
Bakelite molding powder dust	2,500	Limestone dust	3,500
Barrel filling or dumping	3,500–4,000	Lint	2,000
Belt conveyors	3,500	Magnesium dust, coarse	4,000
Bins and hoppers	3,500	Metal turnings	4,000–5,000
Brass turnings	4,000	Packaging, weighing, etc.	3,000
Bucket elevators	3,500	Downdraft grille	3,500
Buffing and polishing		Pharmaceutical coating pans	3,000
Dry	3,000–3,500	Plastics dust (buffing)	3,800
Sticky	3,500–4,500	Plating	2,000
Cast-iron boring dust	4,000	Rubber dust	
Ceramics, general		Fine	2,500
Glaze spraying	2,500	Coarse	4,000
Brushing	3,500	Screens	
Fettling	3,500	Cylindrical	3,500
Dry pan mixing	3,500	Flat deck	3,500

Dry press	3,500
Sagger filling	3,500
Clay dust	3,500
Coal (powdered) dust	4,000
Cocoa dust	3,000
Cork (ground) dust	2,500
Cotton dust	3,000
Crushers	3,000 or higher
Flour dust	2,500
Foundry, general	3,500
Sand mixer	3,500–4,000
Shakeout	3,500–4,000
Swing grinding booth exhaust	3,000
Tumbling mills	4,000–5,000
Grain dust	2,500–3,000
Grinding, general	3,500–4,500
Portable hand grinding	3,500
Jute	
Dust	2,500–3,000
Lint	3,000
Dust shaker waste	3,200
Pickerstock	3,000
Silica dust	3,500–4,500
Soap dust	3,000
Soapstone dust	3,500
Soldering and tinning	2,500
Spray painting	2,000
Starch dust	3,000
Stone cutting and finishing	3,500
Tobacco dust	3,500
Woodworking	
Wood flour, light dry sawdust and shavings	2,500
Heavy shavings, damp sawdust	3,500
Heavy wood chips, waste, green shavings	4,000
Hog waste	3,000
Wool	3,000
Zinc oxide fume	2,000

* *1,000 ft/min ≃ 5 m/s.*

Table 12.9 Classification of transport velocities for dust collection

Material	Minimum transport velocity (ft/min)*
Very fine, light dusts	2,000
Fine, dry dusts and powders	3,000
Average industrial dusts	3,500
Coarse dusts	4,000–4,500
Heavy or moist dust loading	4,500 and up

* *1,000 ft/min ≃ 5 m/s.*

A fan of proper type, size and speed should be selected to give the highest operating efficiency based on total air-flow requirements and the total system resistance or static pressure. A fan curve, or system curve, depicts the possible combinations of volumetric flow and static pressure for a given application as shown in Fig. 12.4. Because the fan and system can each operate only at a single point on its own curve, the combination occurs at the point of intersection. If the flow rate of a fan is inadequate, theoretically one can either reduce the resistance of the system or increase the fan speed. However, while volume is proportional to fan speed, power is proportional to the cube of fan speed with implications for energy costs. A summary of various fan types available and their characteristics is given in Table 12.10.

Ducting before the fan will be at negative pressure and leaks in the system cause air to flow from the room into the duct. Ducting after the fan, however, will be under positive pressure allowing contaminated air to escape via leaks into the room. All positive pressure ducting should therefore be outside the room, preferably by externally locating the fan itself.

Hoods will comprise either total enclosures, booths, captor hoods or receptor hoods. If they are not designed to contain a source of pollution, they should be as close as practicable to it in order to achieve maximum efficiency.

Enclosure

Where possible, equipment or plant handling hazardous chemicals should be totally enclosed. This approach is used, for example, for carcinogenic or radioactive materials. A room is built around the vessel with grilles to allow airflow. The room functions as a hood and operates under slight negative pressure with plant controls located outside the room. Entry into the room is restricted and, when essential, is likely to entail use of complete personal protection, etc. Clearly other safety features are

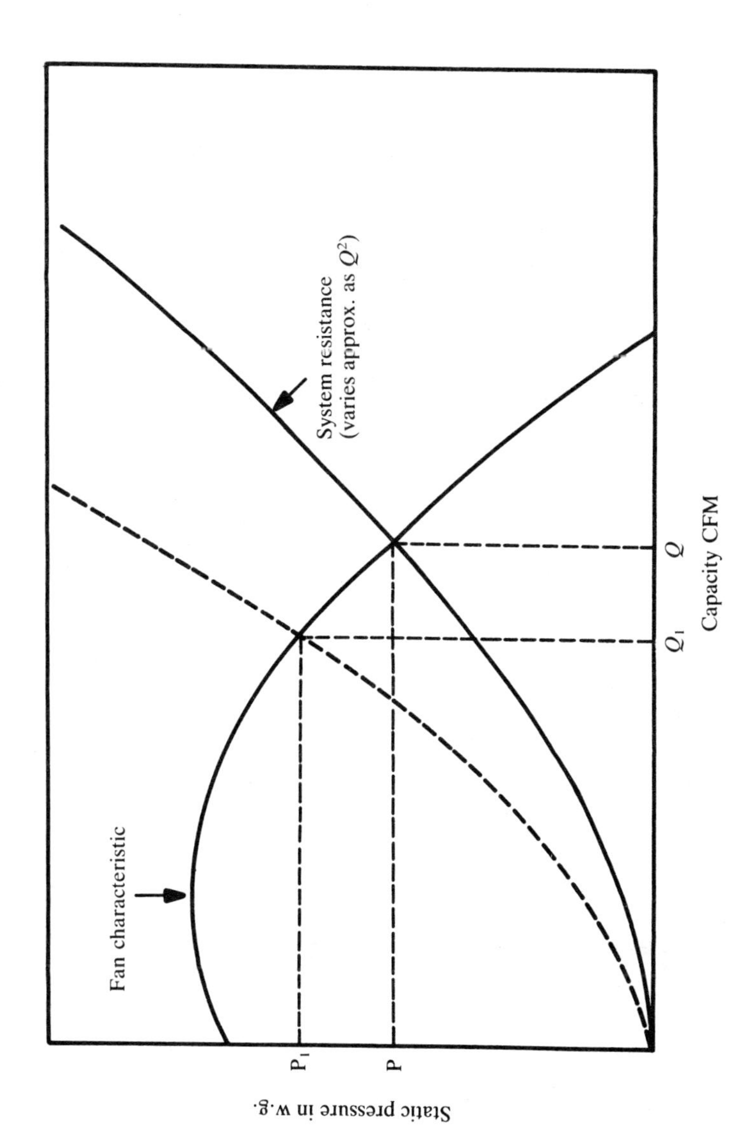

Fig. 12.4 Typical fan characteristic curve

Table 12.10 Summary of fan types

Fan type	Advantages	Disadvantages	Applications
1. Axial-flow (without guide vanes)	Very compact, straight-through flow. Suitable for installing in any position in run of ducting.	High tip speed. Relatively high sound level comparable with 5. Low pressure development.	All low pressure atmospheric air applications.
2. Axial-flow (with guide vanes)	Straight-through flow. Eminently suitable for vertical axis.	Same as 1 but to less extent.	As for 1 and large ventilation schemes such as tunnel ventilation.
3. Forward-curved or multivane centrifugal	Operates with low peripheral speed. Quiet and compact.	Severely rising power characteristic requires large motor margin.	All low and medium pressure atmospheric air and ventilation plants.
4. Straight or paddle-bladed centrifugal	Strong simple impeller, least clogging, easily cleaned and repaired.	Inefficiency. Rising power characteristic.	Material transport systems and any application where dust burden is high.

5. Backwards-curved or backwards-inclined blade centrifugal	Good efficiency. Non-overloading power characteristic.	High tip speed Relatively high sound level compared with 3.	Medium and high pressure applications such as high velocity ventilation schemes.
6. Aerofoil-bladed centrifugal	Highest efficiency of all fans. Non-overloading power characteristic.	Same as 5.	Same as 5 but higher efficiency justifies its use for higher power applications.
7. Propeller	First cost and ease of installation.	Low efficiency and very low pressure development.	Mainly non-ducted low pressure atmospheric air applications. Pressure development can be increased by diaphragm mounting.
8. Mixed-flow	Straight through flow. Suitable for installing in any position in a run of ducting Can be used for higher pressure duties than 2. Lower blade speeds than 1 or 2 hence reduced noise.	Stator vanes are generally highly loaded due to higher pressure ratios. Maximum casing diameter is greater than either inlet or outlet diameters.	Large ventilation schemes where the somewhat higher pressures developed and lower noise levels give an advantage over 2.
9. Cross-flow or tangential-flow	Straight across flow. Long, narrow discharge.	Low efficiency. Very low pressure development.	Fan-coil units. Room conditioners. Domestic heaters.

required, e.g. air monitors inside and outside the room, air filters/scrubbers, permits to work, methods of decontamination, etc. However, since total enclosure is rarely used, partial enclosure is more common, small openings being provided to permit charging or removal of materials. The total area of these openings must be relatively small and ventilating air must be extracted at sufficient rates to prevent dust or fumes from leaking out. The requisite air velocity depends on many factors but, for example, if an enclosure is sufficiently large to contain a naturally occurring dust cloud such that dust does not impinge on the walls, air velocities should be in the range 0.28 to 0.56 m s^{-1}; higher velocities are required if there is significant disturbance inside the enclosure.[54] Fume-cupboards are partial enclosures and are described more fully in Chapter 11.

Booths

A booth should be designed of sufficient size to contain any naturally occurring dust cloud, hence minimising escape from the open face. Air velocities of 0.56 m s^{-1} should be provided over the whole of the open face, but much higher velocities are required if there is considerable air movement in the booth.[54] Special consideration must be given to cope with convection currents, e.g. due to hot equipment or thermal lift of pollutants. As shown in Fig. 12.5 booths should be deep, to contain eddy currents at the back corners; escape of these eddies, with accompanying pollutant, from shallow booths can be minimised by the use of baffle plates or multiple off-takes.

Obviously the design should not require any operator to work between the source of pollutant and the back of the booth, i.e. the front open face is the work position.

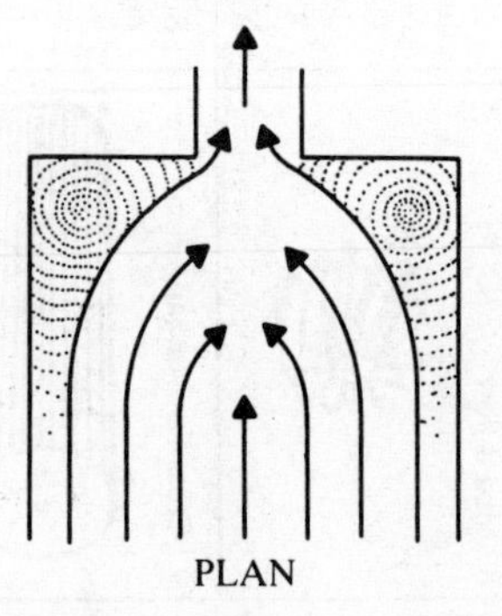

Deep booth – even air flow, no leakage

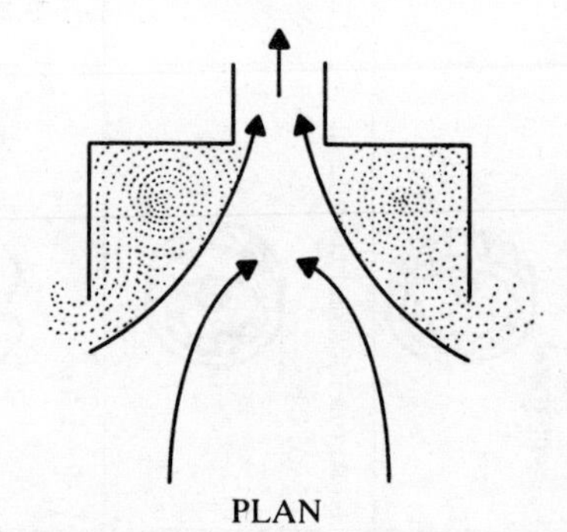

Shallow booth – uneven air flow, leakage from eddies

Fig. 12.5 Effect of booth depth on capture efficiency[54]

Captor hoods

If total enclosure or use of a booth is impracticable, then a hood unit must be used at some distance from the source of dust/fumes. A captor hood is designed to capture dust or fumes which would not enter it naturally, i.e. the movement of polluted air is changed so that it all flows into the hood. The rate of flow of air into the hood must be sufficient to capture contaminant at the furthermost point of origin. Recommended capture velocities for typical applications in manufacturing industries are given in Table 12.11.[51]

An important feature is that air velocity falls off rapidly with distance from the face of the hood as illustrated in Fig. 12.6. Thus in practice a source of dust should not be more than one hood diameter from the hood.[54]

Equipment vents should lead away from working/access areas; in some cases, e.g. with toxic or corrosive gases, they need to be scrubbed in an absorption tower or passed through an adsorber. Flammable gases may either be vented via individual flame arrestors or into a flare header.

The efficiency of a captor system can be significantly improved by the use of flanges and/or by the removal of abrupt changes in direction of the ducting as shown in Fig. 12.7. Air jets may also be used to assist captor hoods.

Receptor hoods

Receptor hoods simply receive contaminant, e.g. dust or fumes, which are driven into it by the source of generation. The flowrate is therefore designed to ensure that the fan empties the hood more rapidly than the

Table 12.11 Range of capture velocities

Condition of dispersion of contaminant	Examples	Capture velocities (m s^{-1})
Released with practically no velocity into quiet air.	Evaporation from tanks; degreasing, etc.	0.25 to 0.51
Released at low velocities into moderately still air.	Spray booths; intermittent container filling; low-speed conveyor transfers; welding; plating; pickling.	0.51 to 1.02
Active generation into zone of rapid air motion.	Spray painting in shallow booths; barrel filling; conveyor loading; crushers.	1.02 to 2.54
Released at high initial velocity into zone of very rapid air motion.	Grinding; abrasive blasting; tumbling.	2.54 to 10.2

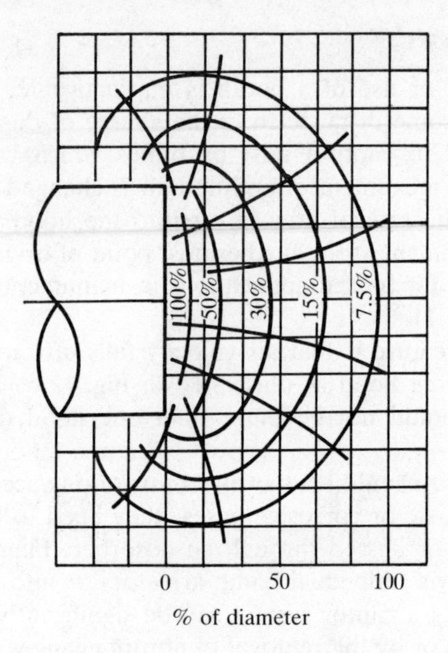

Fig. 12.6 Fall-off in air velocity with distance (expressed as % of diameter) from captor hood

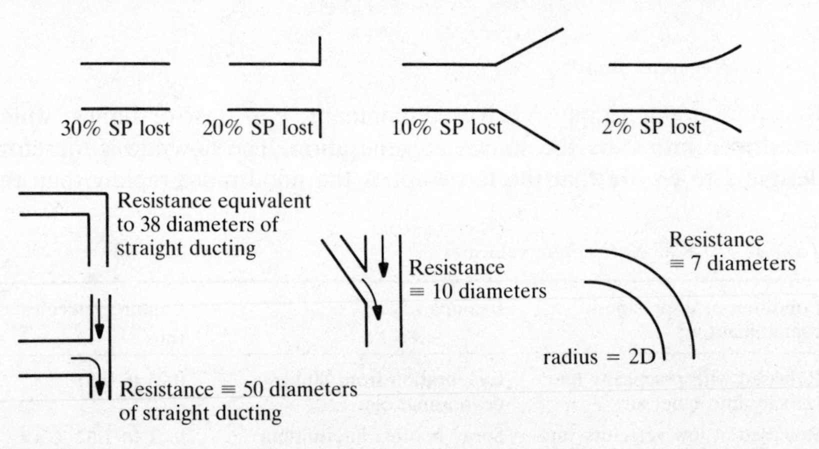

Fig. 12.7 Energy loss associated with hood duct design (SP = static pressure of the hood)

process fills it and to overcome the effect of draughts, etc. The arrangement should avoid the need for operators to work between the hood and the source of contaminants.

Receptor hoods may receive air by convection from hot sources, e.g. furnaces, or by induction from cold processes, e.g. from above material falling down chutes. In either case the hood must be capable of absorbing all the air received, i.e. including entrainment.

Exhausted air

Air removed by local exhaust ventilation has to be replaced via correctly located, adequately sized inlets. Otherwise air will enter indiscriminately via doors, windows, etc., resulting in draughts and possibly redispersion of dust or interference with collection of the contaminant at source.

Occasionally it is permissible to recirculate exhausted air thereby conserving energy. However, the exit concentration must not exceed allowable concentrations which may be calculated for equilibrium concentrations using the following equation:

$$C_R = \tfrac{1}{2}(HS - C_O) \times \frac{Q_T}{Q_R} \times \frac{1}{K}$$

where: C_R = concentration of contaminant in exit air before mixing; Q_T = total ventilation flow through the affected workspace (m^3/min); Q_R = recirculated airflow (m^3/min); K = mixing factor (usually 3–10); HS = hygiene standard; C_O = concentration of contaminant in worker's breathing zone with local exhaust discharge outside.

Dilution ventilation

Reinforcement of the local exhaust provisions may be necessary by means of general, that is dilution, ventilation. This involves flushing the workplace with clean air to reduce the level of contaminants to a minimum. This is a dilution/dissipation process and should be considered secondary to other precautions, since with toxic materials, for example, the contaminant may only be diluted after passage through an operator's breathing zone.

The siting of the extract fan and air inlet require careful consideration in order to provide a suitable air-flow pattern in the room to ensure operator exposure is minimised. A variety of arrangements is summarised in Fig. 12.8.[55] The rules are:

- Site the exhaust fan near to the source of the contaminant.
- Ensure fresh-air movement is from worker to leak source and not vice versa.
- Ensure air-inlet supply is not contaminated with exhaust air.
- Provide back-up handling for air inlet where necessary.

Where a process or operation involves a continuous release of gas or vapour the steady-state dilution ventilation required to reduce the atmo-

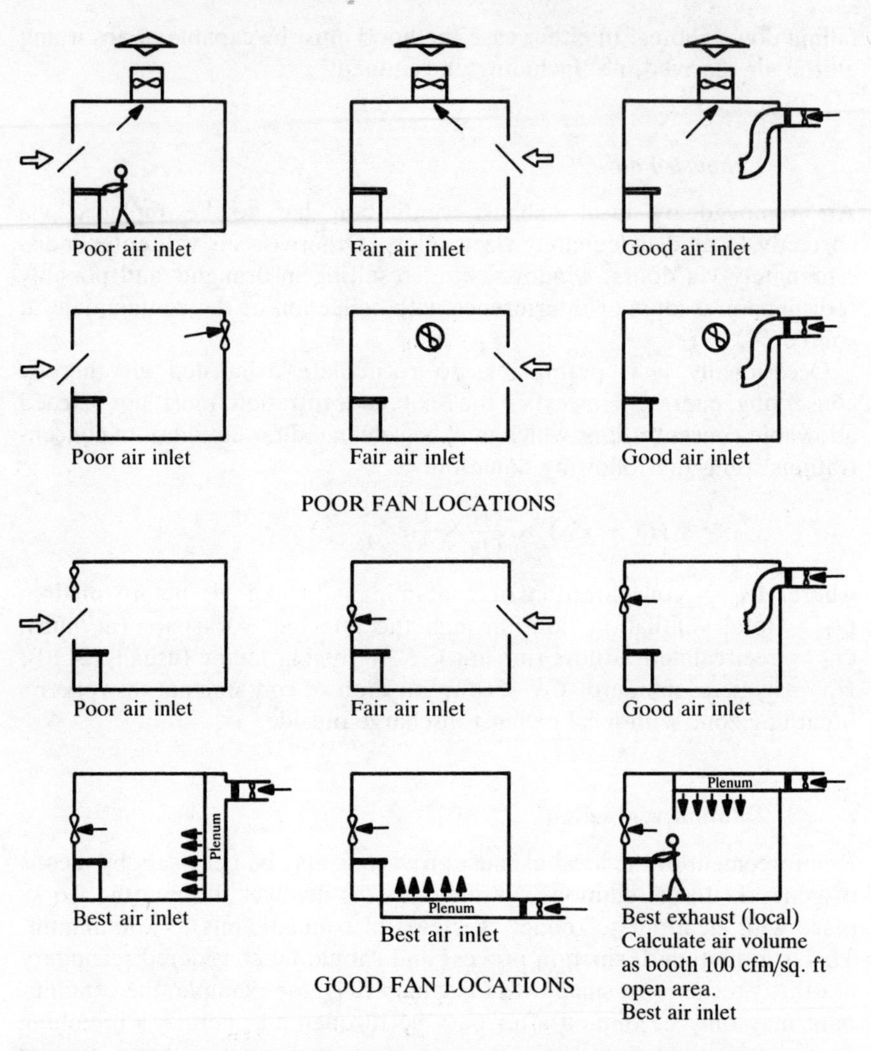

Fig. 12.8 Dilution ventilation. Examples of good and poor fan locations[55]

spheric concentration of a contaminant to a level below the hygiene standard is given by:

$$Q = \frac{3.34 \times 10^5 \times SG \times ER \times K}{MW \times HS}$$

where: Q = ventilation rate (m^3/min);
SG = specific gravity of evaporating liquid;
ER = evaporation rate (l/h);
MW = molecular weight;

K = design factor to account for incomplete mixing (usually between 3 and 10);
HS = relevant hygiene standard (OEL, TLV, MAK, etc.).

1,1,1-trichloroethane evaporates from a tank at a rate of 2 l/h. The OEL (1984) is 350 ppm, the specific gravity is 1.3376 and the molecular weight is 133. Assuming $K = 6$, then the ventilation rate required to maintain concentrations below the OEL is given by:

$$Q = \frac{3.34 \times 10^5 \times 1.34 \times 2 \times 6}{133 \times 350}$$
$$= 115 \text{ m}^3/\text{min}$$

In general, dilution ventilation is inappropriate for substances of high toxicity or OEL < 50 ppm, carcinogens, dusts or fumes, or where concentrations fluctuate, or where an operator works close to a point of release of contaminant. Since hygiene standards are subject to revision (usually downwards) specifications of existing systems may be nullified. Other difficulties surround both the inability to circulate air to conserve heat, and the variations in natural ventilation to account for seasonal changes.

Materials handling

Solids

Solid raw materials or products are received/dispatched in a variety of containers of different designs and capacity, e.g. paper or plastic sacks, metal or fibre-board drums, bulk tanker or lorry load. Depending on the nature of the material and its value it may be stored in piles, or in hoppers or silos, or in the containers as received.

The handling of 'toxic' solids is discussed on pages 323–331. With combustible solids the inherent properties will determine the size of each storage unit and the degree of segregation.

Materials prone to self-heating (see Ch. 4) either in bulk or when stacked in bags require special consideration. The actions to control this phenomenon are[56]:

Breakdown of the stock into numerous smaller units;
or Blanketing of the material, i.e. either preventing the ingress of air to oxidisable material or complete quenching with, e.g. water.

Regular temperature measurements serve to check on the adequacy of control. Special recommendations must be followed when storing significant quantities of sodium chlorate in drums.[57]

Liquids

Liquids are shipped in bottles, plastic cartons, steel or resin-lined steel drums, by road or rail tanker, by ship and on a large scale by pipeline. Bulk storage is in tanks generally in designated areas, i.e. tank farms, with good access. Storage at atmospheric pressure or under slight positive pressure is most common, but the design pressure is determined by the vapour pressure at storage temperatures.

Refrigerated containment of materials with a high vapour pressure at normal ambient temperature is common and avoids the requirement for highly rated pressure vessels. However, it is necessary to consider:

- The effects of failure of refrigeration.
- The potential for low-temperature embrittlement of the vessel walls.

A runaway reaction occurred in an underground 272,000-litre acrolein rundown tank when, due to a power failure, temperature increased and initiated an uncatalysed reaction (see Ch. 10). Rupture of the tank ejected parts of it and concrete fragments over the area.[58]

Table 12.12 Outdoor drum storage recommendations (after ref. 59)

If possible situate drum storage areas remote from any building or plant. Prohibit location beneath pipe bridges or cable runs.
Limit permissible fire load to 75×10^8 BThU (79×10^8 kJ) ≡ 250 tonne of hydrocarbon.
Label all drums. Affix the standard 'Highly Flammable' label if appropriate.
Provide suitable absorbent materials for soaking up spillages.
Limit stack height to 4.5 m, i.e. five high 200 litre drums or four high on pallets, for drums stored on end. Limit to four high if on their sides.
Segregate highly flammable liquids. Mark such stacks. Classify as Zone 2. Space > 15 m from any working building, amenity building or plant, and > 7.5 m from boundary plant, boundary fence.
Space combustible material stacks > 7.5 m from buildings and > 4 m from plant boundary fence.
Limit area for any stack of drums to 230 m^3 with, in general, a maximum length of 18.5 m (use a reduced area for hazardous materials, e.g. ether or carbon disulphide). Limit number of 180 litre drums to 1,500.
Demarkation lines should be marked on the ground which should be appropriately sloped and drained, and have an impervious surface.
Provide > 5 m clear between adjacent stacks (increase for hazardous materials). For the maximum stacking area ensure access for fire-fighting on three sides.
Provide an adequate number of fire hydrants. Consider the provision of fixed monitors for drenching the stacks with water. Stack drums > 2 m from any hydrants. Leave clear access, preferably straight through, on two sides of a hydrant and 4.5 m on fourth side.
Leave space for possible hose runs for a fire on any stack.
Place portable dry powder extinguishers around the area.

With the exception of LPG storage (see Ch. 4) it is normal for bund walls or dikes to be provided around bulk liquid storage vessels.

Whereas fixed tanks are, in general, carefully designed and located and provided with appropriate protection, bulk or intermediate storage of drums may tend to grow in an unplanned manner. Some recommendations for design are summarised in Table 12.12.[59]

Gases

Gases are marketed in either the gas phase (e.g. inert gases or compressed air in cylinders) or the liquid phase (e.g. inert gases or oxygen delivered by road tanker; chlorine or ammonia in cylinders). Liquefaction requires either refrigeration or high pressure. Bulk quantities or fuel gases (e.g. natural gas) are also distributed by pipeline and stored in gas holders at around atmospheric pressure.

Clearly containment is a more important consideration than with the majority of liquids since any leak results in atmospheric dispersion as described in Chapter 3; the potential consequences are summarised in Chapters 4, 5 and 15. Apart from measures to ensure integrity of the storage and distribution piping, as summarised below, gas detection devices of an appropriate design (Ch. 7) may need to be strategically placed and connected to alarm systems.

With LPG it is mainly the failure of storage and transfer systems that has resulted in the largest releases.[60]

Pressure systems

Factors which may singly, or in combination, contribute to failure of a pressurised system include inadequacies in, or lack of control over:

- Design
- Materials of construction
- Manufacture and workmanship
- Inspection, testing and commissioning
- Service conditions
- Maintenance and repair.

Pressure vessel design is covered by codes of construction (BS 5500, ASME VIII, etc.); if these are followed correctly few failures can be attributed to design deficiencies.

Typical ranges of 'safety factors' applied during design are summarised in Table 12.13.[61] From mechanical strength considerations (factors 2–4) this shows that an item of equipment may be 2 to 3 times the strength compared with that if no safety factors were applied. However, inherent safety may be improved by a reduction in vessel size and/or pressure.

Table 12.13 Safety factors in design

Basic design of plant unit	Factor range
1. On basic functional requirements (volume/heating area/residence time, etc.)	1.1–1.2
2. On strength of material (at yield) (Standards and Codes of Practice)	1.5–1.7
3. On corrosion potential	1.1–1.3
4. On design conditions (start-up, shut-down, abnormality)	1.1–1.3
Sub-product (1 × 2 × 3 × 4)	2.0–3.4

Hence stored pressure energy can be minimised by storage at atmospheric pressure, so that the only pressure remaining is that due to the head of liquid.[62] (Thus large quantities of ammonia, or LPG, are stored in fully refrigerated vessels; these are inherently safer – particularly when double containment is used.)

Failure may result from the use of materials which are unsuitable, incorrect or defective. Cases of defective or incorrect materials are now minimised by a high degree of quality control.[63] Adequate corrosion allowances allow for the phenomena summarised in Chapter 3 throughout the life of the equipment (or between replacement of internals, e.g. tubes in heat exchangers for highly corrosive fluids). However, corrosion data from the literature need to be applied with caution since trace contaminants, or erosion, can accelerate the rate of metal loss. An increase in inherent safety may be achieved by the use of higher, more expensive, grades of construction materials. Materials selection may require careful consideration apart from 'corrosion' effects, e.g. copper or any alloy containing >70% copper must not be used with acetylene due to the risk of formation of copper-acetylide (see Ch. 10, page 499).

Manufacture and workmanship cannot generally be perfect, and all welded vessels will contain defects.[62]. Most are benign but some may serve as initiation points for subsequent failure. Inherent safety is improved by thorough examination using non-destructive testing (NDT), and the repair of significant defects. Design codes specify a range of construction categories where the extent of NDT depends on type and thickness of materials, design stress limits, etc., e.g. under BS 5500:

Category 1 100% NDT
Category 2 Spot NDT
Category 3 Visual inspection only

to provide adequate integrity for normal purposes. Additional requirements may apply to special risks, e.g. chlorine vessels come into Category 1.[64]

Table 12.14 Operating deviations

Where appropriate, both static and dynamic effects require consideration.

Major operating limits
Flow
Temperature
Pressure
Level
Chemical reactivity
Mechanical stresses

Other operating limits
Corrosion
Erosion
Resistance
Fouling
Cavitation
Vibration
Hammer
Loadings
Expansion
Contraction
Thermal/mechanical shock
Cycles of activities
Environmental factors

Material physical characteristics
Viscosity
Miscibility
Melting/boiling points
Density
Vapour density
Phase
Appearance
Particle size

Chemical compositions
Hazard characteristics of mixtures

Reactions
Extent and type
Side reactions
Catalyst behaviour, e.g.
  activity
  toxicity
  reaction
  regeneration
  decomposition
Planned/unplanned reactions
Contaminants including materials of construction
Corrosion products
Catalyst
Reactor runaway
Explosion
Combustion

Time
Contact time
Sequence
Design cycle

Local effects
Distribution
Mixing
Hot spots
Overheating
Resonance
Stress on bearings/shafts
Lubricating faults
Vortex generation
Blockage
Slugs
Sedimentation
Stagnation
Adhesion
Crushing
Grinding
Separation
Siphoning
etc.

Failure to contain materials
Spillage
Leakage
Vented material

Construction
Defective materials of construction
Plant incomplete
Plant unsupported
Plant not aligned
Plant not tight
Plant not level
Plant not clean

Major deviations
Start-up/shut-down
Maintenance/inspection
Planned changes in normal operations
Supply/equipment failure
Demand change
Unplanned ignition source
Control disturbance
Loss of communication
Human error
Climatic effects

A well-planned testing procedure sets out to identify and correct potential causes of plant failure prior to production. Hence some failures do occur in the testing and commissioning phase (see page 749).

Service failure may occur due to abnormal conditions of operation compared with design intentions. An improvement in inherent safety can be achieved by designing to allow for abnormal service conditions; where an event is foreseeable it should be taken into account at the design stage.[62] However, in a chemical plant there is wide range of possible deviations from normal operating conditions as exemplified in Table 12.14.[6]

Two men were scalded by live steam, with fatal results, in a hospital laundry. A failure by one of the untrained, unsupervised laundry staff who operated the steam plant to drain the system before allowing steam to enter resulted in water hammer which caused a valve to fracture.[65]

Both static and dynamic effects need to be considered, e.g. temperature and temperature course. For example it may be necessary to take into account the effect of low temperatures on the mechanical strength of normal carbon steel, i.e. low-temperature embrittlement.

The overhead gas from a methaniser column at about −150 °C was discharged to a carbon steel flare line after passage through a heat exchanger; the latter utilised hot propylene vapour to raise the gas temperature to a value acceptable for carbon steel. When the vaporiser failed to operate, the outlet line to the flare ruptured resulting in a fire and explosion.[66]

An example of how accelerated creep due to quite moderate over-temperature excursions can shorten the life of a heater tube is given in Table 12.15.[67]

While furnace design methods enable prediction of the temperature of various parts of heater tubes for a given firing pattern deviations from the calculated tube wall temperature profile of up to 30 °C are not uncommon due to maldistribution of process fluid to the tubes, internal

Table 12.15 Accelerated creep of a heater tube designed to rupture after 10^5 hours

Equivalent age at 500 °C	10,000 hr	20,000 hr	50,000 hr
Duration of overheating	Actual tube wall temperature		
10 hours	610 °C	621 °C	635 °C
100 hours	570 °C	580 °C	595 °C
1,000 hours	534 °C	544 °C	558 °C
10,000 hours	—	510 °C	526 °C
30,000 hours	—	—	506 °C

fouling, differences in local inside heat transfer coefficients, and non-uniform firing of burners.[67]

Piping

The probability of a major leakage of chemicals from a plant is approximately proportional to:

- The length of pipelines.
- The number of drains, bleeds and vents.
- The piping complexity, generally reflected in the number of pump-arounds and other recycle streams.

With regard to pipe lengths a balance is necessary between the added safety which may be provided by increased equipment and vessel spacing, and segregation of hazardous units; and the increased cost and increased inventory within, and probability of leakage from, the pipework.

Simple design rules to reduce the probability of leaks from pipework include (after ref. 68):

- Minimise the number of branches and deadlegs.
- Minimise the number of small drains. Pipe runs should be arranged to facilitate drainage at a few easily accessible, and easily observed, locations.
- Design small diameter branches to the same code, and test them to the same stringency, as the main pipe. Ensure that small branches are reinforced at the junction and adequately supported.
- Provide flexibility to allow for thermal expansion of pipework and vessels. Correctly installed bellows may be necessary on short pipe runs; such bellows should only be subjected to axial movement. An inner sleeve may be required to avoid the convolutions being filled with solid residues.
- Direct the discharges from automatic drains, e.g. lutes and traps, to locations visible to operators. (The system of work should allow for regular checks and reports on these drains.)
- Provide gaskets completely compatible with the internal fluid at the maximum possible operating temperature and capable of compression to seal against the maximum internal pressure.
- Minimise the number of flanges on vacuum lines (e.g. the condenser on a vacuum distillation column).
- Provide removable plugs on valved sample points.
- Provide adequate pipe supports. Check for reaction forces, e.g. from discharges from safety valves.
- Avoid pipe runs where they may be subject to mechanical damage, e.g. from vehicles or from equipment removal/maintenance operations; alternatively, provide shields or barriers.

- Provide adequate walkways, ladders, etc., so that pipework is not clambered on/over.
- Use high-tensile bolts on large flanges subject to high temperatures.

Flexible piping

The selection and use of flexible hoses, e.g. for tanker off-loading, require considerable care. Some considerations are (after ref. 69):

- Conformance of the hose and coupling to appropriate standards, e.g. BSS for the specific duty.
- Provision for rapid isolation in an emergency (e.g. for tanker off-loading an emergency isolation valve at one end of the hose and a non-return valve at the other; excess flow valves have been used instead of remote isolation valves).
- Use of bolted clips, not jubilee clips.
- Provision of adequate protection and support while in use and protection, e.g. from crushing or contamination, when not in use.

All removable flexible pipes operated above atmospheric pressure should be provided with vent valves so that they can be depressurised prior to uncoupling.

> An operator had to disconnect a flexible pipe used to blow a permanent line through with nitrogen. He had shut the nitrogen valve first, allowing process liquid to enter the flex, and there was no provision for venting it prior to disconnection. When it was disconnected a spray of process liquid was ejected with the nitrogen and he sustained corrosive burns.

Glass piping and equipment

Particular care is required with industrial glass piping, valves and equipment.[70] If there is any possibility that it can be pressurised then suitable relief protection is required in the form of lutes or rupture discs. Mains water pressure may burst glass cooling coils if a blockage occurs, or if a downstream valve is shut. Under-pressuring also requires guarding against. In addition, external protection is advisable against the potential for flying fragments.

While adequate for supporting pipework the rubber bushed supports normally provided should not be used for equipment; rigid support plates are needed to take the weight via horizontal joint flanges directly below the centre of gravity. Supports are required adjacent to stopcocks, e.g. one either side. Manufacturer's instructions should be followed precisely regarding joints and gasket materials. PTFE bellows should not be used to join pipes which are misaligned.

Finally, it is always advisable to provide a 'drip tray' of appropriate size, or piped to a drain, to contain any spillage, e.g. due to fracture.

Pressure relief

The provision of over-pressure, or under-pressure, protection is an essential feature of designing for containment. However, the need for normal relief systems (as distinct from fire relief valves) and the associated flare systems may sometimes be avoided by using stronger vessels.[71] For example with a series of vessels with a fall in pressure between them relief valves can be eliminated by designing all the vessels to withstand the full upstream pressure. Similarly if a distillation column is made sufficiently strong to contain the contents at the temperature of the heating medium a relief valve may be avoided.

Some foreseeable causes of over-pressure were described in Chapter 4, e.g.:

- Thermal expansion of liquids
- Temperature excursions
- Melting of solids
- Presence of inert gas with process vapours.

To these must be added:

- Out of balance flows
- Fire conditions
- Failure of heat removal facilities
- Positive displacement pumping.

Even tanks designed for, and intended to operate at, atmospheric pressure require vents to guard against accidental over-pressurisation.

Figure 12.9 illustrates an arrangement in which a tank was provided with an overflow pipe extending down to ground level. An instruction was given to fill the tank as full as possible and in so doing the operators ignored the high-level alarm.

When the tank was overfilled the contents flowed out of the overflow line into the bund faster than the liquid was pumped in. A partial vacuum was produced in the tank and it collapsed.[72] (Low-pressure storage tanks of this type should be provided with vents or pressure/vacuum relief valves on their roofs.)

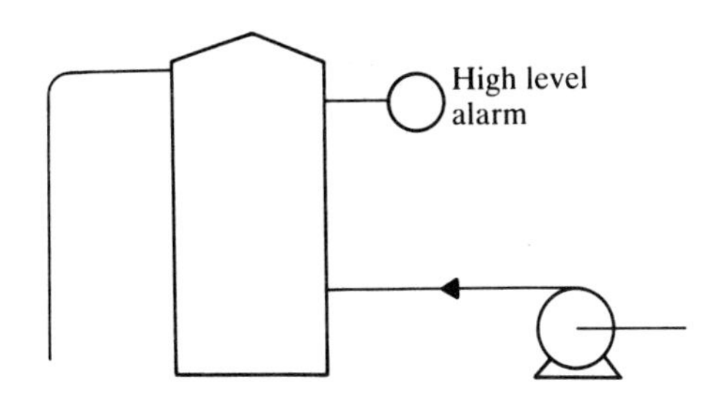

Fig. 12.9 Atmospheric storage tank with overflow to ground level.

Atmospheric storage tanks will normally only withstand pressures of several inches water gauge and therefore an adequately sized vent, not prone to blockage, is essential.

Apart from open vent lines, via seal pots or flame arresters where appropriate, pressure-relief valves or rupture discs are the common alternatives (see Figs 12.10 and 12.11). Trip valves actuated by a pressure controller may also be used. The relief device must be sized to discharge material, at the set pressure and operating temperature, at a rate equal to the net input otherwise the pressure will continue to rise.[8] Relieved gases should be vented so that they are safely dissipated away from buildings and equipment. In summary:

- Gases should be discharged in a safe, remote area.
- Vent lines should be securely anchored to withstand extreme discharge pressures.
- If remote discharge is impracticable vent lines should extend vertically well above buildings and equipment.

The use of flare stacks or flame arrestors for flammable gas/vapour venting is discussed later. Any toxic vapours may need to be vented via a suitable scrubbing system.

On the plant involved in the Seveso incident (page 510) the advice of the suppliers of the bursting disc was:

> '. . . bursting discs may be preferred in the case of fluids of great value and high toxicity where the loss of such fluids should be avoided. In this case a second receiver tank is required to recover the discharged fluid.'

No such tank was provided.[73]

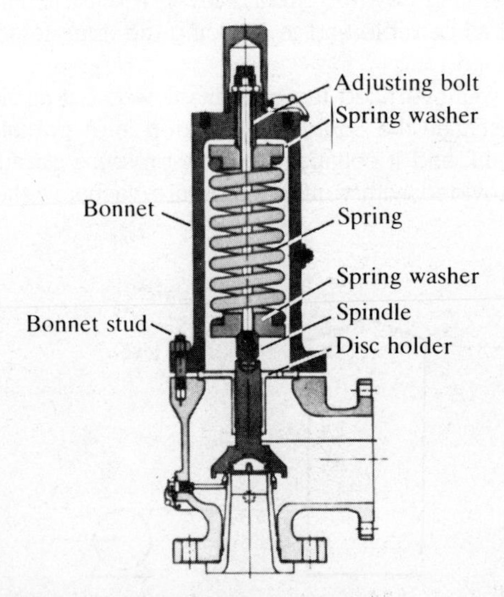

Fig. 12.10 Pressure-relief valve assembly
(Courtesy of Crosby Valve & Engineering Co. Ltd)

(a)

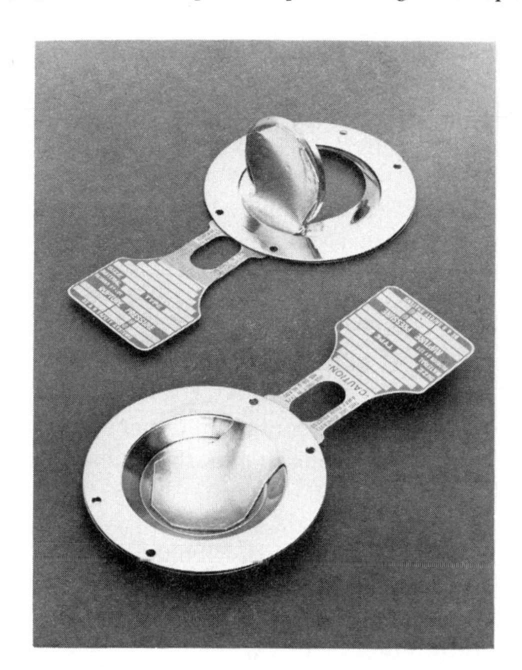

(b)

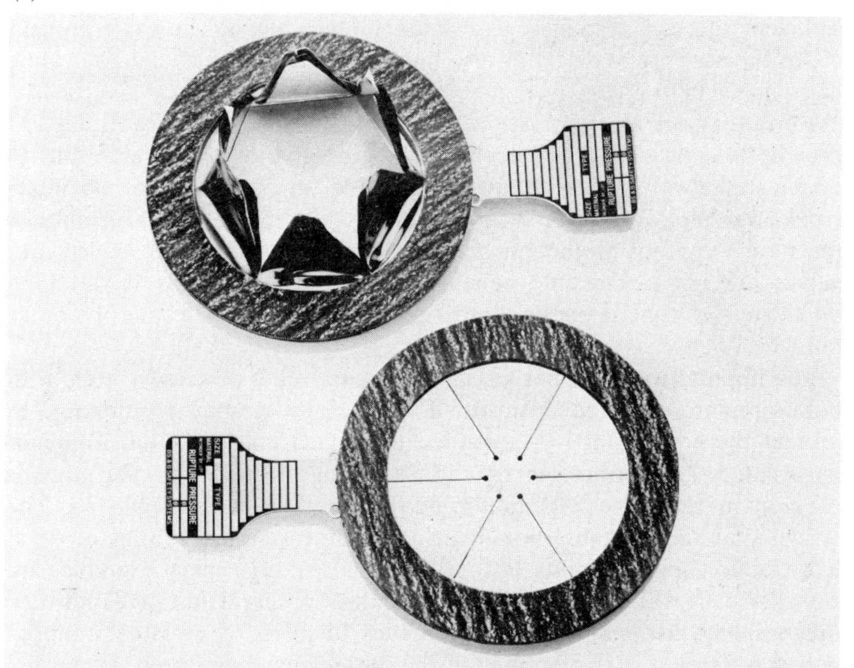

Fig. 12.11 Bursting (rupture) discs[82]: (a) RLS Reverse Buckling Disc for gas and full liquid service. (b) AV Tension Loaded Disc. No holder required, installed between flanges. (Courtesy of BS & B Safety Systems)

The arrangement for toxic liquids such as chlorine involves an absorption plant; on a small scale this may comprise a static tank of caustic soda liquor into which the chlorine gas is vented but on a large plant a fully monitored column is required.[74]

Rupture discs give a more 'positive' guarantee of relief of a reactor than relief valves but the quantity of reactant lost is greater, because once the disc has ruptured there is no control over the release.[75] Failure as designed tends to result in entrainment.

A reactor containing 70% non-volatiles in 30% methanol relieved at about 3.5 bar. More than one-third of the non-volatiles were lost by entrainment over about 10 minutes.

Clearly the venting system must be designed to cope with such effects.[76] Discs may also be prone to abnormal failures, generally partial, e.g. as pin-holing due to[77]:

- Corrosion.
- Atmospheric dirt ingress into the disc/holder niche due to disc flexing under repeated pressure reversal.
- Incorrect disc replacement – coding pins and tags serve to reduce this.[78]
- Discs supplied of too-thin material.

Discs will be preferred to relief valves when a valve does not give sufficient speed of response, or reactants depositing on the valve prevent it closing after use; a disc may also be used upstream of a valve to protect it from reactants.[78]

Rupture discs are generally fitted directly on to equipment, e.g. to provide 'second chance' protection if a trip valve or relief valve fails to open. Relief valves may either be installed on equipment or on interconnecting pipework. The arrangement must preclude any possibility of the item requiring protection being isolated from the relief device, i.e. valves are not permissible beneath a relief device unless linked to a duplicate – so that as one is shut to permit maintenance/testing the other must be open.

The importance of proper sizing of pressure relief devices, to cope with both pressure and the requisite flowrate, and proper positioning, to protect the equipment it is intended to protect and such that it cannot be isolated, is illustrated in Fig. 12.24. Sizing of relief lines for gases is covered in reference 79 and for two-phase flow in reference 76. The design of relief systems should include features which eliminate, or at least reduce, the possibility of their being defeated. Typical examples are shown in Fig. 12.12.[80], the pressure indicator alarm installed between the bursting disc and relief valve serves to warn of pressure build-up between them, due to pinholing of the disc, which might prevent the disc from rupturing at its set pressure. The proper selection and sizing of rupture discs is described in reference 81.

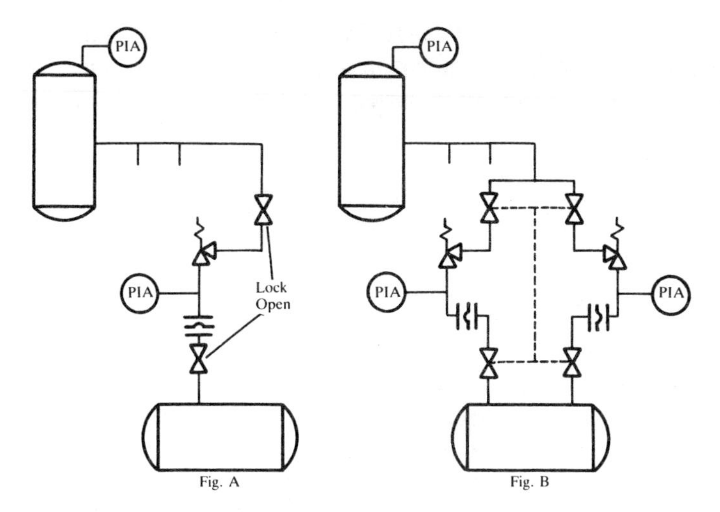

Fig. 12.12 Typical relief and manifold arrangements[80]

Relief valves have the advantage over rupture discs that following a pressure excursion they will reseat, albeit imperfectly; rupture discs require replacement. However, the latter may be preferable on systems prone to blockage, e.g. with polymerisable vapours; in some cases a rupture disc may indeed be installed beneath a relief valve with a pressure indicator between them. The proper inspection and testing of all such systems is of paramount importance.

As the example given in Chapter 3 (page 90) illustrates an open vent on a vessel or tank is, in effect, a relief valve. Therefore it should be treated as such, i.e. it should be checked regularly and not be altered without authorisation.

Flame arresters

Flame arresters are built into equipment or provided on vents or intakes, to prevent the unrestricted propagation of flame through flammable gas (or vapour)–air mixtures. They are unsuitable for flammable dust or other aerosol suspensions.

Flame arresters, or flame traps, comprise an assembly of narrow apertures through which gas can flow but which sub-divide a flame into flamelets; these are quenched before reaching the outlet. Arresters comprise wire gauze, compressed wire, crimped ribbon, perforated metal plates, parallel plates, sintered metal, metal foam, packed towers or hydraulic systems. In a hydraulic flame arrester the mechanism involves division of the gas flow into discrete bubbles. Applications include confining the passage of flame to one part of a plant or preventing flame from entering or leaving. However, while it will restrict an explosion to one part of a plant it does not primarily reduce the pressure; explosion venting is often used in combination for this purpose. Common uses of flame arresters include:

- In pipelines for fuel gas to burners.
- At the base of flare stacks (see page 722).
- On vents from storage tanks.
- In flammable gas pipelines.
- On exhausts from internal combustion engines for flammable areas (see page 155) and on diesel engine crankcases.

Design and location are critical features for safe operation.[83] Choice depends upon gas (ease of ignition, flame speed and corrosion properties). Effective flame arresters are not available for certain materials such as carbon disulphide. In general arresters are also unsuitable for volatile monomers which may polymerise on heating (e.g. in the event of a flame) and block the element. Conventional corrosion protective measures such as epoxy coatings are inappropriate because of the fine nature of the metal mesh coupled with element temperatures in excess of 1,250 °C in the event of a flame: titanium and zirconium have been utilised for speciality applications.

Methanol was stored in three tanks with a total volume of 700 m^3. These were served by a common vent line to a vapour recovery unit; a flame trap was located at about 25 m from this unit.[84] When flame-cutting was being performed near the end of the vent pipe while it was disconnected from the recovery unit (which would have been prevented by a proper permit-to-work system – see Ch. 17), a hot bolt ignited the vapour from it. The flame travelled up the vent line, through the trap and an explosion blew the tops off the storage tanks. The location of the trap allowed the flame to pass through since its velocity increased during travel along the pipeline.

Hence,

- Flame traps should be positioned near the ends of lines and ideally positioned vertically to take advantage of convection effects in the event of burning on the element and to avoid becoming a condensate trap.
- Maintenance is crucial since, with up to 50,000 apertures on a 200 mm dia. arrester, elements can easily become blocked. Condition can be assessed by monitoring pressure drop across the arrester element. Care must be taken during cleaning not to increase the hole sizes in gauze arresters. They can be cleaned by immersion in a bath of a suitable solvent.
- Common vent lines which can allow an explosion to spread to all tanks should be avoided.
- Arresters damaged by previous explosions must be replaced immediately.

Fig. 12.13 Flash-back arrester. For mounting on the outlet of regulators or manifolds of fuel gas (acetylene or propane); an alternative design is available for oxygen (Courtesy AGA Welding Ltd)

One type of flash-back arrester for use with the supply of fuel gas or oxygen is illustrated in Fig. 12.13. This fulfils four functions:

- It prevents reverse flow of gases, e.g. due to blocked nozzles or incorrect purging of hoses.
- It stops and extinguishes, a flash-back (the flame is quenched by a sintered filter and the pressure-wave is stopped by the non-return valve).
- It prevents further supply of gas following a flash-back.
- It prevents further supply of gas following a fire (by activation of a temperature-sensitive cut-off).

Piping and valves – general arrangement

Simplification of piping and valve arrangements, and ease of identification, are important factors in assisting safe operation.

Figure 12.14 illustrates an installation in which three waste-heat boilers shared a common steam drum. Each boiler had to be taken out of service at intervals for clean-up. Twice the wrong valves were closed and an on-line boiler was starved of water and overheated. The first time the damage was serious; a high-temperature alarm was subsequently fitted and prevented serious damage on the second occasion. Then a series of interlocks were installed so that the fuel to each unit had to be isolated before the valves on the steam drum. A better, albeit more expensive, solution followed on newer plants is to provide a single steam drum for each boiler.[71]

When an unstable liquid is being transferred the arrangement of piping, valves and control instrumentation should prevent the liquid becoming stationary in an operating pump.

Explosive decomposition of about 1 tonne of liquid ethylene oxide, at its boiling point and under a pressure of 4 or 5 bar, occurred in a reflux receiver. Decomposition was initiated in a rotary pump following this vessel; on the outlet of the pump was a control valve which was closed for a period by the operation of the level-control system on the vessel.

Fouling in the pump apparently caused overheating and decomposition of the liquid in it. Through a boiling ethylene oxide liquid/vapour mixture an explosive decomposition propagated back and established in the vessel.[85]

With both gases and liquids design must allow for the possibility of flow in the opposite direction to that intended. Case histories have been described of numerous possible backflow scenarios[86] including:

- **Backflow from, e.g. a storage tank or blowdown line, into a plant which has been shut down.**
- **Backflow from equipment into service lines when there is a pressure drop in the latter.**

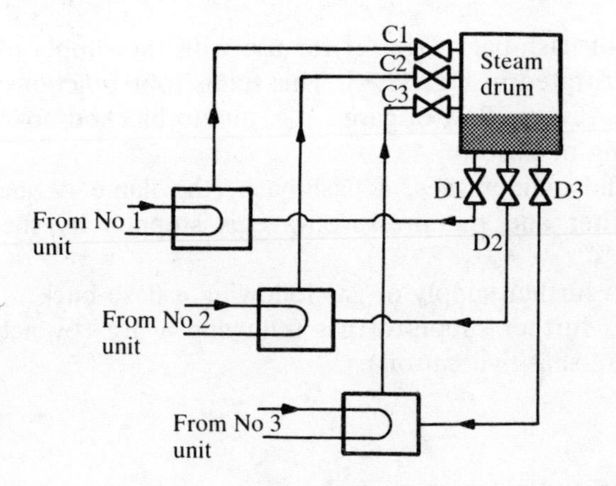

Fig. 12.14 Common steam drum serving three waste-heat boilers

- Backflow caused by failure of a pump.
- Backflow of a reactant up the feed line of a second reactant.

With regard to service lines it is recommended that any service used only intermittently should be connected to process equipment by flexible piping which is disconnected when not in use, or by means of a double block and bleed arrangement. Typical arrangements are shown in Fig. 6.4. For permanent service lines, if the service pressure may fall below the normal process pressure then a low-pressure alarm is desirable on the service line; if the process pressure may rise above normal service pressure then a high-pressure alarm is desirable on the process side. Non-return valves should be provided on service lines.

Non-return valves may be installed to prevent reverse flow in process lines but are seldom relied upon completely since they are prone to 'sticking' open. When failure of a non-return valve on a process line may have serious effects it is recommended that two should be installed, of different types if possible to minimise common mode failures. If the consequences of reverse flow could involve violent reactions or over-pressurisation of equipment then non-return valves provide inadequate reliability.[87] High reliability trip systems and shut-down systems are then required. Reliable shut-off requires double block and bleed arrangements of the type shown in Fig. 6.4.

Consideration should be given to the provision of remotely operated emergency isolation valves in the following cases[88]:

- When a leak may be very large, e.g. from a very large pump.
- When a very large amount of material could flow out of equipment without restriction, e.g. from the bottoms pump on a still containing more than 50 tonnes of flammable material.
- When the equipment is particularly likely to leak, e.g. very hot pumps.

They may also be appropriate where a flashing liquid is pumped.

Self-closing valves, or remotely operated isolation valves, are also recommended on all process drain lines from vessels containing LPG.[6] For ease of operation they may also be fitted on filling lines, on the feed line to the plant, and on large pipelines in preference to ordinary isolation valves.

Reactors

The characteristics of various reaction processes were discussed in Chapter 10.

With exothermic reactions it is very important to analyse the thermal characteristics of the type of reactor selected for a particular process and to identify potential instabilities during operation.[89] The designer must be aware of heats of reactions, rates of heat generation and removal, the effect of temperature changes and the effect of inadequate mixing. Sometimes conditions may change during operation and/or reactants accumulate in the system and it is only with a full understanding of these effects, reinforced by quantitative assessments of deviations from normality possibly using computer simulation, that a safe design can be achieved.[89]

The provisions for removal of heat from a reactor may include any combination of:

- An external jacket or an internal coil or array of tubes with either a flow of cooling media or vaporisation of a liquid.
- An external heat exchanger with recirculation of the reactants.
- Reactants' vaporisation cooling, e.g. with a reflux condenser.

The implications of increased reactor volume in scale-up are explained on page 94.

In the batch reactor system shown in Fig. 12.15, the procedure was for a batch of glycerol to be charged to the reactor and circulated through a heat exchanger which served as either a heater or cooler.[90] It was first used as a heater until the temperature reached 115 °C at which point the feed of ethylene oxide was commenced; since the reaction was exothermic the heat exchanger was switched to cooling. To start the ethylene oxide feed pump required:

Circulating pump operational.
Temperature >115 °C (otherwise no reaction).
Temperature <125 °C (otherwise reaction rate excessive).

On one occasion when the ethylene oxide feed was started the reactor pressure increased, indicative of the ethylene oxide not reacting. The operator deduced that the temperature point was possibly reading low and therefore adjusted the trip setting – to provide more heat to initiate the

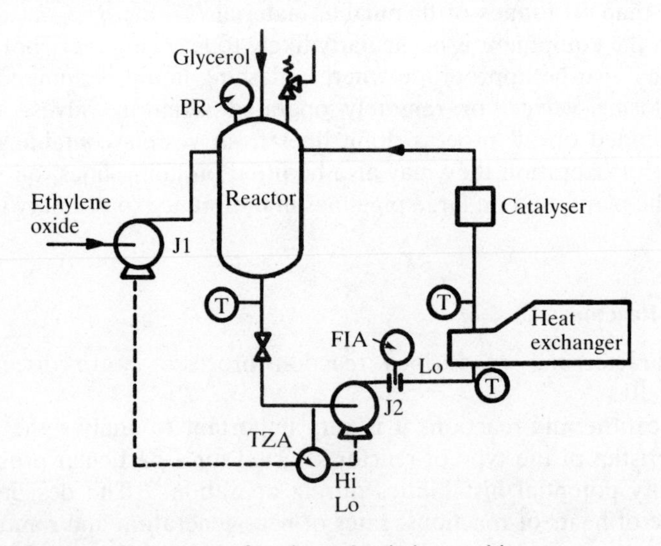

Fig. 12.15 Batch reactor system for glycerol–ethylene oxide

reaction. He allowed the indicated temperature to reach 200 °C but still the pressure did not fall. He suspected his theory was wrong and, on checking, found that the valve at the bottom of the reactor was still closed. He opened the valve and 3 tonnes of ethylene oxide together with the glycerol passed through the heater and catalyser. There was a violent runaway reaction which ruptured the reactor; the escaping gases exploded and two men were injured.[90] (The indicated rise in temperature was unreal. Pump J2 was running with a closed suction valve and the heat generated affected the nearby temperature point. The flow indicator and low-flow alarm were both out of order.)

In general, depending upon the nature of the reactants, the type of reactor, the reactor configuration, the reaction conditions, and whether the process is operated batch-wise or continuously, typical factors which must be considered to ensure safety are listed in Table 12.16.

Considering first the charging or in-flow of reactants, 8 and 10 on Table 12.16 are applicable to both batch and continuous processing and 11 and 12 more particularly to the former. In combustion processes 8 and 10 may rapidly create an explosive mixture in the reaction chamber and this points to a potential requirement for automatic detectors and cut-outs,

Table 12.16 Safety considerations in reactor design

Event

1. Agitator failure (mechanical or electrical).
2. Instrument failure (pressure, flow, temperature, level or some specific reaction parameter, e.g. concentration) or failure of instrument air/electricity.
3. Loss of inert gas blanket (refer to Ch. 4).
4. Failure of relief devices (e.g. pressure-relief valves or rupture discs) or restricted vent.
5. Leakage of materials out (e.g. due to a gasket failure) or air in.
6. Attainment of abnormal reaction conditions (e.g. over-pressure, over-temperature, segregation of reactants, excessive reaction rate, initiation of side reactions).
7. Failure of coolant, refrigerant, or other utilities.
8. Restriction of material flows in/out.
9. Failure of high- or low-pressure alarms/cut-outs.
10. Power failure.

Operator errors

11. Addition of wrong material or wrong quantities (more, or less, or zero).
12. Materials added in incorrect sequence.
13. Failure to add material (e.g. short-stop, or inhibitor) at correct stage.
14. Error in valve or switch operation.
15. Spillage of material.
16. Improper venting to atmosphere (i.e. not via vents with flame arresters, or scrubbers, or via a knock-out drum, or to the correct flare system).
17. Failure to actuate agitation.

standby pumps or compressors and adequate warning devices. The consequences of 11 and 12 may be no more than a ruined batch or conversely may be so serious as to indicate a potential requirement for valve interlocking. In some cases, for example 1 in nitrations, 3 in polymerisations, 6 and 7 in exothermic reactions, and possibly others may be so serious as to necessitate emergency shut-off of in-flows.

A reaction between *N*-substituted aniline and epichlorohydrin involved heating the reagents to 60 °C by the application of steam to the reactor jacket. The reaction then commenced and required cooling water to be passed through the jacket instead.

An operator allowed the temperature of one batch to exceed 70 °C before applying cooling. With cooling full-on the temperature continued to rise slowly. The building was therefore evacuated before, at 120 °C, there was an internal explosion in the reactor.[91]

If the feed system itself incorporates any heat exchange facilities, e.g. precoolers or preheaters, then the effect of event 7 on these, possibly leading to unusually high or low temperatures of the reactor may be important.

An unusual 'runaway' occurred in a resins manufacturing plant when because of sub-zero weather, steam was applied to a catalyst weigh-tank associated with a batch reactor. Excess temperature initiated the reaction in the weigh-tank and, since it had no cooling provisions, the exotherm caused a boil-over of the tank's contents. The vapour cloud ignited and the explosion disabled the sprinkler protection. The reactor area and an adjoining warehouse were destroyed by fire.[58]

Analysis is dependent on data accumulated as to reaction rates, rates of heat evolution and pressure rise, etc. In order to be safe, the reaction should be in a stable area of pressure, temperature and composition and should not be accompanied by undesirable side reactions, possibly leading to fouling or dangerous by-products.

Each event in the list is then analysed, for example, the effect of event 1 must be considered with regard to hot spots, side reactions, decreased heat transfer and possible fouling of the reactor. The possibility of 2 or 4 may indicate a requirement for duplication; 13 or 15 may be so serious as to necessitate the provision of interlocks, 15 may affect the floor drainage provisions and the extent of bunds, fire protection, emergency ventilation and personnel protection. Any of the events may necessitate a procedure for emergency shut-down or in ultra-hazardous processes, lead to consideration of isolation or shielding of the reactors or location in the open. The effects of any event may ultimately be detectable in the transfer manifold from the reactor; thus predicted variations in temperature, pressure or concentration of this point may provide a means of control.

If venting can occur then consideration must be given to the provision of appropriate knock-out drums, venting systems and possibly flare stacks. Figure 12.16 illustrates a typical safety system for a continuous polymerisation reactor.

Numerous measures may be applied to minimise any hazard with exothermic reactors, whether batch or continuous. These include:

- Limiting the size of the reactor. This limits the inventory; further, as described in Chapter 3, the smaller the reactor, generally, the better the ratio of heat transfer area to volume.
- Operating with the reactants diluted, either as solutions, suspensions or emulsions.
- Controlling the rate of addition of one component according to the temperature of the reactants.
- Provision of efficient agitation to distribute reactants homogeneously throughout the reactor volume, to avoid hot spots, and to improve the reactor-side coefficient for heat transfer.
- Provision of a reflux condenser to remove latent heat of vaporisation.
- Limiting the temperature difference between coolants and reactants. One recommendation is that this should be maximum of 10 °C.

Clearly all instrumentation, e.g. for temperature, pressure, flow and level control needs to be actuated by properly located sensors and be highly reliable.

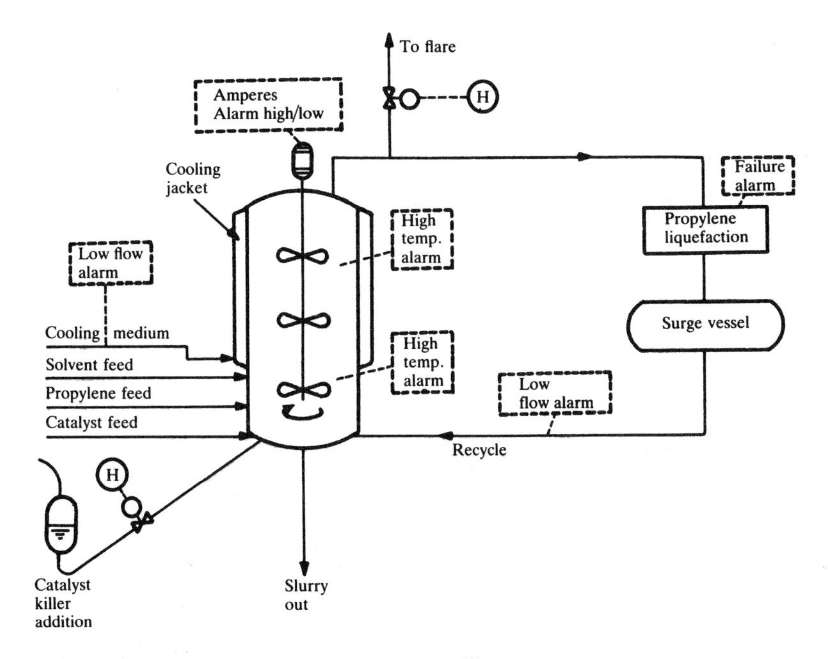

Fig. 12.16 Continuous polymerisation reactor[66]

The agitator on a batch-operated nitration reactor stopped. There was a failure of the instrument provided to alarm and shut off acid feed flow when the agitator ceased to rotate. An explosion occurred when the agitator restarted.[92]

Emergency measures

Some indication of the emergency provisions which may be required to cope with equipment failures/hazardous situations/potential runaway have been mentioned above. Depending on the reaction, reactor type and conditions they include a combination of:

- Vent to blowdown facilities.
- Emergency heat removal, e.g. by supplementary cooling (see also dilution and deluging).
- Dilution with compatible gas or liquid (e.g. water).
- Dumping of contents (e.g. into a flash drum, reservoir or quench tank).
- Shut off feed; increase off-take.
- Destroy catalyst (e.g. add inhibitor).
- Inert gas purging.
- Deluge reactants with compatible liquid (e.g. solvent or water).
- Vary agitation provision.

The selection of the correct combination of emergency measures and their speed of response are clearly important factors.

A batch chlorination reaction was controlled by automatic regulation of the flow of chlorine. One day the thermocouple failed. The coolant (brine) was shut off and the stirrer stopped while the instrument was repaired. Some delay in shutting off the flow of chlorine allowed a high temperature to be reached and the resulting decomposition reactions caused an explosion in the reactor; eight fatalities and extensive damage resulted.[93]

In a catalysed, exothermic, batch polymerisation assume that[15]:

Reaction rate $\propto$ (catalyst concn.)$^{0.5}$

and

Reaction rate doubles per 10 °C rise in temperature.

Also

Reaction temperature in one critical stage is 110 °C maintained by jacket cooling water at 60 °C. Emergency cooling water minimum temperature = 35 °C.

Then

Accidental double charge of catalyst would require cooling water at 40 °C.

If

Reaction temperature is increased to 112 °C to improve reactor usage then average jacket water temperature required $\simeq$ 56 °C.

But

Accidental double charge of catalyst would require cooling water at 32 °C which is not available. Hence there is a potentially uncontrollable scenario for operation at 112 °C.

Instrumentation and control

All operating parameters of relevance to control of the plant will normally be monitored. These will include in many cases:

Temperature
Pressure (and/or pressure differential)
Flow
Level

and depending on the process,

Viscosity
Composition (or proportion of a constituent)
pH.

Indicating or recording instruments only are provided on non-complex operations which can be efficiently controlled manually; as well as display at a specific control area or, on larger plants, a control room the measurements may be displayed locally. The training and response of the operator are then factors in determining the degree of control. The accuracy, and relevance of the information provided by the instruments are other factors. This will depend upon:

- Selection of correct type of instrument.
- Correct location of the instrument, e.g. such that a representative temperature is measured.
- Adequate cleaning and maintenance of the instrument, e.g. such that an orifice plate is not blocked/eroded, or a temperature probe covered in scale or residue, and proper calibration.

Duplication of instruments may be advisable. How, and where, the instrument readings are displayed has an important bearing on operator response. In any event if lack of control can create a significant hazard then protective instrumentation and/or alarms should be provided.

Fully automatic control is increasingly favoured, with process control instruments designed to control process variables within set limits during normal start-up/shut-down and operation. Again protective instrumentation is required as back-up. The de-manning associated with increased automation also has to be taken into account when planning for minor, and particularly major, emergencies. A distinction should be made between control instrumentation, which is by definition intended to control a process variable (e.g. flow, level, temperature), and safety instrumentation which is provided as a safeguard. For example see reference 17.

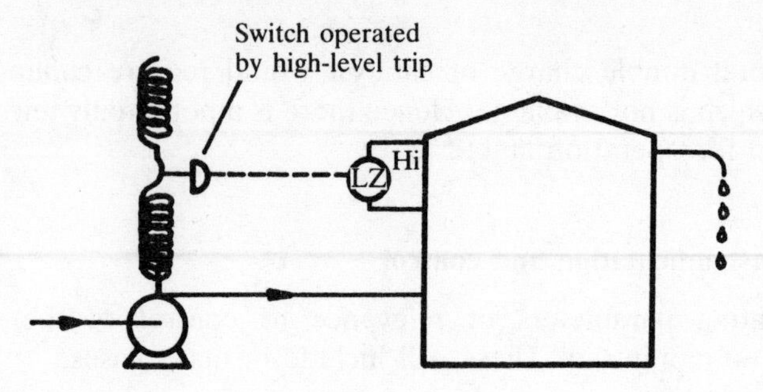

Fig. 12.17 Tank provided with high-level trip

> Figure 12.17 illustrates a tank which was filled once each day. Originally the level was controlled manually but on one occasion, apparently after five years' satisfactory operation, the tank was overfilled. Therefore a high-level trip was installed, as shown in the figure, to guard against operator error.
>
> After a further year the tank was again overfilled. The operator had got into the habit of leaving the trip to control the level in the tank while he did other work. The spillage occurred when the trip failed – as it was likely to about once every two years.

Trip systems

Trip systems are used to open and close valves, to switch off the electric power to drive motors (e.g. on pumps, compressors, or agitators, etc.). A trip system comprises[67]:

- An initiator, which monitors the relevant process variable and provides a signal when a set point is past.
- A link composed of relays which transfer the signal to the activator.
- An activator which receives the signal and performs the required actions.

A control loop may also function as a trip, as illustrated for a pressure trip in Fig. 12.18(a); the system shown in Fig. 12.18(b) is more reliable since it allows redundancy, i.e. a separate pressure detector and vent/shut-off valve are provided. Just how unreliable the original systems were can be shown by a simple hazard analysis.[7] Consider the most likely causes of over-pressure in the vessel:

1. Failure of PT or choking of impulse line.
2. Motor valve stuck open.
3. Failure of PIC.

In the event of (1) the trip cannot be actuated, in the case of (2) the motor valve will not respond to a signal from the trip; only in the case of (3) will the trip work. It is concluded that the trip will operate on less than one-third of the occasions demanded.

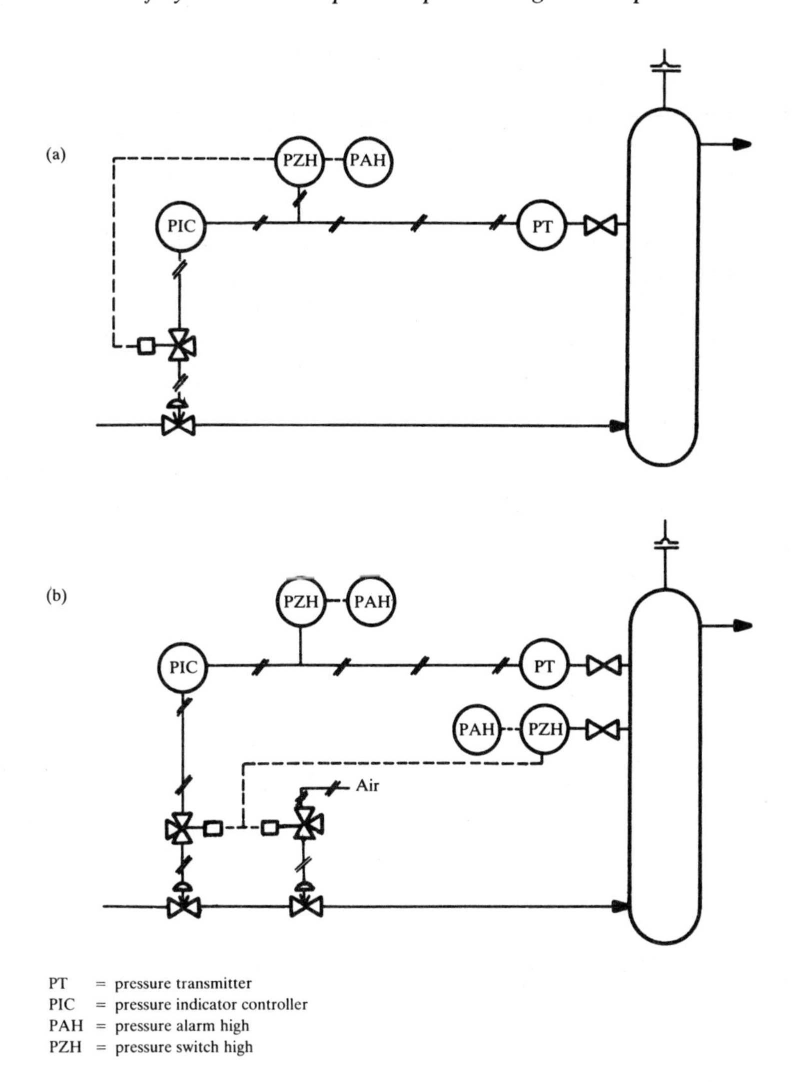

Fig. 12.18 Pressure trip system[67]: (a) Simple system; (b) With redundancy system made more reliable

The activator commonly comprises a solenoid valve which vents a pneumatic control line to a control valve, which fails either open or shut on loss of instrument air depending on the specific application.

Each trip in a plant should be tested at regular intervals while the plant remains in operation. A defect key in the link is generally provided to allow the activator to be tested on-line. In the past there have been incidents because trips were inadequate for the job, or were not tested, or were disarmed without authority. The recommended practice for trip systems includes:

- Regular testing of trips. (Any test which is missed to be carried forward and performed when possible.)
- Clear display of trip settings. (Any alterations to settings to require written authorisation.)
- Disarming of trips should require written authorisation. A clear signal, e.g. a flashing light, should indicate the disarmed state.
- Disarming of trips should not be allowed if any operator's failure to act could have a disastrous effect.

On a particularly hazardous plant where a shut-down is expensive, more than one detecting element may be used for each variable; these are connected to a majority vote relay. This is then a high integrity protective system giving reliability with a significant reduction in the spurious trip rate. (This is particularly relevant to low-manned, continuous operations.)

Electrical equipment

As discussed in Chapter 4, electrical equipment can act as a source of ignition for flammable atmospheres. Therefore assessment is necessary for all electrical equipment for the type of hazard it presents and the degree of risk in various parts of the plant. The following areas are defined[94]:

Zone 0 An explosive gas–air mixture is continuously present or present for long periods.

Zone 1 An explosive gas–air mixture is likely to occur in normal operation.

Zone 2 An explosive gas–air mixture is unlikely to occur in normal operation. If it occurs it will exist only for a short period.

A guide devised by ICI, but not part of BS CP 1003,[95] is:

Division 0 Flammable gas present for more than 1,000 hr/yr.

Division 1 Flammable gas present for more than 10 and less than 1,000 hr/yr.

Division 2 Flammable gas present for less than 10 hr/yr.

Safe area Flammable gas present for less than 1 h in 100 years.

Examples of Division 0 areas would include:

- The ullage space within a fixed roof tank containing a flammable liquid.
- The space above the roof in a floating roof tank containing a flammable liquid.

- A pit below ground level containing an effluent bearing a flammable liquid.

Examples of Division 1 areas include:
- A road tanker loading bay where hoses are frequently disconnected allowing vapour to escape.
- The area surrounding an American Petroleum Institute separator.

Examples of Division 2 areas may arise under abnormal conditions such as a mechanical failure of a gland or pipe joint, or occasional product sampling, e.g.:
- A pump bay where pump glands are well maintained and a leak will occur only infrequently.
- Cooling tower basins, where cooling water may be contaminated by a heat exchanger failure.

Guidance on the selection of methods of safeguarding electrical apparatus in hazardous areas is summarised in Table 12.17.[95]

Flameproof apparatus is achieved by fitting the electrical equipment into an appropriate flameproof enclosure, designed so that if a flammable vapour enters and is ignited by a spark, the flame and products of combustion are contained by the enclosure and do not ignite an external flammable atmosphere. The enclosure must also have sufficient strength to contain the pressure of the internal explosion.

The USA equivalent of flameproof equipment is designated explosion-proof and is similar in design principles. British Standard Spec. 229

Table 12.17 Safeguards for electrical apparatus in hazardous areas[95]

Zone	Gas and vapour risks	Dust risks
Zone 0	Intrinsically safe system category ia (subject to provisions)	Intrinsically safe system category ia (subject to provisions)
Zone 1	Flameproof enclosure	Intrinsically safe system category ia or ib
	Intrinsically safe system category ia or ib	Pressurising or purging
	Approved apparatus and type of protection 's'	Certain types of enclosure
	Pressurising or purging	
	Type of protection 'e' (subject to provisions)	
Zone 2	Any method suitable for Zone 1	Intrinsically safe system category ia or ib
	Type of protection 'N'	Pressurising or purging
	'Division 2 approved'	Certain types of enclosure
	Type of protection 'c'	
	Non-sparking apparatus	
	Totally enclosed apparatus	

defines the different requirements for flameproof enclosures, for the four groups of gases which range from methane to hydrogen.

Division 2 apparatus either produces no sparks in normal operation, or the sparks may be contained in a sealed enclosure. A mercury switch, in which the sparks between the mercury and solid metal contacts are contained by a glass tube and cannot ignite a flammable vapour in the atmosphere, meets this requirement. The requirement is that motors, lighting and other fittings shall be certified for use. If a fault occurs on such apparatus, sparks may be produced, but this is an abnormal condition. The presence of a flammable vapour is also abnormal in Division 2 areas, and the two abnormalities are unlikely to coincide.

With pressurised or purged apparatus the enclosure containing the electrical equipment is supplied with air free from flammable vapours or with inert gas. If a slight positive pressure is maintained in the enclosure, the ingress of a surrounding flammable atmosphere is prevented. Similarly, if a suitable outlet is provided on the enclosure and a sufficient volume of air is introduced, flammable vapours will be swept away. Provision must be made for automatically switching off all electrical supplies to the apparatus if the air or inert gas fails. This type of apparatus may be used in any area.

Intrinsically safe apparatus is designed such that the voltage and current in the electrical circuits are limited, and if a spark is produced in normal working or by failure, the heat energy of the spark is insufficient to ignite a specified flammable vapour. Intrinsically safe apparatus is suitable only for instrumentation, signalling and control circuits. (Low voltage alone does not make equipment safe.) Such apparatus is classified by British Standard Spec. 1259 for use in various gases and vapours. This type of apparatus may be used in Division 0, 1 or 2 areas, in vapours or gases for which it has been certified as being safe.

Certain items mainly portable hand lamps, instruments and signalling systems may be 'approved apparatus' for use in dangerous areas – because they are designed with a high degree of safety. For example, battery operated handlamps have bulbs protected by toughened glass and the switch is suitably enclosed to avoid danger from sparking. Approved apparatus may be used in any dangerous areas, but only in gases or vapours for which it has been approved.

The installation of electrical equipment, particularly cabling, in hazardous areas, the inspection and maintenance of equipment, and electronic instrumentation are covered by several Standards.[96,97]

Consideration must be given to the provision of emergency power for lighting and for some essential services, e.g. instrument air, control instrumentation, agitators and pumps, with automatic changeover from normal to emergency. As with other safeguards, the reliability of standby power supplies (e.g. from a diesel generator) depends upon regular testing and maintenance.

Fencing/covers/guards

Any fixed vessels, sumps or pits (particularly if they contain scalding, corrosive or poisonous liquids)[98] should be securely fenced or covered unless their edge is at least 3 ft (0.91 m) above the ground or adjacent platform.

All moving machinery, i.e. belt or pulley drives, gears, shafts, screw-feeders must be securely guarded. This is a statutory requirement in the UK and is described in more detail in Chapter 2.

Explosion protection

It is important to distinguish between physical over-pressurisation, which is a process occurring over a significant time period, and explosions in which the peak pressure is reached in milliseconds. The measures to avoid, or limit the effects of, explosions are therefore quite different. They include,

- Containment
- Suppression
- Venting (equipment and/or building)
- Automatic isolation
- Partial containment by blast walls or shields

Containment

Because of the large over-pressures generated by explosions it is not usually feasible to design equipment for containment; the exception is small laboratory-scale apparatus.

Specific areas of buildings, e.g. autoclave rooms, may be built to contain explosions. Reinforced concrete walls exceeding 30 cm thick and a roof of similar thickness are necessary for containment of explosive blast impulse and to resist missile penetration.[99] Blast walls or banks may also be used to restrict blast in specific directions; these also contain missiles depending upon trajectory. Heavy netting has also been used to contain missiles.

Explosion relief – buildings

As already mentioned, processes with an inherent explosion hazard should whenever practicable be located in the open air or in open-sided structures.

If the hazard exists in an enclosed building, e.g. with pilot plants or factories filling LPG cylinders or aerosol cans, explosion relief may be desirable. The relevant areas are generally segregated by walls to limit the effects of a vented explosion and the vents themselves have to be sited with regard for the safety of personnel, equipment, etc., outside.

One possible arrangement for a reactor used for an identifiably hazardous reaction is shown in Fig. 12.19.[100] This is located in a building constructed of reinforced concrete with explosion-relief panels and is arranged for remote operation.

Vents are positioned as near as practicable to predictable points of origin of explosions and require careful sizing.[101] The types of vent include:

- Blow-out panels
- Hinged panels
- Unrestricted openings/louvres
- Large, outward-opening, doors with a weak latching device.

It is important that these are installed (and subsequently maintained) in accordance with the design intention/manufacturer's advice since tampering/modification may seriously affect the relief pressure. For example equipment or materials must not subsequently be placed in front of the vents. The effects of venting and the avoidance of missiles must be considered, e.g. access may be restricted and panels restrained.

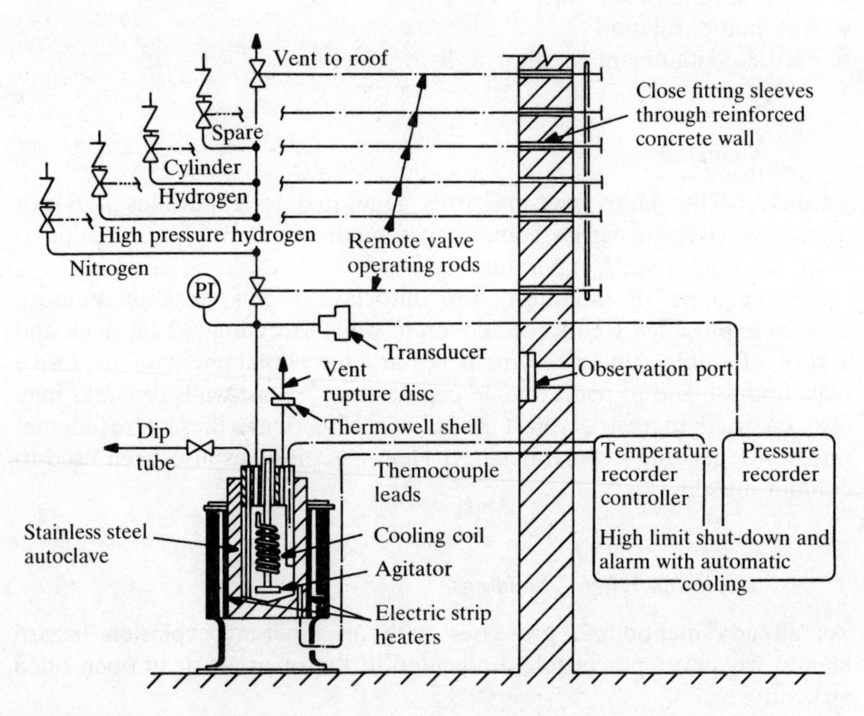

Fig. 12.19 Hazardous reactor layout for remote operation[100]

Explosion relief – equipment

In the UK where there is a danger of a dust explosion in a plant then, unless it is constructed to withstand the over-pressure, there is a statutory duty to take all practicable steps to restrict the spread and effects of such an explosion by means of chokes, baffles, vents or other equally effective device.[102] Blow-off panels or covers should be restrained, e.g. by chains, to prevent them becoming missiles and discharge should be to a safe area in the open air.[103] Vulnerable equipment, e.g. silos or spray driers, are generally located out of doors. Current best practice in the design of vents for dust explosions, including vent closure design and considerations for specific plant items, is reviewed in reference.[104]

Explosion suppression

Explosion suppression systems permit a flammable atmosphere to exist in equipment but if ignition occurs it is detected almost instantaneously and a suppressing agent is quickly injected – during the period of flame propogation when the rate of pressure rise is relatively slow. Hence a hazardous over-pressure is avoided.

This technique is equally applicable to gas, vapour, mist or dust explosion prevention. The most common agent used is chlorobromomethane, but other Halons may be selected for specific uses; powder suppression agents, e.g. ammonium hydrogen phosphate, are also available. Advanced inerting may also be used, i.e. the injection of agents such as nitrogen or carbon dioxide into connecting ducts to prevent flame propagation between different units. A combination of venting and advance inerting is shown in Fig. 12.20.[103]

The design, installation and maintenance of such systems is highly specialised. A strict inspection and servicing routine is essential to ensure

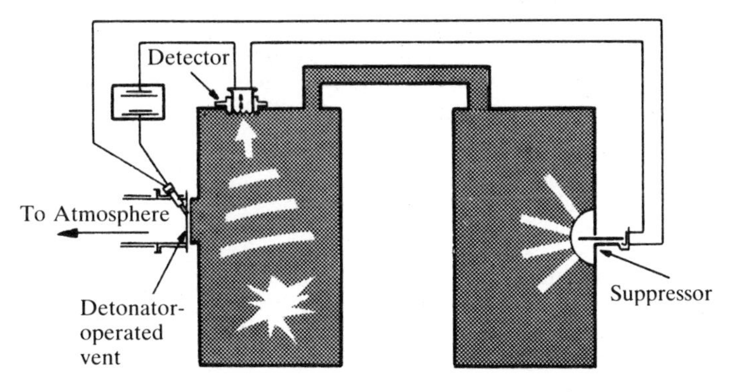

Fig. 12.20 Combined explosion venting and advance inerting

a high level of system reliability, i.e. that the explosion detection system (a pressure/pressure rise or flame reduction detector), control system and suppressant discharge facility all function when called upon. However, potential advantages over other methods may include[70]:

- Use in a congested space or where personnel are likely to be present.
- No significant release of hot combustion products resulting in potential hazards to personnel or neighbouring plant.
- Usually programmed to cause automatic equipment shut-down; hence evidence of origin of explosion and possibly malfunction leading to ignition may be preserved.

Automatic isolation

Passage of an explosion from one section of plant to another may, in a limited number of applications, be avoided by high-speed isolation.

Detection of an incipient explosion triggers a high-speed shut-off valve; such shut-off may be explosive assisted.

Blast walls and shields

Partial containment of plant with a potential explosion hazard or handling reactants at high pressure, e.g. >35 bar, is particularly applicable to pilot-scale operations. The objective is to provide protection against blast and missiles.

Blast walls or shields may vary depending on the scale of the process; designs have included 0.6 cm thick safety glass, 0.6 to 10 cm thick steel plate or 60 cm of reinforced concrete.[105] Two separate plates are more effective than one plate of equivalent thickness and in-fill with sand, etc., can provide added adsorption capacity for missiles.

The design can be extended to form a cubicle in which, for example, the fourth side or roof may be open or arranged to blow out. Any blow-out panels should vent at low over-pressures, e.g. 0.01 bar, and there should be a clear space of 15 m opposite them to minimise damage to nearby buildings by flying debris.[86] Access to this area must be restricted during equipment operation. An earthed bank or hanging steel nets may serve to collect missiles.

Transfer of materials

Container filling

Incidents have been described in which materials have been ejected from drums, or drums have exploded, due to admixture of the intended contents with contaminants either retained in the drums or entering accidentally. The measures to eliminate such incidents will normally include:

- Education of the workforce as to likely chemical reactions with the materials being drummed up or disposed of, e.g. as listed in Chapter 9, page 495–497.
- Thorough internal inspection of drums prior to filling. It may be advisable to keep all drums tightly bunged prior to filling and for the inspection to be just prior to the filling point.
- Avoidance of overhead pipelines at filling stations, e.g. cooling water, condensate or process lines.
- Appropriate shielding of the filling station, e.g. for operator protection and/or to eliminate ingress of contaminants including rainwater.

Solids conveying

Simple rules for the design of pneumatic conveyors include[9]:

- Use of inert gas as the pneumatic fluid if the solid transported is readily combustible. In this case a pressure system is preferable to a vacuum system to reduce the likelihood of air ingress
- Special consideration to fines recovery from exit gas if the solid is significantly toxic (see Ch. 6). Recycle of the carrier gas may obviate this problem.

With combustible powders precautions are necessary against static hazards, e.g. as shown in Fig. 12.21.[106]

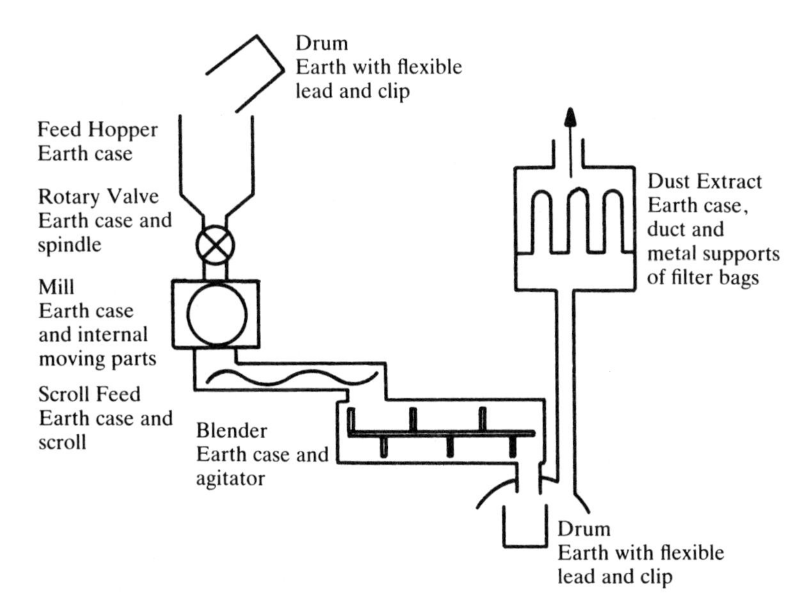

Fig. 12.21 Schematic diagram of typical earthing system in a powder handling plant[106]

Flare systems

Flare systems, of the type illustrated in Fig.12.22, are used for safe emergency disposal of excess flammable gases in circumstances involving[107]:

- Equipment failure
- Process plant upsets, mal-operation, start-up and shut-down
- Emergencies, e.g. plant fire, total power failure
- Depressuring of process systems
- Non-use of process gases as fuel gas.

The system must be provided with an effective liquid knock-out facility to eliminate liquid droplets being discharged. Design will then be based upon considerations of the calculated largest relief quantity, pressure-drop, control of entry of air, thermal radiation, etc.[107]

The flare header into which the various relief lines are connected will normally slope at >1 : 500 and be self-draining to a knock-out drum near the flare. This drum collects any liquid released into, or condensed in, the flare header; it should normally be provided with an automatic

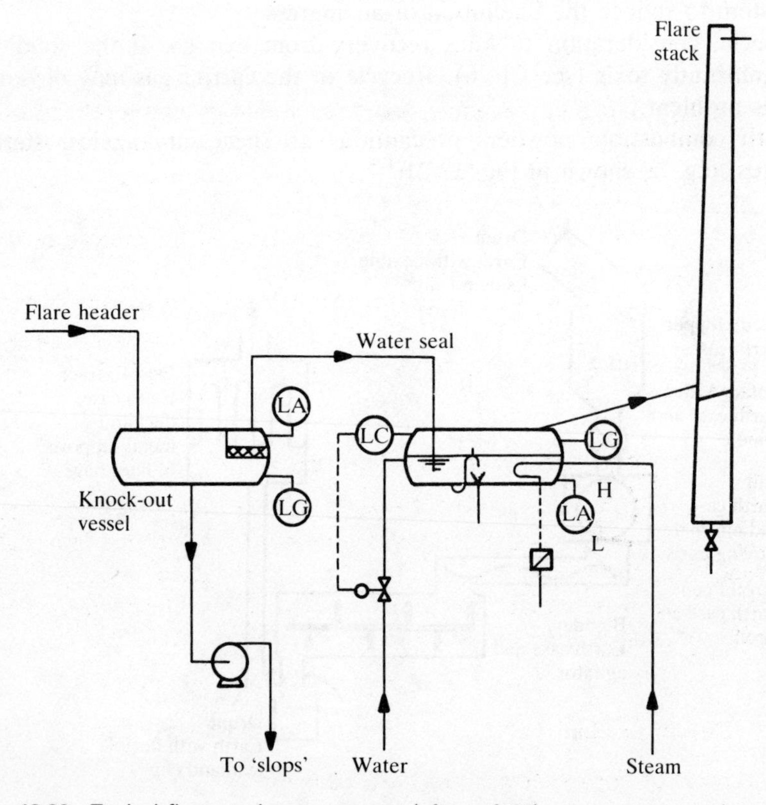

Fig. 12.22 Typical flare stack arrangement (after ref. 66).

pump out and high- and low-level alarms. Steam tracing may be provided on the flare header depending on the gases being disposed of (e.g. methane or hydrogen sulphide) depending on climatic conditions, and provision is made for purge gas injection.

The system is sealed from the flame on the flare by a water seal drum, designed to offer the minimum resistance to flow during emergency venting but ensuring a safe seal to avoid back-flow. The flare should be designed without flanges[108] connections, vents, etc., through which air could be drawn in. Other materials to be avoided include:

- *Low boiling materials* – rapid evaporation can cause heavier materials to solidify, or water to freeze, due to chilling.
- *High boiling materials* – may block the system due to congealing.
- *Excessive amounts of liquid* – may be ejected from the top of the stack.

A continuous flow of purge gas may be used to prevent air penetrating down from the top of the stack. The flare height is such that thermal radiation falling on nearby plant and the surrounding area does not produce a hazard to equipment or workers. Recommended values of radiant heat flux at ground level are:

At the base or within flare zone <6.3 kW/m^2

At boundary or fence line <3.2 kW/m^2

The height, together with the outlet velocity and local meteorological and topographical conditions, must also be such that ground level concentrations of gases do not create a nuisance or hazard within and outside the site. The pilot burners are fitted with flame-failure alarms.

Precautions in operation will generally include:

- Provision and maintenance of a reliable instrument for indicating the flow of inert gas used as a continuous purge.
- Regular testing of the atmosphere within the stack for oxygen.
- On shut-down, extinguishment of the ignition source and purging of the system with inert gas or steam prior to the admission of air.

Preventive and protective features

The basic preventive and protective measures to be found on a chemical plant, depending upon its type and scale and the nature of the materials processed, include those listed in Table 12.18.[25] A useful check-list for the assessment of potential fire hazards and to review the requirements for fire protection is reproduced as Table 12.19 (after ref. 35).

Fire protection

Some aspects of fire protection relevant to factory design and layout were discussed in Chapter 4. Selection of the appropriate fixed fire protection

Table 12.18 Check-list of basic preventive and protective features (after ref. 25)

Adequate water supply for fire protection. Determined by multiplying the length of time that the 'worst possible fire' can be expected to last by its water demand. The supply deemed adequate will vary with different authorities and may range from a 2 hr fire to one lasting 8 hr.
Structural design of vessels, piping, structural steel, etc.
Over-pressure relief devices.
Corrosion resistance and/or allowances.
Segregation of reactive materials in process lines and equipment.
Electrical equipment grounding.
Safe location of auxiliary electrical gear (transformers, breakers, etc.).
Normal protection against utility loss (alternate electrical feeder, spare instrument air compressor, etc.).
Compliance with various applicable codes (ASME, NEC, ASTM, ANSI, Building, Government, etc.).
Fail-safe instrumentation.
Access to area for emergency vehicles and exits for personal evacuation.
Drainage to handle probable spills safely, plus fire-fighting water from hose nozzles and sprinkler heads and/or chemicals.
Insulation of hot surfaces that heat to within 80% of the auto-ignition temperature of any flammable in the area.
Adherence to the National Electric Code, except where variances have been requested/approved.
Limitation of glass devices and expansion joints in flammable or hazardous service. Such devices are not permitted unless absolutely essential. (Where used, they must be registered and approved by the production manager and installed in accordance with in-company standards and specifications.)
Building and equipment layout. Separation of high-hazard area must be recognised especially as it relates to both property damage and interruption of business.
Separation of tanks, at least in accordance with US NFPA No. 30.
Protection of pipe racks and instrument cable trays and their supports from exposure to fire.
Provision of accessible 'battery-limit' block valves.
Cooling tower loss prevention and protection.
Protection of fired-equipment against accidental explosion and resultant fire.
Electrical classification. Division 2 electrical equipment will be required for outside flammable liquid handling where congestion is minimal and natural ventilation is unobstructed. Division 1 equipment is required only for special chemicals and/or special building or process handling conditions, or where ventilation is inadequate.
Process control rooms isolated by 1 hr fire walls from process control laboratories and/or electrical switch-gear and transformers.

depends upon an analysis of the potential fire characteristics. Factors to consider include (after ref.109):

- Type of combustibles, e.g. giving rise to surface fires for which an inerting or inhibiting agent may be suitable; or giving rise to deep-seated combustion requiring a cooling agent.
- Relative importance of speed of detection.
- Degree, if any, of life risk including significance of any toxic combustion or agent-decomposition products.

Table 12.19 Check-list for assessing fire hazards and fire protection requirements

Location

Accessibility from at least two sides.
Traffic – vehicular and pedestrian. Segregation of routes.
Parking areas – entrances, exits, drainage, lighting, enclosures.
Clearances – buildings for railroad traffic and vehicles (overhead width turnarounds).
Drainage, impounding areas and basins.
Road locations, markings, one-way systems.
Entrances exits – pedestrian, vehicular, railroad.
Location of furnaces, units for Dowtherm heat-transfer agent, flare stacks.

Building

Wind pressure, snow loads, floor loads, earthquake design.
Roof material, anchorage.
Roof vents and drains, smoke dispersal.
Stairwells, ramps, lighting.
Elevators and dumbwaiters.
Fire walls, openings, fire doors.
Explosion relief – size, location, venting location.
Exits – fire escapes, identification,
Records storage.
Ventilation – fans, blowers, air conditioning, scrubbing of toxic vapours, location of exhausts/inlets, smoke and heat ventilation dampers, fire curtains.
Lightning protection, structural and equipment grounding for electrical dischargers.
Building heaters (hazardous or non-hazardous area), vents.
Locker rooms including air changes.
Building drainage – inside and out.
Structural steel and equipment fireproofing to approved standards and elevations.
Access ladders to roofs from outside level, escape ladders, fire escapes.
Bearing capacity of subsoil.

Sprinklers, hydrants and mains

Water supply including secondary supplies, pumps, reservoirs and tanks.
Mains – adequate looping, cathodic protection, coated and wrapped when needed, sectional valves.
Hydrants – location.
Automatic sprinklers – occupancy classification, wet systems, dry systems, deluge systems.
Standpipes and tanks.
Type, size, location and number of fire extinguishers needed.
Fixed automatic extinguishing systems, CO_2, N_2, foam, dry powder.
Special fire protection systems – rise in temperature alarms, sprinkler system flow alarms, photoelectric smoke and flame alarms, snuffing steam.

Electrical

Hazard classification (see page 714).
Accessibility of critical circuit breakers.
Polarised outlets and grounded systems.
Switches and breakers for critical equipment and machinery.
Lighting – hazardous or non-hazardous areas, light intensity, approved equipment, emergency lights. Stand-by lighting.
Telephones – hazardous or non-hazardous areas.
Type of electrical distribution system – voltage, grounded or ungrounded, overhead, underground.
Conduit, raceways, enclosures, corrosion considerations.
Motor and circuit protection.
Transformer location and types.

Table 12.19 (continued)

Fail-safe control devices protection against automatic restarting.
Preferred busses for critical loads.
Key interlocks for safety and proper sequencing, duplicate feeders.
Accessibility of critical breakers and switch gear.
Exposure of cable trays to fire damage.

Sewers
Chemical sewers – trapped, accessible clearouts, vents, locations, disposal, explosion hazards, trap tanks, forced ventilation automatic flammable vapour detectors and alarms.
Sanitary sewers – treatment, disposal, traps, plugs, cleanouts, vents.
Storm sewers.
Waste treatment, possible hazards from stream contamination including fire hazard from spills into streams and lakes.
Drain trenches – open, buried, accessible cleanouts, presence of required baffles, exposure to process equipment.
Disposal of wastes, air- and water-pollution safeguards (see Ch. 14).

Storage

General
Accessibility – entrances and exits, sizes.
Sprinklers.
Aisle space.
Floor loading.
Racks.
Height of piles.
Roof venting.

Flammable liquids – gases, dusts and powders, fumes and mists
Closed systems.
'Safe' atmospheres throughout system.
Areas to be sprinklered or provided with water spray.
Emergency vents, flame arresters, relief valves, safe venting location including flares.
Floor drains to chemical sewers, properly trapped.
Ventilation – pressurised controls, etc., and/or equipment.
Tanks, bins, silos – underground, above ground, distances, fireproof supports, dikes and drainage, inert atmospheres.
Special extinguishing systems, explosion suppression – foam, dry chemicals, carbon dioxide.
Dependable refrigeration systems for critical chemicals.

Raw materials
Hazard classification of material including shock sensitivity.
Facilities for receiving and storing.
Identification and quality control tests.
Provisions to prevent materials being placed in wrong tanks, etc.

Finished products (see Ch. 13)
Identification and labelling to protect the customer and transporter.
Conformance with ICC and other shipping regulations.
Segregation of hazardous materials.
Protection from contamination, especially in the filling of tank cars and tank trucks.
Labelling of shipping vehicles.
Routing of hazardous shipments.
Data sheets for safety information for customers and transporters.
Safe storage facilities, piling height.
Safe shipping containers.

Table 12.19 (continued)

Inert gas blanketing of all hazardous products
Consider raw material, intermediates, and products.
Consider storage, material handling and processes.

Materials handling
Truck loading, and unloading, facilities.
Railroad loading and unloading facilities.
Industrial trucks and tractors – gasoline, diesel, liquefied petroleum gas.
Loading and unloading docks for rail, tank trucks and truck trailer – grounding system for flammable liquids.
Cranes – mobile, capacity marking, overload protection, limit switches.
Warehouse area – floor loading and arrangement, sprinklers, height of piles, ventilation.
Conveyors and their location in production areas.
Flammable liquid storage – paints, oils, solvents.
Reactive or explosive storage – quantities, distance separation, limited access.
Disposal of wastes – incinerators, air- and water-pollution safeguards (see Ch. 14).

Machinery
Accessibility, maintenance and operations
  Provision to prevent overheating, including friction heat.
  Possible damage to fire protection equipment from machine failures.
  Protection of pipelines from vehicles, including lift trucks.
Emergency stop switches.
Vibration monitoring.

Process
Chemicals – fire and toxicological hazards (skin and respiratory), instrumentation, operating rules, maintenance, compatibility of chemicals, stability, etc.
Critical pressures, temperatures, flowrates, etc.
Relief devices and flame arresters (sizing and location).
Coded vessels and suitable piping material.
Methods for handling runaway reactions, e.g. dumping, 'short-stopping'.
Fixed fire protection systems – CO_2, foam, deluge.
Vessels properly vented, safe location.
Permanent vacuum cleaning systems or approved portable cleaners.
Explosion barricades and isolation.
Inert gas blanketing systems – list of equipment to be blanketed.
Emergency shut-down valves and switches, location from critical area, action time for relays.
Fireproofing of metal supports to adequate heights.
Safety devices for heat-exchange equipment – vents, valves and drains.
Expansion joints, or expansion loops, for process steam lines.
Steam tracing – provision for relief of thermal expansion in heated lines.
Insulation for personal protection – hot process, steam lines and tracing.
Static grounding for vessels – piping, portable containers.
Cleaning and maintenance of vessels and tanks – adequate manholes, platforms, ladders, cleanout openings and safe entry permit procedures
Provisions for corrosion control and monitoring.
Pipeline identification. Numbers and/or colour coding.
Radiation hazards including personal protection for fire fighters – processes and measuring instruments containing radioisotopes, X-rays, etc.
Redundant instruments with alarms (avoiding possible common-mode failures).
Exposure of process and instrument lines to fire.

Safety equipment general
Medical centre and equipment.
Ambulance (back-up).

Table 12.19 (continued)

Fire engines.
Fire alarm system.
Fire alarms and siren – departments, inside and outside.
Sewage and process waste treatment.
Snow removal and ice control equipment.
Safety ladders and cages.
Emergency equipment locations – gas masks, protective clothing, fire blankets, fire hoses, stretchers, etc.
Laboratory safety shields.
Fire hydrants, location, hose and allied equipment.
Instruments – continuous analysers for flammable vapours and gases, toxic vapours, etc. (see Ch. 7).
Communications – emergency telephones, radio, public address systems, paging systems, safe location and continuous manning of communications centre. Alternative communications centre.
Combustion safeguards on furnaces, e.g. flame-failure devices.
Fuel gas shut-off valves.

- Indoors, for which either a local application or a total flooding system might be chosen, or, outdoors, for which wind effects, exposure of nearby hazards, involvement of other combustibles and difficulties of extinguishment need assessment.
- Probability of fire spread beyond the initial fire area and appropriate measures to contain it (e.g. fire walls, fire-stop doors).
- Requirement for detection and control, or detection and extinguishment, depending upon circumstances and the back-up available.
- Any special features which rule out certain extinguishing agents or make others preferable (e.g. with water-reactive chemicals).
- The possibility of more rapid, positive results with a dual agent system, e.g. water spray and CO_2, or a foam sprinkler system to blanket a liquid fire at floor level, plus dry powder applicators to cover potential running – liquid fires due to process leaks.
- Any explosion risk; this may be inherent, or arise from use of a specific agent, e.g. water spray on metal fires (see Ch. 3) or static discharge due to CO_2 discharge into a flammable vapour zone.
- Any other special features (e.g. operation with personnel present, requirement for rapid 'knock-down').

The characteristics of the various agents are summarised in Chapter 4.

System requirements

A system comprises an extinguishing agent plus any propellant gas needed, together with some combination of storage, pumping, metering, piping and discharge arrangements. It may be activated automatically by a detection system or manually.

In general the requirements are[109]:

- The detection system must be sufficiently sensitive to respond rapidly to any fire situation but should not be unduly subject to spurious operation and unnecessary release of the agent. These objectives are sometimes met by a dual detection system (or indeed in some locations a two-out-of-three voting system); automatic lock-offs may also be required to protect personnel.
- In operation, the system must have the capability to control and extinguish the anticipated fire condition without recourse to outside assistance (unless this is catered for in the planned procedures). The requisite application rates, discharge times and quantities may be assessed on planned spacing of discharge nozzles to cover all the areas at risk.
- The system must be reliable. Its construction should allow for environmental features likely to be detrimental to operation, e.g. dust, tar or other deposits, corrosive atmosphere.
- The potential toxicity of the agent or its breakdown products on the likely fuels affects both the choice of system and the requisite safety measures, e.g. in the case of total-flooding systems.
- The agent(s) used must be compatible with the process and with each other, and with any other installed system(s).
- Visibility in the fire zone following agent discharge is a consideration if manual fire-fighting is also envisaged.

Detailed recommendations have been published regarding fixed protection in factories generally[110] and in chemical plants in particular.[111]

The following features are applicable particularly to chemical process plants and as such are given credits in the Dow Index.[25]

- *Leak detection* – gas detectors may be strategically located to alarm and identify the specific area concerned. In some installations, e.g. on off-shore platforms, an alarm is actuated at 25% of the LEL and a protective system activated at 75% of the LEL.
- *Fire-proofing of structural steel* – fire-proofing, may be applied to all load-bearing steel; this should be to a minimum height of 5 m.
- *Cable protection* – electrical cable and instrument trays are susceptible to damage from fire exposure. Reasonable protection may be provided by 14 or 16 gauge metal sheet beneath the tray with water spray directed on to the top side. Alternatively the cables may be buried below ground.

Steam or water curtains, e.g. activated by strategically located flammable gas detectors, can reduce a flammable gas safely below its LEL in a few seconds.

A water curtain formed by horizontal water-sprays diluted a propane gas cloud below its lower explosive limit; the water pressure required was 300 to 350 kN/m^2 gauge when the nozzles were 1.5 m apart. At 3 m spacing the pressure was 380 to 450 kN/m^2 gauge. A single nozzle fed at 620 kN/m^2 with a 3.6 m diameter spray 3 m above the ground kept an ethylene leak of 0.28 m^3/s below its LEL. In a confined space 2 m × 2 m × 3 m containing a 50% mixture of ethylene in air and an inlet flow of 0.28 m^3/s of pure ethylene, the same spray reduced the concentration in the space to less than 2.7% ethylene in 8 seconds. With no ethylene leak, the time was 6 seconds.[112]

A fire occurred on an olefins plant when a bellows failure released about 57 m^3 of propylene as a large vapour cloud. The sprinkler system on the olefins purification unit, which had tripped prior to the failure, divided the cloud and formed a water curtain around the unit; this limited the damage to around its perimeter.[113]

The explosion of oil mists, containing 5 μm to 10 μm droplets, when oil at ambient temperature is released from high pressure containment, can be prevented by water fog. The water droplets of 80 μm to 90 μm settle about 100 times faster than the oil drops (see Ch. 3) and create a scrubbing action which can reduce the oil mist below the flammable limit in 10 to 15 seconds.[114]

Water sprays can also be effective in reducing the amount of soluble toxic vapours from leaks.

One recommendation relating to a jetty at which liquefied ammonia is unloaded was the installation of an efficient water spray system. If a ship's cargo tanks ruptured the water spray system could be activated quickly and dissolve a high proportion of ammonia vapour from the air thus reducing the amount reaching residential areas.[18]

Dumping/blowdown provision

Emergency venting of gas or vapour may be to a flare system, of the type described at page 722, or into a closed bent receiver. An emergency process dump tank may be provided to receive the complete contents of an upset process condition. This tank is best located outside the vent area. A system which may vent toxic materials in significant quantities should be designed with an appropriate collection/adsorption or scrubbing facility.

Specific recommendations, codes and design practices

Numerous design codes and practices have been recommended for specific plants or the handling of specific chemicals. These can be applied *with caution* by analogy to other chemicals or manufacturing operations.

Fig. 12.23 Water-spray protection of pressurised flammable liquid (Courtesy of Matthew Hall Mechanical & Electrical Engineers Ltd)

Plants handling liquefied hydrocarbons

Design recommendations have been made as listed in Table 12.20.[115]

Design and safety reviews

Various design and safety reviews are performed as design proceeds. The type of review and techniques used depend upon the company and contractor; the options include:

- Fire and Explosion Index developed by Dow[25] or the Mond Index[26]
- Failure analysis and fault trees

Table 12.20 Recommendations for liquefied hydrocarbon plants

Particular care should be taken in the selection of materials of construction of equipment which may at any time contain liquefied hydrocarbons at sub-zero temperatures.

The capacity of fire relief valves should be determined as described in APIRP 2000 and not as in APIRP 520.

Particular care should be taken with the arrangement of vent lines from relief valves on low-pressure refrigerated storage tanks to avoid causing excessive back-pressure.

Liquid-relief valves should be provided on pipelines or other equipment which may be endangered by thermal expansion of locked-in liquid.

Small branches on pressure vessels and major pipelines should be supported mechanically to prevent them from being broken off.

Particular care should be taken to protect equipment against fire exposure by a suitable combination of water cooling, depressuring and fireproof insulation. (Fig. 12.23).

Pumps should in general be fitted with mechanical seals instead of packed glands to reduce leakage.

Process draining and sampling facilities should be designed to withstand mechanical breakage, to minimise the risk of blockage by ice or hydrate and to restrict the quantity of any spillage. There should be a robust connection and first isolation on the plant or storage vessel and a second valve, of > 2 cm size for draining or 0.6 cm for sampling, separated from the first by > 1 m of piping. The discharge pipe from the drain of > 2 cm bore should deliver clear of the vessel and be supported to prevent breakage by jet forces. Both valves should have means of actuation which cannot be readily removed. Samples should be taken only into a bomb through a closed 0.6 cm bore piping system.

Pressure storage vessels should preferably be designed with only one connection below the liquid level, fully welded up to a first remotely operated fire-safe isolation valve located clear of the area of the tank.

Valve connections should be provided on process vessels for disposal of residues of liquefied hydrocarbons, preferably to a closed flare system. No bleed direct to atmosphere should be > 1.9 cm bore.

Remotely controlled isolation valves should be provided on items of equipment which are liable to leak significantly in service.

Discharge of heavy vapour from relief valves and blowdowns should be vented to a closed system, preferably with a flarestack (except when it is possible to discharge to atmosphere at sufficient velocity to ensure safe dilution by jet mixing with air).

Excess flow valves should be installed in liquid and vapour connections which are regularly broken to atmosphere, particularly the flexible hose connections used in tank wagon operations.

Remotely operated isolation valves are better than excess flow-valves.

Flammable-gas detectors should be installed in areas where experience shows there is a significant chance of a leak.

Whenever possible, equipment should be located at the safe distances from sources of ignition determined by the quantitative method of assessment (ref. 109). The horizontal extent of Division 2 areas in electrical classification should be the same as the safe distance.

The ground under pressure storage vessels should be impervious and should slope so that any liquid spillage will flow away from the vessels to a catchment area where it can be safely disposed of, or can burn if it ignites without causing further hazard. Suitable diversion and retaining walls should be provided to prevent uncontrolled spread of the spillage. The height of the walls should be suitably limited in relation to their distance

Table 12.20 (continued)

apart to allow minor leakage to be dispersed by natural air movement. The retention capacity for liquid should be decided in relation to the amount likely to escape allowing for flash-off and boil-off from the ground.

Low-pressure, refrigerated storage tanks should be fully bunded, and the floor of the bund should be sloped so that spillage flows preferentially away from the tank.

The principle of diverting liquid spillage away from equipment should be applied in process areas wherever possible.

In plant or storage areas where safe distances from sources of ignition cannot be met, or in areas near a factory perimeter adjacent to public roads or property the installation of a steam curtain should be considered.

- Hazard analysis
- Hazard and operability study
- Safety audits.

Important reviews which arc mandatory in some companies for many projects are listed in Table 12.21 (after ref. 116).

The value of repeating reviews is shown by the examples of some results of a relief and blowdown review performed one year after start-

Table 12.21 Formal design and safety reviews

- Line diagram review.
  All process and instrument line diagrams.
- Relief and blowdown reviews.
- Hazard and operability study (see page 741).
- Electrical distribution review.
  Consideration of safety consequences of failure at all points in the electrical system.
- Area classification review.
  Ensure that the correct type of electrical and instrument equipment is installed by checking the extent of flammable hazardous areas, referring to Codes of Practice and/or by calculation of the release and dispersion of vapours.
- Layout review
  Ensure conformity with Codes of Practice optimisation of conflicting objectives (e.g. minimum of easy to follow pipe runs, minimum fire loading, minimum inventory of flammable materials overhead, dispersion of flammable vapours, control of fire escalation, ease of isolation for maintenance).
- Paving and drainage review
  Consideration of safe drainage or containment of hazardous spillages.
- Cable routing and protection review.
  Consideration of cable routes and fire protection, including fire hazard sources and electrical safety.
- Means of escape from structures and buildings.
- Fire fighting review.
  Consideration of access for fire-fighting and the provision of fire-fighting equipment.
- Fire protection reviews.

up which, as illustrated in Fig. 12.24, demonstrated the need for additional or larger relief valves, or changes in their position.[117]

These formal reviews, the results and findings of which are documented, will be in addition to informal iterative reviews of special safety features during design.[118]

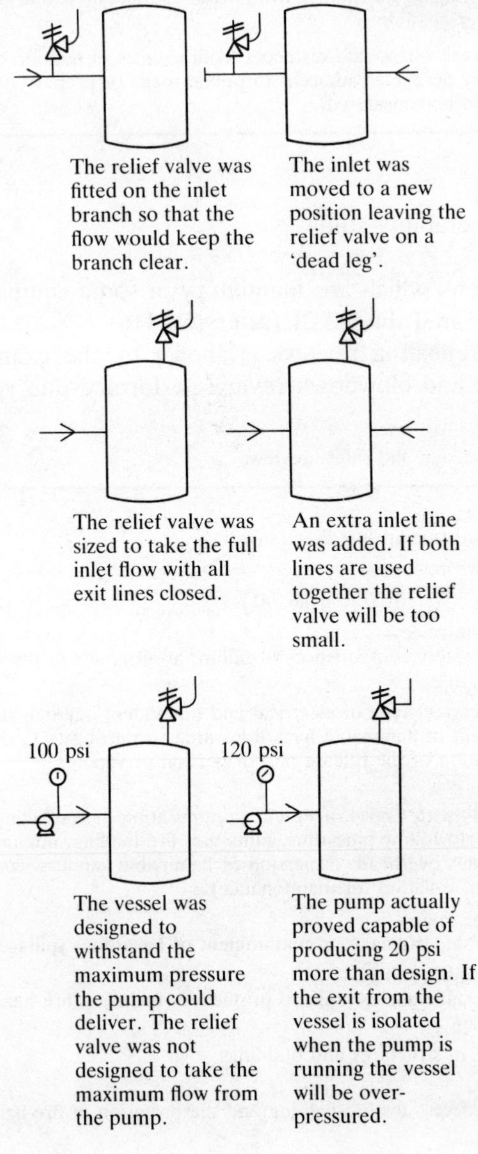

Fig. 12.24 Summary of relief and blowdown review[117]

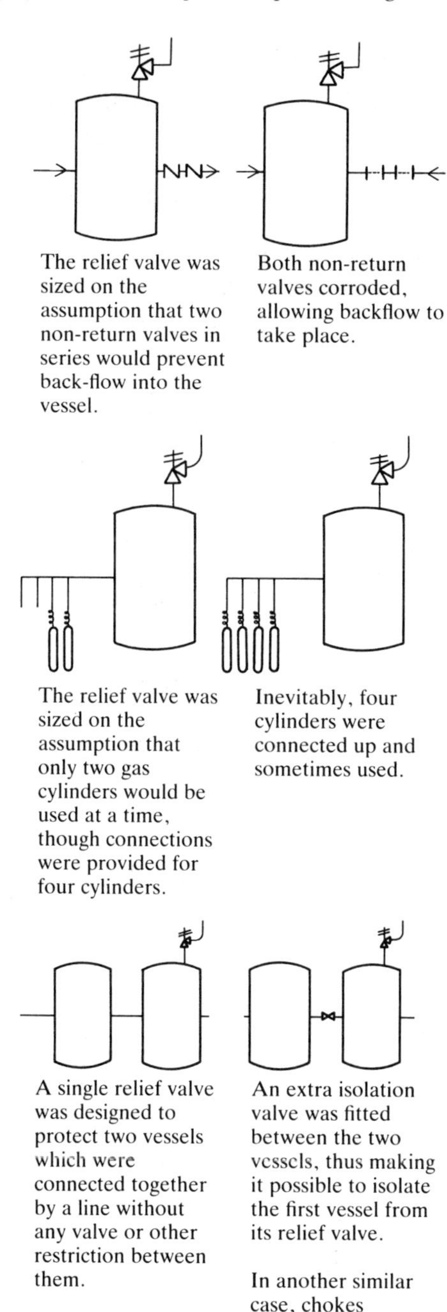

Fig. 12.24 *(continued)*

Dow Fire and Explosion Index[25]

It is frequently an advantage to be able to quantify the relative hazards of a plant, or an individual process unit, in order to assess priorities for special safety precautions, e.g. extra spacing between units, blast-proofing of control rooms, fire-proofing of structures, etc.

The Dow F&EI Guide is used to rate a chemical process unit, e.g. a furnace or a reactor, numerically; the value derived, termed the Fire and Explosion Index, reflects the magnitude of the hazard, albeit in terms of property damage (i.e. no allowance is made for potential toxic effects). The F&EI is calculated on any process units which could have an impact on the process area and at the start of the calculation a Material Factor (MF) is allocated to represent a measure of the intensity of energy release.

Contributing Hazard Factors are then evaluated, either as general process hazards or special hazards, as shown in Fig. 12.25, and appropriate penalties are applied. The guide contains a full explanation of the assessment of penalties but, as an example, General Process Hazards are described in Table 12.22. The product of the 'General Process Hazard Factor' and the 'Special Process Hazard Factor' is the 'Unit Hazard Factor', a measure of the degree of hazard exposure of the process unit. The product of the 'Unit Hazard Factor' and the 'Material Factor' is the F&EI, the range for which is summarised in Table 12.23.

The F&EI can then be used to determine an 'Area of Exposure' around the particular process unit, an indication of which equipment could be exposed to fire or fuel–air explosion generated by this unit.

The 'Unit Hazard Factor' also facilitates determination of the 'damage factor', which is the percentage damage probability caused by fire plus blast damage, taking into account various factors associated with the process unit (i.e. it is the degree of loss exposure). Given the capital value of the equipment and of the product inventory in the Area of Exposure, the 'Base Maximum Probable Property Damage' can be calculated. This is a base value since it incorporates no credits for Loss Control Factors.

The actual Maximum Probable Property Damage (MPPD) is normally less than the Base MPPD because some basic safety design features are included in modern plants or buildings. Additional features which can be utilised to minimise the exposure to an area where an incident occurs and to reduce the probability and magnitude of an incident are categorised as:

- Process control
- Materials isolation
- Fire protection.

FIRE AND EXPLOSION INDEX

LOCATION	DATE

PLANT	UNIT	CHARGE

MATERIALS AND PROCESS

MATERIALS	
CATALYSTS	SOLVENTS

MATERIAL FACTOR			
1. GENERAL PROCESS HAZARDS	PENALTY	PENALTY USED	
BASE FACTOR	1.00	1.00	
A. EXOTHERMIC REACTIONS (FACTOR .30 to 1.25)			
B. ENDOTHERMIC REACTIONS (FACTOR .20 to .40)			
C. MATERIAL HANDLING & TRANSFER (FACTOR .25 to .85)			
D. ENCLOSED PROCESS UNITS (FACTOR .30 to .90)			
E. ACCESS	.35		
F. DRAINAGE (FACTOR .25 to .50)			
GENERAL PROCESS HAZARD FACTOR (F_1)			
2. SPECIAL PROCESS HAZARDS			
BASE FACTOR	1.00	1.00	
A. PROCESS TEMPERATURE (USE ONLY ONE)			
1. ABOVE FLASH POINT	.30		
2. ABOVE BOILING POINT	.60		
3. ABOVE AUTOIGNITION	.75		
B. LOW PRESSURE (SUB-ATOMOSPHERIC)	.50		
C. OPERATION IN OR NEAR FLAMMABLE RANGE			
1. TANK FARM STORAGE FLAMMABLE LIQUIDS	.50		
2. PROCESS UPSET OR PURGE FAILURE	.30		
3. ALWAYS IN FLAMMABLE RANGE	.80		
D. DUST EXPLOSION (FACTOR .25 to 2.00) (SEE TABLE III)			
E. PRESSURE (SEE FIGURE 2)			
F. LOW TEMPERATURE (FACTOR .20 to .50)			
G. QUANTITY OF FLAMMABLE MATERIAL			
1. LIQUIDS OR GASES IN PROCESS (SEE FIG. 3)			
2. LIQUIDS OR GASES IN STORAGE (SEE FIG. 4)			
3. COMBUSTIBLE SOLIDS IN STORAGE (SEE FIG. 5)			
H. CORROSION AND EROSION (FACTOR .10 to .75)			
J. LEAKAGE – JOINTS AND PACKING (FACTOR .10 to 1.50)			
K. USE OF FIRED HEATERS (SEE FIG. 6)			
L. HOT OIL HEAT EXCHANGE SYSTEM (FACTOR .15 to 1.15) (SEE TABLE IV)			
M. ROTATING EQUIPMENT, PUMPS, COMPRESSORS	.50		
SPECIAL PROCESS HAZARD FACTOR (F_2)			
UNIT HAZARD FACTOR ($F_2 \times F_2 = F_3$)			
FIRE AND EXPLOSION INDEX ($F_3 \times$ MF) F & EI)			

Fig. 12.25 F and EI calculation sheet[25] (All figures and tables are given in the full Guide)

Table 12.22 General process hazards

A. *Exothermic reactions*

1. Hydrogenation, hydrolysis, isomerisation, sulphonation and neutralisation = 0.30 penalty.
2. Alkylation, esterification**, oxidation*, polymerisation and condensation = 0.50 penalty.
   * In oxidation reactions involving vigorous oxidising agents such as chlorates, nitric acid, hypochlorous acids and salts, etc., increase the penalty to 1.00.
   ** When the acid is a strong reacting material, increase the penalty to 0.75.
3. Halogenation = 1.00 penalty.
4. Nitration = 1.25 penalty.

B. *Endothermic reactions*

1. Calcination, electrolysis, pyrolysis or cracking = 0.20 penalty.
   (Where the energy source is provided by the combustion of a solid, liquid or gaseous fuel, increase the penalty to 0.40.)

C. *Material handling and transfer*

1. Loading and unloading of Class I flammable liquids or LPG = 0.50 penalty.
2. Use of centrifuges, batch reactions or batch mixing = 0.50 penalty.
3. Warehousing and yard storage:
   (a) Class I flammable liquids or LPG or flammable gases = 0.85 penalty.
   (b) Combustible solids identified as open or open cell (< 40 mm thick) = 0.65 penalty.
   (c) Combustible solids identified as closed cell or dense (> 40 mm thick) = 0.40 penalty.
   (d) Class II combustible liquids = 0.25 penalty.

D. *Enclosed process units*

1. Dust filters or collectors = 0.50
2. Flammable liquids above flash point/below boiling point = 0.30
3. Flammable liquids or LPG above b.p. = 0.60
4. Excess of 10 m. pounds in items 2 or 3 above, penalty is 1.5 times that listed.

E. *Access*

1. Penalty for inadequate access = 0.35

F. *Drainage*

1. Penalty of 0.25 if drainage is directed to an impounding basin.
2. Penalty of 0.50 if diking is used to retain spill around process unit or if spill can be trapped around surrounding process units.

Table 12.23 F&EI range and the corresponding hazard degree

F&EI range		
From	To	Degree of hazard
1	60	Light
61	96	Moderate
97	127	Intermediate
128	158	Heavy
159	UP	Severe

These are assigned credits as in Table 12.24. The product of all factors in each category represents the credit factor for that category, and the product of $C_1 \times C_2 \times C_3$ is converted graphically to an Actual Credit Factor. Then

$$\text{Actual MPPD} = \text{Base MPPD} \times \text{Actual Credit Factor}$$

(Obviously a low credit factor represents a safer process unit and a credit factor of unity means there are no relevant additional safety features.) The technical significance of these values is:

Actual MPPD ≡ Probable loss if an incident of reasonable magnitude occurs *and* the various protective equipment functions.

Base MPPD ≡ Probable loss if an incident of reasonable magnitude occurs and some of the protective equipment fails to function.

(This emphasises the obvious lesson that to be of any benefit protective measures, of whatever type, must be functional when called upon; hence the need for regular inspection and testing; see Ch. 17.)

Finally, provision is made for the calculation of Maximum Probable Days Outage and, hence, given the value of the product manufactured, the Business Interruption (BI) loss.

This is undoubtedly a valuable and rapid procedure for quantifying the magnitude of fire and explosion hazards and the relative advantages of process changes, added safety features, reduced inventories, increased spacing, etc.

Table 12.24 Loss control credit factors for use in Dow F&EI guide[25]

1. Process control (C_1)			
(a) Emergency power	0.97	(e) Computer control	0.89–0.98
(b) Cooling	0.95–0.98	(f) Inert gas	0.90–0.94
(c) Explosion control	0.75–0.96	(g) Operating procedures	0.86–0.99
(d) Emergency shut-down	0.94–0.98	(h) Reactive chemical review	0.85–0.96
2. Material isolation (C_2)			
(a) Remote control valves	0.94	(c) Drainage	0.85–0.95
(b) Dump/blowdown	0.94–0.96	(d) Interlock	0.96
3. Fire protection (C_3)			
(a) Leak detection	0.90–0.97	(f) Sprinkler systems	0.60–0.96
(b) Structural steel	0.92–0.97	(g) Water curtains	0.95–0.97
(c) Buried tanks	0.75–0.85	(h) Foam	0.87–0.98
(d) Water supply	0.90–0.95	(j) Hand extinguishers	0.92–0.97
(e) Special systems	0.85	(k) Cable protection	0.90–0.96

The Mond Index

The Mond Fire, Explosion and Toxicity Index[26,119] is based on a similar methodology to the Dow Index but has useful extensions, e.g. to estimate an Area's Fire Load and a Unit Toxicity Index. Various offsetting values are applied for Safety and Preventative Measures including 'Safety Attitude'. The use of the index is fully explained in the published Guide.[26]

Fault-tree analysis

An accident is generally the outcome of more than one fault in a system. Each fault has a finite probability of occurring in a given time interval and when the appropriate faults occur in a specific sequence an accident results. The sequence of events may be depicted by a fault tree, the branches of which join at nodes or 'gates'. The final outcome, i.e. an accident, will not occur unless the contributing faults combine to provide a route, or routes, through the gates from the lowest branch in the tree to the top. Assigning probabilities to the primary events on the fault tree and combining the probabilities enable the risk of the accident to be estimated. Moreover the impact of precautionary measures and protective devices can be calculated.

A hypothetical fault tree culminating in a major fire is shown in Fig. 12.26.[67] There are numerous AND gates and therefore many possible ways or preventing the fire – by eliminating one of the contributory events at each AND gate. (In practice the aim is to reduce probability to a low, acceptable value.) To achieve the same effect at an OR gate, i.e. zero

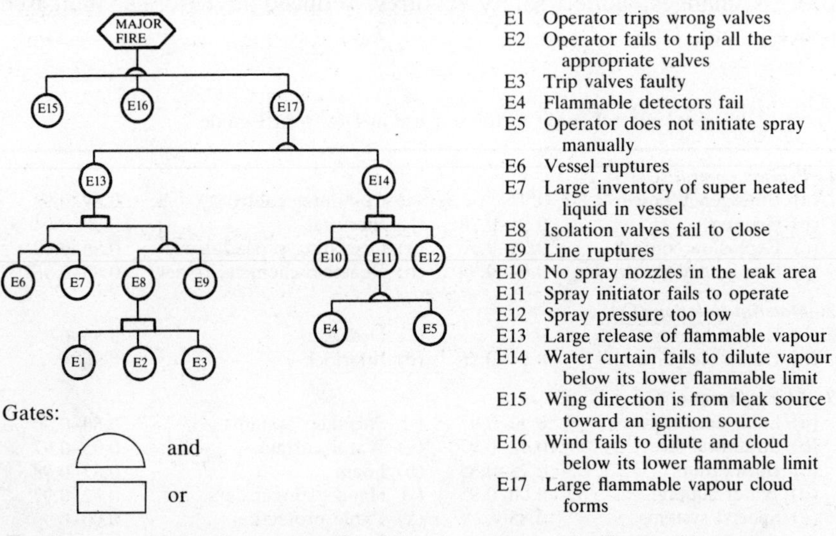

Fig. 12.26 Fault-tree for a major fire following release of a flammable vapour cloud[67]

residual probability, it would be necessary to eliminate all the contributory events – clearly a more difficult task.

Considering failure of the water curtain to dilute vapour below its LEL, E14, it will not be adequate to reduce the probability of the spray initiator failing to operate, E11, by increasing the reliability of the flammable-gas detectors, E4, OR by reducing the probability of the operator failing to initiate the spray manually, E5, *if* the sprinkler system does not cover all areas where a vapour cloud could form, E10, OR the spray pressure is too low, E12, e.g. due to demand elsewhere.

This is a qualitative illustration of the value of a fault tree but the real contribution arises when predicted probabilities of failure are computed.[5]

Hazard and Operability Study

This is a review of the P&I diagrams by a small team of technical specialists using a structured technique based upon key words. Potential malfunctions are identified by considering all possible deviations from design intention. The set of *key words* contains two sub-sets: '*property words*' which focus attention on the design conditions and the design intention; the second sub-set is '*guide words*' which focus attention on to possible deviations.

Typical 'property' words are given in Table 12.25.

A list of 'guide' words which have been recommended is given in Table 12.26.[120,121] The use of this set is not essential but the key words must be carefully selected before starting the study to avoid meaningful deviations being overlooked.

A detailed sequence for the application of an operability study is shown in Fig. 12.27. This represents an extremely thorough method of hazard analysis and assessment. Full details have been published of the method of application[121] and of the use of computer methods to reduce the man-hours involved.[122]

This is a very powerful technique for the identification of hazards and their consequences from P&I diagrams and it leads to the proposal of design safeguards. Several worked examples have been published.[120,123–125]

Table 12.25 Typical property words

Common	Application-dependent
Flow	Viscosity
Temperature	Flash point
Pressure	Vapour pressure
Level	Heat transfer
Concentration	Separate
	Absorb
	etc.

Table 12.26 A list of guide words for HAZOP studies

Guide words	Meanings	Comments
No or Not	The complete negation of these intentions.	No part of the intentions is achieved but nothing else happens.
More Less	Quantitative increases or decreases.	These refer to quantities and properties such as flowrates and temperatures as well as activities like 'heat' and 'react'.
As well as	A qualitative increase.	All the design and operating intentions are achieved together with some additional activity.
Part of	A qualitative decrease.	Only some of the intentions are achieved; some are not.
Reverse	The logical opposite of the intention.	This is mostly applicable to activities, for example reverse flow or chemical reaction. It can also be applied to substances e.g. 'poison' instead of 'antidote' or 'D' instead of 'L' optical isomers.
Other than	Complete substitution.	No part of the original intention is achieved. Something quite different happens.

A reactor into which two feed materials A and B are pumped is shown in Fig. 12.28. Application of the guide word 'No' and property word 'Flow' to flow of material A downstream of the pump suggests the following causes (after ref. 126):

- Stock tank empty (level below off-take).
- Outlet valve on stock tank closed.
- Outlet valve from pump closed.
- Pump not primed.
- Pump not switched on.
- Pump not working (electrically or mechanically).
- Suction line blocked.
- Suction line fractured.
- Material A solidified/too viscous.
- Stock tank not vented, or not under pressure, or not balanced to reactor.

The consequences of no-flow of A into the reactor would then be assessed and if necessary safeguards recommended to prevent, warn of, or cope, with the above causes.

The following is a simple example of an incident in which a **Hazop** would have identified a hazardous situation and lead to the consideration of appropriate safety measures.[127]

In a plant where acidic effluent was neutralised with a slurry of carbonate in an agitated vessel the operator observed that excessively acidic effluent was

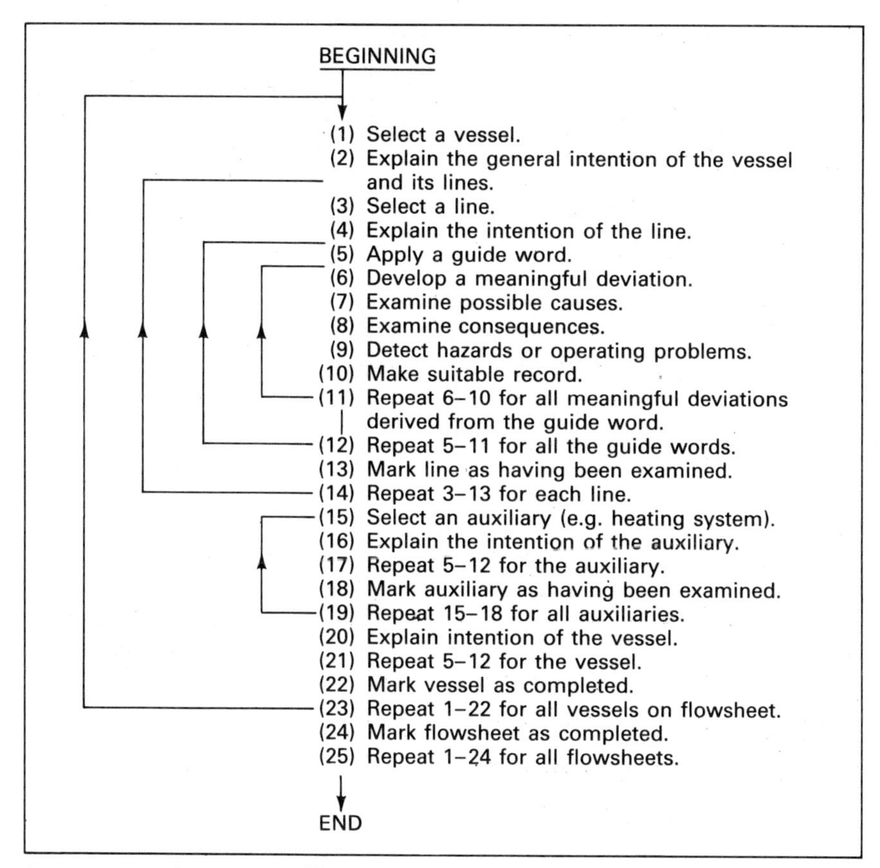

Fig. 12.27 Hazop sequence[121]

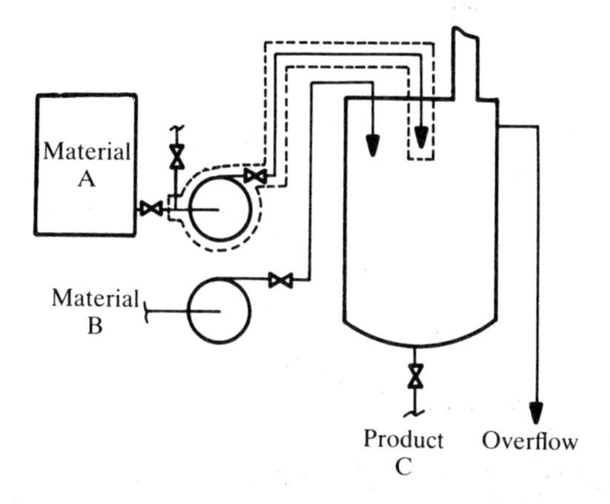

Fig. 12.28 Reaction vessel arrangement[126]

being discharged to drain. The vessel agitator was found to have stopped; he therefore restarted it. A violent reaction followed involving accumulated, unreacted calcium carbonate with a massive evolution of carbon dioxide; this lifted the bolted lid on the vessel and blew off the manhole cover. (Following a similar incident earlier an instruction had been issued as to action in the event of agitator failure; the operator was unaware of this.)

The safety measures for selection comprise:

- Provision for tank venting.
- Provision for automatic shut-off of slurry flow on agitator failure.
- Provision of high/low effluent pH alarm or agitator trip alarm.

Incidentally in such a case it is a check on actual rotation of the agitator which is desirable rather than on power drawn by the motor.

Hazard Analysis

Arising from a Hazop study, or from some other method of hazard assessment, certain areas of a plant (e.g. a pressure storage of a toxic gas) may be clearly identified as a serious hazard. Such an area can be subjected to a detailed, quantitative, Hazard Analysis in which the frequency and consequences of occurrence of hazards are estimated and then compared with some criteria of acceptability.[7,18] Reassessment of the plant is required if the criteria are not met.

Construction safety

Purchasing

As in other factories, it is important that engineering specification and purchasing procedures are devised to ensure that all purchased items of equipment are to the design requirements and comply with in-company and national standards.

In addition to price and delivery, a reputation for equipment reliability and maintainability will generally be taken into account by the company's project engineers, or by a contractor, when placing orders. Audits may also be carried out at appropriate stages in mechanical design and fabrication to ensure that the specified standards and codes are followed; an inspecting authority may be appointed for this purpose. Pressure, running, leak and other tests will be specified. Equipment will also be checked against drawings on delivery to site.

Conditions will normally be applied to purchasing dealing with commercial considerations. However, some of relevance to safety and loss prevention include:

- Notice of testing and inspections off-site to be given to the purchaser, so that they can be witnessed by a representative, e.g. an insurance engineer.

- Acceptance of equipment may be deferred until it has been installed and has performed satisfactorily and safely, for a specified period.
- If installation forms part of the contract, the supplier must comply with all safety procedures on the purchaser's site and make no use of lifting gear, scaffolding, cranes or ladders there without specific permission.

Installation

Good liaison with the installers of the new plant, who are likely to be special contractors, may help check defects at the construction stage. The following features of construction need careful attention otherwise hazards can be introduced:

- Incorrect foundations; settling, improper or insufficient grouting or separating.
- Incorrect materials of construction; defect or flaw, improper treatment, improper material.
- Incorrect fabrication; welding error, improper treatment or hardening, wrong surface finish, imbalance, failure to leave opening, improper tolerance, poor workmanship.
- Incorrect assembly; mechanical damage, parts wrongly fitted or omitted, wrong connections, misalignment, improper bolting, imbalance, piping stress, foreign material present, use of defective or wrong construction material or oil, incorrect or inadequate cleaning.
- Poor shipping and storage resulting in corrosion, contamination with dirt, damage.
- Inadequate supports or bolting.
- Inadequate testing.
- Inadequate monitoring of fabrication and construction.
- Unavailability of item due to scheduled service, replacement, inspection, safety devices temporarily out of action, controls deactivated/not in service.

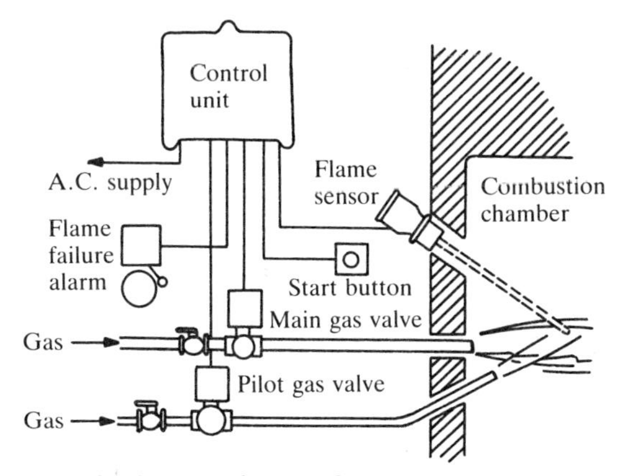

Fig. 12.29 Flame conduction type flame safeguard

- Sources of ignition: open flames, sparking tools, smoking or matches, transport, incomplete lagging of plant items and other hot surfaces.
- Sources of material hazards: flammability, incompatibility, corrosivity.
- Exposed energy source guards removed, manholes lifted and unprotected, tripping hazards.

Contracts must clearly spell out areas of responsibility for health and safety. Where possiblc the whole area should be fenced off and become the responsibility of the contractor until completion and handover. Where this is not possible, or if the project is relatively small, the contractor must be made aware of the owners' health and safety policy and be made to comply with any regulations.

Safety during the construction activities, be it a new plant or an extension/alteration to an existing one, is outside the scope of this text. However, reference to basic precautions is made in Chapter 2 and management considerations are summarised in Table 17.9.

A safety check-list covering the majority of activities is given in Table 12.27 (after ref. 128).

Pressure testing

Pressure testing of equipment and pipework to the specified test pressure is generally performed using liquid, most often water, but in some cases gas or air may be used. Hydrostatic testing is preferable because at low pressures, e.g. < 20 bar, the stored energy is at least 3 orders of magnitude less than with gas.[129,130]

The disaster at Flixborough could have been avoided if either the 50-cm pipe had been tested according to BS 3351 – when the pipe would at least have squirmed – or the bellows manufacturer's recommendations followed. The pipe was pneumatically tested for leaks, which is dangerous unless precautions are taken. The pipe should have been hydrostatically tested to 1.3 times the design pressure.[131]

It is necessary to test sections of plant not merely individual items since, as in the Flixborough case, an item may be satisfactory but the section as installed may not.

Hydrostatic testing may be impracticable in some situations, e.g.:

- If the supporting structure/foundations will not support the equipment weight when full of water.
- If contamination by water is not acceptable.
- If only mains water is available to test stainless steel equipment. This may create problems due to chloride assisted stress corrosion cracking – see page 107; steam condensate or demineralised water is therefore required.

Table 12.27 Safety check-list – construction and alteration projects

Safe access and movements

Vehicles

- Availability of good roads with adequate parking and turning areas and free of mud, obstructions, equipment, etc.
- Provision of separate areas for storage of materials.
- Provision of signs and signals to direct vehicles on site.
- Display of signs indicating vertical clearances and width limits.
- Where appropriate, provision of blocks and sidings for rail traffic.
- Adequate guarding of all powered vehicles.
- Marking of powered vehicles with classification if limited to a specific area and with rated capacities.
- Training of operators, e.g. in loading procedures, traffic rules, fuel handling and switching-off unattended vehicles.
- Establishment of procedures for the safe operation, inspection and maintenance of cranes, hoists, fork-lift trucks, etc.

Workforce

- Provision of adequate walkways, and work areas with proper fencing etc.
- Maintenance of aisles and passages in good condition, free of obstruction or tripping hazards and with marked exits and directions of movement.
- Provision of adequate ladders, stairs or lifts.
- Provision of adequate lighting.
- Good housekeeping practice, e.g. removal of rubbish and construction waste materials from work areas.
- Adequate guarding/covering of floor and roof openings.

Utilities and services

- Convenient and safe location of stores, offices and workshops.
- Location of high-voltage lines so as to allow free access without danger of contacting them.
- Convenient and safe location of toilets, washrooms and temporary power supplies.

Work scheduling

- Planning of work so that all the trades required to work in any area at one time have adequate space, ventilation, etc.
- Provision of an adequate supply of safety equipment, e.g. helmets, safety harnesses, gloves, goggles, boots, etc.

Work practices and procedures

- Arrangement of work such that operator does not have to reach through/over/under dangerous machinery, cannot start machinery accidentally, or be endangered while working near/on it.
- Arrangement of work such that the operator is not endangered by vehicles.
- Provision for audible signals which can be heard by all operators concerned.

Materials, equipment and construction control

- Control to ensure that only materials and equipment specified in the contract, or approved by the site engineers, are used in construction.
- Planning of inspections and overhauls when equipment or machinery is relocated.

Table 12.27 (continued)

- Procedures for testing of high-voltage wiring and apparatus before use.
- Provision for leak-testing of pipes and vessels (see Ch. 7)
- Planning of work so that no fire-resistance rating is affected by improper storage/handling of materials or equipment.
- Arrangements for independent tests of factors affecting integrity of construction, e.g. soil properties, concrete, welds.

Materials handling
- Provision of appropriate equipment for materials handling, i.e. cranes, hoists, lifts, fork-lift trucks, etc.
- Provision of adequate space for loading, unloading and storage of materials.
- Marking of floors with load-bearing capacity.
- Arrangements for secure stacking and storage of materials.
- Arrangements for separate storage of combustible, flammable and highly flammable materials with appropriate precautions.
- Arrangements for separate storage of gas cylinders, chemicals and explosives.
- Establishment of procedures and precautions for the use and filling of storage tanks.
- Display of 'No Smoking' signs in storage areas where appropriate.
- Arrangements for storage and regular disposal of combustible wastes.
- Provision and conspicuous identification of fire extinguishers in work areas.
- If provided, inspection and commissioning of sprinkler systems.

Construction equipment and tools
- Provision of appropriate tools for each stage of the work.
- Arrangements for periodic inspection, maintenance, overhaul or replacement of tools, ladders, scaffolding, etc.
- Checking of electrical equipment for earthing prior to issue (and periodically).
- Maintenance of correct guards on exposed gears, sprockets, pulleys, fly-wheels, belt and chain drives, etc., of all power transmission machinery (see Ch. 2).
- Checking that start, stop and lock-out devices on equipment are within the operator's reach.
- Provision of adequate local exhaust ventilation, e.g. on welding in semi-confined spaces.
- Provision of all appropriate personal protective clothing and equipment (see Ch. 6).
- Procedures for inspection of hired equipment.

Personnel
- Selection of competent workmen, supervisors and management.
- Adequate training – for construction activities and in hazards specific to the site/plant/equipment.
- Site procedures – familiarisation.
- Safety publicity and encouragement, e.g. safety bulletins, posters; regular safety meeting; investigation and reporting of accidents.
- Instruction, and practice in, safety drills and emergency procedures.
- Procedures to cope with changing labour force.

At high pressures hydrostatic testing must be performed cautiously. Care should be taken that any plastic plugs of the type used to seal ports or pipe connections during transit, and possibly painted over, are removed prior to testing on site.[132] Incidents have been reported of significant spillages of flammable liquids from a pump due to a 'shipping-type' plug being lost or blown-out.[133]

The routine precautions include in any event[129]:

- Elimination of trapped air from the equipment; this requires the provision of vents/bleed valves.
- Avoidance of adjustments, e.g. tightening of bolts while under pressure.
- Avoidance, whenever practicable, of temporary threaded joints.
- Avoidance of testing at water temperatures <7 °C, to eliminate any hazards due to brittle failure.
- Leaving the equipment under pressure for some time before approaching it; the pressure indicator must clearly also be accurate.
- Avoidance of freezing while the equipment is full of water and also avoidance of temperature increase.
- Provision of adequate venting capacity during drainage.

Commissioning safety

Special efforts are needed to ensure safety during commissioning. This stage of project development can be fraught with accidents because of the pressures on the team and the unproven nature of the plant. Commissioning involves many operations that may never be performed on the fully operational plant. A Hazard and Operability Study for these irregular procedures must therefore be carried out at the commissioning stage in addition to the analysis at the design stage for the proposed normal operating conditions. Again it is essential to establish good communications and team co-operation between contractors, production and commissioning personnel.

Pre-start-up inspection

The following identifies key questions to ask on a pre-start-up inspection (though the list is not intended to be exhaustive).

- Is the plant built to the design specification?
- Are all hot/refrigerated surfaces or pipes with which personnel can come into contact lagged?
- Have all equipment, piping (including any 'dead-legs'), etc., been pressure tested to the required standards?
- Is it installed *exactly* where it should be?

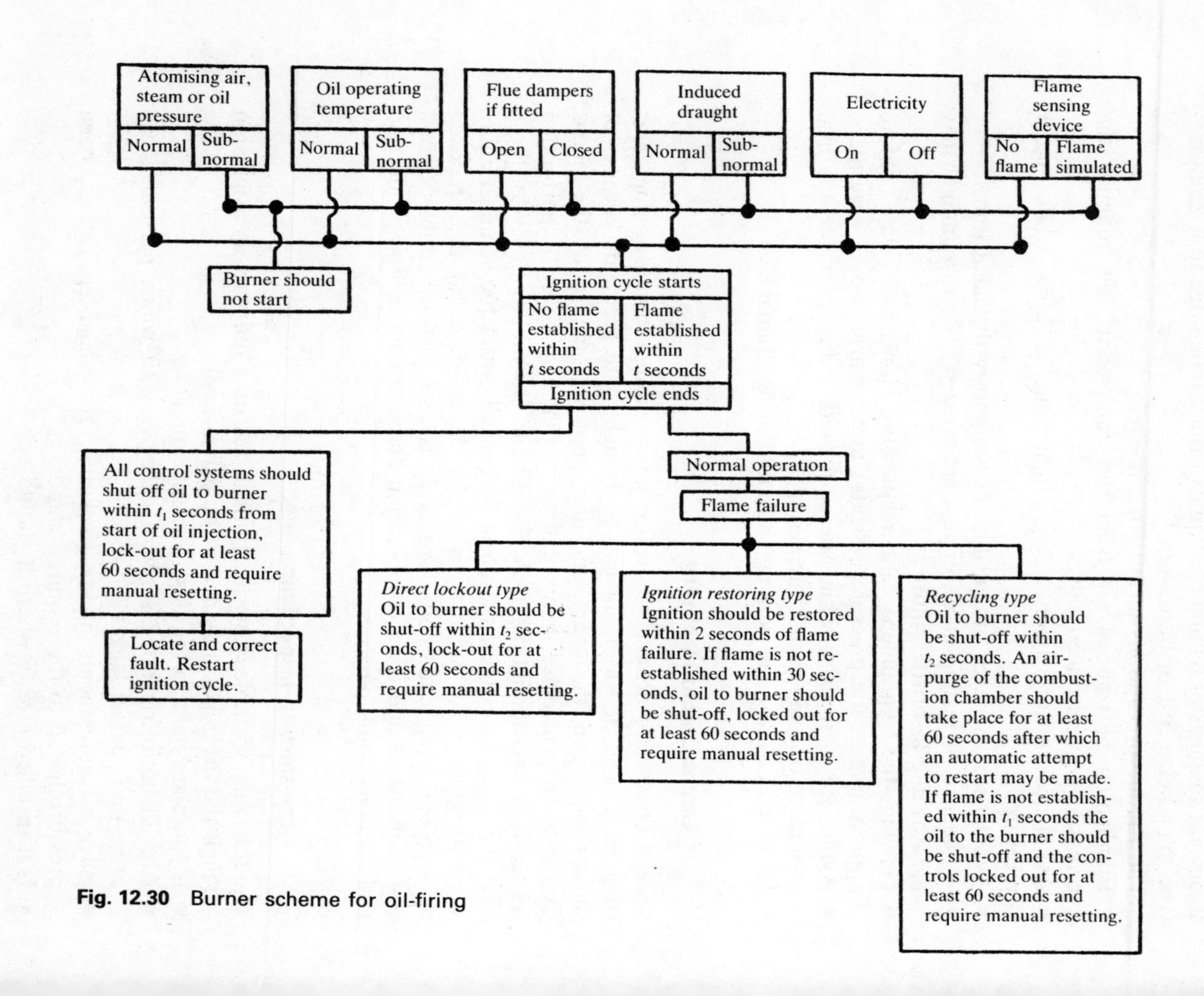

Fig. 12.30 Burner scheme for oil-firing

- Is it complete, supported where necessary, correctly aligned, tight, and level?
- Are the ergonomics satisfactory (pipes, valves, access, etc.)?
- Do all valve positioners work?
- Do instruments sense and have they been calibrated correctly?
- Are pipes and vessels labelled correctly. Blanks removed?
- Are gaskets indeed between flanges?
- Have ground wires and earthing continuity been checked?
- Are moving parts guarded and do interlocks work?
- Are any small components such as bolts, valve wheels plugs, shields for glass components, flexible pipe clips missing?
- Have any construction aids been left in which could cause a serious hazard when the plant starts up?
- Is motorised equipment aligned and has it been tested dry?
- Has the equipment been cleaned with liquid or gas to remove debris, etc., before start-up?
- Are plant areas clean and free from fire hazards, tripping hazards, items obstructing access/egress?
  If appropriate, have all non-flameproof electrical equipment or other sources of ignition been removed?
- Is all plant lighting functional?

Operational safety

Management systems to ensure safety in operation are discussed in Chapter 17. Consideration here is therefore restricted to certain technical aspects particularly relevant to chemical plants. As with other activities, and as emphasised in Chapter 17, safety depends upon a proper combination of hardware and 'software', i.e. operating procedures and training; the one cannot function satisfactorily without the other. For example safe operation of gas- or oil-fired ovens depends upon the considerations listed in Table 12.28 (after ref. 134).

Start-up and shut-down are often the most difficult operations with chemical process equipment. Typical considerations are given in Table 12.29; more details are described later.

Identification

Restricted areas, areas requiring the use of specific personal protection, exits and the locations of safety equipment, etc., should be clearly marked, e.g. using signs of the type shown in Plate 1, after page 771[135].

All piping, including service pipes, should be identified. Colour coding may be used for this purpose.[136,137] Identification is particularly important for vessel filling lines.

Table 12.28 General considerations for safety with gas- or oil-fired appliances[93]

Hardware

- Provision of a good view of the burners (e.g. by the provision of mirrors).
- Provision of efficient means of ignition (e.g. not involving a spill of paper or oil-impregnated cotton waste thrown into the combustion chamber which have caused flash-backs).
- Provision for purging gas or vapour from the combustion chamber.
- Provision of flame-failure safeguards, e.g. simple thermo-electric valves as in Fig. 12.29 or more elaborate electronic devices.
- Ensuring that all parts which must not fail during an explosion, e.g. door fastenings, are of adequate strength *and* provision of explosion relief vènting to a safe area.
- Provision of appropriate fuel, air, ignition source interlocks, e.g. as in Fig. 12.30.
- Design for the minimum number of burners to reduce the probability of error on start-up. (Conversely the flow per burner is inevitably greater.)

Software

- Provision of clear, unambiguous lighting-up and shut-down instructions preferably affixed to the appliance. Prohibition of tampering with cut-outs, interlocks, etc.
- Training of operators in lighting-up and shut-down procedures; in detail, e.g. including the need to close doors gently to avoid extinguishment of the burners by draught.
- Allowance for articles remaining after heating is turned-off which may continue to evolve flammable vapours.
- Provision of maintenance of all controls and safeguards to a high standard.

Table 12.29 Pre-start-up, start-up and shut-down

- Residual flammable materials should be removed before admitting air during shut-down. Similarly air should be removed before admitting flammable material on start-up.
- Water should be disposed of, and guarded against in feed tanks and pipelines, to avoid freezing, or 'steam explosions' on contact with hot process materials (see page 83).
- Checks are required at intervals to ensure that all vents are free before venting, and all drains unplugged before draining.
- Any material suspected of being pyrophoric should be kept wet with appropriate liquid until removed and disposed of (see page 152).
- Procedures for lighting furnaces should be rigidly enforced (see page 156).
- Leak checks should be made prior to start-up.
- Instruments should be commissioned thoroughly in order that they can be relied upon during start-up.
- Operating conditions should be changed slowly during start-up/shut-down to minimise thermal or pressure shocks.
- Permit-to-work systems should be rigidly enforced (see page 975).
- Correct methods of isolation should be followed (see page 978). Slip-plates or spectacle blinds should be installed/removed in the proper sequence and checked off against a list.
- Whenever practicable, any equipment or pipeline should be depressured and drained carefully before any flanges are broken; flanges should be cracked open to check for residual materials/pressure.

Sodium sulphydrate and sulphuric acid were stored in tanks connected by different filling lines to an area where tank trucks were discharged. A tank truck of sulphuric acid was connected in error to the sodium sulphydrate filling line. Evolution of hydrogen sulphide resulted in 3 fatalities and 20 people, including 6 firemen, were hospitalised.[138]

Equipment should also be clearly marked with name and/or number to avoid confusion over operating or maintenance instructions.

Seven pumps were installed in a row and a fitter was issued with a permit to work on No. 7. Assuming that this was the end pump he dismantled it. The pumps were in fact in the order Nos 1,2,3,4,7,5,6 and while he was fortunately out of the area, hot oil was discharged from the open connections.[139] (In circumstances where there are no permanent numbers or labels on equipment requiring maintenance a numbered tag should be tied on; this number should then be quoted on the permit – see Chapter 17.)

Particular attention is needed to mark up similar equipment or vessels which extend through one or more floor levels.

The maintenance department were requested to repair two autoclaves Nos 1 and 3 in a chemical plant. The fitter checked with the chief operator before opening autoclave No. 1. He then proceeded, unknown to the operating staff, to open the top manhole in No. 3 at third floor level; at second floor level he then accidentally opened the manway on autoclave No. 4 instead of No. 3. Autoclave No. 4 contained liquid vinyl chloride under a nitrogen pressure of 70 psig. Formation of polymer around the inside of the manhole initially concealed the fact that the vessel was under pressure but following removal of all the bolts the cover was blown off and hit a nearby railing. Escaping vapour was ignited and three fatalities resulted.[140]

Commissioning and start-up

Generally there is no clear delineation between pre-commissioning tests, commissioning and start-up since different areas of a plant tend to be at different stages of completion. Overlap between the activities also occurs with different equipment items in the same plant area. Therefore a sound communication system is essential, e.g. all commissioning staff must be made aware when leak testing is underway, which items are ‘active’, etc., and all authorities, other plants and services (i.e. fire, emergency, effluent treatment) must be advised of impending start-ups.

Commissioning

Normally safety features will be checked and commissioned before process materials are admitted to the plant. These will include:

- Nitrogen purge systems, blowdown systems, oil–water separator, drains.

- Fire alarms, hoses, mains, extinguishers, drench water, steam hoses, gas detectors, all stations in position and checked.
- Air sets, showers, wash bottles, protective clothing, first aid, resuscitation.
- Perimeter fence, gates, dematching, access/egress for fire-fighting.

Start-up

As with other phases of operation, a planned procedure is necessary on start-up, with built-in flexibility to cope with acceptable parameter variations or operating difficulties. The procedure will involve some combination of the activities listed in Table 12.30.
Some indication of the type of errors possible on start-up is given in Table 12.31.

Table 12.30 Start-up activities

Preparation	Inspection of plant. Removal/installation of blanks and blinds (to documentation).
Activation of services	Activation of compressed air, steam, water, purge gas, etc. Re-commissioning of instruments.
Purging	With water, steam, purge gas, as appropriate.
Leak-testing	Under pressure; by inspection, pressure retention, use of detectors (Ch. 7). Where the process involves hazardous gases a suitable inert alternative may be used.
Drainage/venting	Removal of water, purge gas, etc.
Conditioning	For example atmosphere, temperature, freedom from specific contaminants; possibly drying-out and sterilising.
Bringing-on stream	Staged/gradual adjustments of flow, temperatures, pressures, etc., up to normal conditions of operation.

Table 12.31 Examples of start-up errors

- Wrong route selected for flows – involving failure to ensure that correct valves are open and all other valves are closed.
- Drain valves left open, resulting in release of material.
- Errors of sequence (e.g. water added to concentrated acid instead of acid to water).
- Valves left closed resulting in over-pressurising the system.
- Failure to complete a purging cycle before admission of a fuel–air mixture.
- Mixing of cold and hot layers of liquids with consequent excessive vaporisation (see page 86).
- Admission of steam into a cold line full of condensate, resulting in water hammer.
- Back-flow of material from a high pressure to a low pressure system.
- Setting wrong valves for operating parameters, e.g. agitator speed, jacket temperature, reflux flow.

Purging

A purge flow of air or an inert gas is frequently used to free vessels or confined spaces of toxic or flammable vapours/gases.

In a perfectly mixed system the change in concentration of the atmosphere with time can be predicted from the equation:

$$\frac{c}{c_0} = \exp - \left[\frac{Q\theta}{V}\right]$$

where: θ = time; V = vessel or confined space volume; Q = purge gas flowrate; c = concentration of a specific component in the atmosphere after time θ; c_0 = initial concentration of a specific component in the atmosphere.

All are in consistent units.

Now $Q\,\theta/V = N$, the number of changes of the atmosphere. So that,

$$\log_{10}\left(\frac{c}{c_0}\right) = \frac{N}{-2.3}$$

In theory, therefore, the concentration of a contaminant in a given atmosphere could be reduced by a factor of 100 by about 5 changes.

Plug flow is a more effective method of purging than perfect mixing. This may be approached in[5]:

- 'Vessels' with a high aspect ratio, e.g. tall columns or a pipeline.
- Gravitational displacement of a heavy vapour.
- Bouyant displacement of a cold gas.

If the density difference between purge gas and atmosphere is small then turbulent mixing is essential. In a non-baffled storage tank this requires jets for the purge gas to direct it at a velocity along the axis of about 1.7 m s^{-1} in remote corners.[5]

A rail-tank car was used to transport a heavy inert gas, and after emptying it was pressure tested with nitrogen. A leak was found; therefore the pressure was released and the tank prepared for entry. This was done by supplying air via a 2.5-cm hose to the base of the tank; a 10-cm suction hose connected to a blower extended to the base of the tank. Purging continued for 5 hr. Subsequently, a gas sample taken from 1.2 m below a top runway indicated an acceptable oxygen content. When a man entered the tank he collapsed. The heavy inert gas had a much higher density than either nitrogen or air and had not been dissipated from a layer in the bottom of the tank.[141] [This is a further example of the effect of varying gas densities discussed in Ch. 3.]

Clearly, in such circumstances, predictions are no substitute for proper atmospheric monitoring, and appropriate checks should be written in to the purging procedure. For example on refinery units purging with steam or inert gas is continued until the oxygen content is typically <0.5% by volume.

Materials receipt and dispatch

Clear operating procedures are particularly important with regard to receipt and dispatch of materials.

A driver attempted to drive away a 12.2 m^3 LP gas tanker after filling at a bulk plant. A 5-cm hose was still connected to the loading risers; this and the valve were pulled off the truck leaving a 5-cm discharge orifice.

The ensuing vapour cloud was ignited by a vehicle in a nearby street and fire flashed back to the plant. This incident resulted in six fatalities and a loss of around £170,000, at 1970 costs.

Maintenance

Maintenance is a continuous operation but periodic thorough examinations by a competent person are also appropriate for certain plant, e.g. in the UK (under the Factories Act) lifting machinery, steam boilers and other pressure vessels. Pressure-relief devices, trips, etc., similarly require a formal inspection schedule.

Chemical plant problems

Maintenance which involves dismantling of plant which has been on-stream, clearly has the potential for the release of toxic, corrosive or flammable materials. Therefore, such operations, and maintenance involving entry into confined spaces or hot work, or the types of activity mentioned on page 762, will normally be controlled by a permit to work system.

Corrosion and contamination by process materials are common problems. For example, regular checks are needed for damage to mobile equipment, e.g. lifting gear, ladders and scaffolding, if it is used in adverse conditions.

Maintenance planning

In general breakdown maintenance – in which equipment is simply operated until a fault occurs – is not desirable on chemical plants since failure can be hazardous, or result in unplanned shut-down with loss of production, and/or lead to loss of process materials/products. Thus planned, scheduled maintenance is favoured; equipment and instruments are inspected, adjusted, repaired or replaced at predetermined intervals.

This requires detailed record-keeping and planning since in many cases shut-down of a section of plant is necessary.

Regular inspection

Items likely to require routine inspection and/or testing are listed in Table 12.32. This is a specialist function full details of which are given in stan-

Table 12.32 Items requiring routine inspection/testing*

Items	Including
Pressure vessels	Pressure testing
Steam boilers	Pressure testing, ss. 32–35, 37 and 38. Steam receivers and fittings also
Air receivers	Pressure testing. Fittings also, ss. 36 and 37.2
Pressure relief devices	Relief valves, bursting discs, open vents (relief pressure and capacity)
Local ventilation provision	Efficiency, condition of ducting, efficiency of pollution control devices, e.g. bag filters, scrubbers. Optimum pressures in ducts. s. 63
Lifting equipment	ss. 22–25 Chains, s. 26
Machine guards	s. 16
Trips	Actuators and valves
Stand-by equipment	Diesel generators Emergency lighting
Emergency equipment	Alarms, sprinkler systems Fire-fighting equipment Breathing apparatus and resuscitation apparatus First-aid provisions Emergency showers, eye-wash provisions Foam supplies
Ladders, scaffolding, temporary structures	
Process piping, valves fittings	Insulation condition Leak-tight. Flexible connectors/hoses
Control instruments	Settings, operability
Protective instruments	Alarms and cut-outs Flame detectors Flammable/toxic gas detectors
Means of access/escape	Normal and emergency, s. 40 Fire doors
Portable electric tools	Earthing
Electrical equipment	Earthing Correct classification Bonding continuity Standard tests
Fired heaters	Burners. Flame-failure devices. Pressure trips
Explosion reliefs	Freedom from obstruction or added restraints Flame arresters
General ventilation	s. 4 Efficiency. Make-up air.
Testing apparatus	See Chapter 7

* References are to UK Factories Act 1961.

dard texts.[142,143] Only a selection of aspects of particular relevance to chemical process plants are summarised here.

Pressure vessels, process equipment, piping and valves

It is now almost unknown for the catastrophic failure of a boiler or pressure vessel to result from a gross design or manufacturing error, but explosions of pressure plant still occur either due to a gradual loss of physical strength of the materials of construction usually caused by corrosion or cracking, or to the malfunction, incorrect design or fitting of the control gears.[144]

Periodic pressure-testing is a statutory duty in the UK for steam boilers (every 14 months or in certain specific cases every 26 months), air receivers (every 26 months), steam receivers and condensers (every 26 months) and for water-sealed gas holders (every 2 years). All pressure vessels are also normally tested at suitable intervals.

However, the degree and frequency of inspection, using expensive techniques such as ultrasonics and magnetic particle inspection, will reflect the potential hazard which the vessel represents. For example with small air receivers the damage resulting from failure will be limited but with pressure storage of chlorine, ammonia or similar toxic gases failure could – under certain conditions only – result in multiple off-site casualties. A vessel classified as representing a 'major hazard' should have regular, full inspections and an assessment of its 'fitness for purpose', i.e. certification of maximum load the vessel and its supports can sustain. Process equipment and pipe systems may be similarly re-tested and/or leak-tested. Additional considerations in the inspection of process plant are summarised in Table 12.33 (after ref. 144).

Table 12.33 Considerations in process equipment inspection

Over-stressing	Distortion, change in dimensions
Interior finish	Erosion, corrosion, inclinations Lining condition.
Exterior finish	Painting. Atmospheric protection
Insulation	General condition. Freedom from contamination with oil/process materials. Sealing where appropriate, e.g. vapour seal on low-temperature duties
Gaskets, seals and tube-to-plate joints	General condition. Leak-tightness Signs of creep after high-temperature duty
Contact of dissimilar materials	Corrosion particularly in the presence of liquids
Wear of running surfaces	Surfaces subject to abrasion particularly if unlubricated

Machine guarding

In the UK there is a statutory duty for all fencing or other safeguards provided in connection with the safety of machinery to be of substantial construction and constantly maintained and kept in position while the relevant machinery is in motion.[145]

Electrical equipment

Maintenance of electrical equipment should be performed exclusively by trained staff or contractors. Routine examination and testing is recommended together with the keeping of appropriate records. Records should include the results of specified tests[146] and the general condition of equipment, circuit diagrams, lists of fuse ratings, relay settings, insulation resistances, etc. Earthing, lightning protection and, if applicable, bonding should also be checked regularly.

Hand tools

Proper maintenance of hand-tools is a self-evident precaution; some guidelines are summarised in Table 12.34 (after ref. 147).

Hoists, lifts and cranes

In the UK there is a statutory duty for every hoist or lift to be of good mechanical construction, sound material and adequate strength and to be properly maintained.[101] Also, with regard to cranes and other lifting machines, all parts of working gear shall be of good construction, sound material, adequate strength and free from patent defect and shall be properly maintained.[148]

Table 12.34 Maintenance of hand-tools

Hammers	Avoidance of split/broken/loose shafts Avoidance of worn/chipped heads Proper securing of heads in shafts
Files	Avoidance of use as levers Effectively maintained, proper handle
Chisels	Sharpening of cutting edge to correct angle Avoidance of 'mushroom' heads
Screwdrivers	Avoidance of use as 'chisels' or hammering Avoidance of split handles
Spanners	Avoidance of splay jaws Scrapping of any which are stripped
Cutting tools	Correct heat treatment

Frequent inspections and repair or replacement of defective parts are therefore essential. For cranes the items to be checked include[147]:

- Defects in ropes.
- Cracks, e.g. in cast iron parts, such as wheels and clutches.
- Security of locking pins, cotters, screws and nuts, e.g. on braking mechanisms.
- Lubrication.

For chains there should be a well-organised system for storage and inspection. Tables of safe working loads (SWL) must be marked on the tackle or displayed in the stores, etc.

Pressure relief

Atmospheric vents should be checked regularly to ensure that they have not been reduced in size compared with design intentions and are not blocked, blanked off, covered with plastic sheeting, modified, etc.

Instruments

An LPG storage sphere with a diameter of 14.3 m was found to have increased in diameter by 15 cm. A float from a level controller had come loose and lodged in the line to the relief valve when the vessel was overfilled.

Trip systems

Protective instrumentation and trip systems require regular inspection, testing and maintenance (see Ch. 17).

A slurry of salt in water at 1,000 psig was heated to 300 °C by passage through the tubes of a shell and tube heat exchanger; the heating media was oil at 10 psig. A pressure switch was installed in the oil line to prevent over-pressurisation of the shell in the event of water leakage from the tubes causing rapid steam generation. The signal from the switch served to close the oil and slurry lines and vent the shell to an expansion tank. Although the system operated when a tube leaked, the valves closed slowly and hot oil was blown out of the expansion tank; this ignited and a serious fire resulted.

Subsequent investigation showed that the pilot valves had not been lubricated, the pressure switch setting had been raised from 75 psig to 140 psig, and the trip system had not been tested for 6 months.

The data in Table 12.35 show the effect of test interval on hazard rate for a typical trip[149]; from this it can be deduced that unless a trip system is tested fairly regularly, e.g. at monthly intervals, it may fail to operate when needed.[115]

Operators, and indeed supervisory management, must be strongly discouraged from disarming trips as a matter of convenience.

Table 12.35 Typical trip system; effect of test frequency on hazard rate

Demand rate (The rate at which the parameter operating the trip reaches the set point)	1 per year	1 per 5 years
Hazard rate (The rate at which the parameter exceeds the set point without the trip operating) with,		
Weekly testing	1 per 150 years	1 per 750 years
Monthly testing	1 per 36 years	1 per 180 years
Annual testing	1 per 4 years	1 per 15 years

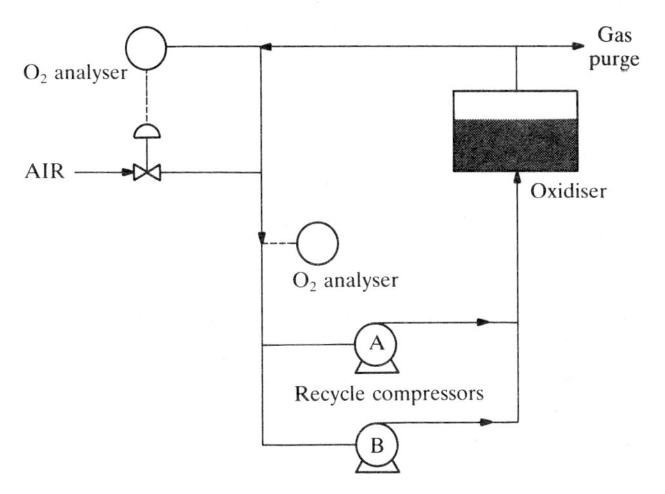

Fig. 12.31 Compressor arrangement on hydrocarbon oxidation plant[150]

Two recycle gas compressors were installed as shown in Fig. 12.31 on a plant in which a hydrocarbon was oxidised with air. If either tripped out the air supply was isolated automatically.[150] One compressor had to be shut down for a few minutes; the operator disarmed the trip but omitted to reduce the air rate. The oxygen analyser was too slow-acting to prevent the oxygen concentration rising to the flammable range and an explosion ensued.

Safety in maintenance operations

It is important that maintenance activities do not in themselves put plant, operators or maintenance workers at risk. The permit-to-work systems described in Chapter 17 are one important facet of this.

Safety lock-out procedures are required whenever accidental/inadvertent operation or energising of equipment may cause personal injury or equipment damage. The 'lock-out device' will comprise a mechanism or arrangement that allows the use of key-operated padlocks to hold a switch lever, or valve handle, in the 'off' position.[112] Lock-out is needed:

- For power isolation before maintenance or repair on any part of a unit.
- For cleaning or oiling movable parts of machinery, e.g. where there are dangerous in-running nips.
- When clearing blocked or jammed conveyors.
- For work on rotating machinery.
- To prevent unauthorised use of machinery.
- For vessel entry.

Advice is available on detailed safety lock-out and electrical isolation procedures.[112] The essential steps in the latter are to switch off the equipment, to lock off the main isolator, and then to attempt to restart the equipment to verify that it is in fact off. Isolation of process equipment and permit-to-work systems are discussed further in Chapter 17.

Maintenance staff need to be aware of the potential hazards from, and measures for dealing with, leaks of process materials or utilities. Spillage of flammable process materials if allowed to accumulate will increase the probability of fire starting; moreover these, or piles of other combustible materials, will result in a more serious fire once started. Fire hazard can be reduced by good housekeeping and simple precautions, e.g.:

- Flammable liquids should not be carried in buckets or other open-topped receptacles, to avoid spillage on to clothing or the ground.
- Factory and storage areas should be maintained in as clean and tidy a state as practicable. Liquid spillages should be dealt with promptly, e.g. by adsorption followed by safe incineration of the selected solid.
- Wastes should be disposed of promptly by authorised routes. Accumulations of open-topped drums full of process wastes, spillages, combustible rubbish, etc., should be avoided near process areas, tank farms or loading bays.
- Unnecessary exposure of flammable liquids should be avoided, e.g. by using non-flammable solvents for cleaning purposes.

References

(*All places of publication are London unless otherwise stated*)

1. Spiegelmann, A., 'Risk evaluation of chemical plants', *Chem. Eng. Prog. Loss Prevention Manual*, 1969, 6, 1.
2. *Smith (or Westwood)* v. *National Coal Board*, 1967, 2 All E.R., 593.
3. Sinnott, R. K., *Chemical Engineering* in J. M. Coulson & J. F. Richardson (eds), *An Introduction to Design*, vol. 6. Pergamon 1983.

4. Perry, J. H. & Chilton, C. H., *Chemical Engineers Handbook*. McGraw-Hill, 1973.
5. Lees, F. P., *Loss Prevention in the Process Industries*. Butterworths 1980.
6. Wells, G. L., *Safety in Process Plant Design*. George Godwin 1979.
7. Kletz, T. A., *Hazop & Hazan, Notes on the Identification and Assessment of Hazards*. Institution of Chemical Engineers, 1983, 53.
8. Institution of Chemical Engineers, *Flowsheeting for Safety*. 1976.
9. Fawcett, H. H. & Wood, W. S., *Safety and Accident Prevention in Chemical Operations*. Interscience: New York, 1964.
10. Vervalin, C. H. (ed.), *Fire Proctection Manual for Hydrocarbon Processing Plants* (2nd edn). Gulf, Houston, Texas, 1973, 82.
11. Anon., *Loss Prevention Bulletin*. Institution of Chemical Engineers, 1977 (014), 2.
12. Anon., *Loss Prevention Bulletin*. Institution of Chemical Engineers, 1980 (035), 1.
13. Maund, J. K., Private Communication.
14. Austin, D. G., *Chemical Engineering Drawing Symbols*. George Godwin 1979.
15. Russell, W. W., *Loss Prevention 10*, 1976 (Chem. Eng. Prog. Tech. Manual). American Instn Chemical Engineers, 80–7
16. Manufacturing Chemists Assoc., *Case History No. 2087*. Washington D.C. 1974.
17. Kletz, T. A., *Chem. Engr*, Mar. 1979, 342, 161–6.
18. Health and Safety Executive, *An Investigation of Potential Hazards from Operations in the Canvey Island/Thurrock Area*, 1978, H.S.E. 1978; *A Second Report on Canvey Island*, H.S.E., 1981.
19. Bell, C. L., 'Planning at Flixborough', in Symposium Design 79, Sept. 12–14, 1979. Institution of Chemical Engineers.
20. Collins, J. H., 'Major hazards – safety factors influencing siting policy', M. Sc. thesis, Univ. of Aston, 1978.
21. Kaloutas, G. J., 'The prediction of hazards from bulk toxic gas releases', M.Sc. thesis, Univ. of Aston, 1980.
22. Purdy, G. and Davies, P. C., *Loss Prevention Bulletin 062*. Institution of Chemical Engineers Apr. 1985, 1–12.
23. Hardman, R., *Process Eng.*, Aug. 1982, 63 (8), 27–9.
24. Mecklenburgh, J. C. (ed.), *Process Plant Layout*. George Godwin/Institution of Chemical Engineers 1985.
25. *Fire & Explosion Index Hazard Classification Guide* (*The Dow Index*) (5th edn), American Institution of Chemical Engineers 1981 (*Chem. Eng. Prog. Tech. Manual*).
26. Imperial Chemical Industries, *The Mond Fire, Explosion and Toxicity Index*, ICI; also Lewis, D. J., American Instn of Chemical Engineers Loss Prevention Symposium, Houston, Texas, 1–5 Apr. 1979.
27. The Petroleum (Consolidation) Act 1928 and S.I. 1929 No. 993 (The Petroleum Mixtures Order 1929).
28. Robertson, R. B., *Proc. of Symposium on Process Industry Hazards*, West Lothian, 14–16 Sept. 1976, Institution of Chemical Engineers.

29. Kletz, T. A., 'Plant layout and location some methods for taking hazardous occurrences into account'. American Instn Chemical Engineers Loss Prevention Symposium, 1–5 Apr. 1979, Houston, Texas.
30. Diaconacolau, G. J., Mumford, C. J. & Lihou, D. A., Symp. on Explosions, Fire Hazards and Relief Venting in Chemical Plant, Institution of Chemical Engineers, 12 Mar. 1980.
31. National Fire Protection Association, *Flammable and Combustible Liquids, Code No. 30*.
32. National Fire Protection Association, *Standard for the Storage and Handling of Liquefied Petroleum Gases, Code No. 58*.
33. Oil Insurance Association, General Recommendations for Spacing in Refineries, Petrochemical Plants, Gasoline Plants, Terminals, Oil Pump Stations and Offshore Properties, No. 631, O. I. Assoc., USA.
34. Stephens, M. M., *Minimizing Damage to Refineries*. US Dept. of the Interior, Office of Oil & Gas, Feb. 1970.
35. Nelson, R. W., *Hydrocarbon Processing*, Aug. 1977, 103–8.
36. Manufacturing Chemists Assoc., *Case History 1887; Loss Prevention*, 1973, 7, 119.
37. Epps, H. M., Paper presented at the 26th American Instn Chemical Engineers. Symposium on Safety in Ammonia Plants and Related Facilities, Montreal, Canada. Oct. 5–8, 1981.
38. Simpson, H. G., *Instn Chem. Engrs Symp. Series*, 1971 (34), 105–10.
39. Kaura, M. L., in C. H. Vervalin (ed.), *Fire Protection Manual for Hydrocarbon Processing Plants*, Vol. 2. Gulf, Houston Texas, 1981, 119.
40. *Factory Steel Stairways, Ladders and Handrails*, Engineering Equipment Users Association Handbook No. 7, 1962.
41. Barker, G. F., *et al.*, *Chem. Eng. Prog.*, Sept. 1977, 64.
42. Kletz, T. A., in *The Technical Lessons of Flixborough*. Institution of Chemical Engineers, 16 Dec. 1975, 7.
43. Barnwell, J., 'Inherent safety design features for an ethylene plant', American Instn Chemical Engineers Meeting, Atlanta, 27 Feb. 1978.
44. Langeveld., J. M., *Proc. of Symposium on Process Industry Hazards*, (*Instn of Chem. Engrs Symp. Series*, 1976, No. 47, 205).
45. Kletz, T. A., *Hydrocarbon Processing*, June 1979, 195.
46. Anon., *Loss Prevention Bulletin*. Institution of Chemical Engineers 1981 (047), 21.
47. *The Flixborough Disaster, Report of the Court of Inquiry*. HMSO 1975.
48. Kletz, T. A., *Accident Case History* 7, 30 May 1972.
49. *Chem. Eng. Prog.*, Apr. 1974, 80.
50. Factories Act 1961, s.63. HMSO.
51. American Conference of Governmental Industrial Hygienists, *Industrial Ventilation – A Manual of Recommended Practice* (13 edn). ACGIH: Lansing, Mich., 1974.
52. National Institute for Occupational Safety and Health, *Recommended Industrial Ventilation Guidelines*. NIOSH: Cincinnati, Ohio 1976.
53. American National Standards Institute, *Fundamentals Governing the Design and Operation of Local Exhaust Systems*. ANSI Z 9.2 Committee, New York, 1971.

54. H.M. Factory Inspectorate, *Health: Dust in Industry*, Technical Data Note 14. HMSO, 1970.
55. Health and Safety Executive, *Principles of Exhaust Ventilation*. HMSO 1975.
56. Gugan, K., *Instn Chem. Engrs Symp. Series*, 1968, No. 25, 8–15.
57. Health & Safety Executive, *The Fire and Explosion at Braehead Container Depot, Renfrew, 4 January 1977*. HMSO 1979.
58. Manuele, F. A., 'One hundred largest losses – a thirty-year review of property damage losses in the hydrocarbon – chemical industries', *Loss Prevention Bulletin*, 1984 (058), 1.
59. Anon., *Loss Prevention Bulletin*. Institution of Chemical Engineers, 1979 (028), 115.
60. Jones, C., & Sands, R. L., *Great Balls of Fire*. Jones & Sands, Coventry, 1981.
61. Maund, J. K., 'Cost effective safety'. Second National Conference Engineering Hazards, London, 1981.
62. Wicks, K. M., *Loss Prevention Bulletin*. Institution of Chemical Engineers, 1983 (053), 7.
63. Hutchings F. R., *Service Failures in the Chemical Industry*. British Engine Insurance Technical Report 1978.
64. *Guidelines for Bulk Handling of Chlorine at Customer Installations*. Chemical Industries Association 1980.
65. Health and Safety Executive, *Health and Safety – Industry & Services*. 1975, 12. HMSO.
66. Klaasen, P. L., *Instn Chem. Engrs Symp. Series*, 1971, No. 34, 111–24.
67. Mumford, C. J., & Lihou, D. A., 'New projects – hazard assessment and control: Part 2 – Assessment and engineering controls', in Design 79 Symposium: Sept. 12–14, 1979. Institution of Chemical Engineers.
68. Lihou, D. A., 'Design and specification for safety', Short Course Lecture Materials, Univ. of Aston.
69. Anon., *Loss Prevention Bulletin*. Institution of Chemical Engineers, 1977 (013), 5.
70. Anon., *Loss Prevention Bulletin*. Institution of Chemical Engineers, 1981 (041), 1.
71. Kletz, T. A., *Chem. Engr.*, Mar. 1972, 342, 161–2.
72. Anon., *Loss Prevention Bulletin*. Institution of Chemical Engineers, 1981 (037), 4.
73. Marshall, V. C., *Chem. Engr*, July, 1980, 499.
74. *Chlorine, Codes of Practice for Chemicals with Major Hazards*, Chemical Industries Association 1975.
75. Scott, D. S., 'Some seldom-considered aspects of pressure-relief systems', Symposium on Explosions, Fire Hazards and Relief Venting in Chemical Plant. Institution of Chemical Engineers, 12 Mar. 1980.
76. Richter, S. H., *Hydrocarbon Processing*, July 1978.
77. Kneale, M., & Binns, J. S., *Instn Chem. Engrs Symp. Series*, 1977, No. 49, 49–54.
78. Torday, J., *Chem. Engr*, July 1977, 525.
79. Duxbury, H. A., *Chem. Engr*, Nov. 1979, **350**, 783.

80. Scott, D. S., *Loss Prevention Bulletin*. Institution of Chemical Engineers, 1981 (048), 1–12.
81. Kayser, D. S., *Loss Prevention* (*Chem. Eng. Prog. Tech. Manual*), 1972, 6, 82.
82. Sarsby, J., *Manuf. Chem.*, Oct. 1984, 35.
83. (a) *Guide to the Use of Flame Arresters and Explosion Reliefs*, Health & Safety at Work No. 34, Health and Safety Executive, 1977. HMSO.
(b) Watson, P. B. 'Flame arresters'; Conference on 'Instrumentation and Safety in the Oil and Natural Gas Industries', Glasgow College of Technology, Mar. 1977.
(c) Bjorklunc, R. A., & Kushida, R. R. *Flashback Flame Arrester Devices for Fuel Cargo Tank Vapour Vents*. U.S. Dept. of Transporation, Mar. 1981.
84. Anon., *F.P.A. Journal*, 1964, 317.
85. Burgoyne, J. K., Bett, K. E. & Lee, R., *Instn Chem. Engrs Symp. Series*, No. 25, 1968, 1–7.
86. Weldon, G. E., *Loss Prevention*, Vol. 6, *Chem. Eng. Prog.*, 105–11.
87. Kletz, T. A., in C. H. Vervalin (ed.) *Fire Protection Manual for Hydrocarbon Processing Plants*, Vol. 2. Gulf, Houston, Texas, 1981, 55–9.
88. Kletz, T.A., *Chem. Eng. Prog.* 1963, **71**, 9.
89. Hearfield, F., *Chem. Engr*, Mar. 1979, 342, 156–60.
90. Rushford, R., *N.E. Coast Instn Eng. Trans.*, 1977, 93, 117.
91. Anon., *Loss Prevention Bulletin*. Institution of Chemical Engineers, 1975 (001), 3.
92. Fritz, E. J., *Loss Prevention*, 1969, 3, 41.
93. Manufacturing Chemists Assoc., *Case History 371*, Vol. 1, 1962.
94. British Standards Institution, BS 5345: Code of practice for the selection, installation and maintenence of electrical apparatus for use in potentially explosive atmospheres (other than mining applications or explosive processing and manufacture), Parts 1–6.
95. Imperial Chemical Industries/Royal Society for Prevention of Accidents, *Electrical Installations in Flammable Atmospheres*, *IS91*. ICI/RoSPA 1972.
96. British Standards Institution, BS 4683.
97. British Standards Institution, BS 5501.
98. The Chemical Works Regulations, 1922, Regn, 1. HMSO.
99. American Instn Chemical Engineers, *Pilot Plant Safety Manual*. A.I.Ch.E. 1972.
100. Shabica, A. C., *Chem. Eng. Prog.*, 1963, **58**(9), 57.
101. Factories Act 1961, s.22. HMSO.
102. Factories Act 1961, s.31.2. HMSO.
103. Dept. of Employment and Productivity, *Dust Explosions in Factories*, Health & Safety at Work. Booklet 22, 1970. HMSO.
104. Schofield, C., *Guide to Dust Explosion Prevention and Protection*: Part 1 – *Venting*. Institution of Chemical Engineering 1984.
105. Ligi, J., in C. H. Vervalin (ed.) *Fire Protection Manual for Hydrocarbon Processing Plants* (2nd edn). Gulf Houston, Texas, 1973, 311.
106. Gibson, N., *Process Eng.*, 11 Sept. 1970, 67–70.
107. Anon., *Loss Prevention Bulletin*. Institution of Chemical Engineers, 1979 (028), 97.
108. Institution of Chemical Engineers, *Fires and Explosions*, Hazards Workshop Module 003, Case History No. 5.

109. Nash, P., I. *Instn Chem. Engrs Symp. Series*, 1977, No. 49, 131.
110. Dept. of Employment and Productivity, *Fire Fighting in Factories*, Health and Safety at Work 10, 1970. HMSO.
111. British Standards Institution, *BS CP3013: Code of Practice for Fire Precautions in Chemical Plant*. July 1974.
112. Watts, J. R. Jr., *Loss Prevention* (*Chem. Eng. Prog. Tech. Manual*), 1976, 10, 48.
113. Saia, S. A., *Loss Prevention* (*Chem. Eng. Prog. Tech. Manual*), 1976, 10, 23.
114. Vincent, G. E., & Howard, W. B., *Loss Prevention* (*Chem. Eng. Prog. Tech. Manual*), 1976, 10, 43 and 55.
115. Kletz, T. A., Second International Symposium on Loss Prevention and Safety Promotion in the Process Industries, Heidelburg. Sept. 1977.
116. Institution of Chemical Engineers, *A Guide to Project Procedure*. 1978.
117. Henderson, J. M., & Kletz, T. A., *Instn Chem. Engrs Symp. Series*, 1976 No. 47.
118. Lihou, D. A., 'Hazard analysis in the design of chemical plants', 139th mtg. of Brit. Assoc. for Adv. of Science, 1977.
119. Lewis, D. J., *Loss Prevention*, 1980, 13.
120. Lawley, H. G., *Chem. Eng. Prog.*, Apr., 1974.
121. Chemical Industries Safety & Health Council, *A Guide to Hazard and Operability Studies*, 1977.
122. Lihou, D. A., *Loss Prevention Bulletin*. Institution of Chemical Engineers, 1983 (051), 19.
123. Lihou, D. A., in D. G. Austin & G. V. Jeffreys *The Manufacture of Methyl Ethyl Ketone From 2-Butanol*, Chap. 12. Institution of Chemical Engineers & George Godwin Ltd, 1979.
124. Knowlton, R. E. & Shipley, D. K., *An Introduction to Hazard and Operability Studies*, Feb. 1976.
125. Kletz, T. A., *Hazop & Hazan – Notes on the Identification and Assessment of Hazards*. Institution of Chemical Engineers, 1983.
126. Menzies, R. M., and Strong, R., *Chem. Engr*, 1975, **342**, 151–60.
127. Anon., *Loss Prevention Bulletin*. Institution of Chemical Engineers, 1977 (014), 2.
128. Bergstraun E. M., *Chem. Engr*, Feb. 27, 1978.
129. Anon., *Loss Prevention Bulletin*. Institution of Chemical Engineers, 1979 (026), 39.
130. Health and Safety Executive, *Guidance Note GS/4: Safety in Pressure Testing*. HMSO.
131. Mecklenburgh, J. C., 'Lessons from the post-inquiry discussion of Flixborough', Design Congress 76. Institution of Chemical Engineers, 1976, A-2-1.
132. Anon., *Loss Prevention Bulletin*. Institution of Chemical Engineers, 1984 (056), 26.
133. Vervalin, C. H., (ed.), *Fire Protection Manual for Hydrocarbon Processing Plants* (2nd edn). Gulf Houston, Texas, 1973, 83.
134. Department of Employment, *Evaporating and Other Ovens*, Health & Safety at Work 46, 1971. HMSO.
135. *Pictograms, RHS 300 Series*, Stimur Ltd, Paisley, Scotland.

136. British Standards Institution, *BS 1710: Identification of Pipelines.*
137. British Standards Institution, *BS 2929: Safety Colours for Use in Industry.*
138. Nailen, R. L., *Fire Engineering*, Oct. 1975.
139. Imperial Chemical Industries, *Safety Newsletter 20*. ICI.
140. Vervalin, C. H., (ed.), *Fire Protection Manual for Hydrocarbon Processing Plants*, (2nd edn). Gulf, 1973, 87.
141. Jenkins, A. J. D., 'Process safety theory and practice', Institution of Chemical Engineer's Course, Teeside Poly., 12–15 July 1976.
142. Pilborough, L., *Inspection of Chemical Plant*. Leonard Hill, 1971.
143. Imperial Chemical Industries/Royal Society for the Prevention of Accidents *Registration and Periodic Inspection of Pressure Vessels Code*. ICI/ RoSPA 1975, IS/107: Institute of Petroleum *Pressure-Vessel Inspection Code*, 1976.
144. *Vigilance*, 1980, **3** (12), 12–14.
145. Factories Act 1961, s.16. HMSO.
146. Hooper, E., *Beckingsale's Safe Use of Electricity*. Royal Society for Prevention of Accidents 1981.
147. *Plant and Machinery Maintenance* Safety, Health & Welfare No. 28. HMSO 1964.
148. Factories Act 1961, s.27. HMSO.
149. Kletz, T. A., *Chemical Processing*, Sept., 1974, 77.
150. Kletz, T. A., *Accidents Illustrated No. 6 – Alarms and Trips. Imperial Chemical Industries. Loss Prevention Bulletin*. Institution of Chemical Engineers, 1981 (049), 7–12.

CHAPTER 13

Safety in marketing and transportation of chemicals

Product liability

There has been a significant increase in 'consumer' legislation in recent years and in many countries greater responsibilities have been imposed on manufacturers. In the UK, s.6 of the Health and Safety at Work etc. Act is particularly important. There is also the Consumer Safety Act 1978, which regulates consumer safety, covering gaps in the Consumer Protection Acts 1961 and 1971 and under which the supply of certain products can be prohibited if they are considered to be unsafe.

However, civil liability arising from the supply of products is an additional problem for manufacturers, i.e. the possibility of a claim for compensation by a consumer or user 'harmed' by the product. Liability may arise in tort, as discussed in Chapter 16, or from a contract.

In the UK liability in contract stems from the Sale of Goods Act 1893. Under this, where a buyer expressly or by implication makes known to the seller the purposes for which goods are required so as to rely on the seller's skill and judgement and these goods are of a description which it is the course of the seller's business to supply, there is an implied condition that the goods shall be reasonably fit for the purpose for which they are intended. There is also an implied condition that the goods shall be of 'merchantable quality', except with regard to defects drawn to the buyer's attention or which ought to have been revealed to the buyer by prior inspection.[1]

> A lady purchased a consignment of smokeless fuel. Unknown to her or the supplier it contained an explosive substance which, in the event, blew up a fireplace and caused considerable damage.[2] She succeeded in a claim for damages on the basis that the goods were not of 'merchantable quality'. A full treatment of the law on the sale of goods is given in reference 3.

Liability is in contract between the buyer and seller and gives no rights to anyone not party to the sale. Important extensions to this Act were made by the Supply of Goods (Implied Terms) Act 1973 including limitation of the operation of exemption clauses.

A firm bought boron tribromide in glass ampoules for use in a manufacturing process. This chemical was perfectly fit for their purposes but, unknown to them was liable to react with great violence on contact with water (see Table 10.2). The process required the labels to be washed off the ampoules and while this was in progress one broke; the ensuing reaction shattered all the others and a violent explosion followed.

The goods were held to be not resonably fit for the purpose for which they were sold, and the defendants were liable because they should have foreseen the possibility of the chemical coming into contact with water.[4]

When any injured party is not the buyer and the manufacturer is not the seller of the goods, liability still exists in tort under the rule in *Donaghue* v. *Stevenson*.[5]

A friend purchased a bottle of ginger beer for Mrs Donaghue in a cafe; the glass of the bottle was opaque. After she had drunk a little of the contents, the remainder was poured out and revealed the presence of a snail in a state of decomposition. Mrs Donaghue suffered from shock and gastroenteritis and sued Stevenson, the manufacturer and bottler. It was held that, although there was no contract between Mrs Donaghue and Stevenson, there was a good cause of action.

Thus where any manufacturer sells a product in the form in which it is to reach the consumer, that manufacturer owes the consumer a duty to take reasonable care in the preparation and 'putting-up' of that product.

The implications of all this for the way a company should manage the various activities in product manufacture and marketing will now be considered. It is convenient for this purpose to base discussion on the duties of manufacturers, designers, importers and suppliers of articles and (with the exception of designers) substances defined by s.6 of the Health and Safety at Work etc. Act. This is reproduced in Table 13.1.

Table 13.1 General duties of manufacturers, etc., as regards articles and substances for use at work

(*Health and Safety at Work etc. Act 1974*)

6. – (1) It shall be the duty of any person who designs, manufactures, imports or supplies any article for use at work –

(*a*) to ensure, so far as is reasonably practicable, that the article is so designed and constructed as to be safe and without risks to health when properly used;

(*b*) to carry out or arrange for the carrying out of such testing and examination as may be necessary for the performance of the duty imposed on him by the preceding paragraph;

(*c*) to take such steps as are necessary to secure that there will be available in connection with the use of the article at work adequate information about the use for which it is designed and has been tested, and about any conditions necessary to ensure that, when put to that use, it will be safe and without risks to health.

(2) It shall be the duty of any person who undertakes the design or manufacture of any article for use at work to carry out or arrange for the carrying out of any necessary

Prohibition signs signifying what must NOT be done (Round with black symbols centralised on a white background with red border and cross bar)

Warning signs identifying particular hazards (Triangular with black symbols against a yellow background with black border)

Mandatory signs signifying what MUST be done (Round with blue background and white symbols)

Safe condition signs identifying safe conditions (Square or oblong with white symbols mounted on a green background).

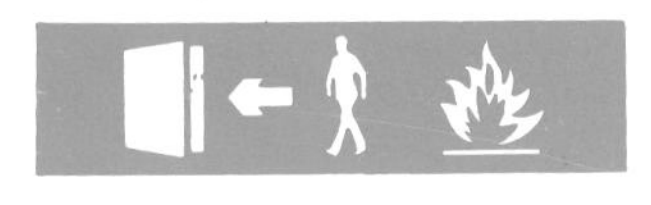

Plate 1 Factory signs
A safety sign combines geometrical shape, colour and pictorial symbol to provide specific health or safety information or instruction. Supplementary text may be used in conjunction with the relevant symbol, provided that it is apart and does not interfere with the symbol. The text should be in an oblong or square box of the same colour as the sign with the text in the relevant contrasting colour.

Air Products Limited Gas cylinder

Standard gases

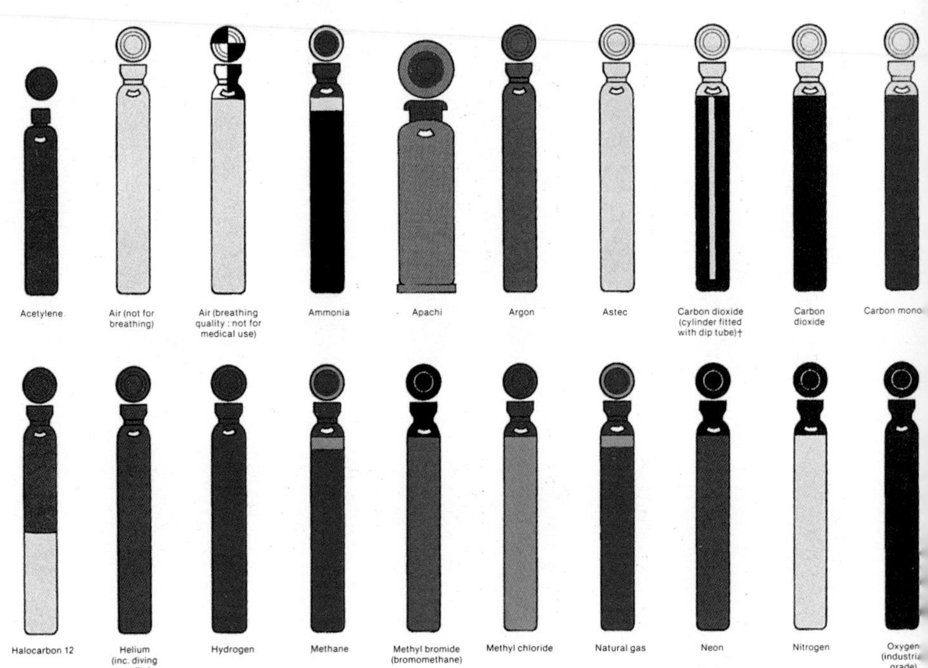

Special gases and mixtures

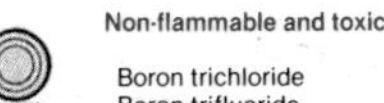

Non-flammable and non-toxic

*Argon
*Carbon dioxide
Halocarbon-11
*Halocarbon-12
Halocarbon-13
Halocarbon-13B1
Halocarbon-14
Halocarbon-21
Halocarbon-22
Halocarbon-23
Halocarbon-113
Halocarbon-114
Halocarbon-114B2
Halocarbon-115
Halocarbon-116
Halocarbon-C318
*Helium
Krypton
*Neon
*Nitrogen
Nitrous oxide
*Oxygen
Perfluoropropane
Sulphur hexafluoride
Xenon

Stencilling

Non-flammable and toxic

Boron trichloride
Boron trifluoride
Bromine pentafluoride
Bromine trifluoride
Carbonyl fluoride
*Chlorine
Chlorine trifluoride
Fluorine
Hexafluoropropene
Hydrogen bromide
Hydrogen chloride
Hydrogen fluoride
Iodine pentafluoride
Nitric oxide
Nitrogen dioxide
Nitrogen trifluoride
Nitrogen trioxide
Nitrosyl chloride
Perfluorobutene-2
*Phosgene
Phosphorus pentafluoride
Silicon tetrachloride
Silicon tetrafluoride
*Sulphur dioxide
Sulphur tetrafluoride
Suphuryl fluoride
Tetrafluorohydrazine
Trifluoromethyliodide
Tungsten hexafluoride

Stencilling

Flammable and non-toxic

*Acetylene
Allene
Butane
1, 3-Butadiene
1-Butene
Cis-2-Butene
Trans-2-Butene
Cis and trans-2-Butene
Deuterium
Dimethyl ether
2, 2-Dimethyl propane
Ethane
Ethylacetylene
*Ethyl chloride
*Ethylene
Halocarbon-142B

Plate 2 Gas cylinder identification – standard gases[40]

identification chart

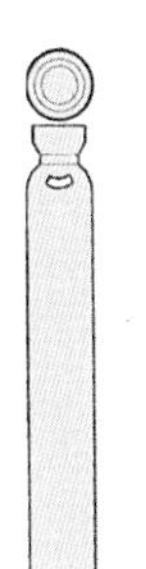
Chlorine

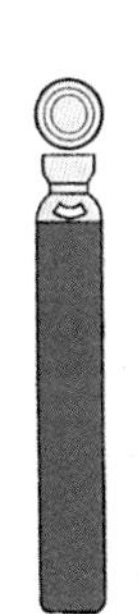
Coogar (carbon dioxide/ oxygen/argon mixtures)

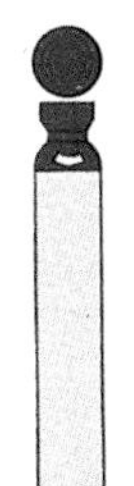
Ethyl chloride

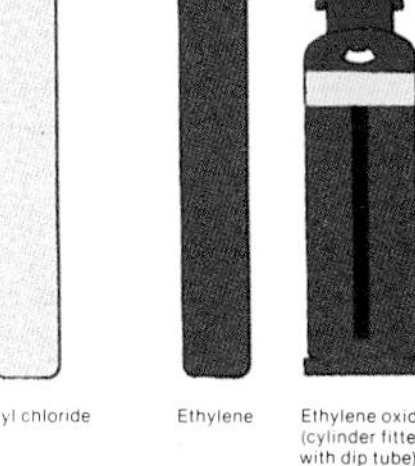
Ethylene

Ethylene oxide (cylinder fitted with dip tube)†

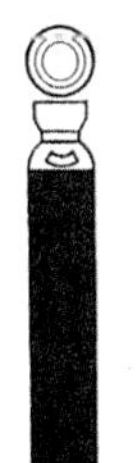
Oxygen (diving quality, not for medical use)

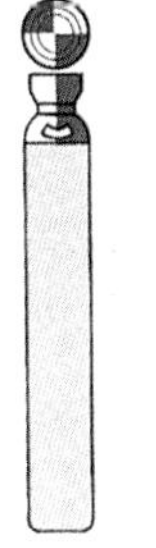
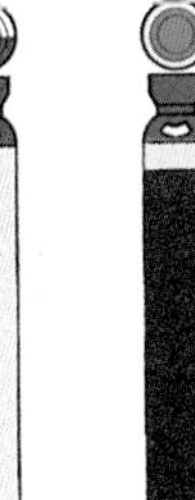
Oxygen/Helium Mixture (diving quality)

Phosgene

Propane

Sulphur dioxide

In the interest of standardisation and safety Air Products are adopting for all gases and mixtures the principles of identification of cylinder contents recommended in BS349:1973.
Primary identification is by means of labelling the name(s) and chemical formula(e) on the shoulder of the cylinder. Secondary identification is by the use of ground colours on the cylinder body and colour bands on the cylinder shoulder to denote the nature of the gas. As an additional aid to identification many Air Products cylinders also have the name, formula and (in the case of mixtures) composition of the contents stencilled on the side of the body (eg 20% argon in carbon dioxide).

On this chart, cylinders containing pure product which are identified by an exclusive colour, are shown left, while cylinders containing other gases and gas mixtures are identified below left. The latter, for which an exclusive identification colour is not provided, have a pastel purple ground colour with further bands of colour on the shoulder to denote the hazard properties (ie. red for flammable and yellow for toxic) of the gas or gas mixture. The colour of the shoulder band indicates the properties of one or more of the components of the mixture, not necessarily the properties of the mixture itself.

Halocarbon-152A
Halocarbon-1132A
Hydrogen
Isobutane
Isobutylene
Methane
Methylacetylene
3-Methylbutene-1
Natural gas
Propane
Propylene
Vinylfluoride
Vinylmethyl ether

Stencilling

Flammable and toxic

*Ammonia
Arsine
Bromotrifluoroethylene
*Carbon monoxide
Carbonyl sulphide
Chlorotrifluoroethylene
Cyanogen
Cyclopropane
Diborane
Dimethylamine
Ethyl fluoride
*Ethylene oxide
Gormanc
Hydrogen cyanide
Hydrogen selenide
Hydrogen sulphide
Hydrogen telluride
*Methyl bromide (bromomethane)
*Methyl chloride
Methyl fluoride
Methyl marcaptan
Monoethylamine
Monomethylamine
Phosphine
Propylene oxide
Silane
Stibine
Trimethylamine
Vinyl bromide
Vinyl chloride

Asphyxiant All gases except oxygen.

Oxidant Oxygen, chlorine, fluorine, iodine pentafluoride, bromine pentafluoride, bromine trifluoride, chlorine trifluoride, nitrogen dioxide and trioxide. These gases may cause flammables, even metals, to inflame instantly.

Pyrophoric Silane and phosphine. Treat these gases as spontaneous flame sources and keep separate from other flammables.

*Pure products contained in cylinders identified with an exclusive colour and shown above left.

†Label indicates that dip tube is fitted.

NB. toxic classification assigned to those gases with tlv <500 ppm.

Table 13.1 (continued)

research with a view to the discovery and, so far as is reasonably practicable, the elimination or minimisation of any risks to health or safety to which the design or article may give rise.

(3) It shall be the duty of any person who erects or installs any article for use at work in any premises where that article is to be used by persons at work to ensure, so far as is reasonably practicable, that nothing about the way in which it is erected or installed makes it unsafe or a risk to health when properly used.

(4) It shall be the duty of any person who manufactures, imports or supplies any substance for use at work –

(*a*) to ensure, so far as is reasonably practicable, that the substance is safe and without risks to health when properly used;

(*b*) to carry out or arrange for the carrying out of such testing and examination as may be necessary for the performance of the duty imposed on him by the preceding paragraph;

(*c*) to take such steps as are necessary to secure that there will be available in connection with the use of the substance at work adequate information about the results of any relevant tests which have been carried out on, or in connection with, the substance and about any conditions necessary to ensure that it will be safe and without risks to health when properly used.

(5) It shall be the duty of any person who undertakes the manufacture of any substance for use at work to carry out or arrange for the carrying out of any necessary research with a view to the discovery and, so far as is reasonably practicable, the elimination or minimisation of any risks to health or safety to which the substance may give rise.

(6) Nothing in the preceding provisions of this section shall be taken to require a person to repeat any testing, examination or research which has been carried out otherwise than by him or at his instance, in so far as it is reasonable for him to rely on the results thereof for the purposes of those provisions.

(7) Any duty imposed on any person by any of the preceding provisions of this section shall extend only to things done in the course of a trade, business or other undertaking carried on by him (whether for profit or not) and to matters within his control.

(8) Where a person designs, manufactures, imports or supplies an article for or to another on the basis of a written undertaking by that other to take specified steps sufficient to ensure, so far as is reasonably practicable, that the article will be safe and without risks to health when properly used, the undertaking shall have the effect of relieving the first-mentioned person from the duty imposed by subsection (1)(*a*) above to such extent as is reasonable having regard to the terms of the undertaking.

(9) Where a person ('the ostensible supplier') supplies any article for use at work or substance for use at work to another ('the customer') under a hire-purchase agreement, conditional sale agreement or credit-sale agreement, and the ostensible supplier –

(*a*) carries on the business of financing the acquisition of goods by others by means of such agreements; and

(*b*) in the course of that business acquired his interest in the article or substance supplied to the customer as a means of financing its acquisition by the customer from a third person ('the effective supplier'),

the effective supplier and not the ostensible supplier shall be treated for the purposes of this section as supplying the article or substance to the customer, and any duty imposed by the preceding provisions of this section on suppliers shall accordingly fall on the effective supplier and not on the ostensible supplier.

(10) For the purposes of this section an article or substance is not to be regarded as properly used where it is used without regard to any relevant information or advice relating to its use which has been made available by a person by whom it was designed, manufactured, imported or supplied.

Manufacturers

A manufacturer has a primary duty to ensure, so far as is reasonably practicable, that any *article* is so designed and constructed that it is safe and without risks to health when properly used. (A designer of an article for use at work has similar duties to the manufacturer. In practice, the 'designer' is the employing body rather than an individual, i.e an employer is vicariously liable.) He must also take steps to ensure that adequate information is available about the designed use and about any necessary conditions of safe use. When it is impracticable to print full safety instructions on the product, or a label permanently affixed to it, a cautionary/advisory leaflet or manual should be provided.[6]

The design of all articles needs to be checked for dangerous features; this requires the use of reasonable foresight to anticipate the likely uses to which a product may be subjected. (In the past in the UK there has been no civil liability for injuries arising from an article being put to a use which could not have been expected, and giving rise to a danger which the manufacturer could not foresee.)[7] While certain dangers may be excluded by design changes, e.g. minor changes in materials of construction, there may be others which even if they could be eliminated, would involve a prohibitive cost; the test to apply is that of reasonable practicability.

The manufacturer, and designer, should carry out (or arrange for) any necessary research (a) to discover and (b) to eliminate or minimise any health or safety risks to which the article (or design) may give rise. There is no requirement to repeat research done by another provided a record of the results is available, and may reasonably be relied upon. In practice, it is important to check that design modifications are not subsequently introduced (e.g. in operating conditions, materials of construction, or dimensions) which invalidate the test results.

The methods used in production and the control of production, especially quality control, need to ensure that the products are all constructed to the approved design. Testing and examination must be carried out unless reliable results are available from someone else. Clearly, part of the final inspection should include a check that all the tests intended for the articles have, in fact, been carried out, that all the ancillaries are provided, and that operating/safety instructions are provided.

A manufacturer of a *substance* (i.e. any natural or artificial substance whether in solid, liquid, gaseous or vapour form, intended for use at work) has a duty to ensure, so far as is reasonably practicable, that it is safe and without risks to health when properly used. Furthermore, they cannot be relieved of their responsibility for research in the way possible for an article.

Importers of articles or substances for use at work have similar duties to manufacturers except as regards research. They should therefore inform all their foreign suppliers of UK standards. In theory they should also check imported products carefully but this clearly presents problems (a) if goods are not actually seen by the importers or (b) when the packing of the product effectively precludes examination and (c) when the importer lacks the expertise to examine a wide range of products for safety.

Suppliers have the same duties as importers. Here 'supply' includes supplying by way of sale, lease, hire or hire-purchase.

Sellers of a jewellery cleaner were held liable when the contents unexpectedly shot out of a plastic container and injured the user's eyes. The sellers had purchased the cleaner from a small manufacturer and had not examined or tested it sufficiently to determine its hazardous qualities or the unsuitability of the container.[8]

Reference has already been made to *Donaghue* v. *Stevenson* in which, by using a sealed and opaque bottle, the defendants had excluded all intermediate interference, i.e. there was no possibility of intermediate examination of the bottle's contents. This is a problem inherent in the supply of chemicals, pharmaceuticals, food products, solvents and cylinder gases. In these cases, it is particularly important to ensure that the substance is of the desired quality, since following *Donaghue* v. *Stevenson*, if someone is injured by consuming a foodstuff or beverage, contaminated because of the negligence of the manufacturer, or by using what should be 'harmless' proprietary substances (e.g. medicines, soap or cleansing powders) the manufacturer can be under a legal liability to compensate for the harm done.

It is also essential for the product to be packed in a properly designed/manufactured container (e.g. drum, cylinder, bottle) which will not leak or burst in transit *and* that confusion does not arise over labelling or colour coding (see packaging and labelling below).

With the sophisticated chemicals and packaging methods now common, it appears to be incumbent on each supplier to perform a hazard analysis on each of his products. Aerosol dispensers are one example of a useful, convenient means of product marketing but which can pose a fire and explosion hazard in storage, use and disposal. The products comprise the essential ingredients dispersed or dissolved in an appropriate solvent, either or both of which may be flammable. Liquefied, compressed, low-boiling-point halogenated hydrocarbons were formerly popular propellants; these are, in general, non-flammable but they can break down when exposed to a flame, or surfaces at a very high temperature, liberating toxic or corrosive gases (see Ch. 5). However, highly flammable LPGs are now favoured as propellants.[9] In any event, irrespective of the flammability of the aerosol dispenser contents an explosion hazard arises if the container is involved in a fire or heated excessively in some other way.

During tests[10] when aerosol containers were heated at the base, the bottom blew off and the remainder of the can became jet propelled; several travelled a distance of over 32.8 metres. When a container was enveloped in flame the walls fragmented and there was a considerable local blast effect. In the comparatively small number of aerosol dispensers examined and tested, there was a surprisingly large incidence of valve failures. In several instances it was difficult to determine whether the propellant was exhausted or whether valve failure had occurred. This may account for the large number of serious explosions which have been reported when containers thought to be empty have been thrown on to a fire.

It follows that basic precautions with an aerosol dispenser include[10]

- Assume that the container contains a liquid of flammability equivalent to that of petrol and treat it as such.
- Do not use an aerosol spray when smoking, or near any other source of ignition, e.g. an electric fire, hot-plate, etc.
- Do not throw used aerosol dispensers on to bonfires.
- Do not let young children have access to aerosol dispensers, full or empty.

It is for the supplier to draw these to the attention of the user in the best way practicable.

A similar example arises with the marketing of cleaning fluids. For example, avoidance of contact with the neat chemical, the proper dilution, and the need to avoid admixture of acid-based and hypochlorite-based agents (which can result in chlorine evolution) should be emphasised. With hydrogen peroxide solutions the importance of avoiding contamination, over-concentration, extraneous contact with metals, etc., is important,[11] and storage should be in a cool place.

It is, of course, vitally important that the quality of material supplied complies in fact with the description applied to it.

In 1973 a routine delivery was made from a chemical factory in Central Michigan to an agricultural feed plant in the same state. A ton of what was believed to be magnesium oxide, in heavy brown paper sacks on which a trade name was crudely stencilled, was unloaded at the plant. Over the following few weeks it was mixed into cattle feed and distributed throughout Michigan. In fact the additive was not Nutrimaster (magnesium oxide) but Firemaster, which was polybrominated biphenyl a chemical used for fire-proofing plastics. As a result, tens of thousands of Michigan cattle were poisoned.[12]

The two chemicals were produced in different buildings at the same factory and should have been stored in separate warehouses and dispatched from different loading areas.

The following is an example of a case, for which damages were recovered but which could have had more serious consequences.[1]

A seller sold some sulphuric acid to a buyer as commercially free from arsenic. The seller did not know that the buyer intended to use it for the manufacture of glucose which he sold to brewers for beer production. The acid was found subsequently to contain arsenic and the buyer recovered the price of the acid and damages for all the goods spoiled.[13] Clearly this defect was not discoverable on reasonable examination.

So far as a manufacturer is concerned then, a failure to take reasonable care can arise at any stage up to and including final inspection, packing and labelling. The customer may be in the position of only being able to describe what happened in the use of a substance or article, that this is not supposed to happen and wouldn't have happened if the manufacturer had taken reasonable care.[1]

A man contracted dermatitis from an excess of chemicals in a pair of new underpants; he could not show how the chemicals got there but proved that, when purchased, the pants were still in their original wrapping. The inference accepted by the court was that the manufacturer had not been sufficiently careful.[14]

The onus of proof has shifted in this way, i.e. such that it was for the manufacturers to show that they had not been negligent, or to give some explanation of the cause of an accident which did not connote negligence, in a case involving a faulty hot-water bottle which split after three months[15] and might logically be extended to, e.g. exploding mineral water bottles (in the absence of excessive impact or intermediate mishandling).

An outline procedure for the re-filling of returnable chemical containers, assuming the quality of the chemical to be charged has already been assured, is illustrated in Fig. 13.1. This should be equally applicable to drums, carboys or – with modifications – to gas cylinders which must by law be tested regularly.

Just after an operator had charged a 45-kg gas cylinder with dimethylamine the bottom of the cylinder ruptured. His ankles were sprayed with liquid. Subsequent investigation found that the bottom had corroded externally.

A 13-year-old gas cylinder used for standard gas mixtures ruptured at a pressure of 120 atm. It came loose from the rack, narrowly missed an operator, demolished a steel cabinet, severely damaged a concrete staircase and shattered neighbouring windows.[16] The cause was found to have been corrosion fatigue because at some time the cylinder had lain horizontally and contained a corrosive solution.

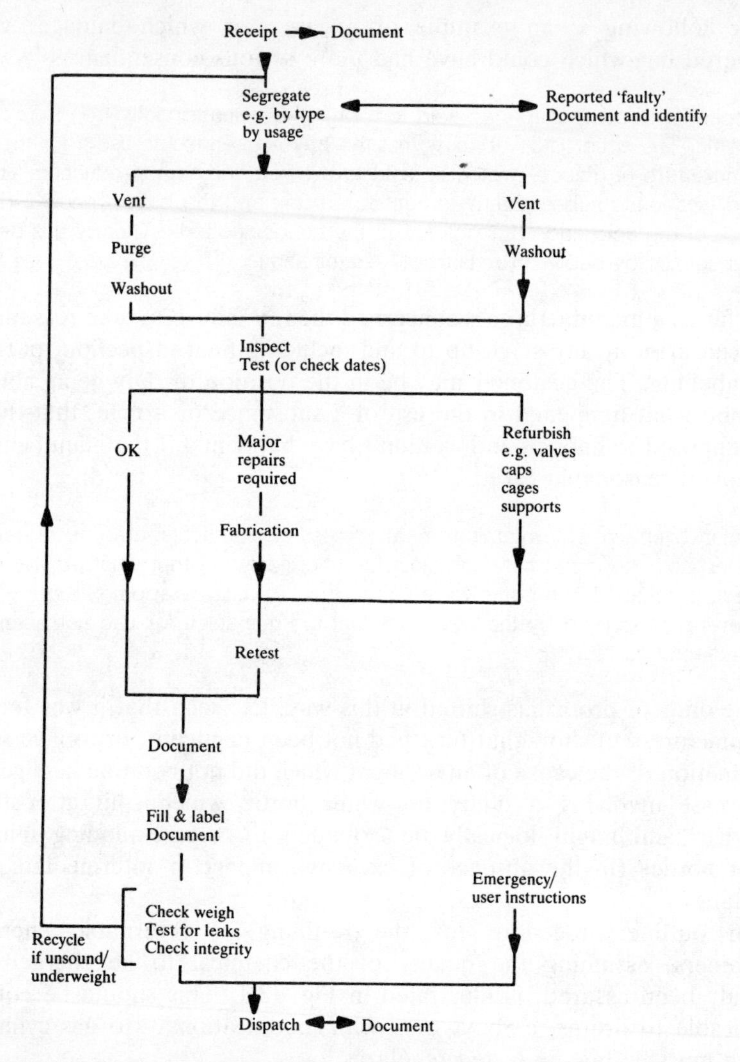

Fig. 13.1 Procedure for re-filling returnable chemical containers

These examples demonstrate the need for thorough external examination on re-filling and periodic proof testing, and restriction of use, of cylinders.

Clearly the integrity, and suitability for their purpose, of bulk containers, e.g road tankers, rail tankers, ships is of paramount importance.

The road tanker involved in the Spanish camp-site disaster (page 816) had its pressure relief valve blanked off from a previous journey.

A $20 $\times$ 10^6 loss occurred at Bantry Bay in 1979 following fires and explosions involving an eleven-year-old 121,000 DWT oil tanker. This ship had completed unloading its first parcel of crude oil at a deep water port. A small fire was noticed on the deck at a time when no transfer operations were in progress. About 10 minutes later fire spread aft along the ship's length and to the sea on both sides. A massive explosion occurred after 30 minutes. In addition to total loss of the ship 340 m of the jetty were damaged or destroyed.

(The initiating event of the disaster was theorised as buckling of the ship's structure at about deck level, followed by explosions in the ballast tanks and the breaking of the ship's back. This was attributed to a seriously weakened hull, due to inadequate maintenance, and an excessive stress due to incorrect ballasting.)[17]

On-site hazards may arise during container preparation and filling as in normal plant operations. During preparation of cylinders, carboys or tankers for receipt of a fresh consignment, depending on the application, testing and inspection may be important between venting, purging and washing operations.

When a tank which had contained phosphorus oxychloride was being filled with water an emission of hydrochloric acid mist necessitated evacuation of works.[18] (Because of very cold weather the phosphorus oxychloride had partially crystallised and some crystals left in the tank reacted with the water.)

Precautions to be taken during the unloading/loading of tankers follow from previous chapters. The most likely causes of a gas escape or liquid spillage include a fractured or failed hose connection, overfilling, faulty metering, pump failure or pump seal leakage. Overfilling may occur for any variety of reasons including incorrect identification, miscalculation, inattention, failure of alarms, etc.

Petroleum spirit (136.2 metres^3) overflowed into a main drainage system and thence into a river running through a heavily populated area. Fortunately, ignition did not occur. The incident was caused by two tanks from a berthed tanker being inadvertantly connected to one of two underground bulk storage tanks each with capacity for one ship's tank only.[19]

Hazardous substances

When it becomes apparent that a substance is not safe to use by the current processes and procedures then the manufacturer should stop supplying it.

Two workers in a rubber factory were found in 1966 to have contracted bladder cancer. In the late 1940s they had been exposed to Nonox S an antioxidant containing carcinogenic substances (β-naphthylamine), and used in the manufacture of tyres and tubes. They sued both their employer and the manufacturer of the antioxidant. It was held that the manufacturer of a chemical product owed to the purchaser's employees a duty of care to satisfy himself that the product was safe, i.e that there was no substantial risk of any substantial injury to health on the part of persons who were likely to use it or to be brought into contact with its use (that use being as reasonably expected). Furthermore, it was held that the manufacturer knew, or ought to have realised, that Nonox S was, or contained, a carcinogen involving a real risk of bladder cancer to rubber workers and was negligent in failing to withraw it before the end of 1946.[20]

As mentioned in Chapter 16, UK law recognises a special duty of care arising in connection with things, e.g. substances, which are dangerous *per se*; this is limited to, for example, explosives, poisons, strong acids and highly flammable materials. Three duties in connection with such things are[7]:

1. They must not be delivered to an irresponsible person. This excludes any possibility of supplying, e.g. petrol or sodium chlorate (weed-killer) to children.
2. They must not be delivered even to a responsible person without due warning. This reinforces the need for adequate marking of containers, e.g. with warning symbols, and supply of information.
3. They must not be left in a situation where they may be tampered with. This illustrates the need for, e.g. acid carboys or LPG cylinders to be delivered and stored under proper control.

Plant erectors and installers

Under s.6(3) of the HASAWA anyone who erects or installs any article for use at work in any premises must ensure, so far as is reasonably practicable, that nothing about the way in which it is erected or installed makes it unsafe or a risk to health when properly used. Thus contractors and companies installing their own products must plan the work, and institute a system for inspection/examination and testing, to eliminate every foreseeable hazard. There have in fact been many actions involving civil liability in this area.

A company designed and installed equipment for another company's factory; this plant was to store and dispense molten stearin. The designer/installers put in equipment which included plastic pipes. The day before it was due to be tested an employee at the factory switched on the heating element so that the stearin would be molten for the tests. Unfortunately, the piping suffered from heat distortion, molten stearin escaped and the factory was burned down.

It was subsequently held that designing and installing the plant with the particular type of plastic piping rendered the equipment wholly unsuitable for its intended purpose; a fundamental breach of contract for which damages were awarded.[1]

Information supply and testing

The duty of manufacturers, suppliers, etc., includes the carrying out of such tests on the substance as are deemed necessary. They are also responsible for taking the necessary steps to ensure that information is made available as to the results of any tests together with any conditions to be met to ensure that the substance will be safe and without risks to health when properly used. The testing of chemicals is discussed later in this chapter.

Suppliers and importers of substances may reduce their commitment to carry out testing by the use of reference data from the manufacturers.[21] However, the wide range of information that may be required on hazardous materials is exemplified in Table 12.2. In any event if suppliers, etc., do rely on data from manufacturers, etc., they must satisfy themselves that the results are adequate and reliable. Users should have rapid and easy access to information. This does not mean that the information needs to be provided with each substance, but that the supplier should be satisfied that it is readily available to the user. Ideally, information should be supplied with the product but if the same product is being supplied frequently, it may be assumed that once the information has been provided it would be unnecessary to do so on each delivery.

A balance should be struck between excessive and inadequate information but it is not sufficient merely to draw attention to the dangerous nature of the substance.[21] A product may be considered defective because it has been marketed without adequate warnings or directions for safe use.[22] It may be argued that any hazardous characteristic of a substance should have been indicated on the packaging or by some other means.

Since it will not always be practicable to predict the use to which a substance will be put, the information provided should contain details of the substance's properties and potential hazards of use so that the user can make informed judgements as to the precautions to be taken depending on the use to which the substance will be put. Information should include the hazardous properties of the substance and the steps that can be taken to avoid the conditions which enhance such properties, together with the proper techniques for storage, handling and use. Further reference should be made to first-aid treatment, fire precautions, etc. Obviously it is in the interests of manufacturers, suppliers, etc., to provide information which reduces the probability of a substance endangering health and safety when used at work. Therefore the information should be as clear and comprehensive as possible.

Attempts to escape responsibility for the information supplied in safety data sheets by disclaimers such as 'The information provided is to the best of our knowledge true and accurate, but all instructions, recommendations and suggestions are made without guarantee. Since the conditions of use are beyond our control we disclaim any liability for loss or damage suffered from use of this information' have no legal standing by virtue of s.2 of the Unfair Contract Terms Act 1977 within the UK, viz.:

2(1) A person cannot by reference to any contract term or to a notice given to persons generally or to particular persons exclude or restrict his liability for death or personal injury resulting from negligence.

2(3) Where a contract term or notice purports to exclude or restrict liability for negligence a person's agreement to or awareness of it is not of itself to be taken as indicating his voluntary acceptance of any risk.

In the UK an employer's duties under s.2 of the HASAWA include the provision of such information . . . as is necessary to ensure so far as is reasonably practicable the health and safety at work of his employees. This involves informing employees of the nature of substances, associated hazards involved and the precautions to be taken in their use. The duty to provide information is complementary to that to provide training so as to ensure that employees understand, and will act upon, the information (see Ch. 17).

The employer is also required to make available to employees' safety representatives information necessary for them to fulfil their functions.[23] This should include 'information of a technical nature about hazards to health and safety and precautions deemed necessary to eliminate or minimise them . . . including any relevant information provided by consultants or designers or by the manufacturer, importer or supplier of any article or substance used or proposed to be used at work by their employees'. Figure 13.2 is an example of a chemical safety information sheet for internal distribution.[24]

Further discussions of the effective management of safety and hygiene information relating to chemicals, and a bibliography, are given in reference.[25]

Packaging and labelling

A widely adopted system of hazard classification for the transport of hazardous materials by land, sea and air follows the recommendations of a United Nations committee of experts on the transport of dangerous goods. Hazard types are divided into nine main classes represented numerically 1–9. Most classes are further sub-divided into hazard divisions and sub-divisions depending on appropriate criteria. A simplified

Roche Products Limited
Chemical Safety Information
STANDARD No 30
DATE: 22 October 1975
Prepared by: Mr K C Ling

Material:
Carbon disulphide

Description:
Clear, colourless liquid – heavier than and insoluble in water. Disagreeable odour.

Health hazard:
Carbon disulphide exhibits both acute and chronic effects.
Inhalation: Threshold limit value for carbon disulphide is 20 ppm. Exposure to levels above 20 ppm will result in initial intoxication followed by depression, stupor and unconsciousness, the degree being dependent on the concentration. Rapid onset of these symptoms would arise with concentrations of 500 ppm. No acute symptoms are shown with levels below 150 ppm. Repeated exposure to levels between 20 and 150 ppm may result in chronic effects – damage to central nervous system characterised by loss of memory, fatigue, headache, disturbance of vision, melancholia, vertigo, loss of reflexes, etc. Provided exposure is ceased when these symptoms first appear, complete recovery is probable.
Skin contact: Defatting action may cause skin irritation and dermatitis.
Ingestion: Carbon disulphide will be absorbed into the system and acute symptoms of carbon disulphide poisoning would follow.

First aid:
Inhalation: Remove person from exposure, patient to rest and keep warm – summon medical aid.
Eye/skin contact: If splash in eye, wash out thoroughly with water. Spill on person – remove affected clothing and allow carbon disulphide to evaporate away. Wash affected areas of body with soapy water.
Ingestion: Wash out mouth with water. Induce vomiting (salt and water emetic).

Fire hazard:
Exceptionally low auto-ignition temperature (temperatures as low as 80 °C are quoted). Thus vapour will ignite on contact with any moderately hot surface (steam pipe, light bulb, vehicle exhaust pipe, etc.). Flash point is −30 °C. Explosive limits are 1%–50% in air. Boiling point is 46.3 °C. Sulphur dioxide formed on combustion, therefore breathing apparatus required.

Fire fighting:
Breathing apparatus must be worn when tackling fire of this material. Water fog or water spray (carbon disulphide is one of the few low flash point liquids heavier than water, thus water is most effective in extinguishing a fire). Dry chemical powder, carbon dioxide and Halons are also effective at extinguishing fire, but do not prevent further evaporation as does water.

Spillage:
1 Clear area of personnel.
2. Cordon off area – notify fire station.
3. If spillage is too large for safe evaporation, blanket with water.
4. Drum spillage and send for disposal.
DO NOT ATTEMPT TO WASH DOWN DRAINS

Handling:
Carbon disulphide should be handled in enclosed systems. Where this is not practical, either the ventilation must be such that the threshold limit value is not exceeded or personnel must be provided with respiratory protection.

Fig. 13.2 Internal chemical safety information sheet[24]

list of hazard types is given in Table 13.2. The criteria adopted for classifying substances vary, depending on the appropriate regulations and mode of transport, e.g. the Intergovernmental Maritime Consultative Organisation Dangerous Goods Code (IMCO),[26] the European Agreement concerning the International Carriage or Dangerous Goods by Road (ADR),[27] the International Regulations concerning the Carriage of Dangerous Goods by Rail (RID)[28] and the International Air Transport Association Restricted Articles Regulations (IATA).[29]

In the UK comprehensive and practical provisions for the classification, packaging and labelling of dangerous substances for both supply and conveyance by road are encompassed by complex legislation.[30,31] The Regulations are supported by Approved Codes of Practice[32,33] and a separate 'Approved List' identifying particulars to be shown on labels for a wide spectrum of specific chemicals. For example, Parts IV and V of the list, reproduced as Tables 13.3 and 13.4, give 'risk ' and 'safety ' phrases to be used on the labels of specific materials. The supply provisions aim to protect those who handle or use dangerous substances whether at work, in education, or in the home. These Regulations and supporting documents serve to implement Treaty obligations laid down by EEC Directives.[35] The packaging and labelling provisions of the Directive on toxic and dangerous waste are also implemented.[36] Separate provisions[37] exist for the bulk conveyance of hazardous substances in road tanks or tank containers, as discussed later.

Table 13.2 UN Hazard Classification

UN class		Hazard type
1	1.1–1.5	*Explosives* Mass explosion hazard – very insensitive substance
2		*Gases* Compressed, liquefied or dissolved under pressure
3	3.1–3.2	*Inflammable liquids* Flash point 23–60.5 °C
4	4.1	*Inflammable solids*
	4.2	Spontaneously combustible substances
	4.3	Substances giving off inflammable gases in contact with water
5	5.1	*Oxidising substances other than organic peroxides*
	5.2	*Organic peroxides*
6	6.1	*Poisonous (toxic) substances*
	6.2	*Infectious substances*
7		*Radioactive substances*
8		*Corrosive substances*
9		*Miscellaneous dangerous substances*

(NB Inflammable ≡ Flammable)

Table 13.3 'Risk' phrases from the Approved List[34]

Indication of particular risks

1: Explosive when dry
2: Risk of explosion by shock, friction, fire or other sources of ignition
3: Extreme risk of explosion by shock, friction, fire or other sources of ignition
4: Forms very sensitive explosive metallic compounds
5: Heating may cause an explosion
6: Explosive with or without contact with air
7: May cause fire
8: Contact with combustible material may cause fire
9: Explosive when mixed with combustible material
10: Flammable
11: Highly flammable
12: Extremely flammable
13: Extremely flammable liquefied gas
14: Reacts violently with water
15: Contact with water liberates highly flammable gases
16: Explosive when mixed with oxidising substances
17: Spontaneously flammable in air
18: In use, may form flammable/explosive vapour–air mixture
19: May form explosive peroxides
20: Harmful by inhalation
21: Harmful in contact with skin
22: Harmful if swallowed
23: Toxic by inhalation
24: Toxic in contact with skin
25: Toxic if swallowed
26: Very toxic by inhalation
27: Very toxic in contact with skin
28: Very toxic if swallowed
29: Contact with water liberates toxic gas
30: Can become highly flammable in use
31: Contact with acids liberates toxic gas
32: Contact with acids liberates very toxic gas
33: Danger of cumulative effects
34: Causes burns
35: Causes severe burns
36: Irritating to eyes
37: Irritating to respiratory system
38: Irritating to skin
39: Danger of very serious irreversible effects
40: Possible risk of irreversible effects
41: Risk of serious damage to eyes
42: May cause sensitisation by inhalation
43: May cause sensitisation by skin contact
44: Risk of explosion if heated under confinement
45: May cause cancer
46: May cause heritable genetic damage
47: May cause birth defects
48: Danger of serious damage to health by prolonged exposure

Combination of particular risks

14/15: Reacts violently with water, liberating highly flammable gases
15/29: Contact with water liberates toxic, highly flammable gas
20/21: Harmful by inhalation and in contact with skin
20/21/22: Harmful by inhalation, in contact with skin and if swallowed
20/22: Harmful by inhalation and if swallowed

Table 13.3 (continued)

21/22: Harmful in contact with skin and if swallowed
23/24: Toxic by inhalation and in contact with skin
23/24/25: Toxic by inhalation, in contact with skin and if swallowed
23/25: Toxic by inhalation and if swallowed
24/25: Toxic in contact with skin and if swallowed
26/27: Very toxic by inhalation and in contact with skin
26/27/28: Very toxic by inhalation, in contact with skin and if swallowed
26/28: Very toxic by inhalation and if swallowed
27/28: Very toxic in contact with skin and if swallowed
36/37: Irritating to eyes and respiratory system
36/37/38: Irritating to eyes, respiratory system and skin
36/38: Irritating to eyes and skin
37/38: Irritating to respiratory system and skin
42/43: May cause sensitisation by inhalation and skin contact

Table 13.4 'Safety' phrases from the Approved List[34]

Indication of safety precautions required
1: Keep locked up
2: Keep out of reach of children
3: Keep in a cool place
4: Keep away from living quarters
5: Keep contents under . . . (appropriate liquid to be specified by the manufacturer)
6: Keep under . . . (inert gas to be specified by the manufacturer)
7: Keep container tightly closed
8: Keep container dry
9: Keep container in a well-ventilated place
12: Do not keep the container sealed
13: Keep away from food, drink and animal feeding stuffs
14: Keep away from . . . (incompatible materials to be indicated by the manufacturer)
15: Keep away from heat
16: Keep away from sources of ignition – No Smoking
17: Keep away from combustible material
18: Handle and open container with care
20: When using do not eat or drink
21: When using do not smoke
22: Do not breathe dust
23: Do not breathe gas/fumes/vapour/spray (appropriate wording to be specified by manufacturer)
24: Avoid contact with skin
25: Avoid contact with eyes
26: In case of contact with eyes, rinse immediately with plenty of water and seek medical advice
27: Take off immediately all contaminated clothing
28: After contact with skin, wash immediately with plenty of . . . (to be specified by the manufacturer)
29: Do not empty into drains
30: Never add water to this product
33: Take precautionary measures against static discharges
34: Avoid shock and friction
35: This material and its container must be disposed of in a safe way
36: Wear suitable protective clothing
37: Wear suitable gloves
38: In case of insufficient ventilation, wear suitable respiratory equipment

Table 13.4 (continued)

39: Wear eye/face protection
40: To clean the floor and all objects contaminated by this material use . . . (to be specified by the manufacturer)
41: In case of fire and/or explosion do not breathe fumes
42: During fumigation/spraying wear suitable respiratory equipment (appropriate wording to be specified)
43: In cases of fire, use . . . (indicate in the space the precise type of fire-fighting equipment. If water increases the risk, add – Never use water)
44: If you feel unwell, seek medical advice (show the label where possible)
45: In case of accident or if you feel unwell, seek medical advice immediately (show the label where possible)
46: If swallowed seek medical advice immediately and show this container or label
47: Keep at temperature not exceeding . . . °C (to be specified by the manufacturer)
48: Keep wetted with . . . (appropriate material to be specified by the manufacturer)
49: Keep only in the original container
50: Do not mix with . . . (to be specified by the manufacturer)
51: Use only in well-ventilated areas
52: Not recommended for interior use on large surface areas

Combination of safety precautions required

1/2: Keep locked up and out of reach of children
3/7/9: Keep container tightly closed, in a cool well-ventilated place
3/9: Keep in a cool, well-ventilated place
3/9/14: Keep in a cool, well-ventilated place away from . . . (incompatible materials to be indicated by the manufacturer)
3/9/14/49: Keep only in the original container in a cool, well-ventilated place away from . . . (incompatible materials to be indicated by the manufacturer)
3/9/49: Keep only in the original container in a cool, well-ventilated place
3/14: Keep in a cool place away from . . . (incompatible materials to be indicated by the manufacturer)
7/8: Keep container tightly closed and dry
7/9: Keep container tightly closed and in a well-ventilated place
20/21: When using do not eat, drink or smoke
24/25: Avoid contact with skin and eyes
36/37: Wear suitable protective clothing and gloves
36/37/39: Wear suitable protective clothing, gloves and eye/face protection
36/39: Wear suitable protective clothing and eye/face protection
37/39: Wear suitable gloves and eye/face protection
47/49: Keep only in the original container at temperature not exceeding . . . °C (to be specified by the manufacturer).

Exclusions apply to substances where existing legislation is already adequate (e.g. medicines, cosmetic preparations, bulk supply and road conveyance, pesticides packaged and labelled in accordance with the Pesticides Safety Precautions Scheme) or where precedent was set by international rules or recommendations (e.g. international journeys in accordance with road/rail/sea/air international model rules, or supply packaging and labelling of gas cylinders). Otherwise any substance is dangerous if in Part 1A of the approved list, or if it exhibits a hazardous property defined in Schedule 1 for supply or Schedule 2 for conveyance as shown in Figs 13.3 and 13.4. The Regulations also apply to a limited

SCHEDULE 1

The classification of and symbols for substances dangerous for supply

PART I

Table of characteristic properties, indications of general nature of risk and symbols

1 Characteristic properties of the substance	2 Classification and indication of general nature of risk	3 Symbol
A substance which may explode under the effect of flame or which is more sensitive to shocks or friction than dinitrobenzene.	Explosive.	
A substance which gives rise to highly exothermic reaction when in contact with other substances, particularly flammable substances.	Oxidizing.	
A liquid having a flash point of less than 0 degrees Celsius and a boiling point of less than or equal to 35 degrees Celsius (see Note 1).	Extremely flammable.	
A substance which— *(a)* may become hot and finally catch fire in contact with air at ambient temperature without any application of energy; *(b)* is a solid and may readily catch fire after brief contact with a source of ignition and which continues to burn or to be consumed after removal of the source of ignition; *(c)* is gaseous and flammable in air at normal pressure (see Note 1); *(d)* in contact with water or damp air, evolves highly flammable gases in dangerous quantities; or *(e)* is a liquid having a flash point below 21 degrees Celsius (see Note 1).	Highly flammable.	

Fig. 13.3 The classification of, and symbols for, substances dangerous for supply

Fig. 13.3 *continued*

1 Characteristic properties of the substance	2 Classification and indication of general nature of risk	3 Symbol
A substance which is a liquid having a flash point equal to or greater than 21 degrees Celsius and less than or equal to 55 degrees Celsius, except a liquid which when tested at 55° in the manner described in Schedule 2 to the Highly Flammable Liquids and Liquefied Petroleum Gases Regulations 1972 (a) does not support combustion. (See Note 1)	Flammable.	No symbol required.
A substance which if it is inhaled or ingested or it penetrates the skin, may involve extremely serious acute or chronic health risks and even death (see Note 2).	Very toxic.	
A substance which if it is inhaled or ingested or it penetrates the skin, may involve serious acute or chronic health risks and even death (see Note 2).	Toxic.	
A substance which if it is inhaled or ingested or if it penetrates the skin, may involve limited health risks (see Note 2).	Harmful.	
A substance which may on contact with living tissues destroy them.	Corrosive.	
A non-corrosive substance which, through immediate, prolonged or repeated contact with the skin or mucous membrane, can cause inflammation.	Irritant.	

Note 1. Preparations packed in aerosol dispensers shall be classified as flammable in accordance with Part III of this Schedule.

Note 2. Substances shall be classified as very toxic, toxic or harmful in accordance with the additional criteria set out in Part II of this Schedule.

(a) S.I. 1972/917.

SCHEDULE 2 Regulations 2(1) and 6(4)

The classification of and hazard warning signs for substances dangerous for conveyance

PART I

Table of characteristic properties, classification and hazard warning signs

1 Characteristic properties of the substance	2 Classification	3 Hazard warning sign
A substance which— (*a*) has a critical temperature below 50°C or which at 50°C has a vapour pressure of more than 3 bars absolute; and (*b*) is conveyed by road at a pressure of more than 500 millibars above atmospheric pressure or in liquefied form, other than a toxic gas or a flammable gas.	Non-flammable compressed gas.	 COMPRESSED GAS
A substance which has a critical temperature below 50°C or which at 50°C has a vapour pressure of more than 3 bars absolute and which is toxic.	Toxic gas.	 TOXIC GAS
A substance which has a critical temperature below 50°C or which at 50°C has a vapour pressure of more than 3 bars absolute and is flammable (see Note 1).	Flammable gas.	 FLAMMABLE GAS

Fig. 13.4 The classification of, and hazard warning signs for, substances dangerous for conveyance

Fig. 13.4 *continued*

1 Characteristic properties of the substance	2 Classification	3 Hazard warning sign
A liquid with a flash point of 55°C or below except a liquid which— *(a)* has a flash point equal to or more than 21°C and less than or equal to 55°C; and *(b)* when tested at 55°C in the manner described in Schedule 2 to the Highly Flammable Liquids and Liquefied Petroleum Gases Regulations 1972 (a) does not support combustion. (See Notes.)	Flammable liquid.	FLAMMABLE LIQUID
A solid which is readily combustible under conditions encountered in conveyance by road or which may cause or contribute to fire through friction.	Flammable solid.	FLAMMABLE SOLID
A substance which is liable to spontaneous heating under conditions encountered in conveyance by road or to heating in contact with air being then liable to catch fire.	Spontaneously combustible substance.	SPONTANEOUSLY COMBUSTIBLE
A substance which in contact with water is liable to become spontaneously combustible or to give off a flammable gas.	Substance which in contact with water emits flammable gas.	DANGEROUS WHEN WET

Fig. 13.4 *continued*

1 Characteristic properties of the substance	2 Classification	3 Hazard warning sign
A substance other than an organic peroxide which, although not itself necessarily combustible, may by yielding oxygen or by a similar process cause or contribute to the combustion of other material.	Oxidizing substance.	OXIDIZING AGENT
A substance which is— *(a)* an organic peroxide; and *(b)* an unstable substance which may undergo exothermic self-accelerating decomposition.	Organic peroxide.	ORGANIC PEROXIDE
A substance known to be so toxic to man as to afford a hazard to health during conveyance or which, in the absence of adequate data on human toxicity, is presumed to be toxic to man.	Toxic substance.	TOXIC
A substance known to be toxic to man or, in the absence of adequate data on human toxicity, is presumed to be toxic to man but which is unlikely to afford a serious acute hazard to health during conveyance.	Harmful substance.	HARMFUL — STOW AWAY FROM FOODSTUFFS

Fig. 13.4 *continued*

1 Characteristic properties of the substance	2 Classification	3 Hazard warning sign
A substance which by chemical action will— *(a)* cause severe damage when in contact with living tissue; *(b)* materially damage other freight or equipment if leakage occurs.	Corrosive substance.	CORROSIVE
A substance which is listed in Part 1A of the approved list and which may create a risk to the health or safety of persons in the conditions encountered in conveyance by road, whether or not it has any of the characteristic properties set out above.	Other dangerous substance.	DANGEROUS SUBSTANCE
Packages containing two or more dangerous substances which have different characteristic properties.	Mixed hazards.	DANGEROUS SUBSTANCE

range of articles containing dangerous substances (Part 1C of the approved list) and to certain products whether or not they are 'dangerous' as defined.

Classification of a substance dangerous for supply is covered by Regulation 5, together with various schedules, and for conveyance by Regulation 6; detailed reference is necessary because of the complexities.

The basic requirements for packaging are that any package should be suitable for its purpose having regard to the substance's properties; receptacles should be so designed, constructed and maintained as to prevent escape of the contents when subjected to the stresses and strains of normal handling. Packages should also be capable of withstanding being dropped from such a height as can be expected in normal handling.[33] The receptacle and its contents must be compatible and any closure intended for repeated re-fastening has to be capable of this duty without leaking. Factors in design and construction of packaging for road conveyance, and special packagings for organic peroxides and other unstable substances, are covered in the Approved Code.[33]

> Plastic materials likely to be softened, rendered brittle or permeable due to the chemical action of a substance should be excluded from contact with it; permeability should also not be such as is likely to create a danger.[33]

In general for supply (Regulation 8) all layers of packaging have to be labelled. The information on the label includes an indication of the general nature of the risk with corresponding symbols, if any; risk phrases – standard phrases used to supplement and complement the symbols and indications of the general nature of the risk and safety phrases – standard phrases setting out sensible precautions to take,[32] plus the name and address of the manufacturer/importer/wholesaler/other supplier.

For conveyance by road (Regulation 9) only the outer layer of packaging likely to be handled need be labelled and there are exemptions for receptacles containing small volumes. The information includes the designation of the substance, the substance identification number, the hazard warning sign and the name and address or telephone number or both of the consigner/other person in Great Britain from whom expert advice on the dangers of the substance can be obtained. For a receptacle which is larger than 25 litres, additional information should indicate the danger to which the substance may give rise and the requisite emergency action, e.g. fire, first aid, spillage instructions; this may be provided in a separate statement accompanying the package. Practical guidance is given in reference.[32] Allowances are made for a single combined form of labelling for substances which are both dangerous for 'supply' and 'conveyance'.

Examples of labels are shown in Fig. 13.5.[31]

Freight containers

Special considerations apply to the booking, packing, transport and unpacking of dangerous goods in freight containers to be carried by sea. Declaration and classification of the substance(s), e.g. to the Merchant Shipping (Dangerous Goods) Rules 1965, labelling, segregation and proper secure stowage are all important. External inspection of packages should be carried out whenever they are handled, and any that are

(a)

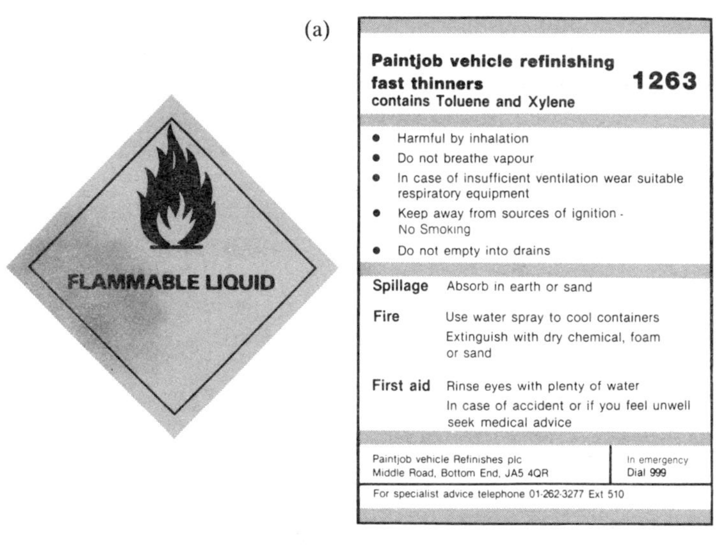

(b)

Paintjob vehicle refinishing fast thinners

contains toluene and xylene

1263

Paintjob Vehicle Refinishes plc,
Middle Road, Bottom End JA5 4QR County

(c)

Paintjob vehicle refinishing fast thinners

contains toluene and xylene

- Harmful by inhalation
- Do not breathe vapour
- In case of insufficient ventilation wear suitable respiratory equipment
- Keep away from sources of ignition — No Smoking
- Do not empty into drains

Paintjob Vehicle Refinishes plc
Middle Road, Bottom End JA5 4QR County

HIGHLY FLAMMABLE

HARMFUL

Fig. 13.5 Examples of labels for packing: (a) Combined supply and conveyance labels including additional information for conveyance; (b) Conveyance labels without additional information (this might be shown when required on a separate statement); (c) Supply labels for each discrete layer of packaging which does not also need to show conveyance labelling information

suspect set aside for examination. Advice on these aspects, and on intermediate storage has been issued by the Chemical Industries Association.[38]

During unpacking of dangerous goods containers:

- The possibility of damage to the cargo in transit should be borne in mind before the doors are opened. Expert advice should be sought before unpacking damaged containers.
- Any container should be ventilated – by leaving the doors open for a short period – before unpacking commences, particularly with toxic products.
- No hazards should be left after unpacking, e.g. special cleaning should be used if a toxic spillage has, or may have, occurred.

Gas cylinders

In the UK cylinders for compressed gases are marked as specified in BS 349:1973.[39] Primary identification is by labelling the name(s) and chemical formulae on the shoulder of the cylinder. Secondary identification is by the use of ground colours on the cylinder body and colour bands on the cylinder shoulder to denote the nature of the gas. Examples for standard gases are shown in Plate 2, before page 772.

Data sheets

While some primary warnings and instructions are carried by labels, it is usually necessary with industrial chemicals to provide more detailed information – generally by the issue of data sheets. One example is reproduced as Fig. 13.6. Many suppliers also produce booklets designed to cover a single product or to provide information on a range of products. More detailed technical reports may be issued for specialist users. Wall-charts of information or a guide to information sources, may be provided for display in laboratories or workrooms.

Information which should *ideally* be supplied to the users of chemicals comprises:

- *Identification*
  Identification of product by trade name. Supplier's name and address plus a telephone number (for emergencies).
- *Chemical composition*
  Chemical composition; details of ingredients, formulae and proportions.

  (The arguments for complete disclosure of composition, which may be contrary to commercial interest, are to enable users to choose the least toxic product capable of doing the job; adopt proper control measures; give proper medical treatment in cases of poisoning; and check independently the accuracy of supplier's health and safety information.[41])

No. 26 Cold Setting Phenolic Resin

Composition
A liquid phenolic resin containing water and small amounts of unreacted formaldehyde and phenol.

Fire Hazard
Products are classed as non flammable but are combustible and should be stored in closed containers, when not in use, and kept away from naked flames in the event of fire.

Explosive Hazard
A violent exothermic reaction occurs if mixed with acid catalyst which must be avoided.

Corrosion Hazard
None.

Health Hazard
Unreacted formaldehyde may exist in small quantities and is an irritant gas, phenol is harmful through skin contact. Skin irritation can therefore occur with sensitive skinned operators and skin contact should be avoided by the use of barrier creams and/or protective clothing.

Decomposition Products
Fumes may arise at the casting, cooling and knock-out stations due to thermal decomposition of resin and catalyst. In addition to formaldehyde and phenol other gases and vapours, e.g. carbon monoxide and sulphur dioxide may be formed. Local exhaust ventilation is likely to be needed to maintain acceptable working conditions. (See Guidance Note EH40/84.)

Storage, Spillage and Leakage Procedures
Store at normal temperature in closed containers away from acid catalysts. In the event of leakage cover with large amounts of sand. Cautiously react the resin-saturated sand with acid catalyst before disposal. If spillage reaches drains inform Regional Water Authority.

Disposal
Unused resin is regarded as notifiable for the purposes of the Deposit of Poisonous Waste Act 1972. Inform the manufacturers. Mixed sand may be disposed of at an approved site away from drinking water.

General Recommendations and First Aid
Eye protection is recommended under all circumstances. If contact occurs immediately irrigate eyes for 15 minutes with clean water. Ingestion must be avoided, accidental ingestion should be treated by an emetic. Inhalation not normally problematic, but where severe exposure occurs remove operator to fresh air. Skin contact should be avoided, if it occurs use cleansing creams and wash with soap and warm water. In all preceding situations medical attention should be sought.

Fig. 13.6 Health and safety data sheet (supplied in this case in a binder with a brief introduction to the safe use of chemical products in foundries) (Courtesy of British Industrial Sand Ltd.)

- *First Aid*
  Including measures to be taken for over-exposure to eyes/skin or by inhalation/ingestion/absorption.
  Information for medical treatment.
- *Disposal – normal and emergency spills*
  General precautions, disposal methods and statutory controls.
  Clean-up, neutralisation, disposal methods for spills.
- *Fire and explosion hazards and precautions*
  Flammable limits, flash point, auto-ignition temperature, etc.[42]
  Reactivity, stability.[43,44]
  Fire-fighting procedures.
- *Health hazards*
  Acute (short-term) effects and chronic (long-term) effects.
  Special hazards.
  First detectable signs of over exposure by all means of entry.
- *Test data*
- *Exposure limits*
  Threshold Limit Values or Occupational Exposure Limits (preferably with reference to their interpretation, i.e. not as 'safe' levels).
  For mixtures – reference to prediction, or substance used for basis.
- *Legal requirements*
  Any legislation applicable to safe handling, storage and use.
- *Control measures*
  Containment, ventilation, means of limiting exposure generally.
- *Personal protection*
  Identify requirements and circumstances for use.
  Emergency requirements.
- *Storage and handling*
  Conditions of storage, segregation, materials of construction.
  Types of container and handling precautions.
  Where relevant, procedures for disposal of containers, e.g. aerosol dispensers, drums when 'empty'.
- *References*
  References consulted and further sources of information.

One company send data sheets by registered or recorded delivery and carefully record mailings. A product is not marketed before a data sheet has been prepared. All data sheets are re-issued periodically and a data sheet is issued for any new sale at the time the order is processed, rather than when the invoice is processed since the product might then have been used before the data sheet arrived.[45]

Testing of chemicals

Testing of chemicals is an inherent part of any safety in use and marketing strategy. Apart from normal quality-control requirements, tests are necessary to determine and quantify the types of hazard reviewed in previous chapters.

With regard to the supply of 'new substances' legal obligations exist within the EEC, under Council Directive No. 67/548/EEC as amended in particular for the sixth time by Council Directive No. 79/831/EEC, for the Authorities to be notified in advance. Member States had until December 1982 to register their products and materials for inclusion on the European Communities Inventory (ECOIN) list. Materials so listed together with supplementary additions will form the European Inventory of Existing Chemical Substances (EINECS) list scheduled to be published in 1985.[46] The EINECS index would serve as the definitive list for use of raw materials in the chemical and allied industries which would not require pre-marketing notification. In the UK these obligations are implemented by the Notification of New Substances Regulations 1982, which require a manufacturer or importer to notify the Executive at least 45 days in advance of supplying the substance where total quantities of 1 tonne or more are involved in any 12-month period. A 'new chemical' is defined as 'any substance except a substance which had been supplied (either alone or as the component of a preparation) to a person in a Member State at any time in the period from 1 January 1971 to 18 September 1981'. The information to be provided to the Executive under the Regulations includes:

1. A technical dossier of information necessary for evaluating the foreseeable risk for man or the environment. Results of studies referred to in Schedule 1 (see Table 13.5) must be given together with details of the studies and methods used.
2. A declaration of any adverse effects of the substance with respect to the uses envisaged.
3. A declaration as to whether the substance is dangerous within the meaning of Part 1 of Schedule 2. If the substance is dangerous additional particulars are required.
4. Proposals for recommended precautions relating to the safe use of the substance.

In addition to the basic information required (Table 13.5) extra data may be required for the purposes of the Regulations (which should be consulted), depending upon tonnages involved as illustrated by Table 13.6.

Guidelines for testing chemicals have been formulated by the OECD.[47] These contain generally formulated procedures for the laboratory testing of a property, or effect, deemed important for the evaluation of health and environmental hazards of a chemical. The subjects covered are summarised in Table 13.7. Sufficient detail is given to enable an operator to carry out the required test, assuming good laboratory practice. Standard methods for flash-point determination are given in Part IV of reference 30. Testing materials for explosive properties is described in Chapter 11 and in reference 48.

Table 13.5 Basic information required for notification of new substances

SCHEDULE 1 — Regulation 4(1) and (3) and 10(3)

(Which reproduces the requirements of Annex VII to the Directive)

INFORMATION REQUIRED FOR THE TECHNICAL DOSSIER REFERRED TO IN REGULATION 4(1)(*a*)

When giving notification the manufacturer or any other person placing a substance on the market shall provide the information set out below.

If it is not technically possible or if it does not appear necessary to give information, the reasons shall be stated.

Tests must be conducted according to methods recognized and recommended by the competent international bodies where such recommendations exist.

The bodies carrying out the tests shall comply with the principles of good current laboratory practice.

When complete studies and the results obtained are submitted, it shall be stated that the tests were conducted using the substance to be marketed. The composition of the sample shall be indicated.

In addition, the description of the methods used or the reference to standardized or internationally recognized methods shall also be mentioned in the technical dossier, together with the name of the body or bodies responsible for carrying out the studies.

1. IDENTITY OF THE SUBSTANCE

1.1 Name

1.1.1. Names in the IUPAC nomenclature

1.1.2. Other names (usual name, trade name, abbreviation)

1.1.3. CAS number (if available)

1.2 Empirical and structural formula

1.3. Composition of the substance

1.3.1. Degree of purity (%)

1.3.2. Nature of impurities, including isomers and by-products

1.3.3. Percentage of (significant) main impurities

1.3.4. If the substance contains a stabilizing agent or an inhibitor or other additives, specify: nature, order of magnitude: . . . ppm; . . . %

1.3.5. Spectral data (UV, IR, NMR)

1.4. Methods of detection and determinations
A full description of the methods used or the appropriate bibliographical references ....

2. INFORMATION ON THE SUBSTANCE

2.1. Proposed uses

2.1.1. Types of use
Describe: the function of the substance ....
the desired effects ....

2.1.2. Fields of application with approximate breakdown
(a) closed system
– industries ....
– farmers and skilled trades ....
– use by the public at large ....

Table 13.5 (continued)

(b) open system
- – industries ....
- – farmers and skilled trades ....
- – use by the public at large ....

2.2. Estimated production and/or imports for each of the anticipated uses or fields of application

2.2.1. Overall production and/or imports in order of tonnes per year 1; 10; 50; 100; 500; 1,000 and 5,000
- – first 12 months .... tonnes/year
- – thereafter .... tonnes/year

2.2.2. Production and/or imports, broken down in accordance with 2.1.1. and 2.1.2., expressed as a percentage
- – first 12 months ....
- – thereafter ....

2.3. Recommended methods and precautions concerning:

2.3.1. handling ....

2.3.2. storage ....

2.3.3. transport ....

2.3.4. fire (nature of combustion gases or pyrolysis, where proposed uses justify this)

2.3.5 other dangers, particularly chemical reaction with water

2.4. Emergency measures in the case of accidental spillage

2.5. Emergency measures in the case of injury to persons (e.g. poisoning)

3. PHYSICO-CHEMICAL PROPERTIES OF THE SUBSTANCE

3.1. Melting point
.... °C

3.2. Boiling point
.... °C............................... Pa

3.3. Relative density
.... (D_4^{20})

3.4. Vapour pressure
.... Pa at .... °C
.... Pa at .... °C

3.5. Surface tension
.... N/m (............................... °C)

3.6. Water solubility
.... mg/litre (............................... °C)

3.7. Fat solubility
Solvent – oil (to be specified)
.... mg/100 g solvent (............................... °C)

3.8 Partition coefficient
n-octanol/water

3.9. Flash point
.... °C ☐ open cup ☐ closed cup

3.10. Flammability (within the meaning of the definition given in Part I of Schedule 2, paragraph 2(*c*), (*d*) and (*e*))

3.11. Explosive properties (within the meaning of the definition given in Part I of Schedule 2, paragraph 2(*a*))

Table 13.5 (continued

3.12. Auto-flammability
.... °C

3.13. Oxidizing properties (within the meaning of the definition given in Part I of Schedule 2, paragraph 2(*b*))

4. TOXICOLOGICAL STUDIES

4.1. Acute toxicity

4.1.1. Administered orally
LD_{50} .... mg/kg
Effects observed, including in the organs ....

4.1.2. Administered by inhalation
LC_{50} .... (ppm) Duration of exposure .... hours
Effects observed, including in the organs ....

4.1.3. Administered cutaneously (percutaneous absorption)
LD_{50} . . . . mg/kg
Effects observed, including in the organs ....

4.1.4. Substances other than gases shall be administered via two routes at least, one of which should be the oral route. The other route will depend on the intended use and on the physical properties of the substance.
Gases and volatile liquids should be administered by inhalation (a minimum period of administration of four hours).
In all cases, observation of the animals should be carried out for at least 14 days.
Unless there are contra-indications, the rat is the preferred species for oral and inhalation experiments.
The experiments in 4.1.1, 4.1.2 and 4.1.3 shall be carried out on both male and female subjects.

4.1.5. Skin irritation
The substance should be applied to the shaved skin of an animal, preferably an albino rabbit.
Duration of exposure .... hours

4.1.6 Eye irritation
The rabbit is the preferred animal.
Duration of exposure .... hours

4.1.7. Skin sensitization
To be determined by a recognized method using a guinea-pig.

4.2. Sub-acute toxicity

4.2.1. Sub-acute toxicity (28 days)
Effects observed on the animal and organs according to the concentrations used, including clinical and laboratory investigations ....
Dose for which no toxic effect is observed ....

4.2.2. A period of daily administration (five to seven days per week) for at least four weeks should be chosen. The route of administration should be the most appropriate having regard to the intended use, the acute toxicity and the physical and chemical properties of the substance.
Unless there are contra-indications, the rat is the preferred species for oral and inhalation experiments.

4.3. Other effects

4.3.1. Mutagenicity (including carcinogenic pre-screening test)

Table 13.5 (continued)

4.3.2. The substance should be examined during a series of two tests, one of which should be bacteriological, with and without metabolic activation, and one non-bacteriological.

5. ECOTOXICOLOGICAL STUDIES

5.1. Effects on organisms

5.1.1. Acute toxicity for fish

LC_{50} .... (ppm)
Duration of exposure ....
Species selected (one or more) ....

5.1.2. Acute toxicity for daphnia
LC_{50} .... (ppm)
Duration of exposure ....

5.2. Degradation
– biotic
– abiotic
The BOD and the BOD/COD ratio should be determined as a minimum

6. POSSIBILITY OF RENDERING THE SUBSTANCE HARMLESS

6.1. For industry/skilled trades

6.1.1. Possibility of recovery ....

6.1.2. Possibility of neutralization ....

6.1.3. Possibility of destruction:
– controlled discharge ....
– incineration ....
– water purification station ....
– others ....

6.2. For the public at large

6.2.1. Possibility of recovery ....

6.2.2. Possibility of neutralization ....

6.2.3. Possibility of destruction:
– controlled discharge ....
– incineration ....
– water purification station ....
– others ....

Note: (This note does not form part of Annex VII)

1. A reference in this Schedule to a substance being placed on the market or to be marketed shall be treated as a reference to that substance being supplied or to be supplied as the case may be.
2. 'IUPAC nomenclature' means the nomenclature of the International Union of Pure and Applied Chemistry details of which are obtainable from the Chemical Nomenclature Abstracts Service, Laboratory of the Government Chemist, Stamford Street, London E.1.
3. "CAS No." means the number assigned by the Chemical Abstracts Service details of which are obtainable from the United Kingdom Chemical Information Service, University of Nottingham, Nottingham.

Table 13.6 Extra information required for notification of new substances

SCHEDULE 5 Regulation 6(2)

(Which reproduces the requirements of Annex VIII to the Directive)

ADDITIONAL INFORMATION AND TESTS WHICH MAY BE REQUIRED FOR THE PURPOSES OF REGULATION 6(2)

Any person who has notified a substance to a competent authority in accordance with the requirements of Article 6 of this Directive shall provide at the request of the authority further information and carry out additional tests as provided for in this Annex.

If it is not technically possible or if it does not appear necessary to give information, the reasons shall be stated.

Tests shall be conducted according to methods recognized and recommended by the competent international bodies where such recommendations exist.

The bodies carrying out the tests shall comply with the principles of good current laboratory practice.

When complete studies and the results obtained are submitted, it shall be stated that the tests were conducted using the substance marketed. The composition of the sample shall be indicated.

In addition, the description of the methods used or the reference to the standardized or internationally recognized methods shall also be mentioned in the technical dossier, together with the name of the body or bodies responsible for carrying out the studies.

LEVEL 1

Taking into account:

- current knowledge of the substance,
- known and planned uses,
- the results of the tests carried out in the context of the base set,

the competent authority may require the following additional studies where the quantity of a substance placed on the market by a notifier reaches a level of 10 tonnes per year or a total of 50 tonnes and if the conditions specified after each of the tests are fulfilled in the case of that substance.

Toxicological studies

- Fertility study (one species, one generation, male and female, most appropriate route of administration)

  If there are equivocal findings in the first generation, study of a second generation is required.

  It is also possible in this study to obtain evidence on teratogenicity.

  If there are indications of teratogenicity, full evaluation of teratogenic potential may require a study in a second species.
- Teratology study (one species, most appropriate route of administration)

  This study is required if teratogenicity has not been examined or evaluated in the preceding fertility study.
- Sub-chronic and/or chronic toxicity study, including special studies (one species, male and female, most appropriate route of administration).

  If the results of the sub-acute study in Schedule 1 or other relevant information demonstrate the need for further investigation, this may take the form of a more detailed examination of certain effects, or more prolonged exposure, e.g. 90 days or longer (even up to two years).

  The effects which would indicate the need for such a study could include for example:

  (*a*) serious or irreversible lesions;

  (*b*) a very low or absence of a 'no effect' level;

  (*c*) a clear relationship in chemical structure between the substance being studied and other substances which have been proved dangerous.

Table 13.6 (continued)

- Additional mutagenesis studies (including screening for carcinogenesis)
  A. If results of the mutagenesis tests are negative, a test to verify mutagenesis and a test to verify carcinogenesis screening are obligatory.
  If the results of the mutagenesis verification test are also negative, further mutagenesis tests are not necessary at this level; if the results are positive, further mutagenesis tests are to be carried out (see B).
  If the results of the carcinogenesis screening verification test are also negative, further carcinogenesis screening verification tests are not necessary at this level; if the results are positive, further carcinogenesis screening verification tests are to be carried out (see B).
  B. If the results of the mutagenesis tests are positive (a single positive test means positive), at least two verification tests are necessary at this level. Both mutagenesis tests and carcinogenesis screening tests should be considered here. A positive result of a carcinogenesis screening test should lead to a carcinogenesis study at this level.

Ecotoxicology studies
- An algal test; one species, growth inhibition test.
- Prolonged toxicity study with Daphnia magna (21 days, this study should also include determination of the 'no-effect level' for reproduction and the 'no-effect level' for lethality).
  The conditions under which this test is carried out shall be determined in accordance with the procedure described in Article 21 of the Directive.
- Test on a higher plant.
- Test on an earthworm.
- Prolonged toxicity study with fish (e.g. Oryzias, Jordanella, etc; at least a period of 14 days; this study should also include determination of the 'threshold level').
- The conditions under which this test is carried out shall be determined in accordance with the procedure described in Article 21 of the Directive.
- Tests for species accumulation; one species, preferably fish (e.g. Poecilla reticulata).
- Prolonged biodegradation study, if sufficient (bio)degradation has not been proved by the studies laid down in Schedule 1, another test (dynamic) shall be chosen with lower concentrations and with a different inoculum (e.g. flow-through system).

In any case, the notifier shall inform the competent authority if the quantity of a substance placed on the market reaches a level of 100 tonnes per year or a total of 500 tonnes.

On receipt of such notification and if the requisite conditions are fulfilled, the competent authority, within a time limit it will determine, shall require the above tests to be carried out unless in any particular case an alternative scientific study would be preferable.

LEVEL 2

If the quantity of a substance placed on the market by a notifier reaches 1,000 tonnes per year or a total of 5,000 tonnes, the notifier shall inform the competent authority. The latter shall then draw up a programme of tests to be carried out by the notifier in order to enable the competent authority to evaluate the risks of the substance for man and the environment. The test programme shall cover the following aspects unless there are strong reasons to the contrary, supported by evidence, that it should not be followed:
- chronic toxicity study,
- carcinogenicity study,
- fertility study (e.g. three-generation study); only if an effect on fertility has been established at level 1,
- teratology study (non-rodent species) study to verify teratology study at level 1 and experiment additional to the level 1 study, if effects on embryos/foetuses have been established,
- acute and sub-acute toxicity study on second species; only if results of level 1 studies indicate a need for this. Also results of biotransformation studies and studies on pharmacokinetics may lead to such studies,

Table 13.6 (continued)

– additional toxicokinetic studies.
Ecotoxicology
– Additional tests for accumulation, degradation and mobility.
The purpose of this study should be to determine any accumulation in the food chain.
For further bioaccumulation studies special attention should be paid to the solubility of the substance in water and to its n-octanol/water partition coefficient.
The results of the level 1 accumulation study and the physiochemical properties may lead to a large-scale flow-through test.
– Prolonged toxicity study with fish (including reproduction).
– Additional toxicity study (acute and sub-acute) with birds (e.g. quails): if accumulation factor is greater than 100.
– Additional toxicity study with other organisms (if this proves necessary).
– Absorption-desorption study where the substance is not particularly degradable.

Table 13.7 OECD guidelines for testing of chemicals

No.	Subject
Section 1 – Physical–chemical properties	
101	UV-VIS Absorption Spectra
102	Melting point/melting range
103	Boiling point/boiling range
104	Vapour pressure curve
105	Water solubility
106	Adsorption/desorption
107	Partition coefficient (n-octanol/water)
108	Complex formation ability in water
109	Density of liquids and solids
110	Particle size distribution/fibre length and diameter distributions
111	Hydrolysis as a function of pH
112	Dissociation constants in water
113	Screening test for thermal stability and stability in air
114	Viscosity of liquids
115	Surface tension of aqueous solutions
116	Fat solubility of solid and liquid substances
Section 2 – Effects on biotic systems	
201	Alga, growth inhibition test
202	*Daphnia* sp., 14-day reproduction test (including an acute immobilisation test)
203	Fish, acute toxicity test
Section 3 – Degradation and accumulation	
Ready biodegradability	
301 A	Modified AFNOR test
301 B	Modified Sturm test
301 C	Modified MITI test (I)
301 D	Closed bottle test
301 E	Modified OECD screening test
Inherent biodegradability	
302 A	Modified SCAS test
302 B	Modified Zahn–Wellens test
302 C	Modified MITI test (II)

Table 13.7 (continued)

No.	Subject
Simulation test – Aerobic sewage treatment	
303 A	Coupled units test
Biodegradability in soil	
304 A	Inherent biodegradability test in soil
Bioaccumulation	
305 A	Sequential static fish test
305 B	Semi-static fish test
305 C	Degree of bioconcentration in fish
305 D	Static fish test
305 E	Flow-through fish test
Section 4 – Health effects	
Short-term toxicology	
401	Acute oral toxicity
402	Acute dermal toxicity
403	Acute inhalation toxicity
404	Acute dermal irritation/corrosion
405	Acute eye irritation/corrosion
406	Skin sensitisation
407	Repeated dose oral toxicity – rodent: 14/28-day
408	Subchronic oral toxicity – rodent: 90-day
409	Subchronic oral toxicity – non-rodent: 90-day
410	Repeated dose dermal toxicity – 21/28-day
411	Subchronic dermal toxicity – 90-day
412	Repeated dose inhalation toxicity – 14/28-day
413	Subchronic inhalation toxicity – 90-day
414	Teratogenicity
Long-term toxicology	
451	Carcinogenicity studies
452	Chronic toxicity studies
453	Combined chronic toxicity/carcinogenicity studies

Two Codes of Practice provide practical guidance relating to the UK Notification of New Substances Regulations 1982, one on methods for physico-chemical properties determination[49] and the other on toxicity determination.[50] A list of agencies providing standards for testing is also given in Chapter 18.

In the USA, under the Toxic Substances Control Act, the standards to be prescribed for the testing of health and environmental effects include carcinogenesis, mutagenesis, teratogenesis, behavioural disorders, cumulative or synergistic effects and any other effect which may pose an unreasonable risk of injury to health or the environment. Characteristics requiring consideration include persistence (a particular problem with polychlorinated biphenyls for example), acute toxicity, sub-acute toxicity and chronic toxicity. Useful guidance on toxicity testing of chemicals, and measures to ensure safety in toxicity-testing laboratories, is given in references 51 and 52.

Transportation in bulk

In November 1979 a goods train derailed in the City of Mississauga in Ontario. Of the 106 rail cars making up the train 24 were derailed including eleven propane, three toluene, three sytrene and one chlorine tank cars. Fire spread through most of the derailled cars and the eighth, twelfth and thirteenth cars which contained propane each BLEVE'd causing considerable damage to neighbouring property. The seventh car in the derailment which contained 90 tonnes of liquid chlorine suffered a hole 0.7 m in diameter and because of the fear of the consequences of the escape of chlorine almost a quarter of a million people were evacuated from their homes and businesses for periods of up to five days.[53]

In May 1976 near Houston, Texas a road tanker carrying 19 tonnes of anhydrous ammonia crashed through a barrier on an elevated section of motorway.[54] The pressurised tank burst on falling to the roadway below; it took off like a rocket and travelled a considerable distance. The initial height of the gas cloud was about 35 m; after about one minute it was 300 m wide by 600 m long. (This gravitational slumping behaviour due to the formation of an ammonia–air mixture denser than air, because of cooling of entrained air by evaporation of liquid ammonia droplets present in the sudden release from a pressure vessel, was referred to on page 96.)

The safe transport of chemicals is a responsibility of the manufacturing company and their employees, any haulier used, and also the customer. Good communications between all parties and the acceptance by each of their responsibilities is vital. The scope of responsibility for the safe transport of chemicals by road is illustrated in Fig. 13.7.[55]

When manufacturers transport their products, the management is responsible for the physical well-being of their transport staff, of the general public and of the customer's employees. Therefore they need to ensure that[55]:

- All the information relevant to the safe handling of products under all foreseeable conditions is available in simple form to the persons involved. Therefore:
  1. Codes of practice, working procedures or general instructions should incorporate all possible safety factors and should be enforced.
  2. Specific instructions should be given to all transport staff, regarding action to be taken in case of emergency involving the spillage or escape of chemicals or other unusual occurrences e.g. fire.

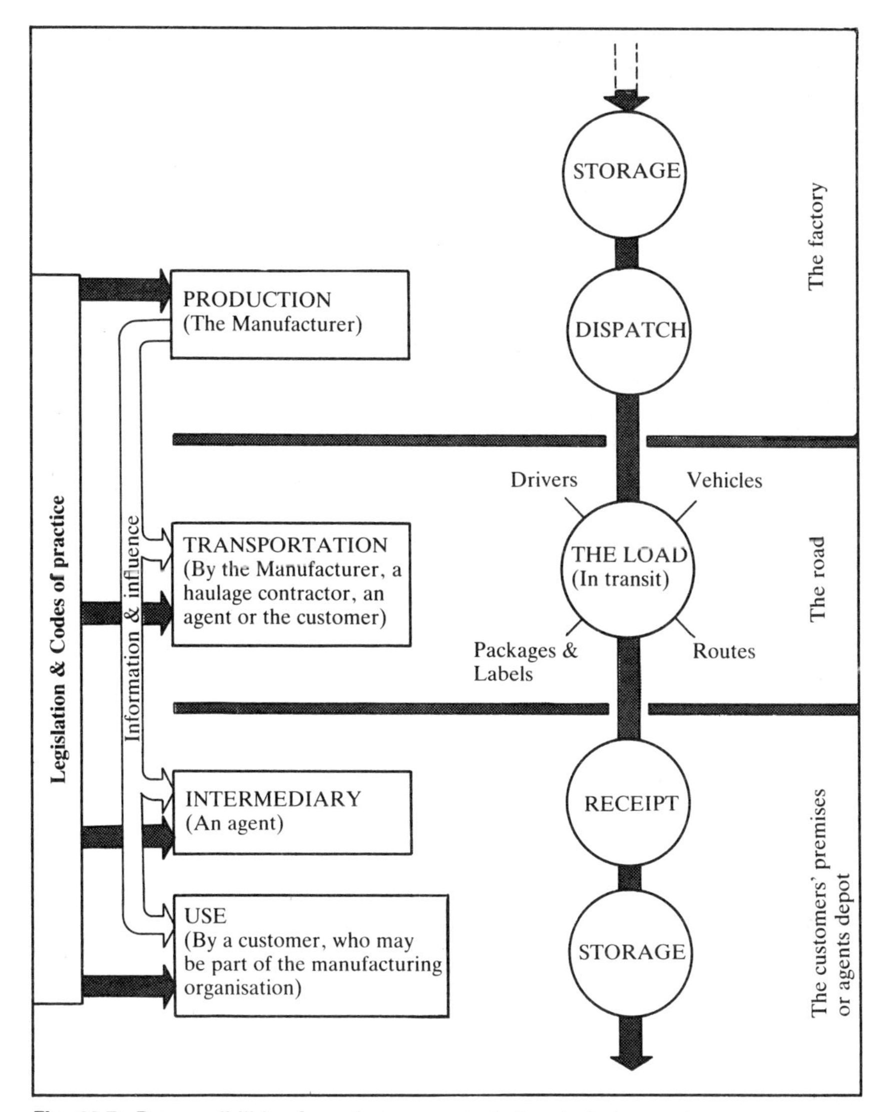

Fig. 13.7 Responsibilities for safe transport of chemicals by road

3. Safe unloading procedures should be followed at a customer's works or depot.
4. All procedures must comply with any statutory requirements applicable to the handling and transport of the product.

- Vehicles are of the correct design and construction for the loads they will carry.

- Local fire and police authorities are given prior information about the intended movement of particularly hazardous loads through their areas of jurisdiction.

The *CIA Manual of Principal Safety Requirements for the Road Transport of Hazardous Chemicals* suggests that safety must ultimately be the responsibility of the manufacturers since they are likely to know more about the specific chemical properties and safe handling and emergency procedures.

In dealing with haulage or warehouse contractors, manufacturers are advised to ensure that the contractor is fully informed of the load to be carried and the precautions to be taken during transit, and to take reasonable steps to assure himself that the contractor is competent to handle hazardous goods. This may in certain circumstances involve inspection of the haulier's facilities before placing or renewing contracts.

Rapid identification of hazards in an emergency

In August 1981 the brakes on a train carrying a large quantity of chlorine between Pajaritos and Tampico, Mexico, failed during the descent of a 3% gradient. The train derailed at over 80 km/h about 350 m beyond Montanas station: twenty-eight of the thirty-two, 55-tonne chlorine tank cars were included in the pile-up; most were badly damaged and an estimated chlorine release exceeding 100 tonnes occurred in the first few minutes. A total of 17 people were killed, 13 as a result of chlorine, and about 1,000 people were treated in hospital.[56]

In the case of accidents involving the release of hazardous chemicals into the environment, or fires, it is important that information is available to enable appropriate action to be taken. In road transport emergencies the first man on the spot is the driver; clearly then drivers with hazardous cargoes require careful selection and training. When the police and fire service are called to the scene they must rely upon the information given to them by the driver, both verbally and from documentation/cards which he carries, and from labels on the vehicle.

Various schemes have been introduced to identify chemicals and their intrinsic properties (toxicity, flammability, etc.) in such situations. The EEC hazard classification system has already been discussed but this is not applicable to bulk supply and road conveyance.

In the UK the conveyance by road of dangerous substances in road tankers of any capacity and in tank containers $>3\ m^3$ capacity is controlled by the Dangerous Substances (Conveyance by Road in Road Tankers and Tank Containers) Regulations 1981.[37] Dangerous substances are either those named in an approved list,[57] or those falling into the classifications in Schedule 1. Except for a slightly different definition for 'other dangerous substance' and 'multi-load' the latter classification is similar to Fig. 13.4.[37]

Exclusions to the Regulations (Regulation 3(1)) include vehicles conveying dangerous substances on an international transport operation provided it complies with the requirements of the CIM (International Convention for the Conveyance of Goods by Rail) or the ADR Agreement (International Carriage of Dangerous Goods by Road).

A check-list of duties arising, where appropriate under these Regulations is given in Table 13.8 (after ref. 58), excluding duties in relation to the unloading of petroleum spirit under Schedule 4. However it is essential in any particular case to refer to the Regulations, Guides and Codes.[59]

A draft Code of Practice on the operational provisions[60] and a guide[58] should be consulted for further details and advice. Pending the issue of an approved code for construction of tankers/tank containers use should be made of existing codes or agreements listed in reference.[61]

The form, dimensions and colour of the hazard warning diamonds for hazard warning panels/compartment labels are specified in Part II of Schedule 1[37] and the make-up of hazard warning labels, etc., in Schedule 3. Specimen panels are shown in Fig. 13.8.

A tanker driver was overcome by fumes and required hospital treatment when several hundred litres of a corrosive liquid were sprayed over a loading bay because of corrosion of a tanker's main filling pipe above the shut-off valve. The accident occurred after 4,500 litres of nitric acid and hydrofluoric acid had been loaded. Subsequent examination revealed that the failed pipe was of unlined, mild steel and initial failure was at a weld.[62].

The firm who had purchased and put into service an eight-year-old tanker without first having it examined and tested by a competent person pleaded guilty to a breach of Regulation 6(1)(c) and was fined £500.[62]

'Conveyance' covers loading, transportation, unloading, cleaning and purging (Regulation 3(2)) and, as referred to in Chapter 11, Fig. 13.9 illustrates the type of facility for unloading a highly toxic, volatile liquid (i.e. chlorine) to a storage vessel (after ref. 63).

The bulk conveyance of anhydrous ammonia by road[64] and by rail[65] is similarly covered by industry codes.

Instruction and training of drivers to deal with emergencies are specific requirements (Regulation 21.1(a)). The actions will normally involve (after ref. 60):

- Arranging for the police and emergency services to be alerted.
- Arranging for assistance to be given to any person injured or in immediate danger.
- If considered safe to do so – having regard to the nature of the emergency, the substance and the emergency equipment available –

Table 13.8 Duties under the Dangerous Substances (Conveyance by Road in Road Tankers and Tank Containers Regulations) 1981

Operator of a road tanker

1. Obtain sufficient information about the substance from the consignor to be able to comply with the Regulations and to be aware of the risks created.
2. Determine whether it is a 'dangerous' substance as defined. If so classify it from the Approved List or from Schedule 1.
3. Check that the substance can lawfully be conveyed.
4. Check whether the regulations apply to the conveyance.
5. Ascertain that the tanker is properly designed – of adequate strength, of good construction from sound and suitable material – suitable for the purpose and will prevent escape of the contents.
   (This does not exclude the fitting of a pressure relief valve but it must be suitable and be positioned/installed to minimise the risk if it operates.)[58]
   For petroleum spirit, a petroleum mixture or carbon disulphide, check the road tanker for compliance with Schedule 2 'Construction of Road Tankers'.
6. Check that there is a suitable written scheme for initial examination and testing, periodic examination and testing of the carrying tank.
7. Check that there is a certificate for the carrying tank in which a competent person has certified it as suitable for the purpose for which it is to be used. These purposes should be specified in the certificate.
8. Ensure that a current report, signed by a competent person, gives details of examinations and tests on the tank, and the interval before which further examinations and tests must be performed. The report should confirm that the tank is suitable for the purposes stated in the certificate or otherwise define its suitability.
9. Ensure that all precautions for preventing fire or explosion are taken throughout the conveyance.
10. Ensure that the tank is not overfilled.
11. Check that the road tanker is fitted with hazard warning labels and compartment labels.
12. Ensure that the driver has received adequate instruction and training to understand the potential dangers, in emergency action and in his duties under the Regulations.
13. Ensure that the driver is provided with written information on the substance conveyed to know its identity, the potential dangers and emergency action.
14. Ensure that hazard warning panels/compartment labels are removed/covered when any/all of the substances have been unloaded and the tank/compartment has been cleaned or purged – such that the dangerous substance remaining is insufficient to create a risk.
15. Ensure that hazard warning panels/compartment labels are displayed at all times as required by the Regulations and that they are kept clean and free from obstruction.

Operator of a tank container

16. As 1 above.
17. As 2 above.
18. As 3 above.
19. As 4 above.
20. Ascertain that the tank container is properly designed – of adequate strength, of good construction from sound and suitable material – suitable for the purpose and will prevent escape of the contents.
21. Check that there is a suitable written scheme for initial examination and testing, periodic examination and testing of the tank container.
22. Check that there is a certificate for the tank container in which a competent person has certified it as suitable for the purpose for which it is to be used. These purposes should be specified in the certificate.
23. Ensure that a current report, signed by a competent person, gives details of examinations and tests on the tank container, and the interval before which further

Table 13.8 (continued)

examinations and tests must be performed. The report should state that the tank container is suitable for the purposes stated in the certificate or otherwise define its suitability.

24. Ensure that the tank container is not overfilled.
25. Check that the tank container is fitted with hazard warning labels and compartment labels.

Operator of a vehicle conveying a tank container

26. As 12 above.
27. As 13 above.
28. As 9 above.
29. As 14 above.
30. As 15 above.

Driver

31. Ensure that written information on the substance conveyed – from the vehicle operator – is kept in the cab and is available at all times.
32. Ensure that this information relates solely to the substance conveyed.
33. As 9 above.
34. Ensure that if conveying a dangerous substance with an emergency action code ending with 'E' (subject to certain exceptions) the vehicle when not being driven is either parked in a safe place or is supervised by himself or some other competent person $>$ 18 years.
35. As 14 above.
36. As 15 above.

Consignor – or other person acting on his behalf

37. Ensure that the information given to the operator concerning substance(s) to be conveyed is accurate and sufficient.
38. If engaged in the conveyance of the substance (e.g. in loading the tank), ensure that all precautions for preventing fire or explosion are taken throughout the conveyance.

All other persons engaged in the conveyance

39. As 9 above.

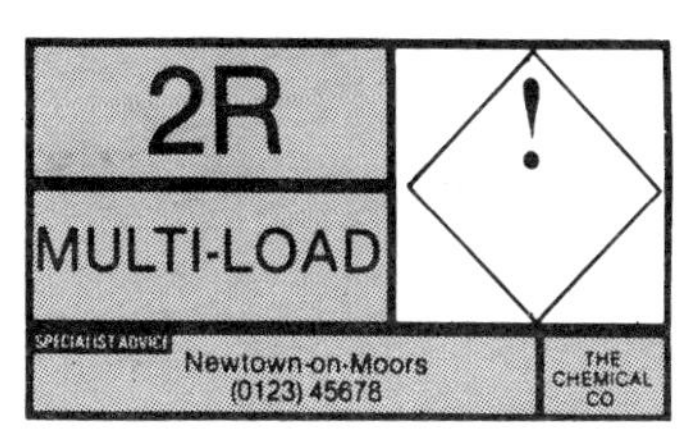

Fig. 13.8 Specimen hazard warning panels

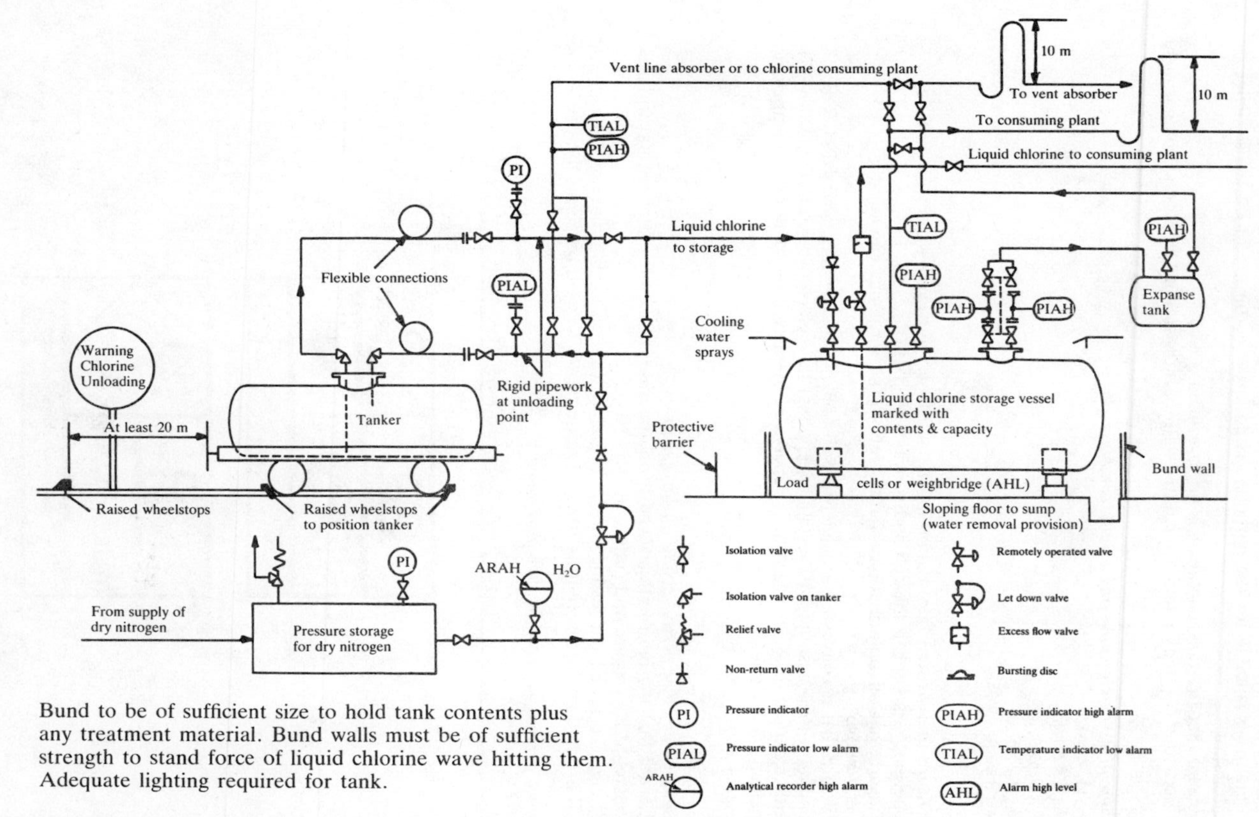

Bund to be of sufficient size to hold tank contents plus any treatment material. Bund walls must be of sufficient strength to stand force of liquid chlorine wave hitting them. Adequate lighting required for tank.

Fig. 13.9 Facility for unloading liquid chlorine from tankers to storage vessel (after ref. 63)

following a selection of the following procedures in an appropriate order:
Stop the engine.
Turn off any battery isolating switch.
If there is no danger of ignition, operate the emergency flashing device.
Move the vehicle to a location where any leakage would cause less harm.
Wear appropriate protective clothing.
Keep onlookers away.
Place a red triangle warning device at the rear of the vehicle and near any spillage.
Prevent smoking and direct other vehicles away from any fire risk area.

- Upon the police/fire brigade taking charge:
  Show the written information, e.g. Tremcard, to them.
  Tell them of action taken and anything helpful about the load, etc.
- At the end of the emergency, inform the operator (for compliance with the Notification of Accidents and Dangerous Occurrences Regulations 1980).

The written information given to the driver should include[60]:

- The name of the substance.
- Its inherent dangers and appropriate safety measures.
- Action and treatment following contact/exposure.
- Action in the event of fire and fire-fighting equipment to be used.
- Action following spillage on the road.
- How and when to use any special safety equipment.

A scheme developed by the European Council of Chemical Manufacturers' Federations (CEFIC) and adopted in the UK by the CIA provides written instructions in the form of transport emergency cards, Tremcards, as Fig. 13.10.

These also give an emergency telephone number of the manufacturer from whom expert advice is available on a 24 hour basis.

With any incident in which the product manufacturer is not known, or cannot be contacted, it is essential to have access to some other sources of information and a back-up advisory service. In the UK this 'service' is provided by CIA in collaboration with the National Chemical Emergency Centre at Harwell, under the Chemsafe scheme. All chemical companies are asked to complete a questionnaire; the information provided on each questionnaire is based on the needs of the emergency services, and is stored in a computerised data bank which covers over 7,000 materials.[66] A document produced from a questionnaire can be obtained from a computer terminal over telephone lines. The form of output from a computer after feeding in the name of the chemical and a code number is shown in Table 13.9.

UN No. 1088

TRANSPORT EMERGENCY CARD (Road)

CEFIC TEC (R)-687
April 1974
Class IIIa ADR
Marg. 2301,1°(a)

Cargo

ACETAL
(Acetaldehyde Diethyl Acetal)
Colourless liquid with perceptible odour
Immiscible with water, lighter than water

Nature of Hazard

Highly inflammable (flashpoint below 21°C)
Volatile
The vapour is invisible, heavier than air and spreads along ground
Can form explosive mixture with air particularly in empty uncleaned receptacles
Heating will cause pressure rise, with risk of bursting and subsequent explosion
The vapour has narcotic effect
The substance has irritant effect on eyes

Protective Devices

Goggles giving complete protection to eyes
Plastic or rubber gloves
Eyewash bottle with clean water

EMERGENCY ACTION – Notify police and fire brigade immediately

- Stop the engine
- No naked lights. No smoking
- Mark roads and warn other road users
- Keep public away from danger area
- Use explosionproof electrical equipment
- Keep upwind

Spillage

- Shut off leaks if without risk
- Absorb in earth or sand and remove to safe place
- Subsequently flush road with water
- Prevent liquid entering sewers, basements and workpits, vapour may create explosive atmosphere
- Warn inhabitants – explosion hazard
- If substance has entered a water course or sewer or contaminated soil or vegetation, advise police

Fire

- Keep containers cool by spraying with water if exposed to fire
- Extinguish with waterspray, dry chemical, alcohol foam or halones
- Do not use water jet

First aid

- Remove soaked clothing immediately and wash affected skin with plenty of water
- If substance has got into the eyes, immediately wash out with plenty of water for several minutes
- Seek medical treatment when anyone has symptoms apparently due to inhalation or contact with eyes

Additional information provided by manufacturer or sender

TELEPHONE

Prepared by CEFIC (CONSEIL EUROPEEN DES FEDERATIONS DE L'INDUSTRIE CHIMIQUE, EUROPEAN COUNCIL OF CHEMICAL MANUFACTURERS' FEDERATIONS) Zürich, from the best knowledge available: no responsibility is accepted that the information is sufficient or correct in all cases. Obtainable from NORPRINT LIMITED, BOSTON, LINCOLNSHIRE
Applies only during road transport

English

Fig. 13.10 CEFIC Tremcard

Table 13.9 Output format of data – Chemsafe Scheme

	PERMANATE
Trade name:	Permanate
Company name:	Parton Chemical Co. Ltd, Northern Road, Parton, Warwickshire
Packaging:	25 kg paper sacks
Code marks:	186KP
Composition:	Potassium permanganate, 100%
Form:	Solid, crystals, dark purple-metallic sheen
Hazards:	Powerful oxidising material. Explosive in contact with sulphuric acid or hydrogen peroxide. Reacts violently with finely divided easily oxidisable substances. Spontaneously flammable on contact with glycerine and ethylene glycol. Increases flammability of combustible materials. A strong irritant due to oxidising properties, use breathing apparatus.
Hazard class:	UN serial no. 1490 UN hazard class 5.1.0 Hazchem code 1P Kemler 50
Spillage:	Wear breathing apparatus at all times with full protective clothing. Flood with water. Beware hazard of contaminated clothing on drying out.
Fire:	Flood with water.
First Aid:	
Knowledge:	Mr H. Weston
Phone:	Parton 49255, available 24 hours
Routes:	Parton to North London and Wales principally
References:	Technical data sheet 1, 2(a) enclosed
Compiler:	Mr J. Jones, Parton 49312
Date:	September 1974

The CIA's Chemsafe scheme has been operating for several years with a country-wide network of response centres run by individual companies incorporating a mutual aid scheme whereby a manufacturer nearest an incident will respond on behalf of the owner of the chemical who may be based some distance away.[67] A long-stop emergency centre is operated by the Harwell emergency centre on a 24-hour basis. Several major companies have, in addition, in-house national or Europe-wide company emergency schemes linking their plants and offices.

US practice

Hazardous materials transported in the USA are subject to the Code of Federal Regulations administered by the Department of Transportation. Pesticides and agricultural chemicals require marking to the specifications of the Environmental Protection Agency. Markings for drugs are regulated by the Food and Drugs Administration. Identification and labelling of materials to these requirements; railroad, pipeline and trucking, airline and marine transportation practice in the USA together with information sources are covered in reference.[68]

Transport controls

With road transport route control is an important aspect of contingency planning. This is the responsibility of the transport manager, who should specify the route with the minimum of traffic hazards. Such routes should avoid crowded areas, areas of high population density and areas especially vulnerable to contamination. The manager should ensure that he obtains up-to-date information about road conditions from the police and/or automobile agencies. In highly industrialised areas, the county authorities, manufacturers and hauliers should collaborate to plan and control the routes taken by vehicles carrying hazardous cargoes.[69]

On 11 July 1978, a road tanker carrying 23 tonnes of liquefied propylene, along a coast road from San Carlos de la Rapita (Spain) developed a 10 cm split along a welded seam. A cloud of vapour and aerosol droplets drifted over a campsite and ignited, producing a fireball about 200 m in diameter. Within seconds, 101 campers were burnt to death; 152 were severely burnt of whom 43 died by the following weekend.[70]

The ultimate death toll was 210 and subsequent analysis[71] suggests that the initial release squirted upwind as a flashing liquid. A small-scale deflagration or flash fire travelled back to the leaking tanker which subsequently BLEVE'd.

Driver instruction and training are clearly important factors. Drivers should be mature with a high sense of responsibility and able to work efficiently without supervision. They must possess the ability, through training and experience, to evaluate emergency situations and to reach sensible decisions, e.g. see page 507.

The inclusion, and location, of hazardous cargoes in goods trains is the responsibility of the railway authority. The parking and shunting of wagons in stations and goods depots is also their concern. In the UK manufacturers must collaborate with British Rail to ensure that tankers carrying hazardous cargoes are correctly marked with a Hazchem label.

Capability of emergency services

The consequences of transportation incidents are often difficult to foresee.

An articulated lorry carrying 19 tonnes of powdered resin in plastic bags left the road and crashed into a cottage. Part of the load was shed and many bags burst generating a cloud of fine powdered resin – a copolymer of acrylic, butadiene and styrene monomers. The mains electricity intake cable to the cottage was damaged in the collision causing arcing between the conductors; this ignited the cloud dispersed on impact to produce a dust explosion which set fire to the resin and sacks on the trailer and roadway. Noxious fumes affected workmen on a building site and fishermen at sea at a range of about 1.5 km.[72]

To cope with incidents involving chemicals may involve a fire service initiating the following actions[73]:

- Appointment of a chemical officer.
- Development of a chemical data bank.
- Improving standards in protective clothing.
- Introduction of positive pressure breathing apparatus.
- Development of decontamination procedures.
- Training of brigade personnel.
- Liaison team of fire, police and ambulance.
- Direct telex reports of chemicals from main importers.

Finally, the reader is referred to reference 74 for detailed advice on classification, licensing, packaging, loading, placarding, transportation, unloading, emergency action, etc., for a wide range of hazardous chemicals. This loose-leaf book is updated periodically.

References

(*All places of publication are London unless otherwise stated*)

1. Broadhurst, A., *Process Engineering*, Feb. 1974, 56–9.
2. *Wilson* v. *Rickett Cockerall & Co Ltd.*, 1954, 1 Q.B. 598.
3. Atiyah, P. S., *The Sale of Goods* (5th edn). Pitman 1975.
4. *Vacwell Engineering Co Ltd* v. *B.D.H Chemicals*, 1969, 3 All E.R. 1681.
5. *Donaghue* v. *Stevenson* (1932) - A.C 562 (HL); (1932). All E.R Rep 1.
6. Broadhurst, A., *The Health and Safety at Work Act in Practice*. Heyden, 1978.
7. Munkman, J., *Employers Liability at Common Law* (8th edn). Butterworths, 1975.
8. *Fisher* v. *Harrods Ltd* (1966) 1 Lloyd's Ref. 500, 85.
9. Sciarra, J. L. & Stoller, L. (eds), *The Science and Technology of Aerosol Packaging*. Wiley 1974.
10. Universities Safety Association, *Safety News*, No. 2, May 1972, 17.
11. Manufacturing Chemists Assoc., *Properties and Essential Information for Handling and Use of Hydrogen Peroxide*, Chemical Safety Data Sheet SD-53, 1969.
12. Egginton, J., *Bitter Harvest*. Secker & Warburg 1980.
13. *Wren* v. *Holt*, 1903, 1 KB. 610.
14. *Grant* v. *Australian Knitting Mills Ltd*, 1936. A.C 85, 100.
15. Borrie, G. & Diamond, A. L., *The Consumer, Society and the Law* (3rd edn). Penguin 1973.

16. Anon., *Loss Prevention Bulletin*. Institution of Chemical Engineers, 1977 (013), 15.
17. Manuele, F. A., *One Hundred Largest Losses – A Thirty Year Review of Property Damage Losses in the Hydrocarbon-Chemical Industries* (7th edn), 1984. Marsh & McLennon Protection Consultants.
18. Department of the Environment, *111th Annual Report on Alkali & Works 1974*, 18.
19. Clarke, A. W., 'Major loss prevention in the process industries', *Instn Chem. Engrs Symp. Series*, 1971 (34), 175–83.
20. *Wright* v. *Dunlop Rubber Co. Ltd and ICI Ltd*, (1972), 13 KIR, 255.
21. Health and Safety Executive, *Articles and Substances for Use at Work*, 1977. HMSO.
22. Miller, C. J., *'Product Liability and Safety Encyclopaedia'*, 1980.
23. Bowes Egan, *Safety Representatives and Safety Committees Regulations 1978 – Apppointment & Functions of Safety Representatives*. New Commercial Publishing 1978.
24. Ling, K. C., *Process Industry Hazards: Accidental Release, Assessment, Containment and Control*. I. Chem. E. Symp. Series, 1976 (47), 109–25.
25. Carson, P. A. & Jones, K., *Journal of Hazardous Materials*, 1984, **9**, 305–14.
26. International Maritime Consultation Organisation, *IMCO International Maritime Dangerous Goods Code*. IMCO 1975.
27. Department of Transport, *European Agreement Concerning the International Carriage of Dangerous Goods by Road (ADR)*. HMSO 1976.
28. Department of Transport, *International Regulations Concerning the Carriage of Dangerous Goods by Rail (RID)*. HMSO 1977.
29. International Air Transport Association, *IATA Restricted Articles Regulations* (19th edn). IATA 1976.
30. The Classification, Packaging and Labelling of Dangerous Substances Regulations 1984.
31. Williamson, G. E., *Chemistry and Industry*, Jan 1985, **21**, 40–4.
32. *Approved Code of Practice: Classification and Labelling of Substances Dangerous for Supply and/or Conveyance by Road*. HMSO.
33. *Approved Code of Practice: Packaging of Dangerous Substances for Conveyance by Road*. HMSO.
34. 'Information Approved for the Classification, Packaging and Labelling of Dangerous Substances for Supply and Conveyance by Road'. HMSO.
35. (a) EEC Directives: 79/831/EEC; *The Sixth Ammendment of Directive 67/548/EEC on Substances*; 73/173/EEC on *Solvent Preparations*; 77/728/EEC on *Paints and Related Products*; 78/631/EEC on *Pesticide Preparations*.
    (b) *British Business*, 1984 (16 Mar.) 517–525.
36. EEC Directive 78/319/EEC.
37. The Dangerous Substances (Conveyance by Road in Road Tankers and Tank Containers) Regulations 1981, SI, 1981, No. 1059.
38. Chemical Industries Association, *Code of Practice for the Safe Carriage of Dangerous Goods in Freight Containers*, CIA, Jan. 1973.
39. British Standards Institution, *BS 349: – Marking of Compressed Gas Cylinders*, 1973.

40. Information Sheet. Air Products Ltd.
41. Frankel, M., *Chemical Risk*. Pluto Press 1982.
42. Meidl, J. H., *Flammable Hazardous Materials* (2nd edn). Glencoe, Encino, California, 1978.
43. Sax, N., *Dangerous Properties of Industrial Materials* (6th edn). Van Nostrand Reinhold, New York 1984.
44. Bretherick, L., *Handbook of Reactive Chemical Hazards*, (2nd edn). Butterworths 1979.
45. Hewstone, R. K., *Designing and Implementing a Product Liability Program*.
46. *Reporting for the EINECS Inventory: Constructing EINECS Basic Documents*. ECSC-EEC-EAEC, Brussels, 1982.
47. Organization for Economic Cooperation and Development, *Guidelines for Testing of Chemicals as Adopted by the OECD Council*, 1981.
48. Field, P., *Explosibility Assessment of Industrial Powders and Dusts*. Department of the Environment, HMSO 1983.
49. *Methods for the Determination of Physico-chemical Properties*, COP9, Health and Safety Commission.
50. *Methods for the Determination of Toxicity*, COPl0, Health and Safety Commission.
51. Jameson, C. W. & Walters, D. B., *Chemistry for Toxicity Testing*. Butterworths 1984.
52. Jameson, C. W. & Walters, D. B., *Health and Safety for Toxicity Testing*. Butterworths 1984.
53. *Report of the Mississauga Railway Accident Inquiry*. The Hon. Mr Justice S. G. M. Grange, Supreme Court of Ontario, Dec. 1980; *Derailment – The Mississauga Miracle*, State of Ontario, 1980.
54. Kaiser, G. D. & Walker, B. C., *Atmospheric Environment*, 1978, **12,** 2289–2300.
55. Chemical Industries Assoc., *The Road Transport of Chemicals: Principal Safety Requirements*. CIA 1967.
56. *Loss Prevention Bulletin*. Institution of Chemical Engineers, Aug. 1983 (052), 7.
57. *Approved Substance Identification Numbers, Emergency Action Codes, and Classifications for Dangerous Substances Conveyed in Road Tankers and Tank Containers (The Approved List)*, HS(R)10. HMSO.
58. Health and Safety Executive *A Guide to the Dangerous Substances (Conveyance by Road in Road Tankers and Tank Containers) Regulations 1981*, Health and Safety Series Booklet HS(R)13, 1981.
59. Health and Safety Commission *Approved Code of Practice: Classification of Dangerous Substances for Conveyance in Road Tankers and Tank Containers*. 1981.
60. Health and Safety Commission *Draft Code of Practice on Operational Provisions of the Dangerous Substances (Conveyance by Road in Road Tankers and Tank Containers) Regulations 1981, Consultative Document*. 1982.
61. (a) Department of Transport, *European Agreement for the International Carriage of Dangerous Goods by Rail* (ADR). HMSO 1976.

(b) Intergovernmental Maritime Consultative Organisation. *International Maritime Dangerous Goods Code.* IMCO 1975.
(c) *Safe Handling and Transport of LPG in Bulk by Road: Code of Practice No. 2, Liquefied Petroleum Gas;*
(d) Industry Technical Association, *U.K Road Tank Wagons Design Code.* Institute of Petroleum;
(e) Chemical Industries Association, *Design and Construction of Vehicles – Guidance Notes for the Design and Construction of Tank Vehicles and Tank Trailers for the Conveyance by Road of Flammable, Corrosive and Toxic Liquids in the UK.* CIA.

62. *ROSPA Bull.*, 1984, Sept. 2.
63. Chemical Industries Association, *Guidelines for Bulk Handling of Chlorine at Customer Installations.* CIA 1980.
64. Chemical Industries Association, *Code of Practice for the Safe Handling and Transport of Anydrous Ammonia in Bulk by Road in the UK.* CIA 1972.
65. Chemical Industries Association. *Code of Practice for the Safe Handling and Transport of Anydrous Ammonia in Bulk by Rail in the UK.* CIA 1975.
66. 'Transport of hazardous materials', *Proceedings of Symposium*, London, 15th Dec., 1977.
67. Anon., 'Distribution safety', *Loss Prevention Bulletin.* Institution of Chemical Engineers, 1981 (039), 17–28.
68. W. E. Isman & G. P. Carlson, *Hazardous Material.* Glencoe, Encino, California, 1980.
69. Hilton, M. R., 'The role of the highway authority', Transchem '79 Symposium, Teeside Polytechnic, UK 1979.
70. Stinton, H. G., 'Spanish campsite disaster', Transchem '79 Symposium, Teeside Polytechnic, UK 1979.
71. Hymes, I., *Loss Prevention Bulletin.* Institution of Chemical Engineers, Feb. 1985 (061), 10–16.
72. *Fire Prevention*, Dec. 1982.
73. Brady, D., 'Experience in dealing with incidents in Humberside', Transchem '79 Symposium, Teeside Polytechnic, UK 1979.
74. Waight, D. C. (ed.), *Dangerous Substances*, Wolters Samson (UK) 1983.

CHAPTER 14

Waste disposal

The control of pollution which can be caused by 'chemical-based' wastes – in either solid, liquid or gaseous forms – is a major concern of industry. Fairly comprehensive legislation to control such pollution exists in most developed countries. In the UK the statutory measures to control environmental hazards and pollution, e.g. under the HASAW Act 1974 and the Control of Pollution Act 1974, are in addition to the strict liability which arises under common law for any damage due to the escape of chemicals brought on to site (see Ch. 16, page 918).

There is an extensive literature on the causes[1–3] and the environmental effects[4–6] of pollution. Current legislation in the UK is described in references 7 and 8 and in the US in references 9 and 13. Industrial practice and procedures for waste disposal and pollution control are also covered by a variety of texts.[10–13] Clearly this is a very wide subject and in this chapter only those aspects of waste disposal relating to safety and loss prevention with 'chemicals' will be reviewed. (However, some aspects of noise control are discussed in Ch. 2.)

While it is convenient to consider wastes as being either solid, liquid or gaseous they may in fact be multi-phase as illustrated in Table 14.1 (after ref. 10). Furthermore, treatment of waste prior to, or as a means of, disposal may involve a phase change and introduce a different range of pollution problems. For example incineration of a solid waste may result in a gaseous emission which requires scrubbing, hence producing liquid effluent, and also some residual ash for disposal.

Recycling and recovery are often proposed as pollution control measures, indeed some proportion of recycling may be inherent in many 'chemical' manufacturing processes. Apart from any hazards associated with loading/transportation/unloading in those cases where recovery is performed at a different site, this approach tends to involve no hazards in addition to those in the original manufacturing process.[14] Mineral oils, solvents, metals (e.g. lead, copper, nickel, iron) are commonly recycled

Table 14.1 Forms of waste

Phase			Examples
Gas	Gas–vapour	—	SO_2; NO_x; HCl CO; hydrocarbons
	Gas–liquid	Mist	Acid mist carryover; chromic acid; oil mists; tar fog
	Gas–solid	Fume	Metal oxides
	Gas–liquid–solid	—	Paint spray
Liquid	Liquid	Solution	Metal plating effluent; 'spent' acids; wash-waters
	Liquid–gas	Foam	Detergent foam
	Liquid–liquid	Emulsions	Oil-in-water (e.g. suds); water-in-oil
	Liquid–solid	Slurry, suspension	Aqueous effluent from fume scrubbing
Solid	Solid	—	Asbestos insulation; heat treatment salts, pulverised fuel ash; refuse
	Solid–liquid	Sludge, wet solid	Filter cake Sewage sludge

but with many other possible recycling options the economics are unfavourable.[10]

Tipping of certain categories of solid waste on land may result in gas generation, mainly methane, and a highly polluted leachate. Any waste treatment may in any event involve hazards *per se*; for example some processes with the potential for the emission of hazardous gases are listed in Table 14.2.[15]

Two different, but related, implications of waste disposal are relevant to the present discussion.

1. Hazard control and loss prevention in the disposal operation – involving collection, segregation, conveyance, treatment, disposal, etc.
2. Control of environmental hazards.

Under (1) so far as liquid and gaseous streams are concerned the measures required do not differ basically from those discussed in other chapters. Differences do arise, however, with solid wastes (which may be heterogeneous) particularly if they are disposed of by tipping on land. Under (2) the need to control 'accidental emissions or discharges' also requires similar measures to those discussed for chemicals in other chapters but requirements for the control of routine, possibly prolonged, dispersal or disposal of wastes involve different considerations, e.g. long-term cumulative effects, persistence in the environment, synergistic effects. This is exemplified by the comparison between occupational and environmental exposures to air pollutants given in Table 14.3.

Table 14.2 Potential for liberation of hazardous gases from waste pretreatment processes[15]

Process	Associated problems	Hazardous airborne contaminants liberated (in addition to the principal reagents)
Neutralisation of strong mineral acids from metal finishing trades (sulphide and hypochlorite contamination common)	Fierce reaction Possibility of mixing with water or organic materials	Chlorine Nitrogen dioxide Sulphur dioxide
Chlorination/oxidation of cyanide wastes from heat treatment plant	Mixing cyanide with acids liberates hydrogen cyanide	Hydrogen cyanide
Separation of oil and water mixtures from engineering and heat raising plant	Emulsion splitting may involve generation of heat, hydrogen and hydrogen sulphide	Hydrogen* Hydrogen sulphide* Phosphine
Detoxification of chromic acid and chromium salts from the plating industry	Use of sulphite	Sulphur dioxide
Detoxification of by-products from smelting	Water and weak acids liberate attack gas	Arsine Phosphine†
Treatment of ammonia bearing waste from chemical industry	Liberation of gaseous ammonia	Ammonium chloride Nitric Acid
Removal of sulphides from leather industry waste	Generation of sulphide gas	Hydrogen sulphide*

* Highly flammable. † Pyrophoric.

Air pollution

Air pollutants

The most important source of air pollutants generally, apart from natural processes, are combustion processes. These may be industrial, domestic or involve motor vehicles. The normal products of combustion of fossil fuel (e.g. coal, oil, natural gas) are carbon dioxide, water vapour and nitrogen. None of these presents a problem, e.g. carbon dioxide is a normal metabolite, the body has a mechanism for its disposal and it is removed from the atmosphere by vegetation. However, the presence of sulphur in fuels results in the formation of sulphur oxides, particularly

Table 14.3 Characteristics of exposure to air pollutants – a comparison between occupational and environmental exposures (after ref. 10)

	Population exposed	Exposure profile Period of exposure	Type of pollutant Dust, fume, gas, vapour, mist (noise)	Level of pollutant	Possible results or effects
Occupational exposure	Adults (16–65) Mainly male Fit for work (health possibly monitored) Easily identifiable	Basic working week e.g. 40 hr/wk, 48 wk/yr + overtime (therefore intermittent elimination and recovery times)	Generally single pollutants Known origin Hazards known Recognised problem Personal protection provided Probably freshly formed/released (e.g. metal fumes)	Possible measurable fractions of Threshold Limit Values which are recognised criteria	Specific effects, e.g. pneumoconiosis, asthma Increased costs (accident rate, compensation, retraining) and lowered efficiency
Environmental exposure	All, including infants, aged and infirm Changing Varying susceptibilities	Continual unless area vacated (therefore elimination and recovery depend on irregular periods of low/zero concentrations)	Mixture of primary pollutants, from different sources, and secondary pollutants Origins may be difficult to prove Hazards not quantified Exposure unheeded	Normally very low At limits of analytical/instrumental sensitivity ($\mu g/m^3$) Variable with dispersion	Decreased well-being Non-specific respiratory troubles Irritation of eyes, nose and throat Damage to property and vegetation Injury to animals Decrease in 'amenity' Long-term ecological effects

sulphur dioxide which in relatively low concentrations acts as a respiratory and sensory irritant, causes discoloration of buildings and corrosion; damage to vegetation may occur at lower concentrations. If nitrogen compounds are present in the fuel, or as a result of the high combustion temperatures, oxides of nitrogen may also be generated (see page 219). Incomplete combustion results in emissions containing carbon monoxide, unburned and partially oxidised hydrocarbons (e.g. aldehydes) all of which may cause adverse health effects if present at sufficient levels. Particulate emissions, viz. smoke, grit and dust, may result in reduced visibility and deposits on buildings; smoke also tends to increase the adverse health effects of other irritants, e.g. sulphur dioxide.

Hence the World Health Organisation air quality criteria and objectives, shown in Table 14.4 (after ref. 7) consider these pollutants in conjunction. These refer to general air quality and if compared with the data for the effects of higher concentrations of sulphur dioxide alone, e.g.

Table 14.4 Air quality criteria and WHO objectives for sulphur dioxide and smoke (WHO, 1972)

Daily average concentration	Effects
500 $\mu g/m^3$ SO_2 with 500 $\mu g/m^3$ smoke*	Increased deaths among people suffering from heart or respiratory disease
500 $\mu g/m^3$ SO_2 with 250 $\mu g/m^3$ smoke*	Worsening in condition of people suffering from bronchitis
250 $\mu g/m^3$ SO_2 with 250 $\mu g/m^3$ smoke*	Increased symptoms in people suffering from respiratory disease
200 $\mu g/m^3$ of sulphur oxides with 120 $\mu g/m^3$ of smoke should not occur on the same day for more than seven to eight days during the year;*†‡	WHO recommended long-term goal
Annual average concentration	*Effects*
100 $\mu g/m^3$ SO_2 with 100 $\mu g/m^3$ smoke*	Increased frequency of respiratory disease and symptoms in children
86 $\mu g/m^3$ SO_2	Chronic plant injury and excessive leaf drop§
80 $\mu g/m^3$ smoke	Complaints of annoyance and reduced visibility
60 $\mu g/m^3$ of sulphur oxides with 40 $\mu g/m^3$ smoke*‡	WHO recommended long-term goal

Notes
* Values for smoke and SO_2 apply only in conjunction with each other.
† The permitted seven to eight days may not include any consecutive days.
‡ Although the criteria refer to SO_2 the goals are actually stated in terms of sulphur oxides.
§ Source: *Air Quality Criteria for Sulphur Oxides*, US Department of Health, Education and Welfare

from an accident or occupational exposures, given in Table 15.8 exemplify the differences referred to in Table 14.3. The WHO has also published air quality criteria and objectives for carbon monoxide.

These are all 'primary pollutants' but some may participate in reactions in the atmosphere producing 'secondary pollutants', e.g. photochemical smogs and acid mists. As already described escaping gas, vapour or dusts may also be associated with the storage or processing of materials, particularly volatile liquids or gases under pressure, or accidents which occur in handling and transportation.

In Table 14.5 emissions are classified as 'persistent' or 'irregular' and by the level of discharge.

Particulate emissions from manufacturing processes consist of any combination of:

- Relatively coarse particles, i.e. a substantial proportion >10 μm, e.g. grit and dust from grinding and screening operations or from solid fuel/waste combustion processes.
- A majority of particles in the range 1 μm to 10 μm, e.g. emissions from smelting processes.

Table 14.5 Continuous/intermittent emissions to atmosphere

High level	
Routine:	Vents – General ventilation (factory atmosphere) – Local extraction (dust, fumes, odour) – Process equipment
	Flare stacks – Continuous Chimneys
Irregular:	Plant maloperation – Process plant – Extraction plant (e.g. low solvent flow on gas scrubbers; excessive gas rate)
	Flare stack – Intermittent Plant failure – Process plant – Extraction/collection plant; cyclones, precipitators, filters, scrubbers (e.g. failure of filter bags)
Low level	
Routine:	Process equipment cleaning Materials handling (discharge, conveying, bagging) dust blowing off accumulations (e.g. 'dusting' of spoil heaps) Waste handling/deposition Tanker, tank container or drum cleaning operations Open effluent treatment tanks, lagoons, basins, channels
Irregular:	Plant maloperation (e.g. unauthorised 'venting') Plant failure Start up/shut-down (i.e. unsteady-state operating conditions) Dismantling/demolition (e.g. lagging removal) Unauthorised waste incineration

- True fumes containing particles $<1\ \mu m$ either alone or in combination with fogs or other contaminants.

Typical efficiencies of arrestment plant at various particle sizes are summarised in Table 6.16. Further details on particulates' control are given in references 10, 11 and 12.

Control

The major items of UK air pollution legislation are the Clean Air Acts 1956 and 1968, the Alkali etc. Act 1906 and the Control of Pollution Act 1974. The Clean Air Acts cover a number of specified fuel-burning operations other than those which present special technical difficulties and registerable under the Alkali Acts. The latter are concerned with the control of emissions of 'noxious or offensive gases' including smoke, grit and dust from specified chemical and industrial processes.

Under the Alkali Acts the general practice in the UK, unlike some other countries, is not to set arbitrary limits for ground-level concentrations. In each case the permitted concentration depends upon the nature of the discharge, its toxicity or nuisance value, the quantities involved, the proposed method of discharge, the pollutants already present or reasonably likely to be present in the atmosphere, and the effects of synergism or poor atmospheric dispersing conditions.[10] One rule-of-thumb method for estimating a limiting concentration at ground level (glc) at which 'no danger' will arise to the health of residents in the area was.

$$\text{Acceptable glc} = \frac{\text{TLV}}{4 \times 10}\ \text{ppm}$$

(The factor of 4 is to allow for 24 hr/day exposure and the factor of 10 to allow for the young, old and infirm in the resident population.) However, this has very serious limitations arising from the incorrect use of TLV criteria, the fact that pollutants do not occur in isolation, and the possible effects of synergism. Thus the best that is possible is to define a limit in terms of a concentration – time relationship at which little or no effect is observed either in isolation or with other pollutants present.[10]

In order to comply with these controls many processes are provided with specially designed pollution control equipment. The processes generally used for the treatment of pollutants in gaseous form are summarised in Table 14.6; further details are given in references 10, 11 and 12. Typical arrestment plant performances are summarised in Table 6.16.

Clearly the proper design, maintenance and continuous operation of such processes is essential. Integrity and reliability of equipment is of prime importance when handling chemicals which are dangerous *per se*.

Table 14.6 Pollution control processes

Process	Examples
• Condensation	On distillation and evaporation generally. On vapour-degreasing, dry-cleaning equipment
• Absorption/scrubbing – in solvent/aqueous solution, possibly followed by stripping	Acid gas removal using alkaline solutions. Hydrogen sulphide removal using ethanolamine solutions
• Adsorption – followed by desorption	Solvents recovery from air streams Odour removal Hydrogen sulphide removal
• Catalytic combustion	Oxides of nitrogen removal Odour removal
• Combustion	Incineration, flaring

Twenty-two people were killed and 320 hospitalised with acute hydrogen sulphide poisoning resulting from an emission from a sulphur recovery plant in Mexico in 1950.[16] A burner on the plant failed under increased hydrogen sulphide flow and gas was emitted over a 25-minute period when atmospheric conditions did not assist rapid dispersion, i.e. weak winds, low inversion layer, fog.

Reference was made to the Bhopal tragedy on page 883 and to the Seveso incident on page 510.

Liquid effluents

In the UK the majority of aqueous effluents are discharged via municipal sewers for treatment prior to discharge to natural waters or the sea. The alternatives involve a higher degree of treatment prior to either discharge to underground streams, or discharge direct to natural waters – rivers, estuaries, lakes or the open sea.

Effluent treatment prior to discharge may involve a combination of physical, chemical and biological methods. Thus some constituents of aqueous effluents and treatment methods are listed in Table 14.7. Clearly the handling and use of chemicals in these processes, e.g. neutralisation with lime or oxidation with chlorine, and the unit operations involved are accompanied by the range of hazards outlined in Chapters 6 and 9 and require similar safeguards.

An operator suffered from benzene poisoning after discharging a 'water layer' from a benzene tank to drain.[17] The cause was equilibration of the benzene in the water with the air above the drain – resulting in benzene vaporisation as described in Chapter 3, page 82.

Table 14.7 Some constituents of aqueous effluents and treatment methods[10]

Pollutant	Example	Treatment method
Alkali	Sodium hydroxide (caustic soda) Potassium hydroxide (caustic potash) Calcium oxide (lime) Calcium hydroxide Sodium, potassium and calcium carbonates Ammonia (q.v.)	Neutralisation if required – see also Total dissolved solids
Ammonia		Air stripping Incineration Neutralisation Nitrification
Biocides		Incineration
Biodegradeable waste	Sewage Food waste Organic chemicals	Incineration Settlement if necessary, biological oxidation and/or biological reduction, tertiary treatment if necessary
Boron, borates, fluoborates	—	Evaporation Ion exchange
Bromine	—	See Chlorine
Chloride	—	See Total dissolved solids
Chlorine	—	Air stripping Reduction
Chromic acid (hexavalent chromium)	—	Adsorption Cementation Dialysis Freeze concentration Ion exchange Oxidation Precipitation as insoluble salt Precipitation, possibly after neutralisation, using barium salts Reduction to trivalent chromium, then as for metal salts in acid solution Reverse osmosis
Cyanide	Copper cyanide Nickel cyanide Potassium cyanide Silver cyanide Sodium cyanide Zinc cyanide	Biological treatment Carbon adsorption Displacement by acid followed by aeration Electrolysis Evaporation Hydrolysis with steam Ion exchange Oxidation to cyanate or carbon dioxide and nitrogen

Table 14.7 (continued)

Pollutant	Example	Treatment method
		Precipitation with ferrous sulphate Reverse osmosis Treatment with hydrogen peroxide and formaldehyde
Emulsified oil	—	Air flotation Centrifugation Chemical emulsion breaking, then as for oil Heating Incineration
Fluoride	—	Adsorption on alumina See also Total dissolved solids Neutralisation with lime and precipitation Precipitation with aluminium sulphate
Fluorine	—	See Chlorine
Metal salts in alkaline solution	Cuprammonium complex Nickel and cobalt ammonia complex Cyanides (q.v.) Copper pyrophosphates Plumbites Zincates	Adsorption Crystallisation Destruction or removal of anion, such as cyanide Electrodialysis Electrolysis Evaporation Freeze separation Ion exchange Oxidation Reverse osmosis Solvent extraction
Metal salts in acid solution	Most metals as acid salts, e.g. chloride nitrate sulphate and others	Adsorption Cementation Crystallisation Electrodialysis Electrolysis Evaporation Freeze separation Ion exchange Neutralisation and precipitation as hydroxide Reverse osmosis Solvent extraction
Mineral acid	Hydrobromic acid Hydrochloric acid Hydrofluoric acid Nitric acid Sulphuric acid	Neutralisation and possibly precipitation of insoluble salt See also Total dissolved solids
Miscible/soluble organic materials	Acetone Alcohol Acetic acid	Carbon adsorption Desorption by air stripping Incineration – see also Biodegradeable waste Solvent extraction

Table 14.7 (continued)

Pollutant	Example	Treatment method
Non-metallic inorganic dissolved compounds	Arsenic Selenium	Adsorption Ion exchange Precipitation as insoluble compound
Oil, grease, wax and immiscible organics	Lubricating oil Animal fat Carbon tetrachloride	Air flotation Centrifugation Coalescence if necessary, separation by flotation then aqueous layer to further treatment if necessary and oil to incineration or recovery Filtration/adsorption
Organometallic compounds	Tetraethyl lead	Ion exchange
pH	—	Neutralisation
Pharmaceuticals	—	See Biodegradeable waste Incineration
Phenols and related compounds	—	Adsorption on carbon See also Biodegradeable waste Incineration Oxidation Solvent extraction
Phosphate	—	Precipitation as insoluble phosphate See also Total dissolved solids
Sewage	—	See Biodegradeable waste
Sulphate	—	Distillation/evaporation Electrodialysis Ion exchange Precipitation as insoluble sulphate Reverse osmosis See Total dissolved solids
Sulphite liquor	—	See Biodegradeable waste
Suspended particles	—	Centrifugation Filtration Flocculation if possible, settlement/clarification/ sedimentation Hydrocyclone
Total dissolved solids	Carbonate Chloride Nitrate Phosphate Sulphate	Distillation Electrodialysis Freezing Ion exchange Reverse osmosis (Desalination)

The similarities between a chemical plant and effluent treatment facilities are illustrated in Fig. 14.1 which is of a plant for treating metal treatment/plating effluents.

Biological oxidation is a process of waste utilisation by bacteria as they multiply. The microbial mass is separated leaving an effluent containing only a small amount of organic material. Aerobic effluent treatment is performed in trickling filters (i.e. a bacteria bed) or in activated sludge tanks, or, in warmer climates, in oxidation ponds.[10]

Disposal of trade effluent to public sewers

In the UK all discharges to a sewer require a 'consent', formerly under the Public Health Acts of 1937 and 1961 but now under the Control of Pollution Act 1974, s.43 and s.44. The relevant water authority will attach conditions to the consent relating to flowrate, total volume discharged, and composition of the effluent; reference may be made to permitted concentrations of specific toxic substances. Typical consent conditions applicable where the effluent represents only a small proportion of the total flow in the sewer are given in Table 14.8.

The purposes of limits or even total prohibition on certain chemicals are to avoid danger to sewermen, obstruction to flow, damage to sewer structures or interference with waste treatment processes. Accumulation of flammable vapours from volatile liquids discharged, or overflowing accidentally, into a factory or local authority drainage system have the potential to cause a fire or even a large confined vapour cloud explosion (see page 875). Organic solvents can cause substantial impairment of sludge sewage digestion processes.[18] Thus there is a prohibition on the

Table 14.8 Typical consent conditions for discharge to municipal sewers – UK

Parameter/substance	Maximum allowed
pH range	6–10 (permitted range)
Sulphate (as SO_3)	500–1,000 mg/l
Free ammonia (as NH_3)	500 mg/l
Suspended solids	500–1,000 mg/l
Tarry and fatty matter	500 mg/l
Sulphide (as S)	10 mg/l
Cyanide (as CN)	10 mg/l
Immiscible organic solvents	nil
Calcium carbide	nil
Temperature	45 °C
Petroleum and petroleum spirit	nil
Total non-ferrous metals	30 mg/l
Soluble non-ferrous metals	10 mg/l
Separable oil and or grease	300–400 mg/l

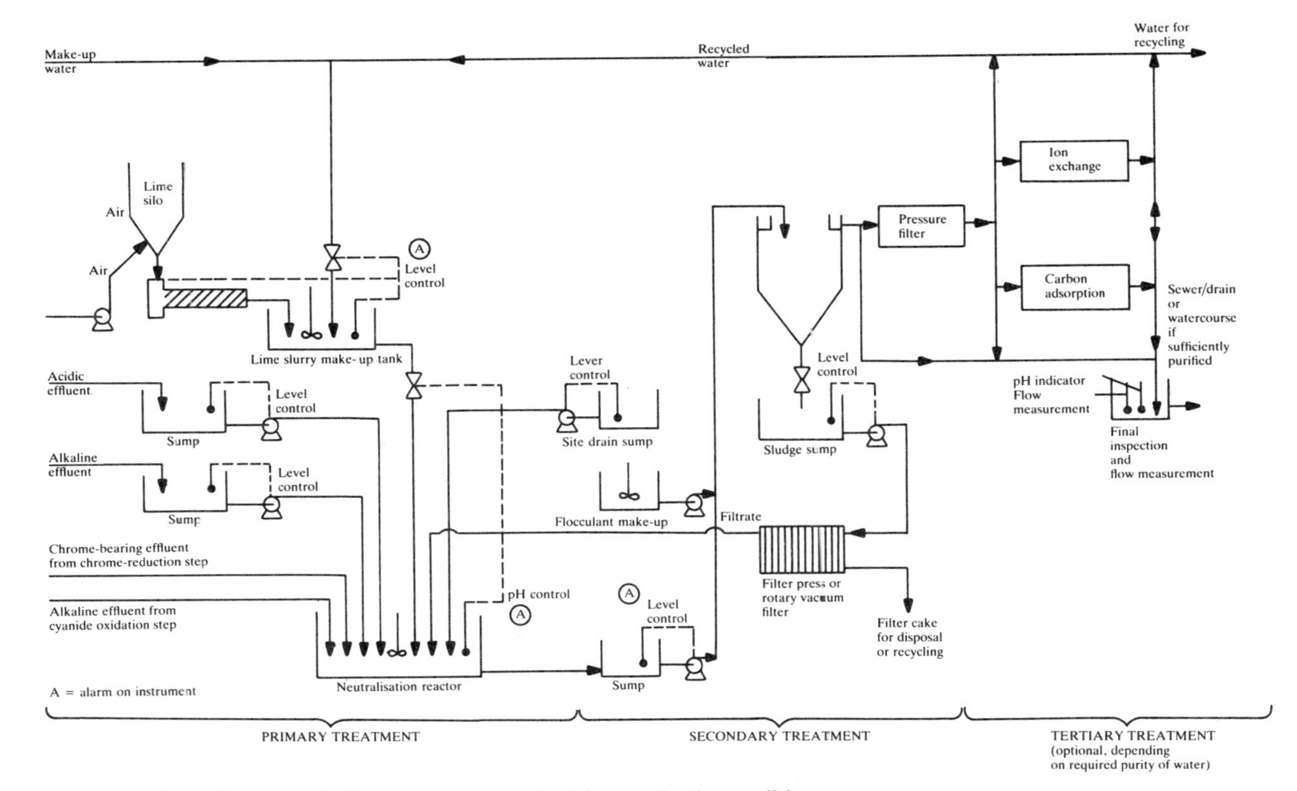

Fig. 14.1 Flow diagram of effluent treatment plant (neutralisation, solids separation and water recycle)[10]

discharge of immiscible organic solvents or petroleum; calcium carbide – which would generate acetylene – is also prohibited.

Limits are also placed upon the quantities of cyanides and sulphides to prevent toxic gas generation – for the safety of effluent treatment plant operatives, sewermen, tanker drivers, etc.

A small plant using high concentrations of alkaline cyanide solution discharged contaminated wastes to a catch basin near a wooded gulley. An operator inspecting the pumping apparatus, on a humid day with no wind movement, was overcome by hydrogen cyanide gas. (The gas was generated due to the action of carbonic acid on the solutions.)[19]

Very hot effluents are excluded partly because they would accelerate chemical degradation and toxic gas generation. A permissible pH range is usually specified and sulphates are limited to avoid structural damage. Since all sewage plants incorporate a biological oxidation unit it is important to avoid any significant concentrations of biochemical poisons which may retard or inhibit these treatment processes, e.g. phenols, cyanides, heavy metals; the latter would in any event pass through and into the surface waters receiving the final discharge.

The consent for discharge normally includes requirements for monitoring the quantity and composition of discharges, and provision for entry to the premises by the authorities to obtain samples. Clearly such control is essential for safe operation.

Within the UK in 1982 some companies were approached by effluent-receiving authorities with a requirement to reduce volatile solvent concentrations in their effluent discharges to a sufficiently low level to ensure that the atmosphere within the drains does not exceed the TLV. However, calculations have shown that this would require their concentration in the liquid phase to be reduced to a few ppm[20] (see also page 83), which is several orders of magnitude lower than the limiting solubility concentrations to which existing consents are generally related. This was considered impracticable and it was recommended that agreed standards should be decided by consultation together with continued provision and use of safety equipment together with suitable breathing apparatus.[20]

Lone entry into any sewer is not permissible and a strict system of work (see 'Entry into confined spaces', Ch. 17) is required for entry. Similar considerations may apply to underground pumping stations. The potential for accumulation of toxic gases, e.g. hydrogen sulphide due to anaerobic decomposition of sulphur-bearing materials, or oxygen deficiency, or pockets of flammable gas requires the provision of appropriate test-apparatus (see Ch. 7).

Two men descended into a stormwater chamber at a sewage works to open a jammed valve to an adjacent sedimentation chamber. At a depth of about

5 m one man indicated they should return to the surface but as they started to climb the vertical access ladder the second man fell off, landed 4 m below, and sustained fatal injuries. Hydrogen sulphide had entered the chamber via a pipe connection from the sedimentation area resulting in a concentration estimated at 5 × TLV.[21]

Discharge to natural waters

Within the UK consents are required for direct discharges to non-tidal waters under the Rivers (Prevention of Pollution) Acts 1951 and 1961. The conditions usually specified relate to both the quantity and quality of the effluent in the final discharge. Certain direct discharges to underground strata also require a consent under the Water Resources Act 1963.

The water pollution problems which controls aim to avoid include:

- Eutrophication, e.g. due to nitrates and phosphates.
- Deleterious effects on river life of heavy-metal salts (e.g. cadmium, mercury).
- Indirect damage due to oxygen depletion.
- Concentration of chemicals by certain organisms.
- Presence of phenolic compounds, cyanides, chlorinated hydrocarbons.

In January 1984 an effluent treatment plant froze up. When it thawed it was overwhelmed and as a result the River Dee was polluted by phenols. This affected the customers of two water authorities.[22,23]

Heavy metals create a particular problem since although generally discharged at very low concentrations they can be concentrated in passage up the food chain. Chemicals which persist in the environment, due to slow rates of biochemical degradation, can also create a hazard because continued discharge may cause a build-up in concentration in the environment.

Examples of typical consent conditions are given in Table 14.9. Again the demand for oxygen in order to decompose organics is expressed as BOD. In the UK this factor, and the ability of a river to support fish life and the presence of toxic pollution, forms part of a Department of the Environment river classification scheme. (Alternatively oxygen demand may be experienced as COD (chemical oxygen demand) – determined with strong oxidants and therefore giving a measure of almost all the organics present.) The water quality in a river is also indicated by the quality and quantity of animal life it will sustain. Thus fish are generally a sensitive indicator of pollution (see LC_{50} discussed in Ch. 6).

Contamination of drinking water supplies due to pollution of a river which serves as a source may arise from[7]:

- Direct discharge of industrial effluent or of sewage works effluent.
- Drain-off of fertilisers and pesticides from agricultural land.

Table 14.9 Typical consent conditions for the discharge of industrial effluent to rivers and streams – UK

Parameter/substance	Maximum allowed
Fishing streams	
BOD (5 days at 20 °C)	20 mg/l
Suspended solids	30 mg/l
pH	5 to 9
Sulphide (as S)	1 mg/l
Cyanide (as CN)	0.1 mg/l
Arsenic, cadmium, chromium, copper, lead, nickel, zinc – either individually or in total	1 mg/l
Free chlorine	0.5 mg/l
Oils and grease	10 mg/l
Temperature	30 °C
Non-fishing streams	
BOD (5 days at 20 °C)	40 mg/l
Suspended solids	40 mg/l
pH	5 to 9
Transparency of settled sample	not less than 100 mm
Sulphide (as S)	1 mg/l
Cyanide (as CN)	0.2 mg/l
Oils and grease	10 mg/l
Formaldehyde	1 mg/l
Phenols (as cresols)	1 mg/l
Free chlorine	1 mg/l
Tar	none
Toxic metals – individually or in total	1 mg/l
Soluble solids	7,500 mg/l
Temperature	32.5 °C
Insecticides or radioactive material	none

BOD = biological oxygen demand (i.e. the amount of oxygen needed to oxidise biologically substances to CO_2), a measure of suspended, colloidal or dissolved organics.

- Accidents involving chemical tankers.
- Illicit waste disposal activities.
- Ingress of contaminated surface water from landfill sites (discussed below).

Clearly, for environmental health considerations, and to comply with the law, technical and managerial control over aqueous effluent discharges and monitoring are essential. To assist this, a check-list is provided at the end of this chapter.

Solid-waste disposal

'Solid' waste comprises those wastes which do not flow from their sources via pipelines, e.g. sewers; it includes sludges and liquids so contaminated with solids that they are unsuitable for treatment as aqueous effluents.

There have in the past been numerous cases of indiscriminate disposal of toxic wastes, resulting in some environmental damage, e.g. land pollution, and high clean-up costs.

In August 1978 some 120,000 litres of waste oil contaminated with PCBs was disposed of illegally by spraying along the verges of 430 km of rural roads in N Carolina, USA.[24] As a result 40,000 tonnes of contaminated soil had to be disposed of.

It has been proposed that waste producers should be made legally responsible for those wastes that are hazardous by the following actions:[25]

- Choice of appropriate transport and disposal methods (unless the use of certain disposal methods is obligatory).
- Avoiding, minimising and recycling wastes (as far as is technically possible and economically reasonable).
- Correct declaration and proper labelling according to the legal requirements.

Within the UK the Deposit of Poisonous Waste Act 1972 first introduced a general prohibition on the tipping of poisonous, noxious or polluting waste where it was liable to give rise to an 'environmental hazard', and a requirement for notification to local and river authorities before such waste was removed or deposited.[10] This Act was repealed and replaced by the Control of Pollution Act 1974 and various Regulations made thereunder. The main relevant provisions which will apply when the Act is fully operational are summarised in Table 14.10.

Table 14.10 Hazardous waste provisions of the Control of Pollution Act 1974 (not yet all in force)

s. 1	Waste Disposal Authorities to ensure that adequate arrangements exist in their areas for the disposal of all controlled waste (household, commercial and industrial wastes).
s. 2	WDAs to survey waste arising and disposal facilities in their areas and to prepare, and periodically revise, a waste disposal plan.
ss. 3–11	All controlled waste to be disposed of at licensed sites. Each licence sets conditions including the types and quantities of waste which can be disposed of, the manner of disposal, the mode of site supervision and operation, the standard of boundary fences, and the provision of site services, e.g. wheel-cleaning equipment, safety showers.[26]
ss. 12–14	Waste to be classified for collection purposes.
s. 16	Local authority empowered to order an occupier to remove controlled waste from any land if it poses, or is likely to pose, an 'environmental hazard'. (The authority may otherwise in certain circumstances remove it and recover reasonable costs.)
s. 17	Provided for the introduction of tighter controls on difficult-to-dispose of or dangerous wastes, termed 'special wastes'.

The Control of Pollution (Special Waste) Regulations 1980, made under s.17, requires waste producers to give advance notification to receiving Waste Disposal Authorities of any intention to dispose of a consignment of special waste. 'Special waste' comprises any controlled waste which consists of, or contains, any of the categories of substances listed in a schedule – reproduced as Table 14.11 – and which is dangerous to life because of its corrosivity, flammability or toxicity. A consignment

Table 14.11 Categories of substances subject to the Control of Pollution (Special Waste) Regulations 1980

1. Acids and alkalis.
2. Antimony and its compounds.
3. Arsenic compounds.
4. Asbestos (all forms).
5. Barium compounds.
6. Beryllium and its compounds.
7. Biocides and phytopharmaceutical substances.
8. Boron compounds.
9. Cadmium and its compounds.
10. Copper compounds.
11. Heterocyclic organic compounds containing oxygen, nitrogen and/or sulphur.
12. Hexavalent chromium compounds.
13. Hydrocarbons and their oxygen, nitrogen and/or sulphur compounds.
14. Inorganic cyanides.
15. Inorganic halogen-containing compounds.
16. Inorganic sulphur-containing compounds.
17. Laboratory chemicals.
18. Lead compounds.
19. Mercury compounds.
20. Nickel and its compounds.
21. Organic halogen compounds, excluding inert polymeric materials.
22. Peroxides, chlorates, perchlorates and azides.
23. Pharmaceutical and veterinary compounds.
24. Phosphorus and its compounds.
25. Selenium and its compounds.
26. Silver compounds.
27. Tarry materials from refining and tar residues from distilling.
28. Tellurium compounds.
29. Thallium and its compounds.
30. Vanadium compounds.
31. Zinc compounds.

note system requires handover against a signature at each stage of the consignment's transfer until it is finally disposed of. Registers of consignment notes have to be maintained by producers, carriers and disposers. Permanent records must also be kept of the locations on sites of deposits of special wastes. Guidance on application of this system is given in references 27 and 28.

In addition to UK legislation four EEC Directives have been adopted with regard to hazardous wastes; these cover waste, the disposal of waste oils, the disposal of PCBs and PCTs (polychlorinated biphenyls and terphenyls, respectively), and toxic and dangerous waste, viz. any waste containing or contaminated with any of 27 categories of substances listed in an Annex in such quantities or concentrations as to present a risk to human health or the environment.[29] Directives have also been issued on waste from the titanium dioxide industry[30] and the protection of groundwater against pollution caused by certain dangerous substances.[27,31]

In the UK the disposal of radioactive wastes is governed separately under the Radioactive Substances Acts already referred to in Chapter 8. An authorisation is required before a user may accumulate or dispose of radioactive waste; this will specify the type of waste, means of disposal, conditions to be observed, and any requisite environmental measurements. Transportation and safe packaging[32] are similarly covered by statutory legislation.

The main provisions of the EEC Directive on toxic and dangerous waste are summarised in Table 14.12.

Table 14.12 Provisions of EEC Directive 78/319/EEC on toxic and dangerous waste

Article	
5	Prohibition of dumping and uncontrolled discharge, tipping or unmanaged transport of toxic and dangerous wastes.
6	Competent authorities to be responsible for planning, organisation, authorisation and supervision of operations for the disposal of toxic and dangerous wastes.
12	Up-to-date plans to be kept by these authorities.
7	Packaging of toxic and dangerous wastes to be labelled appropriately, including nature, composition and quantity of waste.
9	Installations, undertakings or firms providing storage, treatment and/or deposit of toxic and dangerous wastes to hold a licence. Firms transporting these wastes to be controlled by the authorities.
14	Every installation, establishment or undertaking which produces, holds and/or disposes of toxic and dangerous wastes to keep a record of the quantity, nature, physical and chemical characteristics, the origin, method and site of disposal. Toxic and dangerous wastes carried for disposal to be accompanied by a special identification form to the final point of disposal. All relevant documents to be retained.

In the USA the control of hazardous wastes is regulated by the Resource Conservation and Recovery Act (RCRA) 1976 and ensuing Regulations. Here hazardous waste is by definition a solid waste, or combination of solid wastes, which because of its quantity, concentration, or physical, chemical or infectious characteristics may

- cause or significantly contribute to an increase in mortality or an increase in serious irreversible or incapacitating reversible illness; or
- pose a substantial present or potential hazard to human health or the environment when improperly treated, stored, transported, or disposed of, or otherwise managed.

The identification of hazardous wastes is based on measurable characteristics, i.e. ignitability, corrosivity, reactivity and EP (environmental pollution) toxicity using standard tests or by reference to official lists.[33,34] Certain wastes are exempted from full control, including those from premises producing/accumulating <100 kg per month of most hazardous wastes; nevertheless such wastes must be disposed of in approved, environmentally-sound, facilities. For acutely hazardous wastes (e.g. cyanides, many pesticides, arsenic acid) the limit is 1 kg per month; for containers or container liners which hold such wastes the limit is 10 kg per month. The quantity determination excludes hazardous waste that is used/re-used/recycled/reclaimed but some states allow no small-quantity exemptions.[35]

Under the RCRA each hazardous waste handler is given an identification number and, for listed wastes, a hazard code. A waste constitutes a hazardous waste if it contains any concentration of a listed waste – unless subject to the small-quantity exclusion. All generators, transporters and operators are required to participate in a manifest system and producers intending to ship waste outside the USA must keep records and notify the Environmental Protection Agency (EPA); it is the EPA's responsibility to notify recipient countries. Transporters may hold consignments for ≤10 days without an RCRA permit and without complying with all the regulations for treatment, storage and disposal facilities.

The owners and operators of hazardous waste facilities must comply with administrative and operating standards as specified while transporters are responsible for cleaning up any discharge occurring during movement. In addition, under the Toxic Substances Control Act, the use of PCBs is generally prohibited unless they are 'totally enclosed'.

After floods in the Ohio river system in December 1978 hundreds of odorous, corroding 250 litre drums had to be recovered from a creek. Investigation revealed the 'Valley of the Drums', farmland where 25,000 decomposing drums, some leaking chemicals, had been dumped.[24]

Since the RCRA does not cover old or abandoned landfill sites a Superfund was instituted to allow the EPA and Federal agencies to clean up abandoned waste sites. The full implications are discussed in reference.[36]

Clearly from the above a consensus has developed regarding the type of technical and administrative controls required for a variety of hazardous wastes – and there can be no doubt, on the basis of known 'near-misses' and pollution incidents, that these are justified.

The disposal routes[14] for notifiable wastes in the UK comprise: landfill, i.e. deposition of waste in surface depressions 80%; sea disposal, i.e. deposition in any of nine disposal areas in shallow seas off-shore, 9%; treatment and incineration, 11%. Technical memoranda on the treatment and disposal of various wastes with the potential to be hazardous are listed in reference[37]; these include Codes of Practice. (In addition Technical Memorandum No.23 on 'Special Wastes' provides guidance on their definition.)

Landfill is practised either at contained sites, where artificial or natural barriers restrict the movement of liquid, or permeable sites, where liquid can slowly percolate away. Since many of the wastes listed in Table 14.13 may be disposed of by landfill, proper site selection, management and operation are essential for safety and loss prevention; this is discussed further below.

Deposition of hazardous wastes in disused coal or mineral mines is not considered satisfactory since they are usually wet, resulting in easy

Table 14.13 EEC list – toxic and dangerous wastes

1. Arsenic; arsenic compounds
2. Mercury; mercury compounds
3. Cadmium; cadmium compounds
4. Thallium; thallium compounds
5. Beryllium; beryllium compounds
6. Chrome(VI) compounds
7. Lead; lead compounds
8. Antimony; antimony compounds
9. Phenols; phenol compounds
10. Cyanides, organic and inorganic
11. Isocyanates
12. Organic-halogen compounds, excluding inert polymeric materials and other substances referred to in this list or covered by other Directives concerning the disposal of toxic or dangerous waste.
13. Chlorinated solvents
14. Organic solvents
15. Biocides and phytopharmaceutical substances
16. Tarry materials from refining and tar residues from distilling
17. Pharmaceutical compounds
18. Peroxides, chlorates, perchlorates and azides
19. Ethers
20. Chemical laboratory materials, not identifiable and/or new, whose effects on the environment are not known
21. Asbestos (dust and fibres)
22. Selenium; selenium compounds
23. Tellurium; tellurium compounds
24. Aromatic polycyclic compounds (with carcinogenic effects)
25. Metal carbonyls
26. Soluble copper compounds
27. Acids and/or basic substances used in the surface treatment and finishing of metals

Note: The EEC Directive defines toxic and dangerous waste as 'any waste containing or contaminated by the substances or materials listed in the Annex 15 of this Directive of such a nature, in such quantities or in such concentrations as to constitute a risk to health or the environment'.

percolation to acquifiers or low lying surface waters.[38] Deep-well injection of fluid toxic wastes is practised in the USA albeit with some controversy.

Dumping at sea is used for relatively innocuous waste, e.g. pulverised fuel ash, mining spoil and sewage sludge but also for industrial waste – although not necessarily toxic waste.[14] This disposal is subject to licence under the Dumping at Sea Act 1974.

Incineration is used for destroying the most toxic organic chemicals, e.g. pesticides, herbicides and PCBs. (Wastes containing more than trace amounts of specific metals, e.g. zinc, mercury, arsenic, lead, and cadmium, produce toxic vaporisation products and are not therefore incinerated.) Some 'chemical incineration works' of this type, viz. works for the destruction by burning of chemical wastes containing combined chlorine, fluorine, nitrogen, phosphorus or sulphur, are subject to the Alkali Act.

Waste management

Monitoring of wastes as a check against notification documents, to avoid admixture of incompatible chemicals, and to minimise environmental pollution may be prohibitively expensive. This is exemplified by the range of properties from which a selection must be made, as summarised in Table 14.14.[39] A test kit is marketed which will identify hazards (not chemicals) in categories such as poisonous, explosive, flammable and/or corrosive. Any unexpected positive result from a site test performed on

Table 14.14 Waste analyses

	Reaction with water Reaction with acids Reaction with alkalis	Identification of any gases evolved	
For inorganic wastes	Effect of heat pH, total solids Presence of sulphides, total cyanide, ammonium compounds Concentration of metals Pb, Zn, Cd, Hg, Sn, As, Cu, Cr, Ni		For mixtures of inorganic and organic wastes all these analyses should be completed
For organic wastes	Calorific value Flash point Miscibility with water (or other wastes) Viscosity at various temperatures Halogen, sulphur, nitrogen content Ash content Analysis of ash Organic content by biological oxygen demand (BOD), chemical oxygen demand (COD), permanganate value (PV) or total carbon methods		

a small sample of waste can then be followed by more sophisticated analysis or enquiry to the waste producer.

Simple guidelines are applicable to some operations involved in waste disposal.[40] Firstly, the waste should be categorised and certified by the producer and checked by the contractor, who should advise on segregation and storage prior to collection. Admixture of wastes during transportation or disposal should be avoided unless technical evidence shows that no personnel or environmental hazard will be created. In the following example cross-contamination of waste created a serious environmental hazard[41] in the US.

> A load of waste oil was accidentally contaminated in a road tanker by waste chlorinated hydrocarbons from a previous load. The combined waste was sprayed on roads and work areas to suppress dust on a horse ranch. This resulted in ingestion of tetrachlorodibenzodioxin by horses causing serious injuries and deaths and a child was contaminated with development of chloracne.

The correct type of container or tanker should be selected for transportation; loose waste or containers should be adequately checked and secured during transportation.

Notification and monitoring

To handle a waste safely it is necessary to know, in general terms, its toxicity, flammability, stability, corrosivity, dermatitic potential, reactivity and any other properties which could create a hazard to operators (see Chs 4, 5 and 10). Contractors who handle a variety of wastes may be at a disadvantage in this regard, especially if collection and transportation is by a third party. Therefore, it is essential that wastes are properly notified. As summarised above, within the UK every producer must provide information regarding the nature and quantity of hazardous wastes to be disposed of. A model form has been proposed on which the producer is requested to declare the waste's chemical composition; provision is made for an analytical report.[39] Any risks, the best practicable means of disposal and authorisation for such disposal are recorded on the form, a copy of which is carried by the driver of the vehicle used to transport the waste to its disposal point.

Ideally, contractors should ensure that the producer does, in fact, declare all relevant details of the chemical and physical nature of the waste, its 'quality', rate of arising and any special properties.[40] Consideration can then be given to any hazards likely to arise in loading, transportation, unloading, treatment and/or disposal before removal is undertaken. Handling, or pretreatment, at the waste producers prior to disposal also requires the range of precautions discussed in previous chapters.

0.45 kg of sodium hydride was being disposed of in a yard when it came into contact with water.[42] An explosion ensued [see Table 10.2], three men were burned, and the fire spread to the factory.

Some pretreatment processes which are used to reduce or remove toxicity include[38]:

Process	*Reagent/action*
Cyanide → cyanate → $CO_2 + N_2$	Chlorine Hypochlorite under alkaline conditions
Persulphates, chromic acid and chromates → more stable form	Reducing agents, e.g. $NaHSO_3$ soln.
Organophosphorus chemicals (pesticides) → detoxify	Alkaline hydrolysis
Dithiocarbonate fungicides → detoxify	Acid hydrolysis
Strong acids → neutralise Strong alkalis → neutralise	Alkalis Acids (possibly also wastes)
Copper, nickel, cadmium, zinc, tin and aluminium salts → essentially water-insoluble forms	Precipitation as hydroxides

Processes are available for the 'encapsulation' of inorganic wastes using proprietary compounds. The wastes are combined to form insoluble solids; these can be made impervious (and hence non-leaching), chemically inert and non-biodegradable. In general the processes are unsuitable for organic wastes and the relatively high cost has led to their use being restricted, at present, to extremely toxic wastes, e.g. arsenic and cyanide compounds.[38]

Transportation

Whatever the method of disposal wastes need to be segregated, collected, transported and possibly even processed first. A provisional Code of Practice proposed that the waste producer should classify the various wastes, certify their contents and disclose any known hazards.[43] Identification may first be made in terms of physical properties, i.e. liquids, slurries, sludges, thixotropic solids or solids, and then as to hazard using the scheme shown in Table 14.15.

Wastes in different classes, or of different content, should be segregated where reasonably possible to allow for separate collection, trans-

Table 14.15 Classification for hazardous wastes

Class 1	– Explosives
Class 2	– Gases; compressed, liquefied or dissolved under pressure
Class 3	– Inflammable liquids
Class 4(a)	– Inflammable solids
Class 4(b)	– Inflammable solids or substances liable to spontaneous combustion
Class 4(c)	– Inflammable solids or substances which when in contact with water emit inflammable gases
Class 5(a)	– Oxidising substances
Class 5(b)	– Organic peroxides
Class 6(a)	– Poisonous (toxic) substances
Class 6(b)	– Infectious substances
Class 7	– Radioactive substances
Class 8	– Corrosives
Class 9	– Miscellaneous dangerous substances, that is any other substance which experience has shown, or may show, to be of such a dangerous character that these rules should apply to it
Class 10	– Dangerous chemicals in limited quantities

portation and disposal. The producer is recommended to ensure that his employees, and those of any carrier, conform to the Code of Practice. The latter's responsibilities include the provision of equipment, containers and vehicles of suitable design and condition, the instruction and training of his employees, and a system of work to ensure that suitable waste handling equipment and protective clothing are used when necessary. The routes of vehicles should ideally be planned and, in the case of hazardous wastes, personnel should be issued with written instructions, and appropriate emergency services advised of the journey. The vehicles, or containers, should carry bold markings, relevant safety information and some form of transport emergency card, and Hazchem marking if appropriate (see Ch. 13). A particular concern with transport accidents, particularly at sea, is that even if originally the waste containers were labelled with the correct hazard warnings, etc., such directions may be washed off.[44]

A ship carrying potassium methyl sulphate waste from a pharmaceutical company ran aground on the Scottish coast. Due to inadequate labelling, clean-up procedures were not geared to handling the highly toxic, dimethyl sulphate.

Landfill

Disposal methods on landfill sites vary from crude dumping to relatively sophisticated operations involving reception pits, pumps and pipelines to soakage trenches. The waste may first be transferred to site tankers suitable for crossing the rough terrain of the tip. The solid deposits are ideally covered as expeditiously as possible and the liquid waste swiftly soaks away. In general, waste should be deposited in layers of restricted depth and covered with inert, innocuous material at the end of each working day.[45] However, an adequate supply of cover is not always available and complaints may arise regarding the environmental impact associated with some tipping operations.

Any hazard which land deposition may create has to be assessed with regard to the risk of injury or impairment to health, to persons or animals or the pollution or contamination of any water supply above or below ground.

A private waste disposal site was used for the disposal of acids and tarry wastes followed by deposition of inert solid rubble. The site was close to houses and unfenced; as it began to consolidate, children and animals began to use it. Subsequently the acid wastes seeped to the tip surface and children and animals were burned due to contact with them. Eventually the county waste disposal department took over the site and neutralised all the waste with lime, involving a cost of £200,000.[46]

The introduction of site licensing requirements in the UK, such as the provision of security fencing, are beginning to reduce the possibility of trespassers gaining unauthorised access to landfill sites but the presence of totters, who make a living by sorting through wastes, e.g. containing lead, needs to be taken into account.

The potential hazards associated with toxic materials are summarised in Table 14.16. These arise from the nature of the waste and its manner of deposition but there is also the possibility of toxic gas generation due to admixture of 'incompatible wastes', e.g. wastes which undergo chemical reactions of the type exemplified in Tables 10.3(a) and 10.3(b). Some incompatibles are listed in Table 14.17. Since admixture of different chemicals in wastes may allow them to remain in prolonged contact, reactions may follow an 'unexpected' route and result in a fire, explosion, or toxic release.[47]

A number of chemicals cleared from laboratory benches were put into an open-topped 25 litre drum. They included 2 kg of ammonium dichromate and 10 kg of sodium cyanide and, following an 'incubation period' of about 20 minutes, an explosion occurred dispersing a particulates' cloud containing cyanide.[47]

Neutralisation of corrosive wastes was performed in a pit and innocuous wastes were subsequently added to soak up excess liquid. Drummed methyl ethyl ketone had been dealt with in this way but when 4 tonnes of 'general factory waste', comprising mainly scrap oil-filter elements, was tipped into the pit an explosion resulted. The contents of the pit were ejected and the driver's mate, who had been supervising the operation from the rim of the pit, sustained injuries which eventually proved fatal.[48]

Table 14.16 Potential hazards from toxic waste deposition

Air pollution	Dust, effluvia, smoke and fume.
Land pollution	Gross amenity damage; undermining of site stability; sterilisation of surrounding land due to heavy metals, pH changes, etc.
Water pollution	Deposited material or percolate escapes either through surface run-off or by underground movement threatening streams, rivers, aquifers or even the sea. Direct 'poisoning' or eutrophication.

Table 14.17 Potential for toxic gas generation by admixture of incompatible wastes[44]

	Incompatibles		Resulting toxic gas
	A	B	
Inorganic	Arsenical materials	Any reducing agent	Arsine
	Cyanides	Mineral acids	Hydrogen cyanide
	Hypochlorites	Acids	Chlorine
	Nitrates	Sulphuric acid	Nitrogen dioxide and nitrous fumes
	Nitric acid	Copper, brass, many heavy metals	Nitrogen dioxide and nitrous fumes
	Nitrites	Acids	Nitrogen dioxide and nitrous fumes
	Phosphorus	Caustic alkalis or reducing agents	Phosphine
	Selenides	Acids	Hydrogen selenide
	Sulphides	Acids	Hydrogen sulphide
Organic	Dithiocarbamate fungicides	Acids or alkalis	Ethylene thiourea
	Dithiocarbamate fungicides	Water	Carbon disulphide

Even apparently inert wastes may on reaction produce toxic gases, e.g. deposition of calcium sulphate where there is static water can result in hydrogen sulphide formation under anaerobic reducing conditions.[38]

Thus an appreciation of the chemistry, which may be involved, and experience is needed for safe disposal.

Potential operator hazards

Occupational exposure to waste is likely to be most serious in terms of human health.[44]

Few waste disposal routines are fully mechanised and there are numerous ancillary operations in collection, delivery, loading/unloading and debagging/dedrumming which tend to bring operators into close contact with the wastes. 'Hazardous' waste causes most concern; this is considered as any material which could be harmful following ingestion, inhalation or skin contact but clearly fire and explosion risks are also of relevance. However, the normal hazards associated with mechanical handling of materials, are also important.

Operator hazards during waste disposal[49–52] include toxic hazards from inhalation of dust, fumes or gases. Sufficient exposure may occur to exceed the Short Term Exposure Limits during the tipping, or disturbance of waste (e.g. asbestos) despite the operation taking place in the open air. Dedrumming, or debagging, of waste may also result in intermittent exposure to dust clouds or toxic gases. Examples have included a dedrumming operation in which a pneumatic chisel was in use under wet conditions for opening spent cyanide drums, and somewhat rudimentary debagging of lead waste.[53] Toxic emissions may also arise from the wetting of reactive waste, e.g. certain metal drosses, and the mixing of incompatible materials, e.g. hypochlorites, cyanides or sulphides with acids. Some indication of the potential hazard is given by the following calculation (after ref. 54):

$$NaCN + HX \rightarrow HCN + NaX$$

One litre of 10% sodium cyanide solution produces on reaction with an acid 2 gram molecular weights of hydrogen cyanide; this is equivalent to about 50 litres of gas at 25 °C and atmospheric pressure. The short-term exposure limit (10 min TWA value) for this gas is 10 ppm by volume. Thus if evenly dispersed a volume of 5×10^6 litres ≡ 5,000 m^3 could be polluted by HCN at this concentration. Further consideration of such hazards is given in reference 49. Admixture of wastes for neutralisation or other treatment can take place in 'lagoons' rather than process vessels so that control is harder to ensure.

The driver of a road tanker mistakenly discharged 5 tonnes of waste aluminium sulphide liquor into a soakage pit at a landfill site which was normally reserved for acid wastes. The driver of the next tanker to unload quite properly discharged waste sulphuric acid into the same pit. He was subsequently found collapsed in his vehicle, due to hydrogen sulphide poisoning, and was dead on arrival at hospital.[15]

A fatality due to hydrogen cyanide occurred when a man, working with two other men, in the course of disposing of empty 180-litre drums which contained organic wastes drove a tractor close to an industrial waste pond.[55] He was found collapsed in the tractor and despite first-aid treatment, including inhalation of amyl nitrite, and treatment in the medical department with cobalt EDTA he subsequently died.

A fatality, again due to hydrogen sulphide inhalation, occurred in southern Louisiana, USA when a man discharged a truckload of chemical waste into an open pit where it reacted with the contents. This was subsequently found to be one of three open pits operated illegally by a private company that had a permit only to dispose of waste in a deep injection well.[24]

For operation safely the manner of tanker discharge shown in Fig. 14.2 clearly relies upon the proper segregation of wastes to avoid toxic gas generation. This is generally important wherever open waste lagoons or pits are in use.

Exposure may arise in 'confined spaces' during entry for cleaning or inspection of tanker barrels, treatment tanks, soakage trenches or excavations. Finally, toxic gases such as carbon monoxide, hydrogen chloride or hydrogen cyanide could be produced through accidental, or unauth-

Fig. 14.2 Discharge of liquid waste into a concrete reception pit at a landfill site (Courtesy of R. C. Keen)

orised, waste combustion (e.g. some pyrolysis products of plastics are summarised in Table 15.9).

Ingress of toxic substances via intact mucous membranes or skin is a possibility with some material sent for disposal. These include solvents, phosphorous compounds, aniline and alkaloids.

Exposure may arise due to manual operations which either inherently, or by the adoption of 'short-cut' methods, result in physical contact. The significance of the typical minor injuries common to many work situations may be altered in the case of the solid waste worker due to the possible presence of toxic substances or pathogenic organisms. Contamination of any wounds may then have serious consequences.

While significant intake of toxic materials via the digestive system is no longer a problem in factories generally, because better hygiene regulations have prohibited eating and drinking in specific work areas and provide minimum requirements for washing facilities, some waste disposal sites lack these provisions. Hazards may, therefore, arise from classic poisons (e.g. cyanides, arsenic and strychnine), synthetic poisons (e.g. acaricides, insecticides, fungicides and rodenticides), drugs or, in the longer term, heavy metals.

The possibility of eye or skin injuries arises when handling a variety of corrosive chemicals, e.g. alkalis, acids, phenols or cresols. There is also a dermatitic hazard associated with prolonged or intermittent exposure to a wide range of materials, as discussed in Chapter 5.

Unauthorised fires are not uncommon on landfill sites but the direct danger to operators is minimal, because operations are in the open air. The use of landfill machinery to control burning material by smothering or isolation has, however, given rise to hazardous situations.[53] A noise hazard may arise with the operation of some landfill machinery, e.g. bulldozers, scrapers and cranes.

There is a risk of primary entanglement with machinery on some operations, e.g. removal of material fouling the drive on bulldozers or freeing jammed skips. Waste disposal sites tend to be isolated, so that machinery maintenance may be carried out under rudimentary conditions; this may result in 'under-maintenance' and, if facilities for machinery decontamination are rudimentary, enhance any toxic/dermatitic hazards. Where heavy plant is used there is always a possibility of vehicle accidents; added to this there is an overturning risk with bulldozers or compactors on badly planned landfill sites.

A twin-ram tipping discharge municipal freighter rolled over on a landfill site during a conventional discharge operation and crushed a private car being used to deliver household rubbish. A mother and one child in the car were killed; a second child escaped. On this site the rules required that tipping operations should have 20 m of clearance available but clearly this was ignored.[48]

The untidy state of some sites tends to create an increased hazard of 'simple' accidents such as slips and falls (Ch. 2).

Attention has been drawn to the potential role of infective agents, e.g. anthrax and leptospirosis, in affecting operators' health[56]; any leather, wool or bonemeal wastes may be suspect as regards the former and the presence of rodents increases the likelihood of the latter. Any hospital or slaughterhouse wastes or sewage sludge may contain pathogenic organisms which can be dangerous via skin contact. Poor personal hygiene could also result in gastrointestinal infections.

Safety measures advisable on a disposal operation reflect those described for chemical manufacturing or bulk chemical handling factories. However, some are of even more significance because of the variety and admixture of wastes which may be processed. Identification of wastes on site is one example.

> Several hundred gallons of waste nitric acid were left in a demountable tank at a landfill site containing a large lagoon. It was intended to drain it very slowly into the lagoon over a period of two days but an operator dealt with it as 'spent metal finishing acid' and discharged it rapidly into a bunded area already containing a large amount of this acid. A large cloud of nitrogen dioxide was liberated; this drifted towards adjacent buildings but fortunately no one was affected by it.[57]

Ideally there should be operating manuals describing every operation and/or recipe, the correct sequence, potential hazards and specific measures to guard against them. Lone working should be prohibited in especially hazardous situations or in confined space. For work in confined spaces or on contaminated equipment there should be a permit-to-work system (see Ch. 17). Operators should be provided with appropriate personal protective clothing, etc., which should be inspected and maintained on a regular basis. Washing facilities should be provided to, for example, the standards acceptable in the UK under s.58(1) of the Factories Act[58]; changing facilities should be planned to eliminate any cross-contamination between clean and dirty areas. Provision of a mess room reached from the working areas via the changing facilities, and prohibition of its use by operators in overalls, is a useful arrangement. The usual first-aid, and fire-fighting provisions are necessary plus provision for eye-wash facilities, and emergency deluge showers or equipment. Telephones should be available in strategic locations for contacting emergency and fire services when necessary. Rehearsals of emergency drills should be performed at regular intervals.

If possible attention should be given to mechanisation and improved engineering design both to increase operating efficiency and to remove the operator from contact with the waste. If the scale of operation permits, solid waste handling can be completely mechanised.

Environmental hazards

Of the potential problems listed in Table 14.16 chronic environmental exposure is the most serious scenario; it is also the most difficult to evaluate and to remedy.[44] An early example of the complexities involved both chemical interaction of wastes and ground-water migration.[59]

In 1943 manufacture of war materials commenced at an arsenal near Denver, USA. The facilities were leased eight years later to a company for insecticides manufacture. Even prior to this, farmers several miles from the plant complained of extensive crop damage and reported unexplained sickness among livestock. Subsequent investigation revealed that chlorides, phosphates, fluorides, chlorates and arsenic had been discharged from the arsenal into open holding ponds during the years of operation; contamination of ground-water had gradually spread to an area of unknown extent. There was apparently no way to contain the contamination or to stop its advance.

Surprisingly concentrations of the weed-killer 2,4-D (2,4-dichlorophenoxy acetic acid), which had never been manufactured at the arsenal, were found in the holding ponds and in some wells. This was concluded to have been formed spontaneously from other substances discharged into the ponds.

The possible routes of environmental exposure are[44]:

- Consumption of contaminated ground-water due to migration of leachate from a landfill site.
- From pollution of surface water.

Fish from the Hudson river were found to contain PCB concentrations far in excess of the acceptable daily intake due to leaching from wastes. All fishing was therefore banned.[60]

(See also Ch. 5, page 199 for examples of chronic cadmium and mercury poisoning.)

- Inhalation of airborne contaminants, e.g. bacterial spores disseminated during disposal operations[44], or combustion products.
- From development of landfill sites.[61]

It is generally acknowledged that disposal of wastes in landfills produces leachates which can locally give rise to pollution of both surface and ground-water resources unless it is adequately controlled.[62] Phenols are particularly undesirable and therefore it is often recommended that wastes containing high concentrations should be deposited in containment sites where leachates can be collected and, if necessary, treated prior to discharge.[37]

A fire within the body of a landfill site resulted in pyrolysis of a quantity of buried tyres. Phenols were formed among the degradation products and leachate passed into ground-water and thence to an adjacent reservoir. The reservoir was put out of commission because water purification involving chlorine would produce malodorous, maltasting chlorophenols.[63]

The volume of leachate can be estimated with reasonable accuracy using a simple water balance, on the basis of expected inputs of liquid (from liquid wastes, run-in, ground-water ingress and rainfall) and liquid 'sinks' (from run-off, evaporation, evapotranspiration and absorption). However, the composition and 'strength' vary considerably, dependent on the composition of the wastes and upon the complex physical, chemical and biochemical processes involved in waste degradation.

Since these conditions are difficult to control, some countries favour segregation of special wastes from other wastes by depositing them in sealed drums or in cells covered with soils of low permeability rather than domestic solid wastes. Controlled co-disposal of domestic wastes and certain types of industrial hazardous wastes is more favoured in the UK since this can provide significant attenuation of certain water pollutants within the landfill.[62] In a number of cases leachates from co-disposal sites have been comparable with those from domestic wastes but careful control is essential since leaching problems could develop[62] where large volumes of acid waste are deposited, or, where a site is overloaded with liquid waste.

In addition, careful consideration should be given to the hydrogeology of a landfill since under UK climatic conditions generation of a polluting leachate is likely[62]; potential movement of leachate, and attenuation/dilution of contaminants in soil and strata beneath the site require determination.

Precautions

It follows from the above discussion that the avoidance of hazardous materials 'incidents' or 'escapes' from landfill sites depends upon (after ref. 14):

- Selection of a site providing suitable geological and hydrogeological conditions. (Testing methods are under development for assessing the suitability of a specific hazardous waste for disposal at a particular landfill site.)[64]
- Extensive accurate documentation of all wastes.
- A system for segregation of wastes – unless co-disposal is deliberately intended – and for recording the position of all deposited waste. (Segregation of incompatible wastes, and leachate control, can be achieved with inorganic wastes by encapsulation in proprietary compounds – usually based on an inorganic silicate or aluminate and organic polymer; this has been restricted by economics, e.g. for arsenic and cyanide compounds. Metal drums have been used but have a limited life and have caused problems following corrosion.)[38]
- Trained and competent staff with adequate supervision, particularly of new or external (e.g. contractors or delivery) personnel.

- Adequate site security.
- Provisions for dealing with accidental spillages.
- Provision of boreholes and air monitoring facilities (see Ch. 7) as necessary. On-going monitoring is desirable, since cases have been reported of water pollution problems possibly associated with chemicals apparently legally collected and disposed of.[9]
- Regular site inspections to check that no irregularities have occurred (e.g. exposure of buried material, undue waterlogging).
- Taking account of the presence of toxic materials after the site is decommissioned.

With regard to the latter, the general policy of the Department of the Environment in the UK is that former landfill sites should not be redeveloped for housing purposes, both on account of possible land settlement and upward migration of contaminants and organic gases, especially methane.[61,65]

Love Canal in the USA is a former canal in which about 21,000 tonnes of chemicals were disposed of in the 1940s and 1950s. It was sealed with clay and passed over to the Education Authority with a recommendation that the area should not be used for building purposes. Subsequently a school was built there and in the process the capping, which prevented water ingress, was destroyed. In 1976 excessive precipitation resulted in a rising water table which carried the wastes leachate to the surface. This began to ooze out through drains and waste sumps and organic vapours were detected in the basements of adjacent houses. About 80 organic chemicals were detected in the air, soil, and water around the site by the Environmental Protection Agency.[66,67]

Incineration

Effluent incineration is the safest option for the destruction of dangerous organic wastes, e.g. pesticides and herbicides, and pathogenic wastes.[14] However, purpose-built incinerators are generally required and the operation is continually modified to suit the specific waste. The important characteristics[68] are:

- The types and quantities of solid residues.
- Physical and chemical properties of the wastes.
- Behaviour of the wastes in the combustion chamber.
- The composition of the exhaust gases.

A sufficiently high temperature and residence time in the combustion zone must be obtained to achieve decomposition and combustion. For example, polychlorinated biphenyls require temperatures >1,000 °C and such temperatures are recommended for complete destruction of all hazardous chemical wastes.[68]

While inorganic materials are not deliberately burned this is unavoidable in some cases and small quantities of heavy-metal oxides may be formed.[14] Some materials, e.g. mercury and arsenic compounds, should be avoided since their fumes can contaminate the furnace interior and result in a hazard during maintenance.

Provisions are required for waste receipt and storage. Where discharge is from tankers the integrity of the hose and couplings is important to avoid liquid or gas escapes.

A substantial leak in the delivery hose of a road tanker delivering waste hydrocarbons for incineration resulted in a fire at the rear of the tanker. This spread among storage tanks and 20,000 m^2 of plant area and 86 storage tanks were destroyed.[69]

Waste may require preparation, treatment and/or blending. For example liquids may be blended to adjust the calorific value – which may necessitate addition of auxiliary fuel – or to incorporate solid organic residues.[14] Sometimes the waste may be charged to the incinerator in containers, e.g. some incinerators cater for drummed waste. In any event the design of waste loading or injection equipment is an important factor in preventing the escape of toxic gases or dusts – particularly since the materials are being preheated at this stage. The incinerator should also be of integral construction and maintained in good condition to avoid leakages.

Emissions from incineration normally require treatment prior to discharge from an appropriate stack. The pollution problem of solid/liquid wastes must not be merely converted into an air pollution problem; particulates will normally be present in the flue gases and incineration of materials containing nitrogen, sulphur or halogens will inevitably generate acid gases as discussed earlier. Electrostatic precipitation is commonly used to remove particulate matter and water scrubbing to remove acid gases.

Removal of particulates from emissions is considered important because of the presence of trace contaminants in fly ash, a factor which also has to be considered if collected fly ash is deposited or spread while dry.[44]

A large number of organochlorine compounds – in particular PCDDs (polychlorinated dibenzo-p-dioxins) and PCDFs (polychlorinated dibenzofurans) have been identified in the fly ash and flue gas of municipal incinerators. Levels of tetra-CDDs in particulate material from fly ash ex-municipal incinerators are generally <10 ng/g but concentrations from industrial and chemical waste incinerators are $\simeq <100$ ng/g.[44] Since most of the tetra-CDDs are likely to be trapped in the fly ash the risks to the general public are likely to be small.[44]

A modern industrial waste incineration facility, incorporating a vortex-type, high temperature incinerator operating at 1,200 °C with an extended residence time, is illustrated in Fig. 14.3.[70] Under optimum oxidising conditions this has a throughput put of 3 tonnes/hr.

In summary the control of toxic hazards from an incinerator site depends upon[10,14]:

- Extensive, accurate documentation of all waste.
- Checking all waste for conformity to analyses and documentation.
- Adequate waste handling facilities, e.g. suitable lifting gear such as cranes with grab buckets.

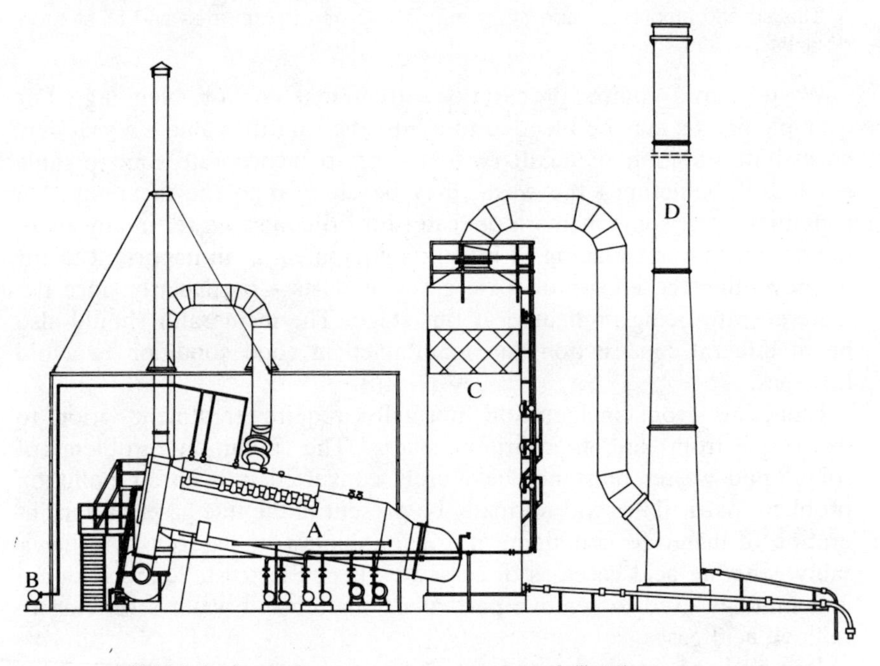

A Slightly inclined combustion chamber in which the high temperature oxidation of the wastes takes place.

B Waste feed pumps.

C The scrubbing tower is provided with large volumes of cooling water to remove the water-soluble impurities produced in the combustion process. The scrubbing tower is constructed in two sections. The lower part is provided with four water sprays to facilitate the primary cleaning of the combustion gases, the upper part comprising a packed section of ceramic intalox saddles.

D The clean gases then pass to atmosphere through the 37.4 m high stack. The liquid discharge from the scrubbing tower and the stack then flows to the acid neutralisation plant.

Fig. 14.3 Industrial waste incineration and chemical treatment facility (Courtesy Cleanaway Ltd)

- Thorough training and supervision of personnel.
- Secure and safe waste storage facilities. These may include tanks, pits, open ground or a covered area for drums and bales. (Drums should preferably be on a concrete pad with provision for drainage to a sump.) Wastes should be segregated according to compatibility. Flammable wastes should be stored in accordance with established practice (see Ch. 4).
- Adequate waste preparation.
- Trouble-free waste loading and injection methods which do not emit dusts or vapours.
- An incinerator which is leak-proof and designed for efficient combustion.
- Enclosure of specific areas to minimise dust and vapour emissions.
- Adequate systems for removal of dust and gaseous combustion products.
- Discharge of all emissions at a sufficient height to ensure dilution and dispersion.
- Siting of plant to minimise the effects from any accident.
- Provision for safe residue removal and subsequent disposal – again requiring mechanical handling.

Sea disposal

Under the Dumping at Sea Act 1974 no waste may be dumped in UK waters (excluding discharges through pipelines) or by British vessels in any seawaters without a licence. Such a licence will include conditions necessary for the protection of the marine environment and the living resources which it supports.

Shallow-sea dumping off the British coast is used for power station ash, colliery spoil, sewage sludge and some industrial waste – mainly thin sludges or dilute aqueous solutions. In theory, biodegradability and dispersion will affect the removal of such wastes. Readily biodegradable and relatively low toxicity organic materials, which soon break down at sea, may also be disposed of well out to sea.

Deep-sea dumping involved the wastes being sealed in drums and encapsulated in concrete. It was used until 1980 for, e.g. solid cyanide residues from heat treatment processes.[38] Mercury and chlorinated antimony compounds were also disposed of in this way; however, this method is now virtually defunct.[14] In any event under an International Convention the substances listed in Table 14.18 (Annex I) are prohibited from being disposed of at sea, other substances (Annex II) require a special permit, and a general permit is required for all other wastes or matter.[71]

Table 14.18 International controls on dumping at sea[71]

Annex I: Wastes and other substances which are prohibited

1. Organohalogen compounds.
2. Mercury and mercury compounds.
3. Cadmium and cadmium compounds.
4. Persistent plastics and other persistent materials, for example, netting and ropes.
5. Crude oil, fuel oil, heavy diesel oil and lubricating oil, hydraulic fluids and any mixtures containing any of these.
6. High-level radioactive wastes or other matter defined by the International Atomic Energy Agency as unsuitable for dumping at sea.
7. Materials in whatever form produced for biological or chemical warfare.
8. The preceding paragraphs do not apply to substances which are rapidly rendered harmless by physical, chemical or biological processes in the sea provided they do not
   (i) make edible marine organisms unpalatable,
   (ii) endanger human health or that of domestic animals.
9. This Annex does not apply to wastes or other materials such as sewage sludges containing the substances referred to in paragraphs 1–5 as trace contaminants.
   Such wastes shall be subject to the provisions of Annexes II or III.

Annex II; Substances for which a prior special permit is required

A. Wastes containing significant amounts of the following:
   arsenic, lead, copper, zinc } and their compounds
   organosilicon compounds.
   cyanides.
   fluorides.
   pesticides and their by-products not covered in Annex I.
B. In the issue of permits for the dumping of large quantities of acids and alkalis, consideration shall be given to the possible presence in such wastes of the substances listed in paragraph A and to the following additional substances:
   beryllium, chromium, nickel, vanadium } and their compounds
C. Containers, scrap metal and other bulky waste liable to sink to sea bottom which may present a serious obstacle to fishing or navigation.
D. Radioactive wastes or other radioactive matter not included in Annex I. In the issue of permits for the dumping of this matter, the Contracting Parties should take full account of the recommendations of the competent body in this field, at present the International Atomic Energy Agency.

Pollution control management

Management for pollution control involves setting objectives and then monitoring, either continually or periodically, how closely they are attained.[10] An initial audit[7] will define the general aims and identify those areas in which performance needs assessment, monitoring and possibly improvement. A summary of some plant features, management controls and monitoring requirements is given in Table 14.19.

Table 14.19 Plant features and management controls

Gaseous emissions

Plant/site features

- Provision of adequately designed, appropriate plant.[11,12]
- Provision of adequate chimney heights.[7]
- Provision and location of instruments, etc., to monitor emissions.

Auditing/control

- List all gaseous pollutants, concentrations, smoke characteristics.
- Assess toxicity and nuisance potential of all pollutants (see Ch. 5). Consider synergistic effects.
- Assess all sources (see Table 14.5).
- Consider adequacy and maintenance of pollution control equipment, i.e. scrubbers, particulates collection equipment.
- Consider effects of poor dispersing conditions, e.g. inversions on dispersion.
- Consider effects of 'prevailing' winds having regard to significant exposed areas (e.g. populated areas, natural waters, agricultural/horticultural land).
- Investigate and report on all alleged pollution episodes (visible, nuisance, damage, etc.).
- Prompt compliance with advice, directives, etc., from statutory authorities.
- Monitoring the environment.[7]
- Compare measured emission levels, ground-level concentrations with statutory requirements.

Liquid effluents

Plant/site features

- Segregation of storm water and foul water drains.
- Bunding of liquid storage tanks.
- Provision of effluent treatment plant able to cope with the full range of flows and concentrations. Adequate instrumentation for control *and* loss prevention (automatic alarm, etc.).

Auditing/control

- List all effluent streams, limits on composition, maximum and minimum flows.
- Consider possible admixture of incompatible effluents.
- Compare recorded analyses, flows, etc., with consents. Act promptly on deviations.
- Assess provisions for disposal of sludges, concentrates, etc., from treatment processes.

Solid wastes

Plant/site features

- Requirement for sumps to collect leachate or special drainage systems.
- Landfill sites, see page 845.
- Incinerators, see page 854.
- Provisions for segregation of wastes.
- Adequate site security features.

Auditing/control

- List all solid wastes, physical forms, analyses.
- Assess 'special' wastes (see Table 14.11) for compliance with approved disposal procedures and routes.

Table 14.19 (continued)

- Consider incompatibility of wastes (see Tables 10.2, 10.3 and 14.17).
- Assess adequacy of drainage systems.
- Assess adequacy of waste identification, labelling, etc.
- Assess accuracy of 'mass balances'
  (i.e. agreement between waste arisings and waste disposal).

References

(*All places of publication are London unless otherwise stated*)

1. Waldbott, G. L., *Health Effects of Environmental Pollutants*. The Mosby Co., St Louis 1973.
2. Wilber, G. C., *The Biological Aspects of Water Pollution*. Charles C. Thomas, Illinois, 1969.
3. Science Research Council, *Combustion Generated Pollution*, 1976.
4. Klein, L., *River Pollution* (3 vols): 1 – *Chemical Analysis*; 2 – *Causes and Effects*; 3 – *Control*. Butterworth 1962.
5. Conway, R. A., & Ross, R. D., *Handbook of Industrial Waste Disposal*. Van Nostrand Reinhold, Wokingham, 1980.
6. Zajic, J. E., *Water Pollution: Disposal and Reuse* (2 vols). Dekker 1971.
7. Frankel, M., *The Social Audit Pollution Handbook*. Macmillan 1978. (& References & Notes therein)
8. Pearce, A. S., 'Legislation for the Control of Industrial Liquid Effluents', in *Effluent Treatment in the Process Industries*, (Instn Chem. Engrs Symp. Series, 77), 71–7.
9. Sell, N. J., *Industrial Pollution Control*. Van Nostrand Reinhold 1981.
10. Bridgwater, A. V., & Mumford, C. J., *Waste Recycling and Pollution Control Handbook*. George Godwin 1980.
11. Nonhebel, G., *Gas Purification Processes for Air Pollution Control*. Newnes-Butterworth 1972.
12. Strauss, W. (ed.), *Air Pollution Control*, Parts I–IV, Wiley-Interscience.
13. Fawcett, H. H., *Hazardous and Toxic Materials – Safe Handling and Disposal*. Wiley-Interscience, New York, 1984.
14. Cook, J. D., *Loss Prevention Bulletin*. Institution of Chemical Engineers, 1984 (055), 17–29.
15. Keen, R. C., 'Disposal of toxic wastes to landfill sites', Training & Technical Symposium on Hazardous Wastes, Univ. of Nottingham, 13–14 Sept. 1982.
16. Chanlett. E. T., *Environmental Protection*. McGraw-Hill 1973, 236.
17. Jennings, A. J. D., *Chem. Engrs*, Oct. 1974, 640.
18. Richardson, M. L., 'Biodegradation and effluent disposal', ibid., 157.
19. Fawcett, H. H., & Wood, W. S., *Safety and Accident Prevention in Chemical Operations*. Wiley 1965, 32.
20. Institution of Chemical Engineers–Technical Response Board, 'Volatiles in Effluent ', *Chemical Engineer*, 1982, **138**, 379.

21. Health & Safety Executive *Health and Safety: Industry & Services*, 1975, 34. HMSO.
22. *New Scientist*, 19 Apr. 1984, 4.
23. *New Scientist*, 10 Jan. 1985, 5.
24. Cookson, C., *New Scientist*, 21 June 1979, 1015.
25. WHO Regional Office for Europe/United Nations Environment Programme, *Hazardous Waste Management*, Interim Document 7, 1982.
26. *The Licensing of Waste Disposal Sites*, Waste Management Paper No. 4. HMSO 1976.
27. Department of the Environment/Welsh Office, Joint Circular 4/81, 8/81. HMSO 1981.
28. Department of the Environment, Waste Management Paper No. 23. HMSO 1981.
29. EEC Directives: 75/439/EEC on *The Disposal of Waste Oils*, 16 June 1975; 75/442/EEC on *Waste*, 15 July 1975; 76/403/EEC on *The Disposal of Polychlorinated Biphenyls(PCBs), and Polychlorinated Terphenyls (PCTs); 18/319/EEC on Toxic and Dangerous Waste*, 20th Mar. 1978.
30. EEC Directive: 78/176/EEC on *Waste from the Titanium Dioxide Industry*, 20 Feb. 1978.
31. EEC Directive: 80/68/EEC on *The Protection of Ground Water Against Pollution Caused by Certain Dangerous Substances*.
32. The Radioactive Substances (Carriage by Road) (Great Britain) Regulations 1974; The Radioactive Substances (Road Transport Workers) (Great Britain) Regulations 1970 & 1975.
33. Sheils, A. K., 'Legislation on hazardous wastes', Training and Technical Symposium on Hazardous Wastes, Univ. of Nottingham, 13–14 Sept. 1982.
34. *Hazardous Waste Regulations under RCRA – A Summary*, E. P.·A. sw- 939, Oct 1981.
35. *Chemical Week*, 9 June 1982, 36–40.
36. Fawcett, H. H., *Hazardous and Toxic Materials – Safe Handling and Disposal*. Wiley 1984.
37. Department of the Environment, Papers in the Waste Management Series: (6) *Polychlorinated Biphenyl (PCB) Wastes – a technical memorandum on reclamation, treatment and disposal*. HMSO 1976; (7) *Mineral Oil Wastes – a technical memorandum on arisings, treatment and disposal*. HMSO 1976; (8) *Heat Treatment Cyanide Wastes – a technical memorandum on arisings, treatment and disposal*. HMSO 1976; (9) *Halogenated Hydrocarbon Solvent Wastes from Cleaning Processes – a technical memorandum on reclamation and disposal*. HMSO 1976; (11) *Metal Finishing Wastes – a technical memorandum on arisings, treatment and disposal*. HMSO 1976; (12) *Mercury-bearing Wastes – a technical memorandum on storage, handling, treatment, disposal and recovery of mercury*. HMSO 1977; (13) *Tarry and Distillation Wastes and Other Chemical-based Residues – a technical memorandum on arisings, treatment and disposal*. HMSO 1977; (14) *Solvent Wastes (Excluding Halogenated Hydrocarbons) – a technical memorandum on reclamation and disposal*. HMSO 1977; (15) *Halogenated Organic Wastes – a technical memorandum on arisings,*

treatment and disposal. HMSO 1978; (16) *Wood-preserving Wastes – a technical memorandum on arisings, treatment and disposal.* HMSO 1980; (17) *Wastes from Tanning, Leather Dressing and Fellmongering – a technical memorandum on recovery, treatment and disposal.* HMSO 1978; (18) *Asbestos Waste – a technical memorandum on arisings and disposal.* HMSO 1979; (19) *Wastes from the Manufacture of Pharmaceuticals, Toiletries and Cosmetics – a technical memorandum on arisings, treatment and disposal.* HMSO 1978; (20) *Arsenic-bearing Wastes – a technical memorandum on recovery, treatment and disposal.* HMSO 1980; (21) *Pesticide Wastes – a technical memorandum on arisings and disposal.* HMSO 1980.

In preparation:
Acid Wastes; Cadmium Wastes; Medical Wastes; Lead-bearing Wastes.

38. Chivers, G. E., *The Disposal of Hazardous Wastes.* Science Reviews 1983.
39. Cope, C. B., Chappell, C. L. & Keen, R. C., *Municipal Engineering*, 9 Jan. 1976, 46–7.
40. National Association of Waste Disposal Contractors, *Code of Practice*, 1976.
41. Kimborough, R. D., Carter, C. D., Liddle, J. A., Chine, R. E. & Phillips, P. E., *Arch. Env. Health*, Mar./Apr. 1977, 77.
42. *RoSPA Bull.*, Mar. 1985, 3.
43. The Institution of Chemical Engineers, *A Provisional Code of Practice for Disposal of Wastes*, 1972.
44. Stevens, C., & Wilson, D. C., Toxicology of Wastes in 'Disposal of toxic wastes to landfill sites', Training and Technical Symposium on Hazardous Wastes, Univ. of Nottingham, 13–14. Sept. 1982.
45. Department of the Environment, Waste Management Papers, Nos. 6–19. HMSO 1976–80.
46. Khan, A. Q., 'Investigation and treatment of Ravenfield tip', *Proc. Conf. Soc. of Chem. Ind.*, 24 Oct. 1979.
47. Cook, J. D., 'Practical methods of laboratory waste disposal including landfill', ibid., 107.
48. Keen, R. C., Private Communication, June 1985.
49. Keen, R. C., 'Operator hazards in toxic waste disposal'. Ph.D. thesis, Univ. of Aston 1980.
50. Keen, R. C., *Some Environmental Health Implications of the Deposit of Toxic Waste on Landfill Sites*, Report to the Environmental Health Officers Association, 1975.
51. Keen, R. C. & Mumford, C. J., *Annals Occup. Hyg.*, 1975, **18,** 213–28.
52. Kinsey, J. S., Keen, R. C. & Mumford, C. J., *Annals Occup. Hyg.*, 1977, **20**, 85–9.
53. Keen, R. C. & Mumford, C. J., 'Waste disposal – reducing the risks involved', *Health and Safety at Work*, Oct. 1978, 70–3.
54. Webster, E. R., 'Occupational health hazards associated with the dispersion of vapour and dust clouds in the open air'. M.Sc. thesis, Univ. of Aston 1976.
55. *Annual Report of H. M. Chief Inspector of Factories, 1971*, 62. HMSO.
56. World Health Organisation Technical Report, *Solid Waste Disposal & Control Series*, No. 484, 1971.

57. Greenfield, M., & Singh, R., *Municipal Engineering*, 30 July 1976, 1167.
58. Health and Safety Executive, *Cloakroom Accommodation and Washing Facilities*, Safety, Health & Welfare Booklet 8, 1968. HMSO.
59. Carson, R., *Silent Spring*. Penguin 1982.
60. 'Heritage', *EPA Journal*, 1981, 4, 8–10.
61. Wilson, D. C., & Stevens, C., *Problems Arising From the Redevelopment of Gas Works and Similar Sites*, Harwell Report, R10366, HMSO.
62. Barber, C., 'Leaching of hazardous wastes in landfills', Training & Technical Symposium on Hazardous Wastes, Univ. of Nottingham, 13–14 Sept. 1982.
63. *Municipal Engineering* 30 July 1975, **152**, 411.
64. Wilson, D. C. & Young, P. J., 'Testing of hazardous wastes to assess their suitability for landfill disposal', presented at Hazardous Wastes Symposium, Univ. of Nottingham, 13–14 Sept. 1982, Instn of Public Health Eng.
65. *The redevelopment of contaminated land: Notes on redevelopment of landfill sites*, ICRCL 17/78 (4), 1978.
66. *In the Matter of the Love Canal Chemical Waste Landfill Site*, Report of the Commissioner of Health of the State of New York, August 1978; *Chemical & Engineering News* 6, Aug. 7, 1978.
67. Glaubinger, R. S., Kohn, P. M. & Remirez, R., *Chemical Engineering*, Oct. 1979, **22**, 86.
68. Ross, R. D., 'Disposal of hazardous materials', in H. W. Fawcett & W. S. Wood *Safety and Accident Prevention in Chemical Operations* (2nd edn). Wiley 1982.
69. Anon., 'Hose and coupling incidents', *Loss Prevention Bulletin*. Institution of Chemical Engineering 1977 (013), 5–13.
70. Grayson, J., 'High temperature incineration', in *The Disposal of Hazardous Waste from Laboratories*. Royal Soc. of Chemistry, 25 Mar. 1983, 16.
71. Convention of the Prevention of Marine Pollution by Dumping of Wastes and other Matter, 29–31 Dec. 1972. HMSO.

CHAPTER 15

Control of hazards from large-scale installations

Introduction

The growth in demand for chemicals, coupled with the economies of scale and advances in technology, has resulted in an increase in the size of many chemical and associated plants. Large chemical complexes have been formed by the integration of related and interdependent processes on single sites. Potential hazards with chemicals are then at the opposite end of the spectrum to those in the laboratory discussed in Chapter 11.

Despite the high standard of care (i.e. in design, construction, inspection and testing, commissioning and operation) normally exercised with such plants 'accidents' may still occur.

The probability and potential scale of any loss of containment is a primary consideration. The theoretical variation of 'disaster potential' with scale has been expressed as follows[1]:

The risk of loss of containment is greater with two X tonne/day plants than with one 2 X tonne/day plant. If the risk is regarded as a function of the total of the periphery of all the joints,

Risk with two X tonne/day plants = $f.2X$

Risk with one 2 X tonne/day plant= $f.\sqrt{2X}$

Except by the fairly remote chance of two X tonne/day plants losing containment simultaneously, the scale of release is in the ratio large plant : small plant = 2 : 1.

Multiplying risk by scale of release yields the disaster potential ratio, i.e.

$$\text{large plant : small plant} = \frac{2\sqrt{2}}{2}$$

Hence disaster potential = $\sqrt{2}$ × scale factor.

Whether or not this is accurate, characteristically because of the scale of operation, the consequences of any accident (including, for example a cataclysmic fire, explosion or release of toxic materials) may affect not only facilities and personnel within the site but also a larger number of people, and possibly buildings and amenities, in the surrounding area.

Examples of 'domino' or 'knock-on' effects include[2]:

- A small leak of flammable gas may ignite and the flame impinge on a large vessel leading to a large spill of hazardous substance.
- Protective systems may be destroyed by flame impingement.
- Missiles from disrupted equipment may breach containment.
- Ship accidents in port areas or coastal waters close to land may adversely affect operations of plant on land or on ships in neighbouring berths.
- Accidents on one industrial site can affect nearby sites.
  At Flixborough in 1974 the UVCE caused a loss of cooling water to blast furnaces at a nearby steelworks.

An outstanding example of a domino effect, and one of the most serious reported fire/explosion incidents in terms of fatalities – over 544 – occurred in Mexico City in November 1984.

Leakage of LPG from a ruptured 20 cm pipeline, followed by ignition, resulted in nine explosions which completely destroyed a large storage facility. About 12,000 m^3 of LPG was involved in fires and explosions with 15 of 48 horizontal vessels becoming missiles and four storage spheres, each of 1,500 m^3 capacity, sustaining BLEVE's. About 7,000 people were injured in a built-up area to the south of the site.

In 1967 and 1972 HM Factory Inspectorate in the UK expressed concern regarding the potential hazards from some large plants.[3,4] Local planning authorities were subsequently reminded[5] to consider safety aspects, and the model application form included a specific question on the use/storage of materials which may constitute a 'major hazard'. Co-operation of the management of existing installations was to be sought in examining ways of reducing risk and attracting attention to potential hazards. Also in 1972 in the UK a committee was set up to review the provision made for the safety and health of persons in the course of their employment and also to consider whether any further steps were required to safeguard the general public from hazards, other than general environmental pollution, stemming from industrial activities. They noted[6]:

1. '. . . the need to protect the public as well as workers from the very large-scale hazards which sometimes accompany modern industrial operations'.
2. '. . . a number of locations in this country where high explosives or flammable substances are kept in such quantities that any failure of control – however remote the possibility – could create situations of disaster potential'.

This potential for harm to the neighbouring community in addition to the local workforce distinguishes major incidents from normal occupational accidents and is the main justification for a different approach towards their control.[7,8]

In 1974 at Flixborough (UK) a temporary bridging pipe failed on a plant producing cyclohexanone (an intermediate in the manufacture of nylon) by the oxidation of cyclohexane at high temperatures and pressures. About 40 tonnes of hydrocarbon were released into the atmosphere and the resulting unconfined vapour cloud ignited and exploded with a force equivalent to 20 tonnes of TNT. The outcome was 28 killed inside the factory; 36 injured inside and 53 outside the plant; 80,000 m^2 of site devasted; over 700 houses damaged up to 8 km from the plant; evacuation of about 3,000 people from their houses downwind (some being unable to return for 9 months); severing of the cooling-water main from the river to a steelworks 2 km away; overhead power lines brought down, etc. Total losses exceeded £40m.[9–11]

This incident, although by no means unique in a global context, had widespread repercussions. In the UK there was an official inquiry and the Health and Safety Commission set up an Advisory Committee on Major Hazards[12]:

'To identify types of installations (excluding nuclear installations) which have the potential to present major hazards to employees or to the public or the environment, and to advise on measures of control, appropriate to the nature and degree of hazard, over the establishment, siting, layout, design, operation, maintenance and development of such installations, as well as overall developments, both industrial and non-industrial, in the vicinity of such installations.'

Initial priority was for installations which could present a major threat to the safety of the workforce or the public from massive escape of volatile liquids or gases forming a large cloud of flammable vapour which may then explode, and/or massive release of toxic substances which could remain lethal for up to 30 km from the point of escape.

Later[13] they made greater mention of potential hazards from pipeline systems and from release of hyper-poisons.

Any factory giving rise to semi-continuous emissions of low concentrations which may have a long-term effect on health but cause no immediate illness (e.g. ranging from particulate lead compounds, or asbestos fibres, to sulphur dioxide), though 'hazardous' by definition is generally excluded from 'major hazards', since the considerations, i.e. reduction of the concentrations to acceptable levels, and the control measures are different. This type of emission (i.e. noxious and offensive gases) and the works involved are generally regulated by specific pollution control legislation, e.g. in the UK the Alkali Act,[14] as discussed in Chapter 14.

Record of major hazard plants

Fortunately incidents involving major hazards in the UK are rare but an analysis of almost 200 reported incidents worldwide with potential for multiple fatalities revealed that on a global scale the frequency of reported major incidents was increasing exponentially[8] as shown in Fig. 15.1.

The upturn in frequency coincides with the start of expansion of the industry in the mid-1950s. Nevertheless the analysis suggests that the trend to increased plant capacity did not necessarily result in more severe accidents. (Thus pre-1950 70% of all reported major incidents studied resulted in five fatalities and the corresponding figure for the period 1951–76 is 77%.) If anything, the data could be indicative of an improvement in safety performance.[8] Additional evidence for an improved safety record is given in a report which shows that in spite of the rise in scale of chlorine production the worst accidents involving failure of storage tanks occurred thirty or more years ago.[13] Figure 15.2 supports this general conclusion.[15]

Possible explanations for the *apparent* reduction in the number of persons killed in present-day major incidents include improvements in plant design and construction, in operating procedures, and in measures to mitigate the consequences of any incident, e.g. plant layout and segregation. Also there is a move towards greater use of automation and to a less labour-intensive industry.[8] However, although most major incidents did not result in their full theoretical potential in terms of deaths, the financial costs of the average major incident have escalated even after normalising for the effects of inflation. There is no correlation between loss of life and material damage.

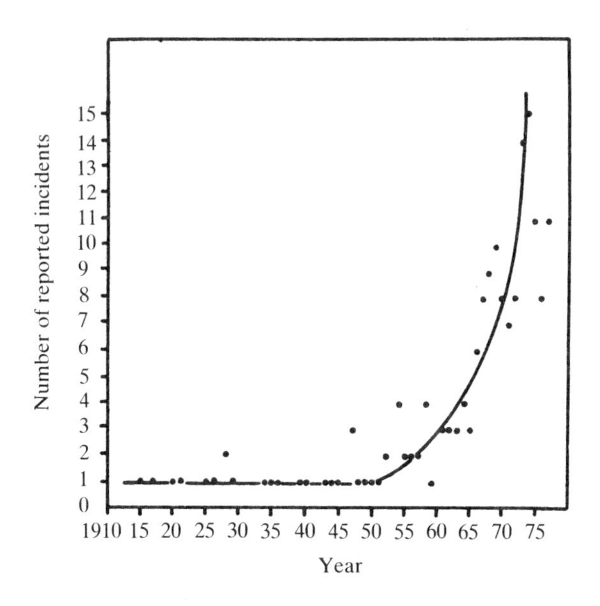

Fig. 15.1 Frequency of major incidents (i.e. with potential for multiple fatalities) worldwide

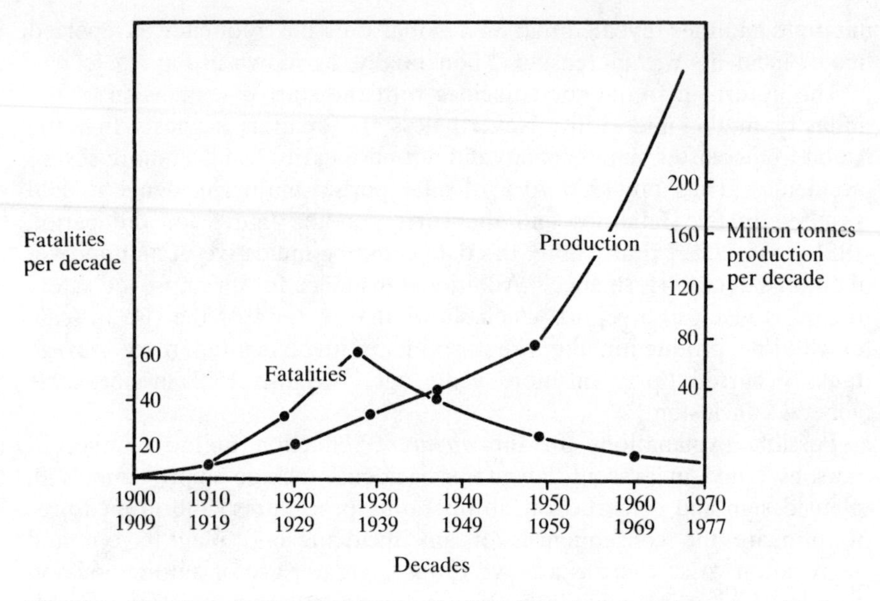

Fig. 15.2 World production and known fatalities from incidents involving chlorine

Large-scale fire hazards

There are more major incidents involving release of flammable material than toxic material.[13] Major fires at UK refineries occur at the rate of 2.5 $\times 10^{-1}$ per year and normally comprise pool fires or running liquid fires.[16] The main hazard associated with the former is secondary fires from radiating heat. The main effects of radiation are shown in Table 15.1.[17]

Since the definition of 'major hazard' embraces the concept of death or serious injury to people outside the confines of the plant, such a risk theoretically exists at radiation fluxes of, or above, 12.6 kW/m^2. The

Table 15.1 Effect of radiant heat

Radiation flux (kW/m^2)	Effect
1.6	Maximum value which will not cause undue discomfort to people going about their normal work.
4.7	Approximate threshold of pain for exposure of any part of the human body.
12.6	Threshold limit above which resin in wood, building roofing felt, oily rags, etc., will start to emit flammable vapour.
38	Maximum permissible flux for uncooled storage tanks.[18]

hazard is reduced by the provision of adequate spacing distances as illustrated in Fig. 15.3,[17] by cooling adjacent plant with water sprays, or by providing water curtains.

The primary danger with running liquid fires is flash-back from the ignition source; the extent of spread depends upon local topography, the presence of drains and trenches, and the restriction by bund walls.

In both cases the greatest potential fire is the largest volume contained in any vessel unless escalation by 'domino effects' is possible in the event of loss of plant integrity.[18] Such escalation may involve failure of a burning vessel or a boil-over such that a larger plant area, e.g. a full bund, and multiple tanks are involved. However, major fires seldom result in 'major hazards' *per se* because the time factor involved is such that the necessary action can generally be taken, adequate means of access and escape can be provided, and suitable fire protection measures can be arranged. Case histories of four major fires between 1972 and 1977 confirm that in each incident some emergency system failed to operate, or was not provided or involved poor design features[19]:

> Because no one was familiar with a specialised valve and no written procedures existed for their overhaul men adopted their own standard. As a consequence clearances were set at 0.63 mm instead of 0.05 mm. Hydrocarbon vapours at their auto-ignition temperature leaked from the gasket and ignited. Frozen water prevented blowdown valves, designed to extinguish fires, from opening. The resulting fire, which took 4 hours to control, caused material damage in excess of $1.3m.

On occasions fires and explosions can take place together, for example primary blast damage caused by an unconfined cloud explosion is invariably followed by a major fire. More serious fires can be encountered if large amounts of flammable vapour can escape rapidly into the atmosphere. Thus breach of containment of liquefied gas stored at, or below,

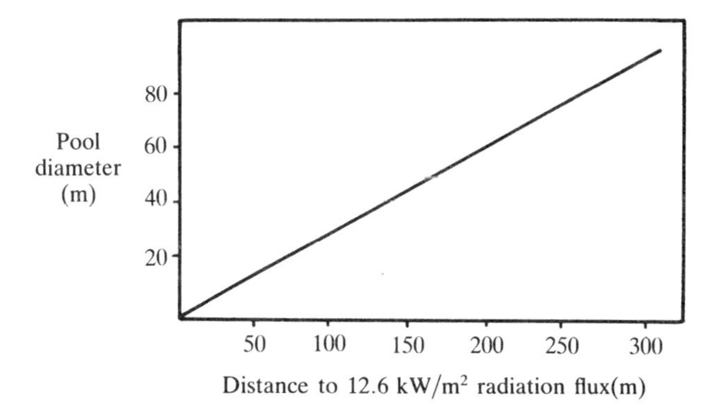

Fig. 15.3 Distance to 12.6 kW/m^2 radiation flux for pool fires of various sizes

ambient temperatures and the boiling point of the liquid, can release significant volumes of vapour into the environment. The initial surge subsides as the liquid cools due to the evaporation process. By way of example, Fig. 15.4 demonstrates the pattern of chlorine escape from a ruptured tanker.[20]

More massive efflux of vapour can occur on rupture of vessels holding large volumes of flammable liquid in equilibrium with its vapour at elevated temperatures, when large vapour clouds can form with tremendous rapidity. Based on experience the radius of unconfined clouds can be estimated from

$$R = 30\sqrt[3]{M}$$

where R is the cloud radius in metres and M is the mass of vapour in tonnes.[21]

The volume of a flammable cloud can be estimated thus[22,23]:

$$V_{(F_1F_2)} = \left[\frac{1{,}000Q}{\pi acU}\right]^{\alpha_1}\left[\frac{\pi ac}{\alpha_1}\right]\left[(F_1)^{-\alpha_2} - (F_2)^{-\alpha_2}\right]$$

where: V = volume of cloud in m^3; F_1 and F_2 = lower and upper explosive limits in mole %; Q = steady rate of release in m^3/s; U = ambient wind speed in m/s.

a, c, α_1 and α_2 depend upon meteorological conditions but pessimistic estimates for clouds up to 500 m from source of emission and U = 2 to 2.5 m/s are:

$ac = 0.0192$ m^2
$\alpha_1 = 1.52$
$\alpha_2 = 0.55$

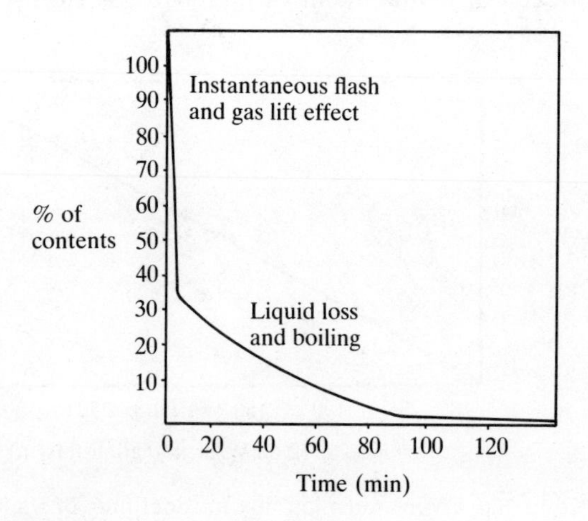

Fig. 15.4 Chlorine escape from punctured tanker

For jets released into still air the mass of flammable vapour contained within the flammable limits is given by:

$$W_{F_1F_2} = 28.84 \left[\frac{Ma^3\ Ta}{MoTo^3}\right]^{\frac{1}{2}} do^3 \left[F_1^{-2} - F_2^{-2}\right]$$

For buoyant plumes the mass is given by:

$$W_{F_1F_2} = 3.93 \quad \frac{Mo}{Ta}\left[\frac{MaToQ^2}{MaTo - MoTa}\right]^{3/5}\left[F_1^{-0.8} - F_2^{-0.8}\right]$$

where: $W_{F_1F_2}$ = mass of flammable vapour within flammable range; M = molecular weight; T = temperature in K; Q = volumetric flow rate in m/s; do = effective diameter of the orifice in m; F_1 and F_2 = lower and upper flammable limits in mole %.

Dispersion of vapour clouds is influenced by many factors, e.g. velocity and direction of discharge; air, vapour and ground temperatures; wind conditions; density of vapour; presence of buildings and plants, etc. Once formed the air/vapour cloud mixture may fail to burn and disperse harmlessly. Some may drift for a considerable time before reaching an ignition source of sufficient energy (e.g. naked flames, electric sparks, internal combustion engines, etc.); others ignite almost immediately.

The loss of 40 tonnes of cyclohexane at Flixborough from a superheated process at 155 °C and 8.8 atmospheres was complete within 45 seconds. The volume of the resulting unconfined vapour cloud, which ignited in less than 1 min, was about 0.5×10^6 m^3 (approximately 2 to 3 times the internal volume of St Paul's Cathedral).[7,13]

Once ignited the cloud may burn with or without explosion. In certain circumstances a cloud too rich in hydrocarbon to explode may commence burning around its perimeter and 'take off' as a fireball. These are particularly hazardous[24] because:

- The hazard becomes mobile, travelling across the site or neighbourhood.
- The intensity of radiated heat can be sufficient to ignite nearby flammable material including property and may cause fatal burns to bystanders.
- Convection currents in the wake of travelling fireballs are capable of drawing in, and igniting, débris.

The following formulae have been proposed[8] to predict the size and duration of fireballs:

$$D = 55 \times \sqrt[3]{M}$$
$$T = 3.8 \times \sqrt[3]{M}$$

(where D is the diameter in metres of the fireball, M is the mass in tonnes of hydrocarbon, and T is the duration in seconds). The following illustrates damage due to heat transfer by radiation.

In Texas in 1956 a spillage occurred while a 2×10^6 litre spheroid tank was being filled with pentane and hexane. Vapour was ignited in a ground fire for an hour. Water was used to cool the tank, on exposed tanks, and to extinguish fires in a nearby floating-roof gasoline tank. Suddenly the top of the spheroid tank ruptured, releasing a huge ball of burning vapours; 19 fire fighters were killed, their bodies being found between 100 m and 130 m from the tank. The fireball ignited three crude-oil tanks 150 m to 180 m away, burned spectators on a road 400 m away, destroyed a trestle on a railway 410 m away and blistered paint on houses 100 m away.[25] (This comprised a BLEVE – see later.)

To put the hazard into perspective, the radiant heat from a large fireball, i.e. $>27 \times 10^3$ kg, can be expected to cause second-degree burns on unprotected skin at a distance of approximately 2 D from the centre of the vapour cloud. This distance decreases to about 1.75 D for 9,000 kg of fuel in the cloud. Dry wood, vegetation and rubbish can be expected to ignite at 2.5 D from the centre of a very large cloud or at 2 D for a 9,000 kg cloud.[26]

Hazards from fires are increased when the air is enriched with oxygen above its normal concentration of 21%. Under these conditions combustion can occur with alarming speed and vigour (as discussed in Ch. 4). The general effect of increased flammability by oxygen enrichment is shown in Table 10.9. It is of interest to note that chlorine affects the flammability in a similar manner to oxygen.[27] Consequently installations using or storing large amounts of liquid oxygen can represent a major hazard.

Large-scale explosion hazards

In an explosion the blast wave travels at such a speed that those caught in the path have no time to protect themselves. However, man's resistance to blast is quite high,[28] and of the 28 fatalities at Flixborough, only 3 were killed by blast.[29] Therefore, the majority of injuries to people outside the plant stem from building collapse or débris. Studies of the response of structures to dynamic loading by shock-waves in air have primarily been done for military purposes. Results are expressed as the damage expected from shocks of specific peak over-pressure. Although the amount of damage is also influenced by the time constant of the pressure decay and the natural vibrational frequency of the structure, for practical purposes the following values may be used:

Pressure (psig)	(kN/m^2)	*Damage*
0.03	0.2	Occasional breakage of glass windows
0.1	0.7	Breakage of some small windows

Pressure (psig)	(kN/m^2)	*Damage*
0.3	2	Probability of serious damage beyond this point = 0.05; 10% glass broken
0.4	3	Minor structural damage to buildings
1.0	7	Partial demolition of houses, uninhabitable, corrugated panels displaced
2.0	14	Partial collapse of house walls and roofs
3.0	20	Steel-frame buildings distorted, pulled from foundations
4.0	28	Oil storage tanks ruptured
5.0	35	Wooden utilities poles snapped
6.0	42	Nearly complete destruction of houses
7.0	50	Loaded wagon trains overturned
10.0	70	Total destruction of buildings, heavy machine tools moved and damaged; very heavy machine tools survived.

Potential major explosion hazards include:

- Confined and unconfined vapour cloud explosions.
- Boiling liquid expanding vapour explosions.
- Dust explosions.
- Thermal, deflagration, or detonation explosions of solids or liquids.
- Explosions due to rapid release of inventories of stored high pressure.

Unconfined vapour cloud explosions

If flammable/air clouds burn in free space with sufficient rapidity to generate pressure waves which propagate both through the vapour cloud and into the surrounding atmosphere then unconfined vapour cloud explosions, UCVCEs, result. Deflagration occurs when the advancing flame front travels subsonically, in most cases <10 m/s. Detonation occurs from the less commonly encountered supersonic advancement of flame fronts.

Conditions leading to a UVCE require a rapid release of a flammable fluid coupled with moderate dispersion to produce a very large flammable air and hydrocarbon cloud, usually with some degree of containment.[30] It has bcen suggested that a minimum of 5 tonnes of vapour is necessary.

In 1970 about 60 tonnes of propane escaped from a ruptured pipeline at Port Hudson, USA. Most people living nearby became aware of the leak because of the noise, and four families evacuated their homes. A white cloud 460 metres long and 3 to 6 metres wide developed in the valley. After about 24 min on reaching an ignition source (probably a refrigeration motor in an outhouse on a nearby farm) the cloud detonated and lit up the valley. There was no observable period of flame propagation but a sudden flash resembling lightning. Almost immediately there was a pressure pulse which knocked down one witness about 800 metres from the centre of the cloud. Eventually a total of 360 tonnes of propane were lost. Fortunately, no one was killed. This is cited as the only example of an UCVCE in which detonation is thought to have been involved.[28]

Shock waves decay as they reach the perimeter of the vapour cloud and it is this decaying wave that causes most of the damage outside the factory. Usually the damage caused by over-pressure is expressed in terms of TNT equivalent, a function of a known amount of explosive and the distance from the blast.

This is not an exact simulation, since UVCEs are 'soft-centred', but in terms of the damage sustained at >300 m gives a reasonable approximation when compared with other models.[31] Most hydrocarbons display a maximum combustion energy of 10,000 to 12,000 kcal/kg compared with 1,100 kcal/kg for TNT. However, the rate of energy release from hydrocarbons is normally too low to create a shock wave. Therefore, an efficiency factor is used to determine the equivalent amount of TNT, e.g.:

Assuming an efficiency factor $\zeta = 2\%$

$$1 \text{ tonne TNT} \equiv \frac{1}{10} \times \frac{100}{2} \equiv 5 \text{ tonne hydrocarbon}$$

Various values have been used for ζ, since it has varied in past real incidents, e.g. 3% to 4% or 10% for conservative estimates, in order to compute the over-pressure and hence the damage at different radii from the epicentre of the explosion.

Since the effect is in any event $\alpha[\text{mass}]^{\frac{1}{3}}$, the accuracy is reasonable[29] but directional effects, degree of confinement by buildings and many other factors determine the damage patterns in real incidents.

Figure 15.5 shows the radii at which over-pressures of 0.8–1.0 psi (0.055–0.07 bar) will exist for various masses of material.[32]

Estimated blast damage for a 20 tonne UCVCE assuming a 10% efficiency factor[32] is shown in Fig. 15.6.

In fact the overall damage may tend to be worse than predicted due to secondary events, i.e. fires, BLEVEs and CVCEs initiated by the explosion.

At Flixborough large quantities of flammable liquids were ejected due to the demolition and rupture of equipment and piping. A fire covered an area of 180 m × 250 m with flames up to 100 m high; smaller fires started in other parts of the site. By the Monday (the accident having occurred on a Saturday) only a few small fires remained but on the following day a benzene tank exploded.[33]

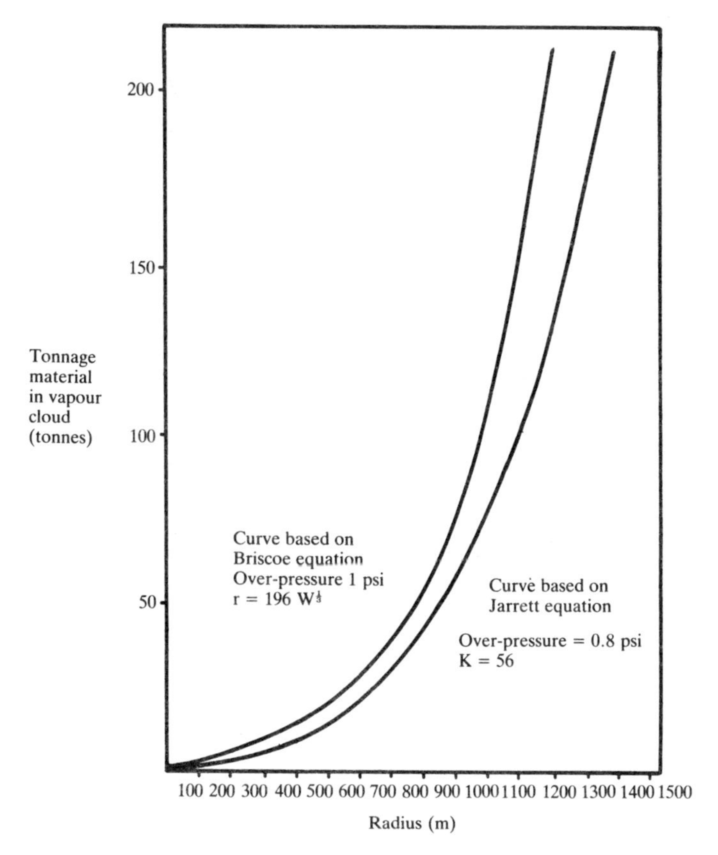

Fig. 15.5 Radius v. tonnage of material for 'acceptable' damage

Hazard-distance relationships for missiles can be related to the blast-wave over-pressure distances. For example a 0.5 psi over-pressure represents the standard separation for protection of personnel in inhabited buildings from the shrapnel following a TNT explosion and a 1-psi over-pressure distance is the normal threshold for serious injury from glass fragments.[26]

Confined vapour cloud explosions

Ignition of a flammable vapour cloud within equipment or a building results in a confined vapour cloud explosion, CVCE. The mass may be relatively small, e.g. <5 kg, if it is confined in a reactor bay or similar. Confinement allows pressure to build up until the containing walls rupture.

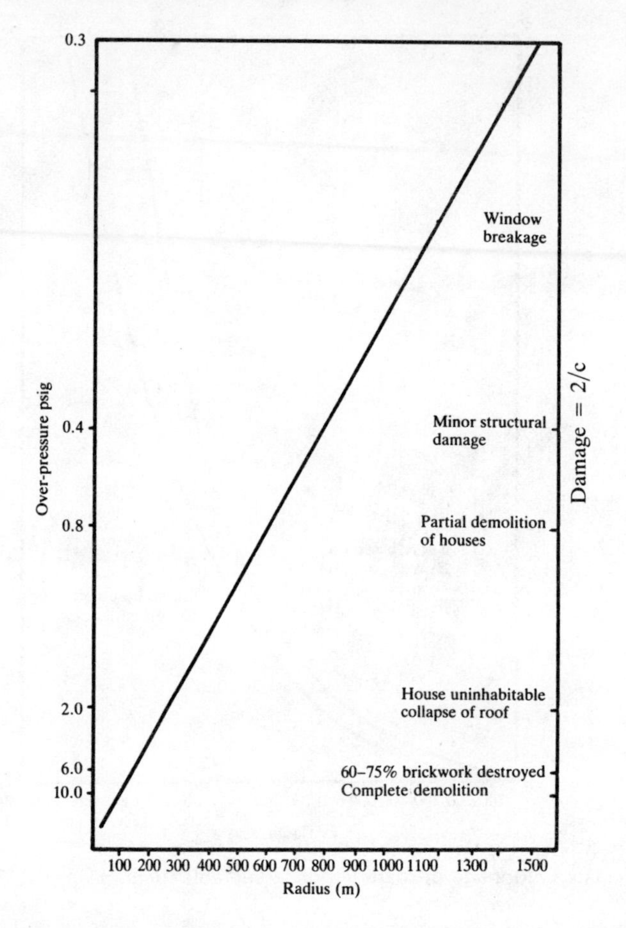

Fig. 15.6 Calculated over-pressure v. radius for 20 tonnes vapour UCVCE

With a pressure vessel, where the bursting pressure is known, the energy resulting from ignition of the mixture is given by the isentropic expansion energy

$$\Delta E = \frac{P_o V_o}{\gamma^{-1}} \left[1 - \frac{(P_o)}{(P_f)}^{\frac{1-\gamma}{\gamma}} \right]$$

where: ΔE = Isentropic expansion energy, J; P_o, P_f = Initial and final absolute pressure, N/m^2; V_o = Original volume, m^3; γ = Ratio of specific heats.

Normally P_f is unknown but is low in the case of confinement within ordinary buildings or structures.

Although these can cause considerable damage, e.g. peak over-pressures of up to 8 bar can be experienced in a fully confined explosion and very much higher in the unlikely event of a detonation,[30] in general they have insufficient energy to have more than localised effects. Indeed much of the major damage found within the cloud confines following a UCVCE probably results from local CVCEs. Clearly, however, if the results of a CVCE can affect nearby plant more serious secondary explosions could follow.

Boiling liquid expanding vapour explosions

BLEVEs, occur when a pressure vessel containing flammable liquefied gas fails on exposure to fire. Failure can be due to either, or a combination of, weakening of the portion of the vessel exposed to the fire, and/or excessive pressure caused by the effect of heat on the vessel contents.

A summary of causes, effects and prevention methods is given in Fig. 15.7.[34]

There are three separate damage-producing effects:

- A blast wave due to relief of internal pressure. This is not a major consideration, except near to the source, e.g. blast over-pressures at the source $\simeq$ 0.05 bar.[30]
- Thermal radiation due to the effects of the fireball resulting from the massive burning of the contents of the vessel in the air; this is preceded by a ground flash.
- Projection of large fragments, scattered for considerable distances by the violent rupture of the tank.

The main effect arises from thermal radiation and distances of 300 metres or more have been cited as having resulted in injury.

However, the hazard range is clearly related to fuel mass, i.e. container size, and the following radii of severe-burn hazard have been quoted as rules of thumb[34]:

Cans and aerosol containers	8 m
Liquid drums and liquefied-gas cylinders	33 m
Large cargo vehicles	164 m
Rail tank-cars	262 m

The secondary hazard is projectiles extending to distances of 300 m for large fragments and 500 to 600 m for small fragments. Again, however, there is a variation with container size, e.g.[34]:

Large pieces, typical non-cryogenic liquefied-gas containers >4,000 US gal.	330 to 650 m
Large pieces, cryogenic containers (rare ∴ estimated)	165 to 325 m
Pieces of flammable liquid drums or liquefied-gas cylinders	66 to 98 m

Missiles generally travel in the direction in which the ends of the container were pointing but deviations often occur.[30]

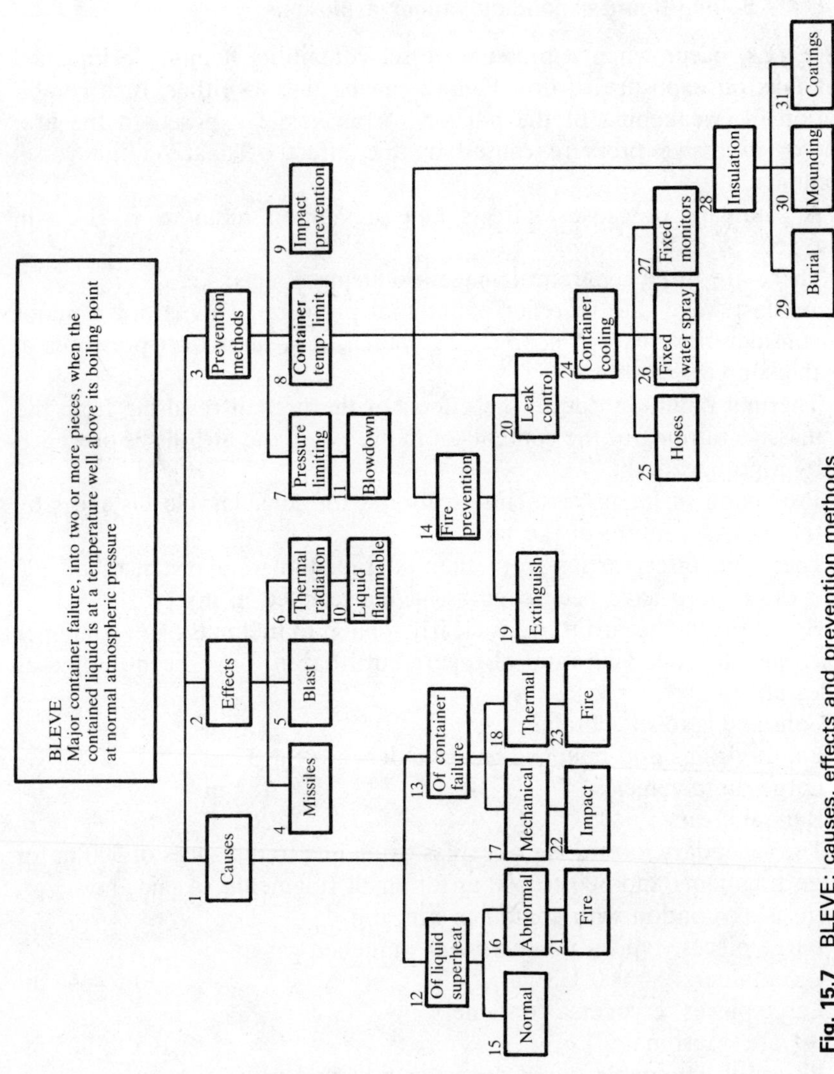

Fig. 15.7 BLEVE: causes, effects and prevention methods

At Feyzin in 1966 an LPG spillage occurred when an operator was draining water from a 2,000-m^3 propane sphere. There was a 1.25-cm and a 5-cm drain valve; since no water came out he opened both valves.

The blockage, probably hydrate, cleared suddenly and a jet of propane (refrigerated by expansion), burned the operator; he was unable to approach close enough to the valves to close them. The cloud of propane vapour spread about 160 m until it was ignited by a motor car on the adjoining motorway.

The liquid spillage burned beneath the sphere and resulted in a BLEVE; the jet of escaping vapour caused the sphere to roll over and débris broke the legs of the next sphere. The firemen were engulfed in burning propane and the fire escalated; in all five spheres and two cylinders containing LPG were destroyed together with a gasoline storage tank. A total of 18 people were killed, 81 injured and the plant seriously damaged.[35]

A similar incident involving water draw-off occurred in Preolo, Sicily.[36]

The control of BLEVEs on fixed installations is discussed in Chapter 4; however, as exemplified in Chapter 13, they may pose significant hazards to onlookers and emergency services at road tanker, or rail tanker, incidents involving LPGs.

Dust explosions

Materials which present a risk of dust explosion, as discussed in Chapter 5, are found in a wide range of industries. The potential hazard exists in premises processing grain into animal or human foodstuffs and fatalities in such incidents have usually resulted from the collapse of buildings or structures.[37] Premises requiring particular attention are:

- Where large quantities of aluminium or magnesium powder are processed – since explosions involving these dusts exhibit the most rapid rate of pressure rise and can have exceptionally severe effects.
- Where sugar, starch or flour are produced – involving large quantities of flammable dusts in processes in tall or multi-storeyed structures of heavy construction.

In any incident fine dust which has escaped from plant and settled within the building can be disturbed and finely dispersed and result in a potentially disastrous secondary explosion. The volume of a dust cloud existing in confined conditions indicates the potential for explosion; in the case of secondary explosions this has resulted in demolition of the building containing the plant.

There have been a significant number of major dust explosions in recent years in US grain handling facilities but conditions in such large silos are considered atypical.[37]

A summary of the few major UK dust explosions is given in Table 15.2.[37]

Table 15.2 Major UK dust explosions

Year	Location	Within premises		Outside premises	
		Killed	Injured*	Killed	Injured*
1911	Glasgow	2	3	3	5
1911	Liverpool	37	100	2	1
1911	Manchester	3	5	—	—
1913	Manchester	3	5	—	—
1930	Liverpool	11	32	—	—
1941	Liverpool	6	40	—	—
1964	Paisley	5	2	—	—
1965	London	5	32	—	—

* These tend to confirm that although enormous destruction has been caused to buildings housing such plants, the risk to persons outside the perimeter of the factory is slight. In the event of a relief venting the slow rate of pressure rise would produce a burning dust cloud rather than an explosion.

The hazard to the environment outside the plant in the case of a massive dust explosion arises from projectiles, notably explosion-relief vents and light roof structures, specifically designed to vent under these circumstances; these may be restrained by chains, etc.

Explosions involving highly reactive or unstable substances

Substances which are highly reactive or unstable when subjected to heat, pressure, mechanical force or on contact with other chemicals represent another potential source of explosive energy. Materials of this class include acetylene, ammonium nitrate, sodium chlorate, nitro cellulose, organic and inorganic peroxides, ethylene and propylene oxide, etc.

Acetylene, used mainly for the manufacture of vinyl chloride, vinyl acetate, chloroprene and trichloroethylene, perchloroethylene and a variety of substituted acetylenes, has a tendency to decompose exothermically (even in the absence of oxygen) and result in explosions or detonations. The character, force and course of the explosion depends upon many factors but pressures in excess of 10 times the initial pressure can result from acetylene explosions and pressures above 50 times the initial pressure have arisen with detonations.

Hazards with ammonium nitrate arise from the ease with which it explodes on detonation, contact with reducing agents, or its sensitivity to thermal stimulus. Many accidents involving this compound have been reviewed.[38]

At Oppau (Germany) in 1921 4,500 tonnes of ammonium nitrate/sulphate mixture detonated with tremendous destructive force – 561 people were killed. The cause of the accident was attributed to the practice of using dynamite to break up caked masses of the salt.[39]

In 1947 at Texas City (USA) two freighters loaded with about 4,000 tonnes of ammonium nitrate exploded, killing 550 people. These explosions, several hours apart followed by extremely vigorous fires aboard the ships, were attributed to the oxidation of the paper packing.[40]

Many organic peroxides, which include hydroperoxides, peroxides, peroxy-acids and peroxy-esters, are also susceptible to facile explosive decomposition under certain conditions as discussed in Chapter 10. The most reactive contain diluents as stabilisers. They are usually employed in small quantities to initiate polymerisations involving ethylene, styrene and other vinyl monomers, and as curing agents for thermoset resins, etc. Some commercially important organic peroxides are listed in Table 15.3.

Ethylene oxide, which is used for producing ethylene glycol, ethanolamines and glycol ethers, is an example of a material which can react violently and with immense explosive force on uncontrolled contact with certain other chemicals.

One person was killed at Doe Run (USA) in 1962 when a 57 m^3 storage tank containing 25 m^3 ethylene oxide became contaminated accidentally with ammonia. The tank ruptured catastrophically and the vapour ignited. Structural damage was extreme within 150 m and minor up to 300 m. Implosion damage was evident within a radius of 200–300 m. Empty tanks at 300 m from the blast centre were pulled 300–450 mm towards the centre by implosion effects. The ground close to the tank was cratered. Tanks weighing 20 tonnes and originally 3 m from the exploding vessel were thrown 120 m. The head of the exploding tank was thrown 800 m and a tank body was lifted 1,200 m. A control room and laboratory about 60 m away from the explosion were destroyed. Other buildings 140 m away were severely damaged. Cooling towers at 180 m were stripped of lagging.[28]

Explosive release of stored pressure

High inventories of stored pressure, e.g. in pressurised reactors and gas collection units, can give rise to incidents due to a catastrophic failure of

Table 15.3 Some commercially important organic peroxides

Dibenzoyl peroxide	t-Butyl peroxyacetate
Dilauroyl peroxide	Diacetyl peroxide
Cumene hydroperoxide	Diisopropyl peroxydicarbonate
Di-t-butyl peroxide	Dicumyl peroxide
Methyl ethyl ketone peroxides	t-Butyl peroxyisobutyrate
t-Butyl peroxybenzoate	t-Butyl hydroperoxide
t-Butyl peroxypivalate	Diethyl peroxide
Didecanoyl peroxide	Dialkyl diperoxides

the pressure shell. If the failure is due to an unforeseen rise in the pressure within the reactor, due for example to a blockage or failure of the pressure relief valve, it will normally occur at or around the designed bursting pressure of the unit, i.e. approximately three times the safe working pressure. This failure will generally be ductile in nature. However, the vessel, if subjected to certain conditions, e.g. low temperature or design weakness, might fail earlier due to brittle failure. The chance of this type of failure is increased by low temperature, rapid rate of pressure rise and use of certain types of constructional materials. Alternatively, the vessel may fail due to corrosion or fatigue. However, these modes of failure will occur below the designed bursting pressure and therefore the pressure energy available will be less than the maximum.

The maximum energy produced from blast waves would arise from brittle, i.e. catastrophic, failure of the vessel at or near to the designed bursting pressure.[32]

In an incident in Hungary, 1969, large storage vessels for liquid carbon dioxide exploded due to brittle fracture. The vessels were 30 m^3 capacity and operated at 15 atmospheres and −30 °C. Nine people were killed and 15 seriously injured. Two similar vessels were turned over and one shot into the process laboratory. Shell and pipe splinters, varying in weight up to several thousand kilograms, were scattered over a 100 m area with other projectiles reaching beyond 400 m. Damage extended beyond the plant area.[41]

With regard to off-site hazards, large inventories of pressure do not generally result in major blast damage and the main hazard is from projectiles launched from the site. For blast effects the energy released can be related directly to the TNT equivalent using a calculated isentropic expansion energy and dividing by the expansion energy of TNT.[32] This gives the equivalent weight of TNT. The maximum pressure should be considered to be the designed bursting pressure of the tank and consideration should be given to the mode of failure of the tank to ensure that does not itself form a missile. This is important in the case of large vulcanisers, or similar vessels, where the most probable mode of failure involves failure of the door giving rise to longitudinal displacement of the pressure shell (see Fig. 3.3).

Some typical examples of high-pressure processes are given in Table 10.10.

Large-scale toxic hazards

Materials capable on release of producing toxic gas clouds outside the plant perimeter can constitute a major hazard. Clearly the substance must be volatile and either be stored in substantial quantities or be produced in large amounts, e.g. on pyrolysis in fires. The main examples include the bulk toxics such as chlorine, ammonia, sulphur dioxide, phosgene, bromine, etc.

By far the worst toxic gas release incident occurred at Bhopal, India in December 1984. Over 2,000 fatalities resulted from an escape of 25 to 30 tonnes of methyl isocyanate – a colourless, lachrymatory liquid with a boiling point of 39 °C and a TLV (skin) of 0.02 ppm – into a shanty town housing 200,000 people.[42] Also fears have been expressed that 20,000 could suffer after-effects.[43,44] The methyl isocyanate (MIC) at Union Carbide was used in the manufacture of 'carbaryl', a carbamate pesticide, according to the following reactions:

$$\underset{}{CO + Cl_2} \longrightarrow \underset{\text{Phosgene}}{COCl_2} \tag{1}$$

$$COCl_2 + \underset{\text{Methyl amine}}{CH_3NH_2} \longrightarrow \underset{\text{Methyl carbamyl chloride}}{CH_3NHCOCl} + HCl \tag{2}$$

$$CH_3NHCOCl \xrightarrow{\text{pyrolysis}} \underset{\text{MIC}}{CH_3NCO} + HCl \tag{3}$$

$$CH_3NCO + \underset{\text{1-Naphthol}}{C_{10}H_7OH} \longrightarrow \underset{\text{1-Naphthyl-}N\text{-methylcarbamate (carbaryl)}}{C_{10}H_7OC(=O)NHCH_3} \tag{4}$$

Chloroform is used as a solvent to quench reaction (2). MIC from an underground storage tank should have vented via scrubbing towers and been neutralised by sodium hydroxide solution but 14,500 litres somehow escaped in a 45-minute period.

A flow diagram of the 57 m^3 storage tank involved is shown in Fig. 15.8.[45] The precise circumstances under which liquid in this tank overheated causing venting via the relief valve, and how the scrubbing system came to be overloaded remains to be proven.[42]

However, the operating company believes that 120 to 240 gal. of water entered the tank and initiated an exothermic reaction.[46] Potential reactions of methyl isocyanate are[47]:

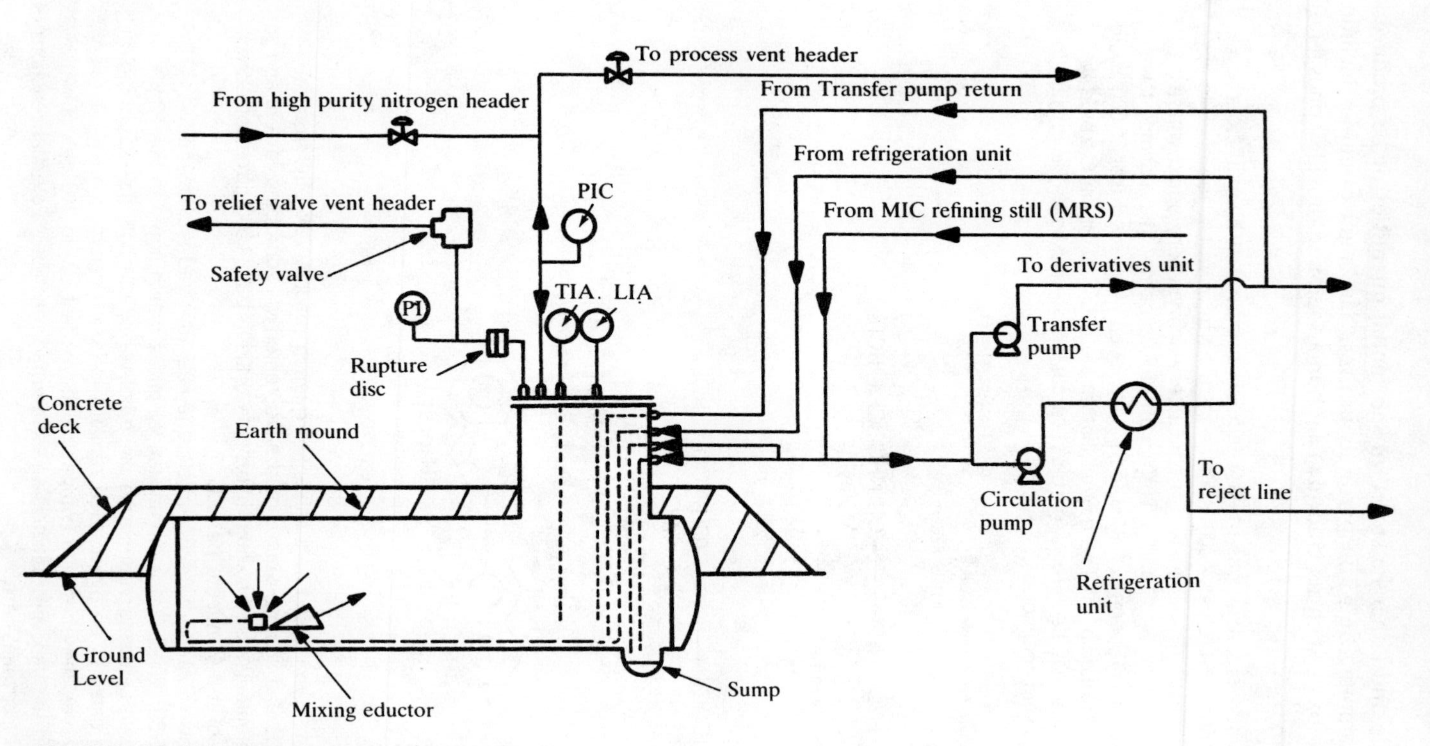

Fig. 15.8 Methyl isocyanate storage tank at Bhopal

$$CH_3N{=}C{=}O + ROH \longrightarrow RO\overset{O}{\overset{\|}{C}}NHCH_3$$

An *N*-methylcarbamate

$$CH_3N{=}C{=}O + H_2O\ (\text{excess}) \longrightarrow CH_3NH\overset{O}{\overset{\|}{C}}NHCH_3 + CO_2$$

1,3-Dimethylurea

$$CH_3N{=}C{=}O\ (\text{excess}) + H_2O \longrightarrow CH_3NH\overset{O}{\overset{\|}{C}}{-}\overset{CH_3}{\overset{|}{N}}{-}\overset{O}{\overset{\|}{C}}NHCH_3 + CO_2$$

1,3,5-Trimethylbiuret

$$3CH_3N{=}C{=}O \xrightarrow{\text{Catalyst}} (CH_3NCO)_3$$

Trimethyl isocyanurate

The hydrolysis occurs slowly at ambient temperature but is exothermic and in the event of a runaway (Ch. 10) would result in the methyl isocyanate boiling – resulting in increased pressure and venting of vapour and carbon dioxide. The refrigeration system, normally used to maintain the methyl isocyanate at 0 to 5 °C[46] is claimed to have been switched off to save electricity.[47] In fact the reaction with water may have been catalysed by the action of Fe ions generated by unusually large amounts of chloroform in the type 304 stainless steel tank.[46]

One explanation of how water entered the tank is that it came via a temporary pipe installed between the relief valve vent header, through which gases were directed for scrubbing, and the process valve vent header.[48] This pipe was apparently installed as a bypass to connect the tank to the relief system while the flare stack was shut down for repair. Four pipelines leading from other equipment into the relief valve vent header were apparently flushed through with water on the evening before the accident but, because of problems with blocked valves, water was able to accumulate in this header. The water was able to flow down the temporary pipe and reportedly entered the tank because the valve on it was either open or leaking. This explanation is not accepted by the operating company which proposed two other routes, i.e. that water could have been added inadvertently or deliberately via the nitrogen pressurisation line.[45]

A review of the human impact, history, scientific background and legal implications of the Bhopal tragedy – in the light of information as at February 1985 – is given in reference 49. Further speculation is not useful pending release of all the facts. However, it is clear that:

- The storage refrigeration plant was shut down.
- The flare stack was inoperative due to maintenance.
- The temperature alarm was disconnected..
- The vent gas scrubber was on stand-by without regular maintenance.

Contributory factors in the scale of the disaster were:

- The juxtaposition of housing and the pesticide plant.
- The tendency of methyl isocyanate, which is twice as heavy as air, to spread near ground level.
- The acute toxicity of methyl isocyanate (e.g. phosgene, which produces similar symptoms, was assigned a TLV of 0.1 ppm).
- The volatility of methyl isocyanate (e.g. TDI and MDI have much higher boiling points).

Historically chlorine has been responsible for most of the world's major toxic releases; some of the most prominent examples are listed in Table 15.4.[8,13]

In 1976 at Louisiana, while a chlorine processing plant was shut down for maintenance, natural gas accidentally entered the plant and storage section along with the inert gas supply. On start-up undetermined conditions triggered an explosion in the storage area. Pressure-relief devices functioned correctly but the force was great enough to dislodge a 160 tonne chlorine storage tank from its foundation. The vessel punctured. Over a 6 hr

Table 15.4 Major releases of chlorine

Location	Date	Source of leakage	Quantity released (tonnes)	Number of fatalities
Newton Alabama	18/11/67	Rail tanker	50	0
La Barre Louisiana	31/01/61	Rail tanker	27.5	1
Griffith Indiana	31/03/35	Rail tanker	27.5	0
Cornwall Ontario	30/11/62	Rail tanker	27.5	0
Chicago Illinois	04/12/47	Rail tanker	16.3	0
Loos, BC	05/03/73	Rail tanker	15.5	0
Niagara Falls NY	28/02/34	Rail tanker	14.5	1
Brandtsville Pennsylvania	28/04/63	Rail tanker	8	0

Table 15.4 (continued)

Location	Date	Source of leakage	Quantity released (tonnes)	Number of fatalities
Mjodalen Norway	26/01/40	Rail tanker	8	3
Chrome New Jersey	1914	Rail tanker	7	0
La Spezia Italy	14/06/66	Rail tanker	7	0
Runcorn UK	19/10/57	Rail tanker	2–3	0
Billingham UK	20/07/50	Rail tanker	0.5	0
Youngstown Florida	Feb. 1978	Rail tanker	50	8
Crestview Florida	April 1979	Rail tanker	Unknown	0
Corbin Louisiana	11/12/71	Rail tanker	Unknown	0
Mississauga Ontario	10/11/79	Rail tanker	75	0
San Luis Potosi Mexico	01/08/81	Rail tanker	300	17
Baton Rouge Louisiana	10/12/76	Storage tank	90	0
Rauma Finland	5/11/47	Storage tank	30	19
St Auban France	13/12/26	Storage tank	24	19
Syracuse N.Y.	10/5/29	Storage tank	24	1
Zarnesti Romania	24/12/39	Storage tank	24	60
Wyandotte Michigan	1917	Storage tank	17	1
Walsum West Germany	4/4/52	Storage tank	15	7
Freeport Texas	1/9/49	Pipeline	4	0
Lake Charles Louisiana	10/3/56	Connecting pipework	3	0
Johnsonburg Pennsylvania	12/11/36	Rail tanker	2	0
Mobile Alabama	12/7/64	Pipeline	Unknown	1

period about 100 tonnes of chlorine escaped. Around 500 people were evacuated shortly after the release and later another 10,000 students and residents were evacuated as a precautionary measure. No one was killed or seriously injured.[50]

Toxicity data of chlorine as a function of exposure time are summarised in Fig. 5.8[51] and in Table 15.5.[52] Based on such data attempts have been made to calculate a mean mortality index.

$$\text{Mean mortality index} = \frac{\text{Mean number of fatalities}}{\text{Mean amount lost}}$$

However, this is of very limited value because:

- The indices were calculated from a low number of incidents.
- The index does not take into account the population density of the area where the accident occurred.

Ammonia has also been responsible for many of the world's large-scale toxic releases,[8,13] as shown in Table 15.6.

At a fertilizer plant at Potchesftroom in 1973 brittle fracture of a dished end of one of 4 × 50-tonne pressure-storage tanks caused it to fail. An estimated 30 tonnes of anhydrous ammonia escaped and a further 8 tonnes were released from a tank car. The blast killed 1 man outright, a further man died immediately from the gas, 3 others died a few days later as a direct result of having been gassed. Outside the plant fence 4 people died immediately and 2 others died several days later. In addition to the 18 deaths about 65

Table 15.5 Human effects from exposure to chlorine gas[52]

Concn. (ppm)	Effects
0.2 to 3.5	Threshold of odour perception in those individuals without chronic exposure to low doses of chlorine.
3 to 5	Tolerated without undue ill effect for 30–60 minutes.
5 to 8	Slight irritation of the mucous membranes of the upper respiratory tract and of the eyes.
15	Effects are immediate. Irritation of nose, throat and eyes with cough and lachrymation.
30	Immediate cough with a choking sensation, retrosternal chest pain and a sense of constriction in the chest. Wheezing due to development of broncho-constriction may occur. Possibility of nausea and vomiting.
40 to 60	Development of a chemical tracheo-bronchitis and pulmonary oedema. The latter may develop after a latent period of several hours with marked respiratory distress, restlessness, possibly cyanosis and frothy sputum. If secondary infection occurs, broncho-pneumonia may develop after a few days.
430	Minimum fatal dose reported for a man after a 30 minute exposure.

Table 15.6 Major releases of ammonia

Year	Location	Quantity released tonnes	Fatalities
1968	Lievin (France)	19	6
1969	Nebraska	90	9
1970	Nebraska	160	0
1970	West Virginia	75	0
1971	Arkansas	600	0
1973	South Africa	38	18
1973	Kansas	277	0
1975	Texas	50	0
1976	Sweden	180	2
1976	Texas	19	6
1976	Oklahoma	500	0
1977	Mexico	Unknown	Unknown
1977	Colombia	Unknown	30

people required hospitalisation and an unknown number were treated by private doctors.[53]

The scale of production and storage capacities for ammonia are of the same order of magnitude as for chlorine. Its main outlets include manufacture of nitric acid, fertilizers, explosives, cleaning agents, etc. The main physiological effects on man are summarised in Table 15.7.

Table 15.8 summarises typical storage arrangements and toxic effects for other key bulk toxics.[54,55] However, a recent trend has been for companies to reduce inventories of stored bulk toxics as far as possible.

Toxic materials of relatively low volatility may create a hazard if a tank is subjected to overheating from an adjacent fire.

Table 15.7 Physical effects of ammonia on man

Concn. (ppm)	Effects
20	Odour threshold
40	Slight eye irritation in a few individuals
100	Noticeable irritation of the eyes and nasal passages after a few minutes' exposure
400	Severe irritation of the throat, nasal passages and upper respiratory tract
700	Severe eye irritation; no permanent effect if the exposure is limited to less than ½ hour
1,700	Serious coughing, bronchial spasms; even exposure of less than ½ hour may prove fatal
5,000	Serious oedema, strangulation asphyxia; fatal almost immediately

Table 15.8 Storage and toxicity details for a selection of bulk toxic materials

Substance	Capacity and type of typical large-scale storage tank	Toxicity to man	
		Level (ppm)	Effect
Sulphur dioxide	Pressurised 100 tonne vessel	6–12	Immediate irritation to nose and throat
		Above 20	Eye irritation
		1%	Irritation to moist skin within a few minutes
Phosgene	Slightly pressurised 35 tonne tank	1	Maximum amount for prolonged exposure
		1.25–2.5	Dangerous to life for a prolonged exposure
		5	Cough or other subjective symptoms within 1 minute
		10	Irritation of eyes and respiratory tract in less than 1 minute
		12.5	Dangerous to life in 30–60 minutes
		20	Severe lung injury within minutes
		25	Dangerous to life for as little as 30 minutes
		90	Fatal within 30 minutes
Hydrogen cyanide	Slightly pressurised 100 tonne tank	18–36	Slight symptoms after several hours
		45–54	Tolerated for ½–1 hour without immediate or later effects
		110–135	Fatal after ½–1 hour or later, or dangerous to life
		135	Fatal after 30 minutes
		181	Fatal after 10 minutes
		270	Immediately fatal
Acrylonitrile	1,000 tonne unpressurised tank	1.4–9	Headaches, irritability
		16–100	After 20–45 minutes' exposure respiratory and nervous system effects observed
Bromine	Slightly pressurised 100 tonne tank	0.1–0.15	Maximal concentration for continuous exposure
		4	Maximal concentration for short exposure
		40–60	Dangerous for short exposure
		1,000	Rapidly fatal for short exposure

Lead anti-knock compounds TEL and TML are liquids at ambient temperature and because their vapour pressures are low, <0.035 bar at 15 °C, in the event of equipment or piping failure the bulk of the liquid would be retained in bunds. Under severe fire conditions, however, failure of a vessel could release a vapour aerosol cloud which could drift and create an on/off-site hazard.

In any fire involving conventional petroleum refinery materials the resulting cloud – of carbon dioxide, carbon monoxide, unburned hydrocarbons and water – rises by thermal lift. In exceptional weather conditions, part of the cloud may return to ground level within a mile from the source, causing irritation and smell; toxic release is not a major problem. However, some substances produce highly toxic products on thermal degradation, therefore major toxic incidents could also arise from certain major fires. Examples include the formation of carbon monoxide from incomplete combustion and the production of acid chlorides and phosgene by the thermal decomposition of halocarbons. Some polymers also pose a threat on over-heating as exemplified in Table 15.9. Organic poly-isocyanates can be additional complications in fires at plants processing polyurethane foam, either as a degradation product or from stocks of the raw material. Moreover, fires involving ammonium nitrate fertiliser tend to result in NO_x emissions.

Major incidents might arise from the smaller scale release of hyper-toxic products, by-products or intermediates, though the number of such substances currently involved in commercial operations is small. Examples include botulinum toxin, staphlococci toxin, aflatoxin and tetra chloro-dibenzo dioxin (TCDD). Important differences between the consequences of release of potently toxic chemicals and release of bulk toxics are:

- With hyper-poisons even if fatalities were few the immediate effects to the community could be devastating in terms of the effects to live-stock, vegetation, water supplies, etc.
- Fallout could contaminate vast areas of land rendering it sterile and hazardous for prolonged periods of time thereby preventing early return of evacuees.
- Because of the trace contaminant character of hyper-poisons, standard decontamination procedures are inappropriate for dealing with these materials.
- There can be serious long-term implications to the health of those exposed to escaped hyper-toxic material.

Table 15.9 Thermal degradation products of polymers

Polymer	Temperature (°C)	Toxic products
Polychloroprene	377	Hydrogen chloride
Polytrifluorethylene	380–800	Hydrogen fluoride
Polyvinyl fluoride	370	Hydrogen fluoride
Polyvinyl chloride	225	Hydrogen chloride
Certain polyamides/urethanes	250–500	Hydrogen cyanide
Certain polyfluoro-olefins	> 250	Perfluoroisobutylene*

* More toxic than phosgene.

The Seveso incident was described in Chapter 10. Though no fatalities were directly attributed to the incident, the drifting poison cloud caused 750 people to be evacuated from their homes, development of chloracne in many of the population exposed, widespread damage to crops and pollution of the Rivers Seveso and Po, and the death of many domestic animals. More than 12 km^2 were contaminated with TCDD at or above a concentration of 5 $\mu g/m^2$.[56]

Identification of major hazards

With notable exceptions e.g. Mexico City (1984), Bhopal (1984), Texas City (1947) incidents involving major hazards seldom realise their full potential in terms of fatalities and damage. The degree of risk associated with major hazards generally can be predicted crudely as a combination of the probability of an incident occurring and the likely consequences of such an event. Thus the risk will vary between installations and will be influenced by the following factors.[27]

Nature of the process

Because of the nature of the process, certain plants must be designed to rigid specifications and incorporate automatic controls, sensors and other safety features. Such processes may include oxidations, polymerisations, chlorinations, hydrogenation, nitrations, etc. The relevance of high pressure has been referred to above and in Chapter 10.

Juxtaposition of plants

Since heat, blast waves or projected débris from a fire or explosion can affect the integrity of adjoining plant and cause an escalation in the severity of an incident, any assessment of the degree of risk associated with a particular plant must take into consideration the potential for 'domino effects'.

Population density

The effect of the neighbouring population density on the outcome of a major incident is self-evident and was a major factor in the catastrophic consequences of the two worst incidents (i.e. Bhopal and Mexico City). Thus an assessment of risk must consider the proximity of housing or industrial estates, hospitals, motorways, railway systems, camp sites, etc.

Topography

The topography surrounding a major installation can influence the outcome of a major incident by, for example, affecting the dispersion of an escaping toxic or flammable gas cloud. Changes in terrain can help determine wind speed and direction, and are most pronounced near coasts where 'normal' temperature and wind profiles can be significantly modified. Differential heating of sea and land masses can cause breezes to blow towards land from the sea for the latter half of the day. Above this landward moving air is a return flow. Similarly differential heating of valleys can bring about local winds deviating from the prevailing wind. A shielding effect may be experienced by the valley walls, resulting in less movement of air inside than outside the valley. A tendency may exist for temperature inversion which could hamper dispersal of atmospheric pollution from the valley floor.

Inventory

The scale of an operation will have some bearing on the outcome of an incident. It has been proposed[57] that the fatality rate in scaling up a process can be correlated by the expression:

$$F_R = K\,S^{0.5 \text{ to } 0.33}$$

where: F_R is the fatality rate; K is a constant; and S is the stream capacity.

Notification of major hazards

Within the UK the Advisory Committee on Major Hazards proposed a notification scheme[12] which formed the basis for the Notification of Installations Handling Hazardous Substances Regulations 1982 operative from 1 January 1983. Notifiable quantities of hazardous substances at any site are as listed in Schedule 1, reproduced as Table 15.10. Quantities in any pipeline within 500 m of the site are included subject to certain exceptions, but not waste at any licensed site. Particulars to be included in the notification are given in Schedule 2 Part I of the Regulations for a site or Part II relating to a pipeline.

EEC Directive on major accident hazards

An EEC Directive deals with a notification scheme[58] applicable to any industrial installation covered by Annex I, reproduced as Table 15.11, and involving a dangerous substance, or other storages covered by Annex II, reproduced as Table 15.12. There is a requirement for a manufacturer 'to take all the measures necessary to prevent major accidents and to limit their consequences for man and the environment' when handling dangerous substances as defined in Annex IV, reproduced as Table 15.13. Under Article 5 of the Directive there is a requirement for notification and supply of information for any relevant industrial installation if a

Table 15.10 Notifiable quantities of hazardous substances in the UK

Part I ***Named Substance*** 1 Substance	2 Notifiable quantity tonnes
Liquefied petroleum gas, such as commercial propane and commercial butane, and any mixture thereof held at a pressure greater than 1.4 bar absolute	25
Liquefied petroleum gas, such as commercial propane and commercial butane, and any mixture thereof held under refrigeration at a pressure of 1.4 bar absolute or less	50
Phosgene	2
Chlorine	10
Hydrogen fluoride	10
Sulphur trioxide	15
Acrylonitrile	20
Hydrogen cyanide	20
Carbon disulphide	20
Sulphur dioxide	20
Bromine	40
Ammonia (anhydrous or as solution containing more than 50% by weight of ammonia)	100
Hydrogen	2
Ethylene oxide	5
Propylene oxide	5
tert-Butyl peroxyacetate	5
tert-Butyl peroxyisobutyrate	5
tert-Butyl peroxymaleate	5
tert-Butyl peroxy isopropyl carbonate	5
Dibenzyl peroxydicarbonate	5
2,2-Bis(tert-butylperoxy)butane	5
1,1-Bis(tert-butylperoxy)cyclohexane	5
Di-sec-butyl peroxydicarbonate	5
2,2-Dihydroperoxypropane	5
Di-n-propyl peroxydicarbonate	5
Methyl ethyl ketone peroxide	5
Sodium chlorate	25
Cellulose nitrate other than – (*a*) cellulose nitrate to which the Explosives Act 1875(a) applies; or (*b*) solutions of cellulose nitrate where the nitrogen content of the cellulose nitrate does not exceed 12.3% by weight and the solution contains not more than 55 parts of cellulose nitrate per 100 parts by weight of solution	50

(a) 1875 c. 17.
(b) OJ No L250. 23.9.80, p. 7.

Table 15.10 (continued)

Part I
Named Substance

1 Substance	2 Notifiable quantity tonnes
Ammonium nitrate and mixtures of ammonium nitrate where the nitrogen content derived from the ammonium nitrate exceeds 28% of the mixture by weight other than – (*a*) mixtures to which the Explosives Act 1875(a) applies; or (*b*) ammonium nitrate based products manufactured chemically for use as fertiliser which comply with Council Directive 80/876/EEC(b)	500
Aqueous solutions containing more than 90 parts by weight of ammonium nitrate per 100 parts by weight of solution	500
Liquid oxygen	500

Part II
Classes of substances not specifically named in Part I

1 Class of substance	2 Notifiable quantity tonnes
1. Gas or any mixture of gases which is flammable in air and is held in the installation as a gas.	15
2. A substance or any mixture of substances which is flammable in air and is normally held in the installation above its boiling point (measured at 1 bar absolute) as a liquid or as a mixture of liquid and gas at a pressure of more than 1.4 bar absolute.	25 being the total quantity of substances above the boiling points whether held singly or in mixtures.
3. A liquefied gas or any mixture of liquefied gases, which is flammable in air, has a boiling point of less than 0 °C (measured at 1 bar absolute) and is normally held in the installation under refrigeration or cooling at a pressure of 1.4 bar absolute or less.	50 being the total quantity of substances having boiling points below 0 °C whether held singly or in mixtures.
4. A liquid or any mixture of liquids not included in items 1 to 3 above, which has a flash point of less than 21 °C.	10.000

dangerous substance listed in Annex III, reproduced as Table 15.14 or Annex II is present in the quantities listed. Under Article 8, persons liable to be affected by a major accident originating in a notified industrial activity should be informed of the safety measures and how to behave in the event of a serious accident. There is also a procedure for reporting major accidents.

Table 15.11 Industrial installations within the meaning of Article 1 [*ANNEX I*]

1. – Installations for the production or processing of organic or inorganic chemicals using for this purpose, in particular:
   - alkylation
   - amination by ammonolysis
   - carbonylation
   - condensation
   - dehydrogenation
   - esterification
   - halogenation and manufacture of halogens
   - hydrogenation
   - hydrolysis
   - oxidation
   - polymerisation
   - sulphonation
   - desulphurisation, manufacture and transformation of sulphur-containing compounds
   - nitration and manufacture of nitrogen-containing compounds
   - manufacture of phosphorus-containing compounds
   - formulation of pesticides and of pharmaceutical products.

   Installations for the processing of organic and inorganic chemical substances, using for this purpose, in particular:
   - distillation
   - extraction
   - solvation
   - mixing.
2. Installations for distillation, refining or other processing of petroleum or petroleum products.
3. Installations for the total or partial disposal of solid or liquid substances by incineration or chemical decomposition.
4. Installations for the production or processing of energy gases, for example, LPG, LNG, SNG.
5. Installations for the dry distillation of coal or lignite.
6. Installations for the production of metals or non-metals by the wet process or by means of electrical energy.

Control of industrial major accident hazards – UK

The EEC Directive referred to above has been implemented in the UK by the Control of Industrial Major Accident Hazards Regulations 1984. Subject to certain exclusions (i.e. nuclear installations, defence installations, licensed explosives sites, mines or quarries, or disposal sites licensed under the Control of Pollution Act), the Regulations are applicable to:

- any operation in an industrial installation specified in Schedule 4 (as Annex I – Table 15.11), which involves one or more dangerous substances, unless that operation is incapable of producing a major accident hazard;
- storage of at least specified quantities of the substances listed in Schedule 2 (as Annex II – Table 15.12), 'isolated storage'.

Table 15.12 Storage at installations other than those covered by Annex I ('Isolated storage') [*ANNEX II*]

The quantities set out below relate to each installation or group of installations belonging to the same manufacturer where the distance between the installations is not sufficient to avoid, in foreseeable circumstances, any aggravation of major-accident hazards. These quantities apply in any case to each group of installations belonging to the same manufacturer where the distance between the installations is less than approximately 500 m.

	Quantities (tonnes) ≥	
Substances or groups of substances	For application of Articles 3 and 4	For application of Article 5
1. Flammable gases as defined in Annex IV (c)(i)	50	300*
2. Highly flammable liquids as defined in Annex IV (c)(ii)	10,000	100,000
3. Acrylonitrile	350	5,000
4. Ammonia	60	600
5. Chlorine	10	200
6. Sulphur dioxide	20	500
7. Ammonium nitrate	500†	5,000†
8. Sodium chlorate	25	250†
9. Liquid oxygen	200	2,000†

* Member States may provisionally apply Article 5 to quantities of at least 500 tonnes until the revisioning of Annex II mentioned in Article 19.
† Where the substance is in a state which gives it properties capable of creating a major accident hazard.

'Dangerous substances' are those which fulfil the criteria in Schedule 1 (as Annex IV – Table 15.13) for very toxic, toxic, explosive, or flammable substances or are listed in Schedule 2 (Table 15.12) or in Schedule 3 (as Annex III – Table 15.14).

Manufacturers controlling such installations are required to be able to demonstrate that they have:

- identified major accident hazards;
- taken adequate steps to prevent or limit the consequences of any major accident;
- provided suitable information, training and equipment for persons working on site to ensure their safety.

Any major accident must be reported to the HSE which is required to send information about it to the Commission of the European Communities.

A manufacturer controlling an activity other than an 'isolated storage' involving at least the quantity specified in Schedule 3 (Table 15.14), or

Table 15.13 Indicative criteria [*ANNEX IV*]

(a) Very toxic substances:
— substances which correspond to the first line of the table below,
— substances which correspond to the second line of the table below and which, owing to their physical and chemical properties, are capable of entailing major-accident hazards similar to those caused by the substance mentioned in the first line:

	LD 50 (oral) (1) mg/kg body weight	LD 50 (cutaneous)(2) mg/kg body weight	LC 50(3) mg/l (inhalation)
1	LD 50 ≤ 5	LD 50 ≤ 10	LC 50 ≤ 0.1
2	5 < LD 50 ≤ 25	10 < LD 50 ≤ 50	0.1 < LC 50 ≤ 0.5

(1) LD 50 oral in rats.
(2) LD 50 cutaneous in rats or rabbits.
(3) LC 50 by inhalation (four hours) in rats.

(b) Other toxic substances:
The substances showing the following values of acute toxicity and having physical and chemical properties capable of entailing major-accident hazards:

LD 50 (oral)(1) mg/kg body weight	LD 50 (cutaneous)(2) mg/kg body weight	LC 50(3) mg/l (inhalation)
25 < LD 50 ≤ 200	50 < LD 50 ≤ 400	0.5 < LC 50 ≤ 2

(1) LD 50 oral in rats.
(2) LD 50 cutaneous in rats or rabbits.
(3) LC 50 by inhalation (four hours) in rats.

(c) Flammable substances
(i) *flammable gases*:
substances which in the gaseous state at normal pressure and mixed with air become flammable and the boiling point of which at normal pressure is 20 °C or below;
(ii) *highly flammable liquids*:
substances which have a flash-point lower than 21 °C and the boiling point of which at normal pressure is above 20 °C;
(iii) *flammable liquids*:
substances which have a flash-point lower than 55 °C and which remain liquid under pressure, where particular processing conditions, such as high pressure and high temperature, may create major-accident hazards.

(d) Explosive substances:
Substances which may explode under the effect of flame or which are more sensitive to shocks or friction than dinitrobenzene.

an isolated storage involving at least the quantity specified in Schedule 2 (Table 15.12), must fulfil the following obligations:

- Send a report to the HSE containing specified information relating to every dangerous substance involved in a relevant quantity, relating to the installation, relating to the management system for controlling the

Table 15.14 List of substances for the application of Article 5 [*ANNEX III*]

The quantities set out below relate to each installation or group of installations belonging to the same manufacturer where the distance between the installations is not sufficient to avoid, in foreseeable circumstances, any aggravation of major-accident hazards. These quantities apply in any case to each group of installations belonging to the same manufacturer where the distance between the installations is less than approximately 500 m.

Name	Quantity (⩾)
1. 4-Aminodiphenyl	1 kg
2. Benzidine	1 kg
3. Benzidine salts	1 kg
4. Dimethylnitrosamine	1 kg
5. 2-Naphthylamine	1 kg
6. Beryllium (powders, compounds)	10 kg
7. Bis(chloromethyl)ether	1 kg
8. 1,3-Propanesultone	1 kg
9. 2,3,7,8-Tetrachlorodibenzo-p-dioxin (TCDD)	1 kg
10. Arsenic pentoxide, Arsenic(V) acid and salts	500 kg
11. Arsenic trioxide. Arsenious(III) acid and salts	100 kg
12. Arsenic hydride (Arsine)	10 kg
13. Dimethylcarbamoyl chloride	1 kg
14. 4-(Chloroformyl) morpholine	1 kg
15. Carbonyl chloride (Phosgene)	20 t
16. Chlorine	50 t
17. Hydrogen sulphide	50 t
18. Acrylonitrile	200 t
19. Hydrogen cyanide	20 t
20. Carbon disulphide	200 t
21. Bromine	500 t
22. Ammonia	500 t
23. Acetylene (Ethyne)	50 t
24. Hydrogen	50 t
25. Ethylene oxide	50 t
26. Propylene oxide	50 t
27. 2-Cyanopropan-2-ol (Acetone cyanohydrin)	200 t
28. 2-Propenal (Acrolein)	200 t
29. 2-Propen-1-ol (Allyl alcohol)	200 t
30. Allylamine	200 t
31. Antimony hydride (Stibine)	100 kg
32. Ethyleneimine	50 t
33. Formaldehyde (concentration ⩾ 90%)	50 t
34. Hydrogen phosphide (Phosphine)	100 kg
35. Bromomethane (Methyl bromide)	200 t

Table 15.14 (continued)

Name	Quantity (⩾)
36. Methyl isocyanate	1 t
37. Nitrogen oxides	50 t
38. Sodium selenite	100 kg
39. Bis(2-chlorocthyl) sulphide	1 kg
40. Phosacetim	100 kg
41. Tetraethyl lead	50 t
42. Tetramethyl lead	50 t
43. Promurit (1-(3,4-Dichlorophenyl)-3-triazenethio-carboxamide)	100 kg
44. Chlorfenvinphos	100 kg
45. Crimidine	100 kg
46. Chloromethyl methyl ether	1 kg
47. Dimethyl phosphoramidocyanidic acid	1 t
48. Carbophenothion	100 kg
49. Dialifos	100 kg
50. Cyanthoate	100 kg
51. Amiton	1 kg
52. Oxydisulfoton	100 kg
53. 00-Diethyl S-ethylsulphinylmethyl phosphorothioate	100 kg
54. 00-Diethyl S-ethylsulphonylmethyl phosphorothioate	100 kg
55. Disulfoton	100 kg
56. Demeton	100 kg
57. Phorate	100 kg
58. 00-Diethyl S-ethylthiomethyl phosphorothioate	100 kg
59. 00-Diethyl S-isopropylthiomethyl phosphorodithioate	100 kg
60. Pyrazoxon	100 kg
61. Pensulfothion	100 kg
62. Paraoxon (Diethyl 4-nitrophenyl phosphate)	100 kg
63. Parathion	100 kg
64. Azinphos-ethyl	100 kg
65. 00-Diethyl S-propylthiomethyl phosphorodithioate	100 kg
66. Thionazin	100 kg
67. Carbofuran	100 kg
68. Phosphamidon	100 kg
69. Tirpate (2,4-Dimethyl-1,3-dithiolane-2-carboxaldehyde 0-methylcarbamoyloxime)	100 kg
70. Mevinphos	100 kg
71. Parathion-methyl	100 kg
72. Azinphos-methyl	100 kg
73. Cycloheximide	100 kg
74. Diphacinone	100 kg
75. Tetramethylenedisulphotetramine	1 kg

Table 15.14 (continued)

Name	Quantity (≥)
76. EPN	100 kg
77. 4-Fluorobutyric acid	1 kg
78. 4-Fluorobutyric acid, salts	1 kg
79. 4-Fluorobutyric acid, esters	1 kg
80. 4-Fluorobutyric acid, amides	1 kg
81. 4-Fluorocrotonic acid	1 kg
82. 4-Fluorocrotonic acid, salts	1 kg
83. 4-Fluorocrotonic acid, esters	1 kg
84. 4-Fluorocrotonic acid, amides	1 kg
85. Fluoroacetic acid	1 kg
86. Fluoroacetic acid, salts	1 kg
87. Fluoroacetic acid, esters	1 kg
88. Fluoroacetic acid, amides	1 kg
89. Fluenetil	100 kg
90. 4-Fluoro-2-hydroxybutyric acid	1 kg
91. 4-Fluoro-2-hydroxyburytic acid, salts	1 kg
92. 4-Fluoro-2-hydroxybutyric acid, esters	1 kg
93. 4-Fluoro-2-hydroxybutyric acid, amides	1 kg
94. Hydrogen fluoride	50 t
95. Hydroxyacetonitrile (Glycolonitrile)	100 kg
96. 1,2,3,7,8,9-Hexachlorodibenzo-p-dioxin	100 kg
97. Isodrin	100 kg
98. Hexamethylphosphoramide	1 kg
99. Juglone (5-Hydroxynaphthalene-1,4-dione)	100 kg
100. Warfarin	100 kg
101. 4,4'-Methylenebis (2-chloroaniline)	10 kg
102. Ethion	100 kg
103. Aldicarb	100 kg
104. Nickel tetracarbonyl	10 kg
105. Isobenzan	100 kg
106. Pentaborane	100 kg
107. I-Propen-2-chloro-1,3-diol-diacetate	10 kg
108. Propyleneimine	50 t
109. Oxygen difluoride	10 kg
110. Sulphur dichloride	1 t
111. Selenium hexafluoride	10 kg
112. Hydrogen selenide	10 kg
113. TEPP	100 kg
114. Sulfotep	100 kg
115. Dimefex	100 kg
116. 1-Tri(cyclohexyl) stannyl-1H-1,2,4-triazole	100 kg

Table 15.14 (continued)

Name	Quantity (≥)
117. Triethylenemelamine	10 kg
118. Cobalt (powders, compounds)	100 kg
119. Nickel (powders, compounds)	100 kg
120. Anabasine	100 kg
121. Tellurium hexafluoride	100 kg
122. Trichloromethanesulphenyl chloride	100 kg
123. 1,2-Dibromoethane (Ethylene dibromide)	50 t
124. Flammable substances as defined in Annex IV (c)(i)	200 t
125. Flammable substances as defined in Annex IV (c)(ii)	50,000 t
126. Diazodinitrophenol	10 t
127. Diethylene glycol dinitrate	10 t
128. Dinitrophenol, salts	50 t
129. 1-Guanyl-4-nitrosaminoguanyl-1-tetrazene	10 t
130. Bis (2,4,6-trinitrophenyl)amine	50 t
131. Hydrazine nitrate	50 t
132. Nitroglycerine	10 t
133. Pentatrythritol tetranitrate	50 t
134. Cyclotrimethylene trinitramine	50 t
135. Trinitroaniline	50 t
136. 2,4,6-Trinitroanisole	50 t
137. Trinitrobenzene	50 t
138. Trinitrobenzoic acid	50 t
139. Chlorotrinitrobenzene	50 t
140. N-Methyl-N,2,4,6-N-tetranitroaniline	50 t
141. 2,4,6-Trinitrophenol (Picric acid)	50 t
142. Trinitrocresol	50 t
143. 2,4,6-Trinitrophenetole	50 t
144. 2,4,6-Trinitroresorcinol (Styphnic acid)	50 t
145. 2,4,6-Trinitrotoluene	50 t
146. Ammonium nitrate*	5,000 t
147. Cellulose nitrate (containing > 12.6% nitrogen)	100 t
148. Sulphur dioxide	1,000 t
149. Hydrogen chloride (liquefied gas)	250 t
150. Flammable substances as defined in Annex IV (c) (iii)	200 t
151. Sodium chlorate*	250 t
152. tert-Butyl peroxyacetate (concentration ≥ 70%)	50 t
153. tert-Butyl peroxyisobutyrate (concentration ≥ 80%)	50 t
154. tert-Butyl peroxymaleate (concentration ≥ 80%)	50 t
155. tert-Butyl peroxy isopropyl carbonate (concentration ≥ 80%)	50 t
156. Dibenzyl peroxydicarbonate (concentration ≥ 90%)	50 t

Table 15.14 (continued)

Name	Quantity (≥)
157. 2,2-Bis (tert-butylperoxy) butane (concentration ≥ 70%)	50 t
158. 1,1-Bis (tert-butylperoxy) cyclohexane (concentration ≥ 80%)	50 t
159. Di-sec-butyl peroxydicarbonate (concentration ≥ 80%)	50 t
160. 2,2-Dihydroperoxypropane (concentration ≥ 30%)	50 t
161. Di-n-propyl peroxydicarbonate (concentration ≥ 80%)	50 t
162. 3,3,6,6,9,9-Hexamethyl-1,2,4,5-tetroxacyclononane (concentration ≥ 75%)	50 t
163. Methyl ethyl ketone peroxide (concentration ≥ 60%)	50 t
164. Methyl isobutyl ketone peroxide (concentration ≥ 60%)	50 t
165. Peracetic acid (concentration ≥ 60%)	50 t
166. Lead azide	50 t
167. Lead 2,4,6-trinitroresorcinoxide (Lead styphnate)	50 t
168. Mercury fulminate	10 t
169. Cyclotetramethylenetetranitramine	50 t
170. 2,2',4,4',6,6'-Hexanitrostilbene	50 t
171. 1,3,5-Triamino-2,4,6-trinitrobenzene	50 t
172. Ethylene glycol dinitrate	10 t
173. Ethyl nitrate	50 t
174. Sodium picramate	50 t
175. Barium azide	50 t
176. Di-isobutyryl peroxide (concentration ≥ 50%)	50 t
177. Diethyl peroxydicarbonate (concentration ≥ 30%)	50 t
178. tert-Butyl peroxypivalate (concentration ≥ 77%)	50 t

* Where this substance is in a state which gives it properties capable of creating a major-accident hazard

industrial activity (i.e. staffing arrangements; arrangements to ensure that the means for safe operation are properly designed, constructed, tested, operated, inspected and maintained; training arrangements) and relating to the potential major accidents.

- Send the above report at least three months before commencing any new activity, or in the case of an existing activity before 8 January 1989. (In the latter case certain preliminary information was required by 1 April 1985 unless information had already been provided under the Notification Regulations summarised on page 893.)
- Keep the report up to date.
- Provide further information to the HSE if required.
- Prepare and keep up to date an on-site emergency plan, detailing how major accidents will be dealt with on site.

- Arrange that persons outside the site, who may be affected by a major accident, are informed of the nature of the hazard, and of safety measures and the correct behaviour to adopt if such an accident occurs.

The local authority in whose area there is a site covered by the Regulations is required to prepare and keep up to date an adequate off-site emergency plan detailing how emergencies relating to a possible major accident will be dealt with. For this purpose they should consult the manufacturers, the HSE and other appropriate persons. (The local authority is enabled to recover from the manufacturer costs reasonably incurred in the preparation and up-dating of this plan.)

Control of major hazards

Following identification of 'major hazard' sites their control requires careful consideration. The crucial rôle of 'management' in securing safety at major hazard installations has been emphasised by the Advisory Committee.[12] Management deficiency has been highlighted as the key cause of the Flixborough disaster, and the UK legislation places specific duties on management and requires reports on management systems.

Since major hazard installations are characterised by a high technical content of the operation and the potential for tremendous harm outside the plant perimeter, the management policy must address itself to two aspects:

1. Reducing the probability of failure of key units to acceptable low levels,
2. Minimising the outcome of a major incident.

Reducing the probability of an incident occurring

Measures available for reducing the chance of incidents occurring include[27]:

1. Good design, e.g. using approved or equivalent Codes of Practice coupled with probability analysis of failure of components and overall hazard assessment by, e.g. hazard and operability studies (HAZOPS – see page 74).
2. Sound construction with attention focused on location and structural integrity. Attention should be devoted to the calibre of contractor and to pre-commissioning checks with reference to suppliers' recommendations (cf. the Flixborough accident).[9]
3. Adequate communication channels and training schedules.
4. Procedures for plant modification and maintenance should be formalised, especially when plant integrity may be affected and in all high-risk areas.

   Tactics include permit-to-work systems, maintenance schedules, non-destructive testing (e.g. visual, optical, ultrasonic, X-ray, etc.). The criteria for modifications and maintenance control should be the potential effect on plant and process safety and not the cost of the work (see Ch. 17).

5. Regular safety audits to identify potential hazards, compilation of operating manuals and detailed in-plant training programmes should receive attention to ensure the safe running of the plant.

Mitigating measures

Aspects to consider in minimising the severity of an accident include[27]:

1. Allowance for a *cordon sanitaire* in the case of new installations. 'Containment' of the hazard in this way by provision of a separation distance relies on estimation of the 'maximum credible effect'. This can only be calculated by reference to historical data and a common-sense approach using the standard 'rules'.
   - Plant or storages at a high pressure, particularly at high or low temperature, have a probability of failure even if correctly designed and maintained.
   - The simultaneous failure of two or more storages is exceedingly unlikely.
   - Secondary effects, i.e. missile or blast damage to other tanks, are unlikely to be more dangerous than the maximum credible primary effect since it would still require damage to two or more storages.[32]

   It is important to note that separation distances are practicable for explosion hazards[31] but predictions from dispersion models suggest that, while separation is desirable, such criteria are less applicable to credible bulk toxic-gas release scenarios.
2. Careful design and choice of location for the control room[13] and other rooms with high occupancy.
3. Arrangements for the rapid assessment of the likely consequences in the event of, e.g. loss of containment of a hazardous material (e.g. to identify the nature and quantity of material lost, to check wind direction, to determine probable areas to be affected from a map and the likely population density threatened).
4. In the case of potential toxic releases provision of escape rooms, local respiratory protection, etc.
5. Establishment of an emergency plan. The development, up-dating and rehearsing of plans for major emergencies – both on- and off-site – are extremely important. Within the UK (Regulations 10(1), 11(1) and 11(2) of the Control of Industrial Major Accidents Hazard Regulations) there is a requirement for an on-site emergency plan to be prepared by a manufacturer, the manufacturer to supply sufficient information to the local authority to prepare an off-site emergency plan, and the authority to prepare such a plan. Some aspects of plan-

ning are summarised below; guidance on the preparation of emergency plans is to be issued by the HSE.

With regard to spacing and location, preliminary calculations with regard to distances based upon predicted overpressures may be based on the following,

Item	*Should not be subjected to incident overpressure of (bar)*
High population densities, e.g. schools, hospitals	0.02
Domestic housing	0.04
Public roads	0.05
Ordinary plant buildings, e.g. offices	0.07
Buildings with shatter-resistant windows. Fixed roof tanks containing highly flammable/toxic materials	0.10
Floating roof tanks. Other fixed roof tanks. Cooling towers. Utility areas. Site roads	0.20
Plants with large atmospheric pressure vessels or units of large superficial area.	0.30
Other hazardous plants	0.40
Non-hazardous (unoccupied) plants. Control rooms designed for blast resistance.	0.70

Major emergency planning

An emergency scheme may be based upon four identifiable phases.[59]

Phase 1 – Preliminary action Specific plan preparation and familiarisation of employees with the details. Purchase and positioning of essential equipment. Training of personnel. Initiation of inspection programmes of hazard areas, testing warning systems and evacuation procedures.

Phase 2 – Action when emergency is imminent A preliminary warning may arise of an emergency, e.g. a potentially dangerous on-plant situation due to accidental damage/maloperation. In this period there should be assembly of key personnel, review of standing emergency arrangements, provision of advance warning to external authorities, and testing of all emergency systems.

Phase 3 – Action during emergency Assuming Phases 1 and 2 are properly performed, Phase 3 may proceed to plan. However, unexpected variations in predicted emergencies (e.g. 'boil-overs' of burning tanks, changes of wind direction, unforeseen escalation) can require rapid decisions and decisive management by appointed personnel (e.g. an emergency controller).

Phase 4 – Ending the emergency A procedure for declaring plants/areas safe and for orderly re-occupation and, if possible, plant close-down/return to normality.

Close liaison with external authorities, e.g. regulatory authorities, local fire, police and medical services is obviously important in major emergency planning. Local authorities are also involved, their possible roles in an emergency being to support the emergency services (e.g. with engineering problems, traffic problems, volunteer coordination), care of the homeless (i.e. temporary accommodation, feeding, rehousing, transport and information) and co-ordination of information transfer.[60,61] An outstanding example of this has been described in relation to the Mississauga incident.[62] Other public authorities, e.g. electricity, gas, water and telephone communications may have roles in an emergency in performing repairs and re-establishing supplies in accordance with priorities and prior discussion is therefore relevant. A mutual aid scheme whereby neighbours stand ready with advice, trained personnel and equipment to help one another and the surrounding community in the event of a disaster, or the threat of it, is also advisable.[63,64]

A low-level explosion and fire occurred in a 4,500 litre carbon disulphide tank and the contents spilled into the surrounding bund. Three fire trucks were required to fight the fire and to cool adjoining equipment.[63]

Extensive damage was caused as the tank, loading facilities and adjacent operating areas became involved. The emergency control programme was activated and within minutes the plant's fire-fighting equipment was in operation. Explosions occurred in a captive sewer system.

The mutual aid consortium was alerted and responded with generators and portable lights, a laboratory to check for river pollution, fire trucks, foam and fire-fighting equipment. After several hours the emergency, which could have resulted in a major disaster involving the surrounding community, had been satisfactorily dealt with.[63]

The plan should cover the full range of possible incidents; other incidents may be 'so unlikely' that they can be disregarded. (The assessment of probability in this case may require a thorough hazard analysis.) Account must be taken of on-site and off-site areas likely to be affected. Realistic estimates of the potential effects, e.g. using computer models[65,66] will assist planning as will consideration of past mistakes.

The emergency plan at Flixborough called for plant and laboratory personnel to assemble in the control room. In the event everyone in the control room was killed in the UVCE. Those in the adjacent laboratory who ignored the plan, and got out of the building and ran as fast as they could from the vapour cloud, all survived despite being caught in the open by the blast.[67]

The plan should include the establishment of lines of communication between key personnel (e.g. management, laboratory staff, plant supervision, medical officer, safety and security officers) and police, fire authorities, and personnel in adjacent factories. This necessitates:

- Appointment of a central communication co-ordinator.
- Closely identified and agreed duties/responsibilities for each key person.
- Identification of persons to be contacted in advance, including those in local authorities and neighbouring works.

Advance arrangements are also necessary for foreseeable contingencies such as:

- The need to transfer switchboard to night service in the event of evacuation.
- The need to use neighbouring facilities if the works medical centre becomes untenable.
- The incident occurring during holidays or at night (i.e. on a weekday there is obviously a 2 in 3 chance of an incident occurring outside the 9 a.m.–5 p.m. period on Mondays to Fridays).

Provision should also be made for the cancellation of signals as soon possible in the event of a false alarm.

Emergency control

A senior member of staff is normally designated as emergency controller. A deputy should be appointed to cover periods when the controller is unavailable, and outside normal working hours initial control is vested in the senior member of site management until the controller arrives.

An emergency control centre will normally be sited in an area of minimum risk; an alternative centre may also need to be provided on some sites. The centre will contain maps, lists of materials on site, protective equipment and lists of key personnel for immediate and for subsequent call-out. It will serve:

- To receive information from forward control and assembly points.
- To transmit calls for assistance to external authorities.
- To call out key personnel.
- To operate a check to account for all personnel at each assembly point.
- To release information via an appointed public relations officer.

Alarms

The type of internal alarm system will depend upon the size of the works, the inter-dependence of plants, the rapidity with which escalation can occur and the existence of alarms to signal other emergencies, e.g. fires.

The on-site alarm should be selective on large works; a two-stage warning is used on some sites.[59] In some situations there may also be a selective evacuation alarm to be sounded only on the instruction of the emergency controller.

Provision of warnings to the general public requires particularly careful planning due to the changing population around a site.

Training

Detailed instructions should be issued to all those required to act to control or minimise the effects of any major incident. Short summaries serve as useful reminders when an emergency occurs. Thorough training, including realistic practical exercises, with such participation by outside services as possible, is important. All personnel should be familiar with the actions to take, escape routes, assembly areas, etc. This should be reinforced by training exercises – to increase the probability that the right actions are followed in a real emergency, i.e. to minimise irrational behaviour under stressful conditions. The latter is illustrated by the ammonia tank explosion at Potchefstroom which resulted in 18 fatalities. Analysis of the escape routes followed by personnel, shown in Fig. 15.9, reveals that several of those who died actually ran towards the zones of highest toxic concentrations.[53]

Full advice on emergency planning is given in reference 59 to serve as a basis for a plan necessarily specific to each particular site. Published case histories of major emergencies such as those at Doe Run[68] or given in reference[69] indicate the range of problems and contingencies which may arise.

Associated hazards

Because of the scale of operations it is equally necessary to consider potential hazards from:

1. Transport of raw materials to, and products from, the site by road, rail or sea, or via pipelines.[37]
2. The disposal of liquid effluents and solid wastes.

Factors to consider under (1) are discussed in Chapter 14. Hazards associated with (2) are well documented (Ch. 14) and with strict adherence to legislative controls applicable to the particular waste and location, long-term problems can be avoided. Again, however, it is the 'sudden event' which may not be foreseen, e.g. potential pollution of water supplies by spillage of effluent into a river.

In February 1982 a fire occurred at a transit warehouse storing paraquat- and diquat-based herbicides contained in drums, plastic bottles and cardboard cartons and also octylphenol in paper sacks. As a result, a

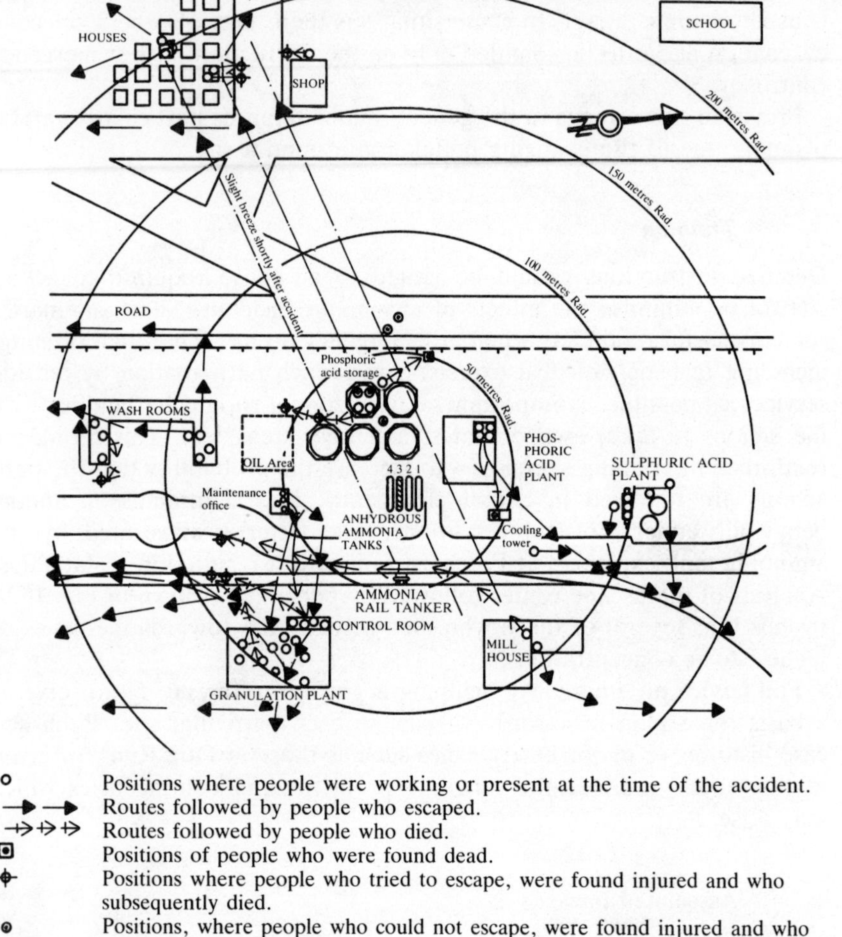

Fig. 15.9 Potchefstroom plant layout and escape routes followed by personnel

quarter of the herbicide stock was liberated; this, together with fire-fighting water, drained into a water course near housing. The River Calder became contaminated, resulting in widespread environmental pollution, and the herbicide-saturated warehouse foundations continued as a source of contamination in subsequent rain storms.[37]

In addition to factors discussed in Chapter 14, the lessons from this incident include the need for appropriate site location and construction of warehouses containing hazardous substances with good access and the availability of someone on site with chemical expertise.[37]

Fumes containing phenolic compounds with small quantities of organic bromine escaped into the atmosphere from a plant near Basle in 1986. Although air sample tests indicated no public health hazard, some residents in surrounding communities complained of eye irritation and annoyance from odours.[70]

This plant was near to where a warehouse fire, at the Sandos site in November 1986, resulted in large quantities of toxic chemicals entering the Rhine causing severe river pollution extending through five countries. This fire which involved agricultural chemicals including insecticides, burned for six hours during which time it destroyed a 6,5000 m^2 warehouse and some 800 tonnes of chemicals. About 30 tonnes, including mercury derivatives (e.g. ethoxyethyl-mercury hydroxide) were washed into the river by fire-water.

Apart from the fact that different regulations allowed a warehouse to be operated with a different standard of fire protection to that required throughout the EEC, two lessons arising from these incidents – which have been emphasised earlier – are:[71]

- When designing emergency systems, and planning for incidents, consideration must be given to the drainage and disposal of contaminated fire-water.
- Emergency response systems depend upon rapid activation (The official Rhine warning system designed to mitigate the consequences of such an incident was only activated after some time had elapsed.)

Major incidents from other sources are fortunately rare.

A massive explosion in central Los Angeles was followed by fires which burned for four days. One major store was demolished and damage was valued at millions of dollars; 200 businesses and homes were evacuated.[72] Methane gas seeping from oil fields drilled and abandoned in the 1920s and 1930s had collected in a basement or underground lot and been ignited by a stray spark.

General guidelines for those planning to deal with the consequences of major industrial emergencies are given in reference 73.

References

(*All places of publication are London unless otherwise stated*)

1. Marshall, V. C., *Chemical Engineering*, 22 Dec. 1975.
2. Health and Safety Commission, *The Control of Major Hazards*. Advisory Committee on Major Hazards, 3rd Report, 1984.
3. Great Britain: Department of Employment. *Annual Report of HM Chief Inspector of Factories, 1967*. HMSO.
4. Great Britain: Department of Employment. *Annual Report of HM Chief Inspector of Factories, 1972*. HMSO.

5. Department of Employment, *Development Involving the Use or Storage in Bulk of Hazardous Materials*, 1972, PBI/766/6. HMSO.
6. Department of Employment, *The Report of the Robens' Committee, 1972*. HMSO.
7. Carson, P. A. & Mumford, C. J., *J. Occupational Accidents*, 1978, **2**, 1.
8. Carson, P. A. and Mumford, C. J., *J. Hazardous Materials*, 1979, **3**, 149.
9. Department of Employment, *The Flixborough Disaster: Report of the Court of Inquiry*. HMSO 1975.
10. Bell, C. L., 'Flixborough and the Health and Safety at Work Act'. Paper presented at Symposium on Health in Research and Chemical Process Industries, University of Leeds, Feb. 1976. Institution of Chemical Engineers.
11. Marshall, V. C., 'Chemical conurbations – the domino effect'. Paper presented to Eurochem Conf. on Chemical Engineering in a Hostile World, Birmingham, June 1977.
12. Health and Safety Commission, *The Advisory Committee on Major Hazards, First Report, 1976*. HMSO.
13. Health and Safety Commission, *The Advisory Committee on Major Hazards, Second Report, 1979*. HMSO.
14. Bridgwater, A. V. & Mumford, C. J., *Waste Recycling and Pollution Control Handbook*. George Godwin 1979.
15. Harris, N. C., 'Analysis of chlorine accident reports'. The Chlorine Institute, 21st Plant Managers Seminar, Houston, 1978.
16. Health and Safety Executive, *An Investigation of Potential Hazards from Operations in the Canvey Island/Thurrock Area, 1978*. HMSO.
17. Robertson, 'Spacing in chemical plant design against loss by fire' in *Proceedings on Process Industry Hazards*, West Lothian, 14–16 Sept. 1976. Institution of Chemical Engineers.
18. Kletz, T. A., 'Plant layout and location – some methods for taking hazardous occurrences into account', American Instn Chemical Engineers Loss Prevention Symposium, 1–5 Apr. 1979.
19. Kletz, T. A., *Hydrocarbon Processing*, 1979, 5, 243.
20. Westbrook, G. W., 'Carriage of liquid chlorine', in C. H. Buschmann (ed.), *Proceedings of the First International Loss Prevention Symposium*, The Hague, May 1974. Elsevier Science: Oxford.
21. Marshall, V. C., 'The siting and construction of control buildings – A strategic approach', *Instn Chem. Engrs Symp. Series*, 1976, No. 47.
22. Pasquill, F., *Metereological Mag*., 1961, **90**.
23. Burgess, D. *et al.*, 14th Combustion Symposium. Combustion Institute, 1974, 289.
24. Slater, D. H., *Chem. and Ind.*, 1978, 295.
25. National Fire Protection Association, *Fireman's Magazine* (USA), Sept. 1956.
26. Campbell, J. A., *Estimating the Magnitude of Macro-hazards*, SFPE Technology Report 81–2. Society of Fire Protection Engineers.
27. Carson, P. A. & Mumford, C. J., *J. Occupational Accidents*, 1979, **2**, 85.
28. Gugan, K., *Unconfined Vapour Cloud Explosions*. Institution of Chemical Engineers in association with George Godwin 1978.

29. Briscoe, F., 'A review of current information on the causes and effects of explosions of unconfined vapour clouds', in *Canvey; an Investigation of Potential Hazards from Operations in the Canvey Island/Thurrock area*. HMSO, 1978.
30. Crawley, F. K., 'The effects of the ignition of a major fuel spillage', *Instn Chem. Engrs Symp. Series*, No. 71, 1982, 125–45.
31. Diaconicolau, G. J., Mumford, C. J. & Lihou, D. A., 'Prediction of acceptable spacing distances for plants involving an UVCE hazard', Symposium of Institute of Chemical Engineers, 12 Mar. 1980.
32. Collins, J., MSc thesis, Univ. of Aston 1978.
33. Tucker, D. M., The explosion and fire at Nypro (UK) Ltd., Flixborough on 1 June, 1974, in 'The Technical Lessons of Flixborough'. Instn of Chem. Eng., Nottingham, 16 Dec. 1975.
34. Walls, W. L., *Fire Command*, May 1979, 22–4; June 1979, 35–7.
35. *Fire International*, 1966, 12, 16; *Fire* – Special Supplement, Feb. 1966; *Petroleum Times*, 21.1.66, 132; *Paris Match*, 15.1.66.
36. Anon., *Loss Prevention Bulletin*, Institution of Chemical Engineers, Feb. 1985 (061) 33.
37. Health and Safety Commission, *The Control of Major Hazards*. Advisory Committee on Major Hazards, Third Report. HSC 1984.
38. Sykes, W. G. *et al.*, *Chem. Eng. Prog.*, 1963, **59**, 66.
39. Holmes, W. C., 'Ammonium compounds', in D. F. Othmer (ed.), *Encyclopaedia of Chemical Technology*, Vol. 2. Interscience 328.
40. National Board of Fire Underwriters, *Chem. Eng. News*, 1947, **25**, 1594.
41. Vörös. M. & Honti, Gy, *Proc. 1st International Loss Prevention Symposium*, 28–30 May 1974, 337–46.
42. MacKenzie, D., *New Scientist*, 13 Dec. 1984, 3.
43. *The Observer*, 9 Dec. 1984.
44. *Sunday Times*, 9 Dec. 1984.
45. *Bhopal Methyl Isocyanate Incident Investigation Team Report, March 1985*. Union Carbide Corporation, Danbury, Connecticut.
46. *Chemical Engineering*, 1 Apr. 1985, 9.
47. Worthy, W., *Chemical & Engineering News*, 11 Feb. 1985, 27.
48. MacKenzie, D., *New Scientist*, 11 Apr. 1985.
49. 'Bhopal, The continuing story', *Chemical & Engineering News* (Special Issue), 11 Feb. 1985, 3.
50. Baton Rouge, *Morning Advocate*, Dec. 1976.
51. Sellers, J. G., 'Quantification of toxic gas emission hazards', in *Proceedings of Symposium on Process Hazards*, Sept. 1976, West Lothian, Institution of Chemical Engineers.
52. *First Report of the Toxicity Working Party of the Institution of Chemical Engineers Advisory Panel on Major Hazards*, 1985.
53. Lonsdale, H., 'Ammonia tank failure in South Africa', *Ammonia Plant Safety*, 1975, **17**, 126.
54. American Conference of Governmental Industrial Hygienists, *Documentation of Threshold Limit Values For Substances in Work Room Air* (3rd edn) ACGIH, 1971 (3rd printing 1976).

55. *IARC Monographs on the Evaluation of the Carcinogenic Risk of Chemicals to Humans*, 1979, **19**, 73.
56. Anon 'Seveso: Causes; Prevention', *Loss Prevention Bulletin*, Instn. of Chem. Eng., Oct. 1983 (053), 27
57. Marshall, V. C., *Safety Survey*, 1977.
58. European Economic Community, *Council Directive on Major Accident Hazards of Certain Industrial Activities*, 82/501/EEC, June 24 1982.
59. Chemical Industries Association, *Recommended Procedures for Handling Major Emergencies*. Nov. 1970.
60. Diggle, W. M., *Instn Chem. Engrs Symp. Series*, No. 47, 1976.
61. Hill, H. R., Bruce, D. J. & Diggle, W. M., 'Loss Prevention and Safety Promotion in the Process Industries, 2nd Int. Symposium, Heidelberg, 1979.
62. State of Ontario, Canada, *Derailment – The Mississauga Miracle*, 1980.
63. Underwood, H. C., Jr., Sourwine, R. E. & Johnson, C. D., *Chem. Engr*, 11 Oct. 1976.
64. Webb, H. E., *Chem. Engr*, 1 Aug. 1977.
65. Kolios, E. L., 'An analysis of hazards from bulk toxic gas releases in transportation incidents'. M.Sc. thesis, Univ. of Aston, 1981.
66. Kaloutas, G. J., 'The prediction of hazards from bulk toxic gas releases. M.Sc. thesis, University of Aston, 1980.
67. King, R. & Magid, I., *Industrial Hazard and Safety Handbook 179*. Newnes-Butterworth, 1979.
68. Troyen, J. E. & Le Vine, R. Y., *Loss Prevention*, (*Chem. Eng. Prog. Tech. Manual 2*), 125, 1968.
69. Zajic, J. E. & Himmelman, W. A., *Highly Hazardous Material Spills and Emergency Planning*. Marcel Dekker, 1979.
70. Anon., *Chemical Week*, 26 Nov. 1986, 15.
71. Editorial, *Loss Prevention Bulletin*. Institution of Chemical Engineers, Dec. 1986 (072).
72. Scobie, W., *The Observer*, 31 Mar. 1985.
73. Soc. of. Ind. Emergency Services Officers, *A Guide to Emergency Planning*. Paramount Publishing 1986.

CHAPTER 16

Legislative controls

In all developed countries chemicals handling and use are controlled by law.

The two most significant sources of law are legislation and case (or common) law. While the majority of this chapter relates to UK legislative controls, the range of hazards which they seek to ensure protection against are common to all industrial activities. Therefore the value of reference to this chapter – and in particular to the measures necessary to provide a safe workplace, safe equipment and safe working procedures – is not restricted to the UK situation.

United Kingdom

The present legal system can be traced to the reign of Henry II (1154–89). Justice for the most part was administered in local courts. Local lords or County sheriffs administered law in their respective areas and decided upon cases on the basis of local custom. Henry II ensured that royal justice was open to all. In the reign of Richard II (1337–99) a system of circuit judges from the King's Bench was introduced. These judges selected the best customary rulings and applied them outside their county of origin, which resulted in a more uniform law 'common' throughout the kingdom.

In early times there were few statutes and the bulk of law was case law, 'law by precedent'. In the reign of Henry II legislation (also called constitutions, charters etc.) was generally made by the King in Council, but sometimes by a sort of Parliament which consisted of a meeting of nobles and clergy from the counties. From the Tudor period onwards Parliament became more independent and the practice of making law by statute increased. Legislation comprises Acts of Parliament plus legislation passed by lower bodies; examples include local authority by-laws and Statutory Instruments. Statute law can be used to abolish outdated common law rules. An Act of Parliament is binding on everyone within the sphere of the jurisdiction but can be repealed by the same, or a subsequent, Parliaments. Contraventions are a criminal offence and sanctions such as fines and imprisonment are intended as a punishment, a deterrent, and to reform, as opposed to generally providing compensation to injured parties. In criminal charges, brought by the police or other officials such as factory inspectors, the burden of proof rests with the prosecution to prove the case beyond all reasonable doubt.

Case law is often required to interpret legislation and civil cases are brought by injured parties to seek compensation.

Civil liability

Civil actions are proved on a balance of probabilities. The two torts most encountered in safety cases are those of negligence and breach of statutory duty. The tort of negligence has three components and to succeed in an action the plaintiff must show:

1. The existence of a duty to take care which was owed to him by the defendant.
2. Breach of such duty by the defendant (the standard of care a defendant must take is that of the average prudent man and not necessarily that of a particularly conscientious person).
3. Resulting damage to the plaintiff.

With regard to (1), and of particular relevance to chemicals, what can be foreseen depends on knowledge (i.e. what the defendant knew, or should have known, at the material time); if with such knowledge no risk can be foreseen then there is no duty.

> A workman contracted dermatitis from the dust produced by sanding a rare African hardwood [see Table 5.9]. Since the connection between the two was not generally known at the time the dermatitis was not reasonably foreseeable and the employer was therefore held not to be liable.[1]

A special duty of care arises with 'things dangerous in themselves' including explosives, poisons, acids, chemicals, and highly flammable substances. In summary, they must not be delivered to an irresponsible person (e.g. an irresponsible apprentice or juvenile worker), or to a responsible person without due warning, and they must not be left in a situation where they may be tampered with (see also Ch. 13).

Although the burden of proof in negligence normally rests with the plaintiff there is a principle called *res ipsa loquitur* (the thing speaks for itself). This applies when it is so unlikely that such an accident would have happened without negligence. However, two conditions must be satisfied:

1. The activity causing harm must be wholly under the control of the defendant.

2. The accident would not have happened if proper care had been exercised.

The presumption may be rebutted by an alternative explanation not involving negligence, by positive disproof of negligence, or by complete evidence of the facts.

Res ipsa loquitur was applicable in an explosion involving oxygen fed to a furnace, probably due to impurities in the pipe [see Ch. 4, oxygen enrichment]. Negligence was not excluded since the defendants had not examined the filters.[2]

Sometimes both parties are deemed to have been negligent and the doctrine of contributory negligence applies: liability is apportioned between plaintiff and defendant and damages may be reduced on a percentage basis according to the degree of negligence by the plaintiff.

Substantial contributory negligence, 75%, was attributed when a foundry worker failed to wear protective boots or spats against splashes of metal. He knew that spats were available free and boots at a cost.[3] Indeed on appeal there was held to be no negligence by the defendant.[4]

Damages for personal injury may be awarded under pain and suffering, loss of enjoyment of life, loss of expectation of life, loss of earnings, both actual and prospective.

So far as the employer–employee relationship is concerned, the employer has a duty to take reasonable care for the safety of his employees in the course of their employment. This includes, but is not restricted to, the provision of 'a safe place of work', the provision of 'safe plant and appliances' and the provision of 'a safe system of work'. The latter includes adequate supervision, instructions and training, etc., as outlined in Chapter 17. The employer is also responsible for the acts of his servants in the course of their employment, as where a workman injures another, e.g. by handling chemicals carelessly; this is termed vicarious liability.

Generally once the existence and extent of the duty (of care) is established the employer must take the workman as he finds him, e.g. with any predisposition to more severe effects from an accident, or prolonged exposure to chemicals, that happen to arise.

A galvaniser was engaged in lowering articles into a bath of molten metal using a crane. The articles had previously been immersed in hydrochloric acid and if this, or air bubbles, were trapped it caused molten metal to splash out of the bath [see Ch. 3]. Sheets of corrugated iron, curved at the top, were provided as temporary shelters, but when operating the crane – with his back to the bath – the man probably looked out from behind the shelter and was burned on the lower lip by molten metal. He had a pre-malignant condition in the tissues of the lips due to contact with tar over a ten-year period in the gas industry [see Ch. 5, page 210]; as a result the burn ulcerated, became cancerous and ultimately had fatal consequences.[5] The accident was held to be foreseeable, the shelter inadequate, and the employer responsible for *all* the consequences of the injury.

Where injury to a worker results from a breach of statutory duty (e.g. failure to fence a dangerous machine) he may be able to sue his employer for damages by using the breach of statutory duty to establish this duty of care; this statutory obligation is absolute.

A job entailed taking a small coke-carrying vehicle to a lift, the operator accompanying it to the first floor of a building, emptying it and returning to the ground floor by the lift. The operator had deposited one load of coke and began to move the vehicle back into the lift. He had not noticed that the lift platform had moved upwards. Both he and the vehicle plunged down the shaft. It was found as a fact that the reason why the lift platform had moved was a failure of the braking system. This had never happened before and the lift was found to be in perfect working order when examined by experts. Thus there was nothing the employers could have done to prevent the accident. The man's wife sued the company for breach of statutory duty.

It was held that the duty to maintain equipment when this is specified in legislation, is absolute. 'Maintain' means 'kept in perfect working order at all times'. The failure of a component to work properly is proof that it has not been maintained. Being an absolute duty, it makes no difference that the defect is one which the employer could not have discovered.[6]

There are numerous statutes which are designed to protect the health, and provide for the welfare and safety of employees (see below). A manufacturer's or supplier's liability for his products is discussed in Chapter 13.

In the UK the Employer's Liability (Defective Equipment) Act 1969 makes an employer responsible for defective equipment, so that an injured worker can sue for compensation. Provided the accident occurs in the course of employment, the employer becomes liable even when no negligence can be attributed to him; all that is necessary is that negligence is attributable to somebody. It is then up to the employer, if appropriate, to seek recovery from the equipment manufacturer. Useful general texts on law are given in the references.[7–13]

Finally there are two cases relevant to this text where the common law recognises 'strict liability', i.e. negligence need not be proved (although certain defences are applicable)[9]:

1. Liability for the 'escape' of a fire from the defendant's premises, other than a fire started accidentally.
2. Liability for damage caused by the escape of things brought on to his premises by the defendant and which, either inherently or because of

accumulation in quantity, are likely to cause damage if they escape.[14] This rule has been applied to 'things' likely to give off noxious gas or fumes, explosives and gas.

Clearly only the general tenor of common law liability has been illustrated and reference is recommended to specific texts.[7–13]

Safety legislation in the UK

In ancient and medieval days little was done to control accidents in the workplace. With the onset of the industrial revolution, following the introduction of the steam engine and the water frame in the mid-1700s, increasing numbers of people were exposed to a wider variety of hazards at work with little attention being paid to health, safety or welfare aspects. Eventually conditions became intolerable and legislation was deemed necessary for their control. Through the long history of industrial safety legislation a few milestones are worthy of mention.

In 1802 Sir Robert Peel introduced to Parliament the first Factory Bill which became the Health and Morals of Apprentices Act. This Act sought to impose certain standards to protect young persons employed in cotton and other mills and factories. Fundamental hygiene standards were identified. It provided for very general standards of environmental conditions such as heating, lighting, ventilation, lime-washing of looms, and general cleaning, in cotton mills and textile factories; limiting the work day for apprentices in cotton mills to 12 hours; the religious and general education of apprentices; and introduced the principle of inspection of factories by volunteers appointed by the local magistrates. Though largely ineffective the principle of government participation in industrial legislation was now established.

Through enlightened employers a new bill to control hours of work for women and children was introduced to Parliament in 1819 but was rejected by the Lords. Even so, Robert Owen by refusing to employ persons under the age of 10 years, shortening the working day, improving conditions and providing for adult and child education, applied the principles to his New Lanark Mills.

In 1833 a Royal Commission on employment of child labour was followed by the passing of the Factory Act. This included in its general provisions: retention of 12-hour day for young persons but extended the scope to cover the woollen and linen industry; educational facilities; and inauguration of the Factory Department with four inspectors with full power of entry into a factory or mill, powers to make rules, regulations and orders to implement the Act with authority to ensure enforcement.

The first provision for the fencing of machinery was introduced in 1844, resulting from the recognition of risks of injury to women and children through trapping of clothing in moving parts of machinery. The 1844 Act also resulted in doctors being appointed as certifying surgeons to investigate, and report to factory inspectors, on factory accidents.

The Ten-Hour Act of 1847 restricted the hours of work of women and young persons in factories to 58 each week.

The 1856 Act was the only piece of retrograde legislation made in connection with factory safety by restricting the 1844 Act's fencing requirements to those parts of machines with which children and women were liable to come into contact.

In 1864 and 1867 certain factories where employees were engaged in engineering processes were introduced to legislative control. Inspectors were given authority to enforce employers to control fumes and dusts by means of fans and other ventilating systems. The Workshops Regulations Act of 1867 provided protection for workers in smaller establishments. This was limited in scope but valuable advancement nevertheless. The following is indicative of conditions at that time:

> The saltcake men (in the first stage of the Leblanc process for soda manufacture) were the roughest men in the works; their job was to charge, rake and empty the saltcake furnaces (in which salt was decomposed with sulphuric acid), in the course of which they were exposed to sulphuric acid gas. A saltcake man could be recognised by his toothless mouth, and the 'decomposing house' was characterised by the heaps of bread crusts which the men could not eat.[15]

By 1875 the law covering factories and workshops was set out in a patchwork of statutes and regulations. A common characteristic of legislation thus far was that the legal controls were created to meet specific needs that had become apparent, often as a result of serious accidents. No general view of employee safety had been taken by Parliament. In 1876 a Royal Commission set up to make a considered analysis reported, and the first piece of legislation designed to create comprehensive control of factory activities became law as the Factory and Workshop Act 1878.

The Office of Chief Inspector was created in 1878 and the first medical inspector was appointed in 1898. The Employer's Liability Act was passed in 1880 to give employers additional incentive to fence machinery, even though the Act contained no safety legislation as such.

In the last quarter of the nineteenth century further legislation was promulgated and the former chaos with a variety of laws and regulations was restored. A fresh endeavour to provide comprehensive control was made by the enactment of the Factories and Workshops Act of 1901; this remained the principal law until repealed by the Factories Act 1937. Again, particular happenings and immediate needs led to supplementary laws and regulations. The Factories Act 1937 repealed the existing factories and workshops laws, but maintained statutory regulations made under them. It exhibited a greater concern with the amenities of factory life and made a considerable number of detailed requirements for the safety, health and welfare of employees.

In the 1950s a review was undertaken of law covering the safety and management of mines and quarries. This led to the passage of the Mines and Quarries Act in 1954 which, for these premises, remains the principal Act today. The 1937 Act was itself updated in 1948 and 1959 and repealed by the Factories Act of 1961. This major law, subject to a number of amendments, remains in force today. Its main provisions are given in Table 16.1. The Act is applicable only to a 'factory' as defined and the duties rest upon the 'occupier'.

Like all the principal laws passed previously the 1961 Act contained an open-ended power to make regulations that would extend legal controls and govern dangerous processes, plants and substances.

The Offices Shops and Railways Premises Act 1963 structurally parallels the Factories Act of 1961.

Table 16.1 Some relevant sections of the Factories Act 1961

Section	Provision	Section	Provision
1	Cleanliness and regular painting or whitewashing of work areas	22	Provision, maintenance and testing of hoists and lifts.
2	Prevention of overcrowding with minimum cubic space per employee specified	26	Provision of chains, ropes and lifting tackle.
3	Maintenance of a reasonable temperature in the workplace	28	Provision and maintenance of floors passage ways and stairs.
4	Adequate ventilation to prevent accumulation of dust or fumes, etc., in the work area	29	Provision of safe access to place of work (including working at heights).
5	Adequacy of lighting and the cleaning of skylights	30	A system of work in confined space in which dangerous fumes are liable to be present.
6	Effective drainage	31	Precautions with respect to explosive or flammable dust, gas, vapour or substance.
7	Provision, maintenance and cleaning sanitary conveniences	32–39	Use, maintenance, examination, etc., of steam boilers.
12–16	Provision and maintenance of machinery guarding	40–52	Means of escape in case of fire.
18	Protection against dangerous substances	58	Washing facilities.
20	Cleaning of machinery by women and young persons	63	Removal of dust and fumes.
21	Operation of machinery by young persons	72	Manhandling of excessive weights

In summary, detailed Acts of Parliament had sought for 175 years to protect health and safety and to impose precise safety standards in respect of particular kinds of factories, other workplaces and in relation to particular machines, processes and chemicals. However, these attempts were often ineffective with the system showing deficiencies as new hazards were identified.

In 1970 the Robens Committee was appointed with wide terms of reference. Included in their prime recommendations were that the piecemeal development of special laws should be supplemented by general duties of an all-embracing nature so as to create duties even where no specific regulations or standards exist. Safety legislation should be extended to cover all work places and not just factories, and protection should be afforded to employees, visitors and the general public. Management organisations should ensure the involvement in safety of senior executives, line managers and employees.

In 1974 the Health and Safety at Work etc. Act (HASAWA) was published. This enabling Act, which is based heavily on the findings of the Robens Committee,[16] is divided into four parts. The objectives of Part I are to secure the health, safety and welfare of persons at work; to protect persons other than persons at work against risks to health or safety arising out of, or in connection with, the activities of persons at work; to control the keeping and use of explosive or highly flammable or otherwise dangerous substances and generally to prevent the unlawful acquisition, possession and use of such substances; and to control the emission into the atmosphere of noxious or offensive substances from prescribed premises. This part also deals with the making of regulations, etc, and their enforcement. Part II concentrates on the Employment Advisory Service, while Part III modifies law relating to building regulations under ss.61 and 62 of the Public Health Act of 1936. Part IV contains miscellaneous and general provisions and includes sweeping power to repeal or modify, by regulation, any provision of the 1974 Act or any other Act passed before or in the same section as the 1974 Act if it appears expedient to do so in consequence of, or in connection with, any provision made by or under Part I.

Duties of employers, plus those of employees and manufacturers or suppliers under the HASAWA are summarised in Table 16.2.

For those sections of the Act of greatest importance to employers with regard to in-factory operations the implications for practices, procedures and equipment are exemplified, but not circumscribed, by the series of questions, i.e. a check-list reproduced below.[17] Many of these items are also the subject of specific items of UK legislation. Where relevant to the handling of 'chemicals' these questions are dealt with in detail in other parts of the text.

Table 16.2 Duties of employers, employees and manufacturers under HASAWA

Section	Duty
Employers	
2 (3)	To prepare and up-date as often as necessary a written statement of his general policy with respect to health and safety at work of his employees. This must also describe the organisation and arrangements for implementing the policy, and it must be brought to the attention of all employees.[18]
2 (1)	To ensure the health, safety and welfare of all employees.
2 (2) (a)	To provide and maintain safe plant and systems of work.
2 (2) (b)	To ensure the safety and absence of risk in connection with the use, handling, storage and transportation of articles and substances.
2 (2) (c)	To provide the necessary information, instruction, training, and supervision for the health and safety of employees.
2 (2) (d)	To ensure the place of work is maintained in a safe condition with safe access and egress.
2 (2) (e)	To provide and maintain a safe working environment which is without risk to health and adequate for their welfare at work.
2 (6)	To consult with safety representatives.[19,20]
2 (7)	To set up a safety committee.
3	To conduct his undertaking so as not to expose the general public to risks.
4	To prevent those, who are on his premises but who are not his employees (e.g. visitors, contractors), from being exposed to risks in their health and safety.
5	To use best practicable means for preventing noxious or offensive dusts or fumes from being exhausted into the atmosphere, or that such are harmless.
Employees	
7 (a)	To take reasonable care of his own safety and health and of those who may be affected by his acts or omissions.
	A shift manager in charge of a conveyor belt instructed two operators to clean the belt of oil contamination. This they did after climbing over a wire-mesh guard while the conveyor was running. The shift manager visited the line after the two men climbed back out and he told them to dry the belt using an air-line. In doing so one operator slipped and was driven into an inrunning nip. The manager was prosecuted under s. 7 of the HASAWA for not having taken reasonable care for the safety of other persons and was fined £150.[21]
(b)	To co-operate with the employer in discharging his duties.
	A 29-year-old employee was fined £50 and made to pay costs for consistently refusing to wear ear defenders in a noisy mill. This was the first case brought by HSE against a worker for failing to co-operate with his employer.[22]
8	Not to interfere or misuse anything provided for health safety or welfare reasons.
Designers, manufacturing importers or suppliers	
6	To ensure their products are designed and installed so as to be safe when used and to test and provide necessary data on the product for safe usage.

The duties imposed under s.2. are:

2.2(a) 'The provision and maintenance of plant and systems of work, that are, so far as is reasonably practicable, safe and without risks to health.'

- Is all plant up to the necessary standards with respect to safety and risk to health?
- When new plant is installed is latest good practice taken into account?
- Is there provision by regular inspection, examination and, where necessary, testing to ensure that plant and its safety devices have not deteriorated?
- In such csses would the examination, etc., be more suitably assigned to specialists?
- Do all the systems of work provide adequately for safety?
- Are they properly enforced?
- Has a thorough examination been made of all operations undertaken in the workplace (especially those carried out only infrequently) to minimise danger or risk to health?
- What attention has been paid to the safety of cleaning, repair and maintenance operations?
- Should special safety systems such as 'permits to work' be considered?
- Is the work environment regularly monitored to ensure that, where known toxic contaminants are present, the protection conforms to current hygiene standards?
- Is monitoring also carried out to check the adequacy of control measures?
- Have arrangements been made for regular inspection of all equipment and appliances used for safety and health (for example, dust and fume extraction equipment, guards, safe arrangements for access and monitoring and testing appliances)?
- What personal protective equipment is required (e.g. protective boots, helmets, goggles, respirators, ear protectors)? Has it been issued? Are adequate arrangements made for its storage, maintenance, cleaning and renewal? Have those who need it been trained in its use?
- Have arrangements been made for regular maintenance and testing of electrical installations and equipment?
- Have emergency procedures and contingency plans been formulated, to cover, for example escapes or spillages of toxic or dangerous materials, fire, escapes of gases, etc., and also emergencies due to hazards arising in adjacent premises, or sabotage?

2.2(b) 'Arrangements for ensuring, so far as is reasonably practicable, safety and absence of risks to health in connection with the use, handling, storage and transport of articles and substances.'

- Have the methods of manufacture been examined carefully of every substance likely to give rise to risk, to ensure that every necessary precaution has been taken?

- Has an audit been carried out to list every substance used at work to identify the specific health and safety risks to which any substance may give rise?
- Are the containers and handling devices suitable or should expert advice be sought?
- Are the containers of all substances correctly labelled?
- Has particular attention been given to the manipulation of molten metal?
- Can storage and transport arrangements be improved to make them safer and reduce health risks?
- If there is mechanical transport on the premises have the operating procedures been critically appraised? Are the transport rules adequate and are they properly enforced?
- Are structural modifications to plant, buildings or operating areas necessary to achieve safety from the use of transport?
- Have proper procedures been established for assessing new proposals for handling materials or using transport?
- Could safety be increased or working conditions improved by substitution of less toxic or less dangerous substances or by improvements in enclosure or by institution of remote handling methods?
- Has particular attention been given to the safety of systems of work and of the handling of substances undertaken outside the normal production processes?
- Has special attention been given to the precautions in the carriage and transport of dangerous materials, such as those of high toxicity, or with explosive properties or radiation emitters?
- Should the safety of handling articles in use be again reviewed?
- Is further training in the use of handtools, machines and mobile plant necessary?
- Is all equipment safely stored?
- Are any processes being undertaken using unsuitable machines or equipment?

2.2(c) 'The provision of such information, instruction, training and supervision as is necessary to ensure, so far as is reasonably practicable, the health and safety at work of his employees.' (This duty can extend to third parties, e.g. contractors.)

During work on a ship a fire was caused after a welder struck an arc in an oxygen-enriched atmosphere which had been building up all night because an oxygen hose had been left connected by sub-contractors installing a chill water supply system. Eight men died after being trapped below deck. Charges were brought under ss.2 and 3 of HASAWA.

While the shipbuilders were aware of the dangers of oxygen enrichment and had published a manual on the subject they had not distributed it to the sub-contractor. In the Appeal Court it was held that 'if the provision of a safe system of work for the benefit of his own employees involves information and instruction as to the potential dangers being given to persons other than his own employees, then an employer is under a duty to provide such a system'.

The shipbuilders were fined £3,000 and the sub-contractor £15,000.[23]

- What can be done to bring within the procedures for planning for health and safety the co-operation of the workpeople themselves?
- Is adequate information and guidance given to all employees on the hazards of the work activities and the methods for avoiding them and on any other matters affecting health or safety?
- Has every worker exposed to a health hazard been informed of the risks and the precautions? Have arrangements been made for him to be told of the results of any relevant monitoring carried out?
- Is any further information necessary with respect to special equipment or substances or processes?
- Are there arrangements for a proper flow of information to employees and also for rapid and unhindered communication on safety and health matters from employees to management?
- Has information been provided to employees on legal requirements?
- Has adequate technical information been provided at suitable levels?
- Are arrangements such that advisory literature (including any new publications) on health and safety is available for all whom it concerns?
- Have adequate arrangements been made for training in safe practices, in procedures for avoiding risks to health, and in the use of equipment for safety? Does such training take into account the levels at which it must be approached and the capabilities of the recipients? Is there proper training for supervisory staff? Is the effectiveness of the training and its retention by the trainees monitored regularly? Are there arrangements for retraining of those in post as well as for training new employees?
- Has action been taken to foster a high standard of safety awareness in all employees?
- Has a high standard of skill been achieved in all supervisors in their role in the management of safe procedures and systems of work?
- Has a system been instituted for identifying the particular training needs within the organisation and of any special needs of individual employees?
- Are there processes with special hazards which have particular training needs?
- Are safe methods of working receiving full emphasis in all training given?

2.2(d) 'So far as is reasonably practicable as regards any place of work under the employer's control, the maintenance of it in a condition that is safe and without risks to health and the provision and maintenance of means of access to and egress from it that are safe and without such risks.'

- Has consideration been given to the safety of all places of work and the means of access to, and egress from, them?
- Do all buildings comply adequately with safety standards? Are they so maintained? Are professional surveys required for any buildings?
- Has adequate consideration been given to special safety requirements of all buildings, such as fire escapes, fixture points for window cleaners' harnesses, and preplanned arrangements for building maintenance?
- Are there specially awkward places within the buildings which must be reached at particular times, for example for observation or for control of processes of plant? If so can the means of access be improved?
- Do persons have to enter plant for maintenance or cleaning, where there may be special hazards and, if so, have proper procedures and precautions been taken?
- Do good housekeeping and cleanliness receive the attention due to them?
- Are fire exits identified and are they checked to ensure that they are maintained free from obstructions?
- Is fire-fighting equipment provided and is it maintained? Are persons adequately trained in its use?
- Is there a fire certificate for the premises and are its conditions being observed?
- Has the fire alarm been tested? Are fire drills held?

2.2(e) 'The provision and maintenance of a working environment for his employees that is, so far as is reasonably practicable, safe, without risks to health, and adequate as regards facilities and arrangements for their welfare at work.'

- Are there any problems in the premises in connection with heating, lighting, ventilation or noise?
- Are the welfare arrangements – seating, washing accommodation, clothing accommodation and lavatories satisfactory?
- Are first-aid arrangements satisfactory?
- Are there any problems arising from dangerous waste?
- Are there any potential environmental hazards? Have the risks been assessed eg. by atmospheric analysis?

2.3 'Except in such cases as may be prescribed, it shall be the duty of every employer to prepare and, as often as may be appropriate, revise a written statement of his general policy with respect to the health and safety at work of his employees and the organisation and arrangements for the time being in force for carrying out that policy, and to bring the statement and any revision of it to the notice of all his employees.'

- Has the written statement of safety policy, organisation and arrangements been prepared in accordance with this section?
- Has a system of supervision been organised which will ensure that sustained attention is given to all duties placed upon the employer by this legislation?
- Have measures been taken to encourage employees at all levels to be aware of and work in accordance with the safety policy?
- Have the relative responsibilities for safety and all that is involved with it been assigned clearly at all levels of the organisation and particularly within the management structure?
- Has there been full acceptance of responsibility for health and safety by all managers and do they regard this responsibility as no less important than their responsibilities for production?
- If the organisation is a large one, have safety responsibilities been spread throughout the intermediate management structure?

The duties imposed under **s.3**. are:

'It shall be the duty of every employer to conduct his undertaking in such a way as to ensure, so far as is reasonably practicable, that persons not in his employment who may be affected thereby are not thereby exposed to risks to their health or safety.'

'In such cases as may be prescribed, it shall be the duty of every employer . . ., in the prescribed circumstances and in the prescribed manner, to give to persons (not being his employees) who may be affected by the way in which he conducts his undertaking the prescribed information about such aspects of the way in which he conducts his undertaking as might effect their health or safety.'

- Could any activities undertaken within your organisation cause danger of injury or to health to members of the public, or other workpeople not employed by you?
- Are toxic gases or dust emitted into the atmosphere from your premises? (s.5 may also be relevant).
- Is any machinery or plant adequately protected against risks to the public or, for example, to children, even when trespassing?
- If you use or produce radiations could they cause danger at any place to which the public or other employed persons have access?
- Could there be danger to the public from any conveyance of toxic or otherwise dangerous substances?

With regard to other activities, **s.6** imposes duties on designers, importers, manufacturers, suppliers and erectors and installers. These duties include:

(a) 'to ensure, so far as is reasonably practicable' that articles and substances for use at work are 'safe and without risks to health when properly used'.

(b) 'to carry out, or arrange for the carrying out of, such testing and examination as may be necessary for the performance of the duty imposed on him by the preceding paragraph'.
(c) to ensure 'that there will be available in connection with the use of the article at work adequate information about the use for which it is designed and has been tested and about any conditions necessary to ensure safety and absence of risks to health when put to that use'.
(d) to ensure 'that there will be available in connection with the use of the substance at work adequate information about the results of any relevant tests which have been carried out', and 'any conditions necessary to ensure that it will be safe and without risks to health when properly used'.
(e) 'to carry out or arrange for the carrying out of any necessary research' in connection with articles and substances 'with a view to the discovery and, so far as is reasonably practicable, the elimination or minimisation of any risks to health or safety to which' they may give rise.
(f) 'the duty of any person who erects or installs any article for use at work in any premises where that article is to be used by persons at work to ensure, so far as is reasonably practicable, that nothing about the way in which it is erected or installed makes it unsafe or a risk to health when properly used'.

Some indication of the types of measures necessary in order to comply with s.6 is given by the following check-list.[24]

- Have steps been taken that it would be reasonably practicable to take to ensure that the article or substance will be
  (a) safe when properly used
  (b) without risk to health when properly used?
- Have adequate tests and examination been carried out to ensure compliance with the above?
- Has adequate research been carried out by the designer or manufacturer?
- Is there any evidence from previous use of similar articles or substances that particular hazards or potentially dangerous circumstances may arise?
- Are there any particular hazards or potentially dangerous circumstances which may arise to which the user's attention should be drawn?
- What precautions must be taken to ensure the safe use of the article during: (a) setting; (b) operating; (c) adjusting; (d) cleaning; (e) maintenance; (f) other?
- What precautions must be taken to ensure safe use having regard to the inherent properties of the substance?

- Is enough information available about the article or substance to enable it to be safely used at work?
- Have practical methods of securing that this information is available been considered, e.g. appropriate literature, such as brochures, pamphlets, operating manuals, data sheets, etc.; labels, markings, any other?
- Are there any special conditions the user should be advised to observe?
- Have steps that it would be reasonably practicable to take to ensure that the article has been correctly erected or installed been taken, (including carrying out any tests and examinations), so that it is
  (a) safe
  (b) without risks to health?
- If the customer is to provide a written undertaking to take specified steps, is it adequate and does it show the specified steps to be taken to ensure that the article will be safe and without risks to health when properly used?
- Have any necessary steps been taken to comply with other legal provisions which lay duties on designers, manufacturers, importers and suppliers?
- Has consideration been given to other legal provisions with which the user may have to comply?
- Does the article or substance conform to existing recommendations for health and safety, i.e.
  (a) Advisory literature produced by the Health and Safety Commission and Executive (see Ch. 18)
  (b) Reports of Industry Advisory Committees
  (c) Reports of Joint Standing or Joint Advisory Committees
  (d) British and International Standards
  (e) Courts of Inquiry
  (f) Any other?

A man was spray painting the inside of a vessel using a twin-pack epoxy resin base system. The paint was obtained from a small paint manufacturer on the advice of the agent for the vessel's original manufacturer. It was supplied with no information on technical specification, although the label indicated the contents had a flash-point below 32 °C. The agent was unaware of any hazards as were the purchasers, who made no special enquiries. Spraying continued over a number of days during which time the operator's health progressively deteriorated and he eventually suffered a severe asthmatic attack. Six weeks after the incident he was still suffering from a 50% reduction in lung function and appeared to be permanently sensitised to epoxy systems. Spraying was allegedly performed under appalling conditions with no fume control, and no proper respiratory protection or other protective equipment. Informations were laid against the man's employer under s.63 of the Factories Act and under s.2(1) of the HASAWA. An information was also laid on the supplier under s.6(4) of the HASAWA for failing to provide information.

The employer was fined £300 for the s.63 offence and £200 for the s.2(1) case. The paint suppliers were fined £150 with £30 costs.[25]

The HASAWA resulted in the establishment of the Health and Safety Commission and the Health and Safety Executive. The former under its chairmanship consists of representatives from both sides of industry and local authorities. The functions of the HSC, which include policy formulation, are described under s.11 of the Act. The HSE is responsible for implementation of policy.

Under the Act, Regulations can be drafted. This usually involves extensive discussions and often results in the proposals being issued as a consultative document for comment by all interested parties.

Under the Act, documents such as British Standard Specifications can be approved. Approved Codes of Practice can be prepared by the HSC under s.16. These documents are supplemented by Guidance Notes issued by the HSE. Enforcement of the Act is described under s.18. An inspector has wide powers. If he is of the opinion that a breach has, or may, occur under s.21 of the Act he may serve an Improvement Notice on the employer or employee. Where immediate risk of serious personal injury exists under s.22 he may serve a Prohibition Notice requiring immediate cessation of the activity. Appeals against notices may be made through tribunals.

Under s.34, if an inspector decides to bring about a prosecution, he has six months to do so. If cases are heard summarily fines up to £1,000 can result upon conviction. On indictment penalties include imprisonment and unlimited fines.

Though responsibility for an offence usually rests with the employer or employee, a director, manager, company secretary or other officer can under certain circumstances also be liable as described in s.37.

A workman employed in repainting the bridge over the river Clyde in Glasgow fell to his death. The Council were prosecuted for a breach of a number of safety provisions relating to the lack of a safe system of work, and failure to notify the local inspector that the work was being done.

The Director of Roads was held personally liable for breach of safety laws resulting in the fatality, and was prosecuted under s.37 of HASAWA. His department had failed to prepare and carry out a safety policy. This neglect of duty led ultimately to the breaches of safety provision resulting in the employee's death. He was fined £125.[26]

When a firm's German factory closed, the multi-spindle drilling machines and horizontal milling machines were transferred to the company's Clydebank factory in 1976. The machines were installed between May and

July of that year but were still operating without guards the following December. One of the company's UK directors was prosecuted under s.37 of HASAWA.[27]

Many of the duties in the Act are qualified by the term 'so far as is reasonably practicable'. Whereas 'practicable' imposes a strict standard of what is feasible 'reasonably practicable' is a narrower term than physically possible. It implies that computation must be made in which the quantum of risk is placed on one scale and the sacrifice involved in measures necessary for averting the risk (whether money, time, or trouble) is placed on the other.

The sweeping obligation to provide and maintain 'plant and systems of work that are, so far as is reasonably practicable, safe and without risk' extends the duty given in rules under the Factories Act:

1. To those machines and workplaces not, previously covered.
2. To those situations whereby employers' adherence to regulations is deemed as failing to take all reasonably practicable steps or if the regulations are outdated or deficient.

There are hundreds of other Acts and Regulations relating to occupational health and safety and product liability as illustrated in Appendices 16.1 and 16.2. Each set of Regulations deals specifically with the precautions, etc., required in the factories, situations and processes as strictly defined. For example, the stringent requirements of the Asbestos Regulations 1969 emcompass all industrial processes involving asbestos or any article composed wholly or partly of asbestos, except a process in connection with which asbestos dust cannot be given off. The measures included are summarised in Table 16.3 but, as with all statutory legislation, reference must be made to the full text.[28]

Two workers started to remove an old boiler, unaware that the lagging was 60% asbestos. A factory inspector on a routine visit found them in the process. The asbestos level in the boiler was up to 100 times the maximum acceptable level. The men's employers pleaded guilty to three charges under the Factories Act 1961 for failing to provide the men with approved protective breathing masks, and clothing, and failing to give 28 days' notice to the Factory Inspectorate of dismantling operations. The firm was fined £500.[29]

The legislation applicable to 'chemicals' handling, etc., covers a broad spectrum. For example, in any chemical works, as defined, there is a requirement, under Regulation 10[30]:

In all places where strong acids or dangerous corrosive liquids are used:

(a) There shall be provided, for use in an emergency:

(i) Adequate and readily accessible means of drenching with cold water persons, and the clothing of persons, who have become splashed with such liquids.

Table 16.3 Measures encompassed by the Asbestos Regulations 1969

Regulation	Scope
1–6	Application, interpretation and control.
7	Exhaust ventilation to be provided, maintained and used. Provision for inspection and testing.
8	Provision of approved respiratory equipment and protective clothing (where the requirements of 7 are impracticable).
9	Cleanliness of machinery, apparatus . . . floors, ledges . . ., etc.
10	Cleaning by suitable vacuum cleaning equipment or other suitable method such that 'asbestos dust neither escapes nor is discharged into the air of any workplace'.
11	Cleaning provisions (e.g. protective clothing and respiratory equipment) where 10 is impracticable.
12	Maintenance of vacuum cleaning equipment.
13	Requirements for construction of buildings to be used for scheduled processes involving, e.g. asbestos manipulation and manufacturing processes.
14	Cleanliness of accommodation for clothing.
15	Storage of loose asbestos or asbestos waste in suitable closed receptacles.
16	Distribution of loose asbestos or asbestos waste from/into a factory in suitable closed receptacles or in a totally enclosed system.
17	Marking of crocidolite (blue asbestos) receptacles.
18	Accommodation, etc., for protective clothing and respiratory protective equipment.
19	Cleaning of protective clothing either within the factory or involving dispatch in sealed, marked containers.
20	Restrictions on the employment of young persons ($<$ 18 years of age).

(ii) A sufficient number of eye-wash bottles, filled with distilled water or other suitable liquid, kept in boxes or cupboards conveniently situated and clearly indicated by a distinctive sign which shall be visible at all times.

A summary of the main provisions of the Chemical Works Regulations is given in Table 16.4.

In order to reach an electrical junction box an electrician placed a plank over an open-topped vat containing hot caustic soda. As he stood on the plank it swivelled round causing him to overbalance and he fell into one end of the vat. His legs were immersed in the hot caustic and he required extensive hospitalisation.

His employers were found guilty of two offences under the 1922 Chemical Works Regulations. The first was for allowing a plank of less than 18 inches (46 cm) width to be placed across the vat containing a dangerous substance, and the second for not securely fencing the vat. They were fined £500 on each account and ordered to pay costs.[31]

Table 16.4 Relevant provisions of the Chemical Works Regulations 1922

Section	Scope
1	Fencing of pots, vessels, containers, etc.
2	Ventilation
3	Dust containment
4	Lighting
5	Pressure-relief valves
6	Breathing apparatus, lifebelts
7,8	Entry into confined spaces
10	Emergency showers, eye-wash facilities
12	Ambulance room
14	Training
17,31	Employees' responsibilities
19,20	Entry into vessels
21	Nitro and amido processes
22	Grinding and crushing of caustic
23	Chlorate handling
24	Employment of persons over 18 years
25	Personal protection
26,27,28	Cloakroom, mess room and lavatory provisions
30	Health register, medical examinations

The Regulations contain other common-sense measures, albeit not now catering entirely for technological advances. The Carcinogenic Substances Regulations 1967[32] control or prohibit entirely the use of a limited number of chemicals (see Ch. 5). Further reference to legislation is made earlier for flammable liquids and gases (Ch. 4), for marketing and transportation (Ch. 13), for electricity (Ch. 2) and for large-scale process hazards (Ch. 15).

Much of this legislation is inter-dependent. For example the provision of eye protection may be essential in complying with s.2(2)(a) of the HASAWA in which case the Protection of Eyes Regulations 1974 must be consulted. The main requirements of these Regulations are given in Table 2.5.

Similarly The Safety Signs Regulations, 1980 should be consulted. These Regulations meet an EEC requirement to harmonise safety signs within the Community. Thus, new signs must conform to BS 5378 and existing signs must be replaced by January 1986 with signs complying with the BS. (The Regulations do not apply to the labelling of packages or containers or vessels; these are discussed in Ch. 13). In brief, signs must be designed as follows:

Prohibition – white background with circular band and cross bar in red. The symbol must be black and placed centrally. At least 35% of the sign area will be red.

Warning – yellow background with black triangular band. Symbols in black should be located centrally and 50% of the sign area will be yellow.

Safe condition – white symbol on a green background so that at least 50% of sign area is green. Signs should be square or oblong.

Mandatory – circular signs with white symbols on a blue background such that at least 50% of sign area is blue.

Expert advice is often required in identification and interpretation of relevant laws. Though some regulations have been repealed they are still used as an illustration of minimum standards.

Some recent developments

Two recent developments on legislation relevant to exposure to hazardous chemicals include the introduction of the Poisonous Substances in Agriculture Regulations 1984 and publication of draft Regulations and draft approved Code of Practice on the Control of Substances Hazardous to Health.

The Poisonous Substances in Agriculture Regulations 1984

These list scheduled operations involving poisonous substances which can only be undertaken by persons over the age of 18. Employees on scheduled operations should not work for more than 7 hours in any day and 40 hours in any 7 consecutive days or 80 hours in 21 consecutive days. They indicate specific duties of employers. Key examples include providing and maintaining personal protection (see Ch. 6, page 294), providing washing facilities, drinking water, ensuring equipment is free of contamination and that vessels containing specified substances are securely closed, training of employees, notification of cases of sickness and absence, and the keeping of a register. The latter must include details of all employees carrying out scheduled operations, their hours of work with specified substances and any matters requiring notification of inspectors or relating to exemption certificates.

Employees obligations require them to co-operate with the employer in discharging of their duties particularly with respect to scheduled operations.

The control of substances hazardous to health

Existing Regulations intended to protect workpeople from health risks resulting from exposure to substances are usually limited in scope to a particular substance or group of substances or to a particular process or processes as indicated by the UK Regulations shown in Table 16.5. They often prescribe in detail the methods by which control is to be achieved which vary according to the particular substance or process regulated.

Table 16.5 The Protection of Eyes Regulations 1974

The main requirements of these Regulations are:

- Employees shall be provided with eye protectors or a shield or a fixed shield, according to the nature of the process, if employed on a specified process.
- There are 35 different processes listed in Schedule 1 of the Regulations and, as with all statutory legislation, when in doubt reference should be made to the original Statute. However, the following processes are clearly of relevance here:
  - The pouring or skimming of molten metal in foundries.
  - Work at a molten salt bath when the molten salt surface is exposed.
  - The operation, maintenance, dismantling or demolition of plant or any part of plant, being plant or part of plant which contains or has contained acids, alkalis, dangerous corrosive substances, whether liquid or solid, or other substances which are similarly injurious to the eyes, and which has not been so prepared (by isolation, reduction of pressure, emptying or otherwise), treated or designed and constructed as to prevent any reasonable foreseeable risk or injury to the eyes of any person engaged in such work from any of the said contents.
  - The handling in open vessels or manipulation of acids, alkalis, dangerous corrosive materials, whether liquid or solid, and other substances which are similarly injurious to the eyes, where in any of the foregoing cases there is a reasonably foreseeable risk of injury to the eyes of any person engaged in any such work from drops splashed or particles thrown off.
  - Work at a furnace containing molten metal, and the pouring or skimming of molten metal in places other than foundries, where there is a reasonably foreseeable risk of injury to the eyes of any person engaged in any such work from molten metal.
  - Processes in foundries where there is a reasonably foreseeable risk of injury to the eyes of any person engaged in any such work from hot sand thrown off.
  - Welding of metals
  - Spraying
- Eye protectors shall be given into the possession of the workers, and sufficient eye protectors should be provided for use by occasional workers.
  - The employer shall replace lost, defective or unsuitable eye protectors, and shall keep available a sufficient number to enable him to do this.
  - Eye protectors and shields must be suitable for the person for whose use they are provided.
  - Eye protectors and shields shall be made in conformity with an appropriate and approved specification and shall be marked to indicate the use for which they are intended.
  - Persons provided with eye protectors and shields shall use them while engaged in a specified process or in any place where there is a reasonably foreseeable risk of eye injury.
  - The user shall take reasonable care of eye protectors and shields and report any loss or defect.

Moreover, most Regulations apply only to work activities which take place in factories.

The HASAWA on the other hand covers all work activities and substances hazardous to health but the requirements are so general that the obligations on employers and employees are not always clearly understood. Thus it has been argued that the existing legislative framework is unnecessarily complex, inhibits the introduction of new technology to control the risks, is inadequate to allow ratification of the ILO Convention on Carcinogenic Substances and Agents and does not provide a suitable infrastructure to implement EEC Directives dealing with the protection of workers against the risks of substances hazardous to health, except by way of regulation on a substance by substance basis. Further, as mentioned in Chapter 1, it is known that ill health is caused or exacerbated by occupational exposure to toxic substances, albeit of unknown magnitude. For these reasons the HSC in the UK has published a consultative document which contains proposals for:

- Comprehensive Regulations on the control of substances hazardous to health.
- An Approved Code of Practice for the control of substances hazardous to health.
- An Approved Code of Practice on the control of carcinogenic substances.

The Codes of Practice provide practical guidance on the requirements of the proposed Regulations. The Regulations will apply, with a few exceptions, to all chemicals defined by s.53 of the HASAWA which are hazardous to health by virtue of being very toxic, toxic, harmful, corrosive or irritant, and to dusts of any kind which are present in substantial concentrations. There will be a legal requirement for employers to:

1. Undertake an assessment before starting any work which may involve exposure to a substance hazardous to health. This is the key to compliance with all remaining regulations.
2. Control exposure through selection, use and maintenance of appropriate measures with regard to materials, plant, and processes. Personal protection will only be permitted as a last line of defence.
3. Provide general health surveillances of employees exposed to hazardous substances.
4. Monitor the workplace to determine employee exposure.
5. Provide employees with information (including results of risk assessments, environmental monitoring, health records plus suppliers data sheets, etc.), instruction and training.
6. Maintain records relating to risk assessments, health surveillance, environmental monitoring, and maintenance of control measures.

(Employees have duties to co-operate with employers.)

In-depth discussion of these detailed documents is superfluous since the details are likely to change after the end of the consultation period. The principles and regulations proposed are widely accepted as good occupational health and hygiene practice and the debate surrounds the scope (and hence cost of compliance) and timing of the regulations, and certain definitions.

Table 16.6 summarises the proposed main duties of employers.

Table 16.6 Employers' duties under proposed Control of Substances Hazardous to Health Regulations

Duty of employer relating to:	Employees	Other persons at the premises	Other persons not on premises
Assessment (Regulation 6)	✓	SFRP	SFRP
Control (Regulation 7)	✓	SFRP	SFRP
Use of control measures Maintenance of control measures (Regulations 8 and 9)	✓	SFRP	SFRP
Monitoring (Regulation 10)	✓	SFRP	X
Health surveillance (Regulation 11)	✓	X	X
Information, etc. (Regulation 12)	✓	SFRP	X
Fumigation (Regulation 13)	✓	✓	✓

✓ = absolute requirement.
X = no requirement.
SFRP = So far as is reasonably practicable.

Injury benefits

Any employee injured at work in the UK may be able to claim for benefit under the National Insurance (Industrial Injuries) Act 1946. This is quite apart from his/her right to sue the employer under common law for damages.

The Act provides for insurance against loss of earning power in respect of personal injury caused by accident arising out of, and in the course of, employment; or a prescribed disease or personal injury due to the nature of the person's employment. The periods covered with regard to accidents include general periods at work but not time spent travelling to and from work (except for representatives or on a special works bus); accidents arising incidentally at work, e.g. while visiting the canteen or in the washroom are covered if the work involved was done for the purpose of, or in connection with, the employer's business. Most accidents which occur in an emergency would be covered, the emphasis being on the need to save lives and property.

Prescribed diseases are listed (see Table 16.7) in the National Insurance (Industrial Injuries) (Prescribed Diseases) Regulations; the claimant must

Table 16.7 Prescribed occupational diseases in the UK

Prescribed disease or injury	Nature of occupation
A. ***Conditions due to physical agents***	Any occupation involving:
1 Inflammation, ulceration or malignant disease of the skin or subcutaneous tissues or of the bones, or blood dyscrasia, or cataract, due to electromagnetic radiations (other than radiant heat), or to ionising particles	Exposure to electro-magnetic radiations other than radiant heat, or to ionising particles.
2 Heat cataract	Frequent or prolonged exposure to rays from molten or red-hot material.
3 Dysbarism, including decompression sickness, barotrauma and osteonecrosis	Subjection to compressed or rarefied air or other respirable gases or gaseous mixtures.
4 Cramp of the hand or forearm due to repetitive movements	Prolonged periods of handwriting, typing or other repetitive movements of the fingers, hand or arm.
5 Subcutaneous cellulitis of the hand (Beat hand)	Manual labour causing severe or prolonged friction or pressure on the hand.
6 Bursitis or subcutaneous cellulitis arising at or about the knee due to severe or prolonged external friction or pressure at or about the knee (Beat knee)	Manual labour causing severe or prolonged external friction or pressure at or about the knee.
7 Bursitis or subcutaneous cellulitis arising at or about the elbow due to severe or prolonged external friction or pressure at or about the elbow (Beat elbow)	Manual labour causing severe or prolonged external friction or pressure at or about the elbow.
8 Traumatic inflammation of the tendons of the hand or forearm, or of the associated tendon sheaths	Manual labour, or frequent or repeated movements of the hand or wrist.
9 Miner's nystagmus	Work in or about a mine.
10 Occupational deafness	
11 Vibration white finger. Episodic blanching, occurring throughout the year, affecting the middle or proximal phalanges, or in the case of a thumb the proximal phalanx, of: (a) in the case of a person with 5 fingers (including thumb) on one hand, any 3 of those fingers, or (b) in the case of a person with only 4 such fingers, any 2 of those fingers, or	Any occupation involving: (a) The use of hand-held chain saws in forestry; or (b) the use of hand-held rotary tools in grinding or in the sanding or polishing of metal, or the holding of material being ground, or metal being sanded or polished, by rotary tools; or (c) the use of hand-held percussive metal-working tools, or the holding of metal being worked upon by percussive tools, in riveting, caulking, chipping, hammering, fettling or swaging; or

Table 16.7 (continued)

Prescribed disease or injury	Nature of occupation
(c) in the case of a person with less than 4 such fingers, any one of those fingers or, as the case may be, the one remaining finger. (Vibration white finger)	(d) the use of hand-held powered percussive drills or hand-held powered percussive hammers in mining, quarrying, demolition, or on roads or footpaths, including road construction; or (e) the holding of material being worked upon by pounding machines in shoe manufacture.
B. Conditions due to biological agents	
1 Anthrax	Contact with animals infected with anthrax or the handling (including the loading and unloading or transport) of animal products or residues.
2 Glanders	Contact with equine animals or their carcases.
3 Infection by Leptospira	(a) Work in places which are, or are liable to be, infested by rats, field mice or voles, or other small mammals; (b) work at dog kennels or the care or handling of dogs; (c) contact with bovine animals or their meat products or pigs or their meat products.
4 Ankylostomiasis	Work in or about a mine.
5 Tuberculosis	Contact with a source of tuberculous infection.
6 Extrinsic allergic alveolitis (including Farmer's Lung)	Exposure to moulds or fungal spores or heteorologous proteins by reason of employment in:- Any occupation involving: (a) agriculture, horticulture, forestry, cultivation of edible fungi or maltworking; or (b) loading or unloading or handling in storage mouldy vegetable matter or edible fungi; or (c) caring for or handling birds; or (d) handling bagasse.
7 Infection by organisms of the genus Brucella	Contact with: (a) animals infected by brucella, or their carcases or parts thereof, or their untreated products; or with (b) laboratory specimens or vaccines of, or containing, brucella.
8 Viral hepatitis	Contact with: (a) human blood or human blood products; or (b) a source of viral hepatitis.

Table 16.7 (continued)

Prescribed disease or injury	Nature of occupation
9 Infection by Streptococcus suis	Contact with pigs infected by streptococcus suis, or with the carcases, products or residues of pigs so infected.
C. Conditions due to chemical agents	
1 Poisoning by lead or a compound of lead	The use or handling of, or exposure to the fumes, dust or vapour of, lead or a compound of lead, or a substance containing lead.
2 Poisoning by manganese or a compound of manganese	The use or handling of, or exposure to the fumes, dust or vapour of, manganese, or a substance containing manganese.
	Any occupation involving:
3 Poisoning by phosphorus or an inorganic compound of phosphorus or poisoning due to the anti-cholinesterase or pseudo anti-cholinesterase action of organic phosphorous compounds	The use or handling of, or exposure to the fumes, dust or vapour of, phosphorus or a compound of phosphorus or a substance containing phosphorus.
4 Poisoning by arsenic or a compound of arsenic	The use or handling of, or exposure to the fumes, dust or vapour of, arsenic or a compound of arsenic, or a substance containing arsenic.
5 Poisoning by mercury or a compound of mercury	The use or handling of, or exposure to the fumes, dust or vapour of, mercury or a compound of mercury, or a substance containing mercury.
6 Poisoning by carbon bisulphide	The use or handling of, or exposure to the fumes or vapour of, carbon bisulphide or a compound of carbon bisulphide, or a substance containing carbon bisulphide.
7 Poisoning by benzene or a homologue	The use or handling of, or exposure to the fumes of, or vapour containing benzene or any of its homologues.
8 Poisoning by a nitro- or amino- or chloro-derivative of benzene or of a homologue of benzene, or poisoning by nitrochlorbenzene	The use or handling of, or exposure to the fumes of, or vapour containing a nitro- or amino- or chloro-derivative of benzene or nitrochlorbenzene.
9 Poisoning by dinitrophenol or a homologue or by substituted dinitrophenols or by the salts of such substances	The use or handling of, or exposure to the fumes of, or vapour containing, dinitrophenol or a homologue or substituted dinitrophenols or the salts of such substances.
10 Poisoning by tetrachloroethane	The use or handling of, or exposure to the fumes of, or vapour containing tetrachloroethane.
11 Poisoning by diethylene dioxide (dioxan)	The use or handling of, or exposure to the fumes of, or vapour containing diethylene dioxide (dioxan).

Table 16.7 (continued)

Prescribed disease or injury	Nature of occupation
	Any occupation involving:
12 Poisoning by methyl bromide	The use or handling of, or exposure to the fumes of, or vapour containing, methyl bromide.
13 Poisoning by chlorinated naphthalene	The use or handling of, or exposure to the fumes of, or dust or vapour containing chlorinated naphthalene.
14 Poisoning by nickel carbonyl	Exposure to nickel carbonyl gas.
15 Poisoning by oxides of nitrogen	Exposure to oxides of nitrogen.
16 Poisoning by Gonioma kamassi (African boxwood)	The manipulation of gonioma kamassi or any process in or incidental to the manufacture of articles therefrom.
17 Poisoning by beryllium or a compound of beryllium	The use or handling of, or exposure to the fumes, dust or vapour of, beryllium, or a substance containing beryllium.
18 Poisoning by cadmium	Exposure to cadmium dust or fumes.
19 Poisoning by acrylamide monomer	The use or handling of, or exposure to, acrylamide monomer.
20 Dystrophy of the cornea (including ulceration of the corneal surface) of the eye	(a) The use or handling of, or exposure to, arsenic, tar, pitch, bitumen, mineral oil (including paraffin), soot or any compound, product or residue of any of these substances except quinone or hydroquinone; (b) Exposure to quinone or hydroquinone during their manufacture.
21 (a) Localised new growth of the skin, papillomatous or keratotic (b) Squamous-celled carcinoma of the skin	The use or handling of, or exposure to, arsenic, tar, pitch, bitumen, mineral oil (including paraffin), soot or any compound, product or residue of these substances except quinone or hydroquinone.
22 (a) Carcinoma of the mucous membrane of the nose or associated air sinuses (b) Primary carcinoma of a bronchus or of a lung	Work in a factory where nickel is produced by decomposition of a gaseous nickel compound which necessitates working in or about a building or buildings where that process or any other industrial process ancillary or incidental thereto is carried on.
	Any occupation involving:
23 Primary neoplasm (including papilloma, carinoma-in-situ and invasive carcinoma) of the epithelial lining of the urinary tract (renal pelvis, ureter, bladder and urethra)	(a) Work in a building in which any of the following substances is produced for commercial purposes:- (i) alpha-naphthylamine, beta-naphthylamine or methylene-bis-orthochloroaniline;

Table 16.7 (continued)

Prescribed disease or injury	Nature of occupation
	(ii) diphenyl substituted by at least one nitro or primary amino group or by at least one nitro and primary amino group (including benzidine); (iii) any of the substances mentioned in sub-paragraph (ii) above if further ring substituted by halogeno, methyl or methoxy groups, but not by other groups; (iv) the salts of any of the substances mentioned in subparagraph (i) to (iii) above; (v) Auramine or magenta; (b) The use or handling of any of the substances mentioned in paragraph *a*, except auramine and magenta (sub-paragraph v.), or work in a process in which any such substance is used or handled or is liberated; (c) The maintenance or cleaning of any plant or machinery used in any such process as is mentioned in paragraph *b* or the cleaning of clothing used in any such building as is mentioned in paragraph *a* if such clothing is cleaned within the works of which the building forms a part or in a laundry maintained and used solely in connection with such works.
24 (a) Angiosarcoma of the liver; (b) Osteolysis of the terminal phalanges of the fingers; (c) Non-cirrhotic portal fibrosis	Any occupation involving: (a) Work in or about machinery or apparatus used for the polymerisation of vinyl chloride monomer, a process which, for the purposes of this provision, comprises all operation up to and including the drying of the slurry produced by the polymerisation and the packaging of the dried product; or (b) Work in a building or structure in which any part of that process takes place.
25 Occupational vitiligo	The use or handling of, or exposure to, para-tertiary-butylphenol, para-tertiary-butylcatechol, para-amylphenol, hydroquinone or the monobenzyl or monobutyl ether of hydroquinone.
D. Miscellaneous conditions not included elsewhere in this Schedule	
1 Pneumoconiosis	Employed earner's employment of a type specified in Regulation 2(b) and in Part II of this Schedule.

Table 16.7 (continued)

Prescribed disease or injury	Nature of occupation
2 Byssinosis	Work in any room in a factory where any process up to and including the weaving of cotton or flax is performed.
3 Diffuse mesothelioma (primary neoplasm of the mesothelium of the pleura or of the pericardium or of the peritoneum)	(a) The working or handling of asbestos or any admixture of asbestos; (b) The manufacture or repair of asbestos textiles or other articles containing or composed of asbestos; (c) The cleaning of any machinery or plant used in any of the foregoing operations and of any chambers, fixtures and appliances for the collection of asbestos dust; (d) Substantial exposure to the dust arising from any of the foregoing operations.
	Any occupation involving:
4 Inflammation or ulceration of the mucous membrane of the upper respiratory passages or mouth produced by dust, liquid or vapour	Exposure to dust, liquid or vapour.
5 Non-infective dermatitis of external origin (including chrome ulceration of the skin but excluding dermatitis due to ionising or electromagnetic radiations other than radiant heat)	Exposure to dust, liquid or vapour or any other external agent capable of irritating the skin (including friction or heat but excluding ionising particles or electromagnetic radiations other than radiant heat).
6 Carcinoma of the nasal cavity or associated air sinuses (nasal carcinoma)	(a) Attendance for work in or about a building where wooden goods are manufactured or repaired; or (b) Attendance for work in a building used for the manufacture of footwear or components of footwear made wholly or partly of leather or fibre board; or (c) Attendance for work at a place used wholly or mainly for the repair of footwear made wholly or partly of leather or fibre board.
7 Asthma which is due to exposure to any of the following agents: (a) isocyanates; (b) platinum salts; (c) fumes or dusts arising from the manufacture, transport or use of hardening agents (including epoxy resin curing agents) based on phthalic anhydride, tetrachlorophthalic anhydride, trimellitic anhydride or triethylenetetramine; (d) fumes arising from the use of rosin as a soldering flux;	Exposure to any of the agents set out in column 1 of this paragraph.

Table 16.7 (continued)

Prescribed disease or injury	Nature of occupation
(e) proteolytic enzymes; (f) animals or insects used for the purposes of research or education or in laboratories; (g) dusts arising from the sowing, cultivation, harvesting, drying, handling, milling, transport or storage of barley, oats, rye, wheat or maize, or the handling, milling, transport or storage of meal or flour made therefrom (occupational asthma).	
8 Primary carcinoma of the lung where there is accompanying evidence of one or both of the following: (a) asbestosis; (b) bilateral diffuse pleural thickening. 9 Bilateral diffuse pleural thickening	Any occupation involving: (a) The working or handling of asbestos or any admixture of asbestos; or (b) the manufacture or repair of asbestos textiles or other articles containing or composed of asbestos; or (c) the cleaning of any machinery or plant used in any of the foregoing operations and of any chambers, fixtures and appliances for the collection of asbestos dust; or (d) substantial exposure to the dust arising from any of the foregoing operations.

show that he is suffering indeed from a specific prescribed disease and has for a minimum time period to sustaining it, been employed in an occupation listed.

USA legislation

Industrialisation developed on much the same lines in mainland Europe and the United States as in the UK. The industrial revolution started in the USA much later than in this country. The vastness of the country, and the freedom of each state to follow its own policy, resulted in differences of application. The first Child Labor Law was passed in the State of Massachussetts in 1836 and they subsequently appointed a police officer in 1867 to enforce the law relating to the employment of children under the age of 10 years. The developments followed the same pattern as in the UK – concern for adverse environmental conditions, long hours of work and generally hazardous processes. The Federal Government created a Bureau of Labor in 1884, a Bureau of Mines in 1910 and the Office of Industrial Hygiene in 1914. The latter was part of the Public

Health Service. This organisation has done a great deal over the years to foster and develop the occupational Health Service in the USA.

In the first decade of the twentieth century progressive industrial safety was practically non-existent in the USA. With no workmen's compensation laws all states dealt with industrial injuries under 'common law'. This necessitated the injured person suing employers for recompense. Even then the company did not have to pay if the employee (or a fellow employee) contributed to the causes of the accident, if the employee knew of the hazard before the accident and agreed to work under these conditions for payment, or if the employer was not negligent. New York State in 1908 passed the first workmen's compensation law which required employers to pay compensation to injured employees regardless of fault. Though this was held to be unconstitutional a similar law passed in Wisconsin in 1911 was accepted. All other states followed suit by 1947. Some other important developments in health and safety protection in the USA are listed in the first appendix to this chapter.

The Toxic Substances Control Act of 1976 is of particular relevance and a summary of its technical implications is given in reference 33. The TSCA inventory lists over 63,000 chemicals whose manufacture, importation, or processing for commercial purposes in the US has occurred since 1 January 1975. The *Chemicals-in-Progress Bulletin*, by the Office of Toxic Substances, is a useful current awareness publication in this field.

EEC

The EEC has an ambitious programme in both occupational health and safety and in consumer protection. Directives emanating from the Commission have a profound influence on developments within member states. A small selection of relevant EEC publications is included in the references.[35–46] Chapter 13 also mentions relevant EEC documents relating to marketing of chemicals.

References

(*All places of publication are London unless otherwise stated*)

1. *Ebbs* v. *Whitson*, 1951, 2.Q.B.877, 311.
2. *Colvilles Ltd* v. *Devine* 1969, 2 All E.R. 53.
3. *Haynes* v. *Qualcast (Wolverhampton) Ltd*; 1958, 1, All E.R. 441.
4. *Haynes* v. *Qualcast (Wolverhampton) Ltd*; 1959, 2, All E.R. 38.
5. *Smith* v. *Leach Brain & Co.*, 1962, 2, Q.B. 405, 313.
6. *Galashields Gas Co Ltd* v. *O'Donnell*; (1949) 1 All E.R. 319.
7. Smith, K. & Koenan, D, J., *English Law* (5th edn). Pitman 1975.
8. Jackson, J., *Health and Safety – The Law*, New Commercial Publishing 1979.

9. Munkman, J., *Employers Liability at Common Law* (8th edn). Butterworth 1975.
10. Jackson, J., 'Employment protection', *The Complete Guide to Employee Rights*. New Commercial Publishing 1979.
11. Fife, I. & Machin, E. A., *Redgraves' Health and Safety in Factories*. Butterworth 1976 (plus later supplements).
12. Tye, J. & Egan, B., *Management Guide to Product Liability*. New Commercial Publishing 1979.
13. Martin. D., *Consumer Safety*. New Commercial Publishing 1979.
14. *Rylands* v. *Fletcher*, 1868, L.R. 3HL 330; 37 L.J. Ex. 161.
15. Campbell, W. A., *The Chemical Industry*. Longman, 1971, 48. (from a description in *The White Slaves of England*, Sherard, R. H.).
16. *Health and Safety at Work Report of the Committee 1970–1972* (Chairman Lord Robens) HMSO, 1972.
17. HM Factory Inspectorate, HSE, 1974.
18. Egan, B., *Safety Policies*. New Commercial Publishing 1979.
19. Egan, B., *Appointment and Functions of Safety Representatives* (2nd edn). New Commercial Publishing 1979.
20. Egan, B., *Safety Inspections*. New Commercial Publishing 1978.
21. Royal Society for the Prevention of Accidents, *RoSPA Bull.*, 1984, **14** (2), 2.
22. Janner, G., *Health and Safety at Work*, 1983, **6** (3), 12. HMSO.
23. Anon., *Safety*, 1981 (Aug.), 12.
24. Health and Safety Commission, *Guidance Note GS8: Articles and Substances for Use at Work*. HMSO 1977.
25. The Royal Society for the Prevention of Accidents, *RoSPA Bull.*, 1984 (Apr.), 2.
26. Industrial Relations Services, *Health and Safety Information Bull.*, 1977 (20), 15.
27. Anon., *Industrial Safety*, 1979 (Feb.), 27.
28. The Asbestos Regulations, 1969. HMSO.
29. The Royal Society for the Prevention of Accidents, *RoSPA Bull.*, 1984 (Apr.), 2.
30. The Chemical Works Regulations, 1922 [S.R. & O 1922, No. 731 as amended by SI 1961, No. 2435]. HMSO.
31. *South Wales Evening Post*, 1979 (18 Oct).
32. The Carcinogenic Substances Regulations 1967, HMSO.
33. Fawcett, H. W., *Hazardous and Toxic Materials – Safe Handling and Disposal* Wiley Interscience, New York, 1984.
34. Health and Safety Commission, *Guidance Note GS8: Articles and substances for use at work*. HSC 1977.
35. 'Council Resolution of 28 June 1978 regarding an Action Programme on Safety and Health at Work', *Official Journal of the European Communities*, 11.7.1978 C165, 1.
36. 'Proposal for a Council Resolution on a Second Programme of Action of the European Communities on Safety and Health at Work', *Official Journal of the European Communities*, 25.11.1982 C308, 11.

37. 'Council Directive (EEC/1107/80) of 27 November 1980 for the Protection of Workers from the Risks of Exposure to Chemical, Physical and Biological Agents at Work', *Official Journal of the European Communities*, 3.12.1980, L327, 8.
38. 'Council Directive (77/576/EEC) of 25 July 1977 on the Approximation of the Laws Regulations and Administrative Provisions of Safety Signs at Places of Work', *Official Journal of the European Communities*, 7.9.1977, L299, 12.
39. 'Council Directive (78/610/EEC) of 29 June 1978 on the Approximation of the Laws, Regulations and Administrative Provisions of the Member States on the Protection of Health of Workers Exposed to Vinyl Chloride Monomer', *Official Journal of the European Communities*, 22.7.1978, L197 12, and 7.9.1977, 12.
40. 'Council Directive (EEC/605/82) of 28 July 1982 on the Protection of Workers from Harmful Exposure to Metallic Lead and its Ionic Compounds at Work'. *Official Journal of the European Communities*, 23.8.1982, L247, 12.
41. 'Proposal for a Second Council Directive on the Protection of Workers from the Risks Related to Exposure to Agents at Work: Asbestos', *Official Journal of the European Communities,* 9.10.1980, C262, 7.
42. 'Council Directive (82/501/EEC) of 24 June 1982 on the Major Accident Hazards of Certain Industrial Activities', *Official Journal of the European Communities*, 5.8.1982, L230, 1.
43. 'Second Action Programme of the European Communities on the Environment of 17 May 1977', *Official Journal of the European Communities*, 13.6.1977, C139, 1.
44. 'Council Directive (79/831/EEC) of 18 September 1979 Amending for the Sixth Time Directive 67/548/EEC on the Approximation of the Laws, Regulations and Administrative Provisions Relating to the Classification, Packaging and Labelling of Dangerous Substances', *Official Journal of the European Communities*, 15.10.1979, L259, 10.
45. 'Proposal for a Council Directive Laying Down Basic Standards for the Health Protection of Workers and the General Public Against the Dangers of Microwave Radiation', *Official Journal of the European Communities*, 16.9.1980, C249, 6.
46. 'Proposal for a Council Directive on the Protection of Workers from the Risks Related to Exposure to Chemical, Physical and Biological Agents at Work: Noise', *Official Journal of the European Communities*, 5.11.1980, C289, 1.

Appendix 16.1
Some relevant developments in safety and health related legislation in the USA

1899 River and Harbour Act.
1906 Federal Food, Drug and Cosmetic Act (amended 1938 and 1962).
Prohibits detectable amounts of carcinogens in foods or cosmetics (excluding hair dyes).
1927 Federal Hazardous Substances Act (amended 1976).
Controls flammable, toxic, corrosive or allergenic material in consumer products.
1947 Federal Insecticide, Fungicide, and Rodenticide Act.
1952 Dangerous Cargo Act.
1952 Federal Water Pollution Control Act.
1953 Flammable Fabrics Act.
1954 Atomic Energy Act.
1956 Fish and Wild Life Act.
1960 Federal Hazardous Substances Labelling Act.
1965 Water Quality Act.
1965 Solid Waste Disposal Act.
1966 Metal and Non-Metallic Mine Safety Act.
1967 Air Quality Act.
1969 Coal Mine Safety and Health Act.
1969 National Environmental Policy Act.
1969 Construction Safety Act.
1970 Consolidation of all Federal Pollution Control Groups into one Agency, the Environmental Protection Agency.
1970 Clean Air Act. Amendments.
Allow the Environmental Protection Agency to set national emission standards for hazardous air pollutants. (Currently asbestos, beryllium, mercury and vinyl chloride are covered.)
1970 Poison Prevention Packaging Act.
1970 Water Quality Improvement Act.
1970 Federal Railroad Safety Act.
1970 Resource Recovery Act.
1970 Occupational Safety and Health Act.
Controls hazardous materials in the workplace or materials purchased for use therein.
1972 Noise Control Act.
1972 Federal Environmental Pollution Control Act.
1972 Hazardous Materials Transportation Act.
1972 Consumer Product Safety Act.
1972 Marine Protection Research and Sanctuary Act.
1972 Clean Water Act.
1972 Coastal Zone Management Act.
1973 Mining Enforcement and Safety Administration.
1974 Safe Drinking Water Act.
Controls carcinogens and toxic substances in public drinking water supplies.

1974 Transport Safety Act.
1974 Energy Supply and Environmental Co-ordination Act.
1976 Resource Conservation and Recovery Act.
Controls disposal of toxic and other hazardous waste in landfills, by incineration, etc.
1976 Toxic Substances Control Act.
Required the EPA to order animal testing of any chemical or mixture for which there is insufficient data or experience – in effect to determine whether manufacture, distribution, processing, use or disposal presents an unreasonable risk of injury to health or the environment. Applies to all new chemical substances and to new uses of existing substances.
1977 Federal Mine Safety and Health Act.
1977 Surface Mine Control and Reclamation Act.
1978 Uranium Mill Tailings Control Act.
1978 Port and Tankers Safety Act.
1980 Comprehensive Environmental Response Compensation and Liability Act.

Appendix 16.2
UK Legislation on health, safety and related topics

Under the Health and Safety at Work etc. Act 1974

Health and Safety at Work etc. Act 1974
The Health and Safety at Work etc. Act 1974 (Application outside Great Britain) Order 1977
The Boiler Explosions Legislation (Repeals and Modifications) Regulations (Northern Ireland) 1979
The Building (Prescribed Fees) Regulations 1982
The Classification and Labelling of Explosives Regulations 1983
The Health and Safety at Work etc. Act 1974 (Commencement No. 1) Order 1974
The Health and Safety at Work etc. Act 1974 (Commencement No. 2) Order 1975
The Health and Safety at Work etc. Act 1974 (Commencement No. 3) Order 1975
The Health and Safety at Work etc. Act 1974 (Commencement No. 4) Order 1977
The Health and Safety at Work etc. Act 1974 (Commencement No. 5) Order 1980
The Health and Safety at Work etc. Act 1974 (Commencement No. 6) Order 1980
The Compressed Acetylene (Importation) Regulations 1978
The Dangerous Substances (Conveyance by Road in Road Tankers and Tank Containers) Regulations 1981
Fire Certificates (Special Premises) Regulations 1976
The Gasholders and Steam Boilers Regulations (Metrication) Regulations 1981
The Health and Safety (Animal Products) (Metrication) Regulations 1980
The Health and Safety (Emissions into the Atmosphere) Regulations 1983

The Health and Safety (Fees for Medical Examination) Regulations 1984
The Health and Safety (Genetic Manipulation) Regulations 1978
The Health and Safety Inquiries (Procedure) Regulations 1975
The Health and Safety (Leasing Arrangements) Regulations 1980
The Health and Safety Licensing Appeals (Hearings Procedure) Rules 1974
The Health and Safety Licensing Appeals (Hearings Procedure) (Scotland) Rules 1974
The Health and Safety (Miscellaneous Fees for Approvals) Regulations 1983
The Health and Safety (Youth Training Scheme) Regulations 1983
The Industrial Tribunals (Improvement and Prohibition Notices Appeals) Regulations 1974
The Industrial Tribunals (Improvement and Prohibition Notices Appeals) Regulations (Northern Ireland) 1979
The Industrial Tribunals (Improvement and Prohibition Notices Appeals) (Scotland) Regulations 1974
The Notification of Installations Handling Hazardous Substances Regulations 1982
The Notification of New Substances Regulations 1982
The Packaging and Labelling of Dangerous Substances (Amendment) Regulations 1983.
The Safety Representatives and Safety Committees Regulations 1977
The Safety Signs Regulations 1980
The Asbestos (Licensing) Regulations 1983
The Explosives and Related Matters (Fees) Regulations 1983
The Anthrax Prevention Act 1919 (Repeals and Modifications) Regulations 1974
The Chemical Works (Metrication) Regulations 1981
The Diving Operations at Work Regulations 1981
The Employer's Health and Safety Policy Statements (Exception) Regulations 1975
The Health and Safety (Dangerous Pathogens) Regulations 1981
The Health and Safety (Enforcing Authority) (Amendment) Regulations 1980
The Health and Safety (Enforcing Authority) Regulations 1977
The Health and Safety (First Aid) Regulations 1981
The Health and Safety Inquiries (Procedure) (Amendment) Regulations 1976
The Notification of Accidents and Dangerous Occurrences Regulations 1980
The Packaging and Labelling of Dangerous Substances (Amendment) Regulations 1981
The Packaging and Labelling of Dangerous Substances Regulations 1978
The Safety Representatives and Safety Committees Regulations (Northern Ireland) 1979

The Health and Safety at Work (Northern Ireland) Order 1978

Under this Order:

The Agriculture (Power Take-off) Regulations (Northern Ireland) 1981

The Health and Safety (Dangerous Pathogens) Regulations (Northern Ireland) 1982

The Health and Safety Inquiries (Procedure) Regulations (Northern Ireland) 1980

The Notification of Accidents and Dangerous Occurrences Regulations (Northern Ireland) 1981

The Packaging and Labelling of Dangerous Substances Regulations (Northern Ireland) 1981

The Petroleum-Spirit (Plastic Containers) Regulations (Northern Ireland) 1983

The Safety Signs Regulations (Northern Ireland) 1981

The Agriculture (First-Aid) Regulations (Northern Ireland) 1979

The Agriculture (Safety of Children) Regulations (Northern Ireland 1981

The Agriculture (Tractor Cabs) Regulations (Northern Ireland) 1981

The Employers' Health and Safety Policy Statements (Exceptions) Regulations (Northern Ireland) 1980

The Health and Safety (Animal Products) (Metrication) Regulations (Northern Ireland) 1980

The Office and Shop Premise Act (Repeals and Modifications) Regulations (Northern Ireland) 1979

Under the Factories Act 1937

The Blasting (Castings and Other Articles) Special Regulations 1949
The Chains, Ropes and Lifting Tackle (Register) Order 1938
The Cinematograph Film Stripping Regulations 1939
The Clay Works (Welfare) Special Regulations 1948
The Construction (General Provisions) Regulations 1961
The Construction (General Provisions) Reports Order 1962
The Control of Lead at Work Regulations 1980
The Dangerous Machines (Training of Young Persons) Order 1954
The Diving Operations (Diver's Fitness Register) Order 1972
The Diving Operations Special Regulations 1960
The Docks, Shipbuilding etc. (Metrication) Regulations 1983
The Dry Cleaning Special Regulations 1949
The Employment of Young Persons (Glass Containers) Regulations 1955
The Engineering Construction (Extension of Definition) Regulations 1960
The Factories (Cleanliness of Walls and Ceilings) (Amendment) Order 1974
The Factories (Cleanliness of Walls and Ceilings) Order 1960
The Factories (Cotton Shuttles) Special Regulations 1952
The Factories (Notification of Diseases) Regulations 1938
The Factories (Notification of Diseases) Regulations 1942
The Factories (Testing of Aircraft Engines and Accessories) Special Regulations 1952
The Foundries (Parting Materials) Special Regulations 1950
The Gasholders and Steam Boilers Regulations (Metrication) Regulations 1981
The Health and Safety (Foundries etc.) (Metrication) Regulations 1981
The Indiarubber Regulations 1955
The Iron and Steel Foundries Regulations 1953

The Jute (Safety, Health and Welfare) Regulations 1948
The Magnesium (Grinding of Castings and Other Articles) Special Regulations 1946
The Mule Spinning and Pottery Appointed Doctor (Amendment) Regulations 1963
The Mule Spinning (Health Special Regulations 1953
The Operations at Unfenced Machinery (Amended Schedule) Regulations 1946
The Operations at Unfenced Machinery (Amendment) Regulations 1976
The Operations at Unfenced Machinery Regulations 1938
The Patent Fuel Manufacture (Health and Welfare) Special Regulations 1946
The Pottery (Health etc.) (Metrication) Regulations 1982
The Pottery (Health) Special Regulations 1947
The Railway Running Sheds (No. 1) Regulations 1961
The Railway Running Sheds (No. 2) Regulations 1961
The Railway Running Sheds Order 1961
The Sanitary Accommodation Regulations 1938
The Sanitary Accommodation (Amendment) Regulations 1974
The Shipbuilding (Air Receivers) Order 1961
The Shipbuilding and Shiprepairing Regulations 1960
The Shipbuilding (Lifting Appliances etc. Forms) (Amendment) Order 1964
The Shipbuilding (Lifting Appliances etc. Forms) Order 1961
The Shipbuilding (Particulars of Annealing) Order 1961
The Shipbuilding (Reports on Breathing Apparatus, etc.) Order 1961
The Shipbuilding (Reports on Chains and Lifting Gear) Order 1961
The Shipbuilding (Reports on Lifting Appliances) Order 1961
The Shipbuilding (Reports on Ropes and Rope Slings) Order 1961
The Work in Compressed Air (Amendment) Regulations 1960
The Work in Compressed Air Special Regulations 1958
The Work in Underground Rooms (Form of Notice) Order 1946
The Breathing Apparatus, etc. (Report on Examination) Order 1961
The Electricity (Factories Act) Special Regulations 1944
The Gasholders (Record of Examinations) Order 1938
The Grinding of Metals (Miscellaneous Industries) (Amendment) Special Regulations 1950
The Health and Safety (First Aid) Regulations 1981
The Work in Compressed Air (Health Register) Order 1973

Under the Factories Act 1948

The Control of Lead at Work Regulations 1980
The Mule Spinning and Pottery Appointed Doctor (Amendment) Regulations 1963
The Pottery (Health etc.) (Metrication) Regulations 1982
The Pottery (Health and Welfare) Special Regulations 1950

Under the Factories Act 1959

The Construction (Lifting Operations) Certificates (Amendment) Order 1964

The Construction (Lifting Operations) Certificates Order 1962
The Construction (Lifting Operations) Prescribed Particulars (Amendment) Order 1962
The Construction (Lifting Operations) Prescribed Particulars Order 1962
The Construction (Lifting Operations) Regulations 1961
The Construction (Lifting Operations) Reports Order 1962
The Washing Facilities (Miscellaneous Industries) Regulations 1960
The Washing Facilities (Running Water) Exemption Regulations 1960

Under the Factories Act 1961

Factories Act 1961
The Abstract of Factories Act Order 1973
The Abstract of Special Regulations (Highly Flammable Liquids and Liquefied Petroleum Gases) Order 1974
The Aerated Water Regulations (Metrication) Regulations (Northern Ireland) 1982
The Asbestos Regulations 1969
The Carcinogenic Substances Regulations 1967
The Construction (Health and Welfare) Regulations 1966
The Construction (Notice of Operations and Works) Order 1965
The Control of Lead at Work Regulations 1980
The Employment Medical Advisory Service (Factories Act Orders etc. Amendment) Order 1973
The Engineering Construction (Extension of Definition) (No. 2) Regulations 1968
The Factories Act 1961 etc. (Metrication) Regulations 1983
The Factories Act 1961 etc. (Repeals and Modifications) Regulations 1974
The Factories Act 1961 etc. (Repeals) Regulations 1976
The Examination of Steam Boilers Regulations 1964
The Factories (Notice of Accident etc.) Order 1965
The Factories (Notification of Diseases) Regulations 1966
The Factories (Standards of Lighting) (Revocation) Regulations 1978 [makes amendment to the Woodworking Machines Regulations 1974]
The Foundries (Protective Footwear and Gaiters) Regulations 1971
The Health and Safety (Foundries etc.) (Metrication) Regulations 1981
The Highly Flammable Liquids and Liquefied Petroleum Gases Regulations 1972
The Hoists and Lifts (Reports of Examinations) Order 1963
The Hoists Examption (Amendment) Order 1967
The Hoists Exemption Order 1962
The Ionising Radiations (Sealed Sources) (Health Register) Order 1961
The Ionising Radiations (Sealed Sources) (Laboratory Certificate) Order 1961
The Ionising Radiations (Sealed Sources) (Leakage Test) Order 1961
The Ionising Radiations (Sealed Sources) (Radiation Dose Record) Order 1961
The Ionising Radiations (Sealed Sources) (Radiation Dosemeter and Dose Rate Meter) Order 1961
The Ionising Radiations (Sealed Sources) Regulations 1969
The Ionising Radiations (Sealed Sources) (Transfer Record) Order 1961
The Ionising Radiations (Unsealed Radiactive Substances) Regulations 1968

The Lifting Machines (Particulars of Examinations) Order 1963
The Non-Ferrous Metals (Melting and Founding) Regulations 1962
The Notice of Industrial Diseases Order 1973
The Power Presses (Amendment) Regulations 1972
The Power Presses Regulations 1965
The Protection of Eyes Regulations 1974
The Factories Act 1961 (Repeals) Regulations 1975
The Slaughterhouses (No. 1) Regulations 1962
The Slaughterhouses (No. 2) Regulations 1962
The Slaughterhouses Order 1962
The Woodworking Machines Regulations 1974
The Hoist and Lifts (Metrication) Regulations 1983
The Abrasive Wheels Regulations 1970
The Abstract of Special Regulations (Aerated Water) Order 1963
The Anthrax (Cautionary Notice) Order 1968
The Construction (Notice of Accident, etc.) Order 1964
The Construction (Working Places) Regulations 1966
The Examination of Steam Boilers Reports (No. 1) Order 1964
The Factories Act General Register Order 1973
The Health and Safety (First Aid) Regulations 1981
The Protection of Eyes (Amendment) Regulations 1975
The Work in Compressed Air (Prescribed Leaflet) Order 1967

Factories Act (Northern Ireland) 1965

Under this Act:

The Health and Safety at Work (Northern Ireland) Order 1978
The Factories Legislation (Repeals and Modifications) Regulations (Northern Ireland) 1979

Under Factory and Workshop Act 1901

Factory and Workshop Act 1901
The Aerated Water Regulations 1921
The Aerated Water Regulations (Metrication) Regulations (Northern Ireland) 1982
The Celluloid Regulations 1921
The Chemical Works Regulations 1922
The Chromium Plating (Amendment) Regulations 1973
The Chromium Plating Regulations 1931
The Control of Lead at Work Regulations 1980
The Docks Regulations 1925
The Docks, Shipbuilding etc. (Metrication) Regulations 1983
The East India Wool Regulations 1908
The Electric Accumulator Regulations 1925
The Electricity Regulations 1908
The Electricity Regulations 1908 (Competent Persons Exemption) Order 1968
The Factory and Workshop (Notification of Diseases) (Epitheliomatous and Chrome Ulceration) Order 1919
The Factory and Workshop (Notification of Diseases) Order 1924
The Factory and Workshop (Notification of Diseases) Order 1936
The Flax and Tow Spinning and Weaving Regulations 1906
The Grinding of Cutlery and Edge Tools (Amendment) Special Regulations 1950

The Grinding of Cutlery and Edge Tools Regulations 1925
The Grinding of Metals etc. (Metrication) Regulations 1981
The Hides and Skins Regulations 1921
The Horizontal Milling Machines Regulations 1928
The Kiers Regulations 1938
The Manufacture of Cinematograph Film Regulations 1928
The Refractory Materials Regulations 1931
The Regulations for the Heading of Yarn Dyed by Means of a Lead Compound 1907
The Regulations for the Manufacture of Paints and Colours 1907
The Regulations for the Smelting of Materials containing Lead, the Manufacture of Red or Orange Lead, and the Manufacture of Flaked Litharge
The Regulations for Spinning and Weaving Hemp, or Jute, or Hemp or Jute Tow 1907
The Regulations for Spinning by Self-acting Mules 1901
The Regulations for the Tinning of Metal Hollow-ware, Iron Drums and Harness Furniture 1909
The Regulations for the Use of Horsehair from China, Siberia or Russia 1907
The Wool, Goat Hair and Camel Hair Regulations 1905
The Woollen and Worsted Textiles (Lifting of Heavy Weights) Regulations 1926
The Docks Regulations 1934
The Electricity Regulations 1908 (Portable Apparatus Exemption) Order 1968
The Factory and Workshop (Notification of Diseases) (Toxic Jaundice) Order 1915
The Grinding of Metals (Miscellaneous Industries) Regulations 1925
The Health and Safety (First Aid) Regulations 1981
The Horizontal Milling Machines (Amendment) Regulation 1934
The Kiers Regulations 1938 (Metrication) Regulations 1981
The Regulations for Vitreous Enamelling Metal or Glass 1908
The Indiarubber Regulations 1922

Under Petroleum (Consolidation) Act 1928

Petroleum (Consolidation) Act 1928 (Conveyance by Road Regulations Exemptions) Regulations 1980
The Petroleum (Consolidation) Act 1928 (Enforcement) Regulations 1979
The Gas Cylinders (Conveyance) Regulations 1931
The Petroleum (Compressed Gases) Order 1930
The Petroleum (Corrosive Substances) Order 1970
The Petroleum (Inflammable Liquids and Other Dangerous Substances) Order 1947
The Petroleum (Inflammable Liquids) Order 1971
The Gas Cylinders (Conveyance) Regulations 1959
The Petroleum (Liquid Methane) Order 1957
The Petroleum (Mixtures) Order 1929
The Petroleum (Organic Peroxides) Order 1973
The Petroleum (Regulations) Acts 1928 and 1936 (Repeals and Modifications) Regulations 1974
The Petroleum-Spirit (Motor Vehicles, & c.) Regulations 1929
The Petroleum-Spirit (Plastic Containers) Regulations 1982
The Carbon Disulphide (Conveyance by Road) Regulations 1958

The Carbon Disulphide (Conveyance by Road) Regulations 1962
The Gas Cylinders (Conveyance) Regulations 1947

Under Offices, Shops and Railway Premises Act 1963

Offices, Shops and Railway Premises Act 1963
The Offices, Shops and Railway Premises Act 1963 (Commencement No. 1) Order 1964
The Offices, Shops and Railway Premises Act 1963 (Commencement No. 2) Order 1964
The Offices, Shops and Railway Premises Act 1963 etc. (Metrication) Regulations 1982
The Offices, Shops and Railway Premises Act 1963 (Exemption No. 1) Order 1964
The Offices, Shops and Railway Premises Act 1963 (Exemption No. 7) Order 1968
The Information for Employees Regulations 1965
The Offices, Shops and Railway Premises Forms Order 1964
The Offices, Shops and Railway Premises Act 1963 (Repeals and Modifications) Regulations 1974
The Offices, Shops and Railway Premises Act 1963 (Repeals) Regulations 1975
The Health and Safety (First-Aid) Regulations 1981
The Notification of Employment of Persons Order 1964
The Offices, Shops and Railway Premises (Hoists and Lifts) Regulations 1968
The Offices, Shops and Railway Premises (Hoists and Lifts) Reports Order 1968
The Prescribed Dangerous Machines Order 1964
The Sanitary Conveniences Regulations 1964
The Washing Facilities Regulations 1964
The Offices, Shops and Railway Premises Act 1963 etc. (Repeals) Regulations 1976

Under Control of Pollution Act 1974

Control of Pollution Act 1974
The Control of Pollution Act 1974 (Appointed Day) Order 1976
The Control of Pollution Act 1974 (Commencement No. 1) Order 1974
The Control of Pollution Act 1974 (Commencement No. 2) Order 1974
The Control of Pollution Act 1974 (Commencement No. 4) Order 1975
The Control of Pollution Act 1974 (Commencement No. 5) Order 1976
The Control of Pollution Act 1974 (Commencement No. 6) Order 1976
The Control of Pollution Act 1974 (Commencement No. 8) Order 1977
The Control of Pollution Act 1974 (Commencement No. 9) Order 1977
The Control of Pollution Act 1974 (Commencement No. 11) Order 1977
The Control of Pollution Act 1974 (Commencement No. 13) Order 1978
The Control of Pollution Act 1974 (Commencement No. 14) Order 1981
The Control of Atmospheric Pollution (Appeals) Regulations 1977
The Control of Atmospheric Pollution (Research and Publicity) Regulations 1977
The Control of Noise (Appeals) Regulations 1975
The Control of Noise (Code of Practice for Construction Sites) Order 1975

The Control of Noise (Code of Practice for Construction Sites (Scotland) Order 1982
The Control of Noise (Code of Practice on Noise from Audible Intruder Alarms) Order 1981
The Control of Noise (Measurement and Registers) Regulations 1976
The Control of Pollution (Discharges into Sewers) Regulations 1976
The Control of Pollution (Licensing of Waste Disposal) (Amendment) Regulations 1977
The Control of Pollution (Licensing of Waste Disposal) Regulations 1976
The Control of Pollution (Radioactive Waste) Regulations 1976
The Health and Safety (Emissions into the Atmosphere) Regulations 1983
The Noise Levels (Measurements and Registers) (Scotland) Regulations 1982
The Clean Air Enactments (Repeals and Modifications) Regulations 1974
The Control of Pollution (Special Waste) Regulations 1980

Under Explosives Act 1875 and 1923

Explosives Act 1875
The Explosives Act 1875 etc. (Metrication and Miscellaneous Amendments) Regulations 1984
The Explosives Act 1875 (Exemptions) Regulations 1979
The Magazines for Explosives Order 1951
Order in Council (No. 12) relating to the keeping of explosive for private use and not for sale
Order in Council (No. 16) repealing and consolidating the previous Orders relating to premises registered for mixed explosives
Order in Council (No. 3) relating to magazines for explosives other than gunpowder whether with or without gunpowder
Order in Council (No. 3A) amending the Order in Council (No. 3) as to magazines for explosives other than gunpowder
Order in Council (No. 5) relating to stores for gunpowder exclusively
Order in Council (No. 6) relating to stores licensed for mixed explosives
Order in Council (No. 6A) amending Order in Council (No. 6) relating to stores licensed for mixed explosives
Order in Council (No. 6D) amending the Order in Council (No. 6) as to stores licensed for mixed explosives
The Stores for Explosives Order 1951
The Stores for Explosives Order 1953
The Acetylene (Exemption) Order 1977
The Explosives (Licensing of Stores and Registration of Premises) Fees Regulations 1983
Explosives Act 1923
The Explosives Acts 1875 and 1923 etc. (Repeals and Modifications) (Amendment) Regulations 1974
The Explosives Acts 1875 and 1923 etc. (Repeals and Modifications) Regulations 1974

Under Clean Air Acts

Clean Air Act 1956
The Clean Air Act 1956 (Appointed Day) Order 1956
The Clean Air Act 1956 (Appointed Day) Order 1958
The Clean Air (Measurement of Grit and Dust from Furnaces) Regulations 1917
The Dark Smoke (Permitted Periods) Regulations 1958

The Clean Air Enactments (Repeals and Modifications) Regulations 1974
Clean Air Act 1968
The Clean Air (Arrestment Plant) (Exemption) Regulations 1969
The Clean Air (Emission of Dark Smoke) (Exemption) Regulations 1969
The Clean Air (Emission of Grit and Dust from Furnaces) Regulations 1971
The Clean Air (Height of Chimneys) (Exemption) Regulations 1969
The Clean Air (Height of Chimneys) (Prescribed Form) Regulations 1969
The Clean Air Act 1968 (Commencement No. 1) Order 1968
The Clean Air Act 1968 (Commencement No. 2) Order 1969
The Health and Safety (Emissions into the Atmosphere) Regulations 1983
The Clean Air Enactments (Repeals and Modifications) Regulations 1974

Under Agriculture (Poisonous Substances) Act 1952

Agriculture (Poisonous Substances) Act 1952
The Agriculture (Poisonous Substances) (Extension) Order 1960
The Agriculture (Poisonous Substances) (Extension) Order 1965
The Agriculture (Poisonous Substances) (Extension) Order 1966
The Health and Safety (Agriculture) (Poisonous Substances) Regulations 1975
The Agriculture (Poisonous Substances) Act 1952 (Repeals and Modification) Regulations 1975
The Health and Safety (Agriculture) (Miscellaneous) Repeals and Modifications) Regulations 1976

Under Agriculture (Safety, Health and Welfare Provisions) Act 1956

The Agriculture (Field Machinery) Regulations 1962
The Agriculture (Metrication) Regulations 1981
The Agriculture (Safeguarding of Workplaces) Regulations 1959
The Agriculture (Stationary Machinery) Regulations 1959
The Agriculture (Threshers and Balers) Regulations 1960
The Agriculture (Power Take-Off) Regulations 1957
The Health and Safety (First-Aid) Regulations 1981
The Agriculture (Lifting of Heavy Weights) Regulations 1959
The Agriculture (Tractor Cabs) Regulations 1974
The Agriculture (Safety, Health and Welfare Provisions) Act 1956 (Repeals and Modifications) Regulations 1975
The Agriculture (Avoidance of Accidents to Children) Regulations 1958
The Agriculture (Circular Saws) Regulations 1959
The Agriculture (Ladders) Regulations 1957
The Agriculture (Tractor Cabs) (Amendment) Regulations 1984

Under Alkali, etc. Works Act 1906

The Health and Safety (Emissions into the Atmosphere) Regulations 1983
The Alkali, etc. Works (Registration) Order 1957
The Alkali, etc. Works Regulation Order (Scotland) 1933 (SR & O 1933/878 (S48))

The Clean Air Enactments (Repeals and Modifications) Regulations 1974

Alkali, etc. Works Regulation (Scotland) Act 1951

Under the Hydrogen Cyanide (Fumigation) Act 1937

Hydrogen Cyanide (Fumigation) Act 1937

The Hydrogen Cyanide (Fumigation of Buildings) (Amendment) Regulations 1982

The Hydrogen Cyanide (Fumigation) Act 1937 (Repeals and Modifications) Regulations 1974

The Hydrogen Cyanide (Fumigation of Buildings) Regulations 1951

The Hydrogen Cyanide (Fumigation of Ships) Regulations 1951

Under Mineral Workings (Offshore Installations) Act 1971

Mineral Workings (Offshore Installations) Act 1971

The Offshore Installations (Emergency Procedures) Regulations 1976

The Offshore Installations (Fire-Fighting Equipment) Regulations 1978

The Offshore Installations (Life-saving Appliances and Fire-fighting Equipment) (Amendment) Regulations 1982

The Offshore Installations (Logbooks and Registration of Death) Regulations 1972

The Offshore Installations (Managers) Regulations 1972

The Offshore Installations (Operational Safety, Health and Welfare) Regulations 1976

The Offshore Installations (Registration) Regulations 1972

The Offshore Installations (Inspectors and Casualties) Regulations 1973

The Offshore Installations (Life-saving Appliances and Fire-fighting Equipment) (Amendment) Regulations 1981

The Offshore Installations (Life-saving Appliances) Regulations 1977

The Offshore Installations (Well Control) Regulations 1980

Under Mines and Quarries Act 1954

The Mines and Quarries (Fees for Approvals) (Amendment) Regulations 1984

Mines and Quarries Act 1954

The Coal and Other Mines (Electricity) (Third Amendment) Regulations 1977

The Coal Mines (Respirable Dust) Regulations 1975

The Mines and Quarries Acts 1954 to 1971 (Repeals and Modifications) Regulations 1974

The Mines and Quarries Act 1954 to 1971 (Repeals and Modifications) Regulations 1975

The Mines and Quarries (Metrication) Regulations 1976

The Mines and Quarries Act 1954 (Modification) Regulations 1978

The Coal and Other Mines (Electric Lighting for Filming) Regulations 1979

The Coal and Other Mines (Metrication) Regulations 1978

The Coal Mines (Precautions against Inflammable Dust) Amendment Regulations 1977

The Coal Mines (Respirable Dust) (Amendment) Regulations 1978

The Health and Safety (First Aid) Regulations 1981

The Mines (Precautions Against Inrushes) Regulations 1979

Under Fire Precautions Act 1971

Fire Precautions Act 1971 (Commencement No. 1) Order 1972
Fire Precautions Act 1971 (Commencement No. 2) Order 1976
The Fire Precautions (Application for Certificate) Regulations 1976
The Fire Precautions (Factories, Offices, Shops and Railway Premises) Order 1976
The Fire Precautions (Non-Certificated Factory, Offices, Shop and Railway Premises) Regulations 1976
The Fire Precautions Act 1971 (Modifications) Regulations 1976

Under Police, Factories, &c. (Miscellaneous Provisions) Act 1916

Police, Factories, &c. (Miscellaneous Provisions) Act 1916
The Cement Works Welfare Order 1930
The Dyeing (Use of Bichromate of Potassium or Sodium) Welfare Order 1918
The Glass Bottle, etc. Manufacture Welfare Order 1918
The Gut Scraping, Tripe Dressing etc. Welfare Order 1920
The Herring Curing (Norfolk and Suffolk) Welfare Order 1920
The Herring Curing (Scotland) Welfare Order 1926
The Herring Curing Welfare Order 1927
The Hollow-ware and Galvanising Welfare Order 1921
The Biscuit Factories Welfare Order 1927
The Laundries Welfare Order 1920
The Oil Cake Welfare Order 1929
Order for Securing the Welfare of the Workers Employed in the Bevelling of Glass 1921
Orders for Securing the Welfare of the Workers Employed in Fruit Preserving Factories 1919
The Sacks (Cleaning and Repairing) Welfare Order 1927
The Sugar Factories Welfare Order 1931
The Tanning (Two-Bath Process) Welfare Order 1918
The Tanning Welfare Order 1930
The Health and Safety (First-Aid) Regulations 1981
The Tin or Terne Plate Manufacture Welfare Order 1917
The Bakehouses Welfare Order 1927

Under Pneumoconiosis etc. (Workers Compensation) Act 1979

Pneumoconiosis etc. (Workers' Compensation) Act 1979
The Pneumoconiosis etc. (Workers' Compensation) (Determination of Claims) Regulations 1979
The Pneumoconiosis etc. (Workers' Compensation) (Payment of Claims) (Amendment) Regulations 1982
The Pneumoconiosis etc. (Workers' Compensation) (Payment of Claims) (Regulations) 1979

Under Radioactive Substances Act 1948

The Radioactive Substances Act 1948 (Modification) Regulations 1974
The Radioactive Substances (Road Transport Workers) (Great Britain) (Amendment) Regulations 1975
The Radioactive Substances (Carriage by Road) (Great Britain) Regulations 1974

The Radioactive Substances (Road Transport Workers) (Great Britain) Regulations 1970

Under Radioactive Substances Act 1960

Radioactive Substances Act 1960
The Radioactive Substances (Hospitals' Waste) Exemption Order 1963
The Radioactive Substances (Smoke Detectors) Exemption Order 1980
The Radioactive Substances (Smoke Detectors) Exemption (Scotland) Order 1980
Radiological Protection Act 1970
Rights of Entry (Gas and Electricity Boards) Act 1954

Under Safety of Sports Grounds Act 1975

Safety of Sports Grounds Act 1975
The Safety of Sports Grounds (Designation) Order 1979
The Safety of Sports Grounds (Designation) Order 1980
The Safety of Sports Grounds (Designation) Order 1982
The Safety of Sports Grounds (Designation) (Scotland) Order 1979
The Safety of Sports Grounds (Designation) (Scotland) Order 1980
The Safety of Sports Grounds (Designation) Order 1981

Under Social Security Act 1975

Social Security Act 1975
The Social Security (Industrial Injuries) (Prescribed Diseases) Amendment (No. 2) Regulations 1982
The Social Security (Industrial Injuries) (Prescribed Diseases) Amendment Regulations 1980
The Social Security (Industrial Injuries) (Prescribed Diseases) Regulations 1980 (revokes and consolidates all previous 'prescribed diseases' regulations)

Miscellaneous (i.e. not included above)

Atomic Energy Authority Act 1971
Boiler Explosions Act 1882
The Boiler Explosions Act 1882 and 1890 (Repeals and Modifications) Regulations 1974

The Building Regulations 1976
Under these Regulations:
The Building (Third Amendment) Regulations 1983

The Building (First Amendment) Regulations 1978

The Building (Second Amendment) Regulations 1981

The Building Regulation (Northern Ireland) 1977

The Building Standards (Scotland) (Consolidation) Regulations 1971

The Building Standards (Scotland) Amendment Regulations 1973

The Building Standards (Scotland) Amendment Regulations 1975

The Building Standards (Scotland) Amendment Regulations 1980

Celluloid and Cinematograph Film Act 1922
The Celluloid and Cinematograph Film Act 1922 (Examptions) Regulations 1980
The Celluloid and Cinematograph Film Act 1922 (Repeals and Modifications) Regulations 1974
Chronically Sick and Disabled Persons Act 1970
Chronically Sick and Disabled Persons (Amendment) Act 1976
Criminal Law Act 1977
Docks and Harbours Act 1966
The Docks and Harbours Act 1966 (Modification) Regulations 1974

[Electricity Act 1957]
The Electricity (Overhead Lines) Regulations 1970

[Electricity (Supply) Act 1882 to 1936]
The Nuclear Installations Act 1965 etc. (Repeals and Modifications) Regulations 1974
Nuclear Installations Act 1969
Nuclear Installations (Amendment) Act 1965
Occupiers' Liability Act 1957
Office and Shop Premises Act (Northern Ireland) 1966

[Offices and Shop Premises Act (Northern Ireland) 1966]
The Office and Shop Premises Act (Repeals and Modifications) Regulations (Northern Ireland) 1979

*[Health and Safety]
The Classification, Packaging and Labelling of Dangerous Substances Regulations 1984

*[Health and Safety]
The Construction (Metrication) Regulations 1984

*[Health and Safety] The
Poisonous Substances in Agriculture Regulations 1984
Highways Act 1980

Industrial Diseases (Notification) Act 1981 Industrial Injuries and Diseases (Old Cases) Act 1975

[Industrial Injuries and Diseases (Old Cases) Act 1975]
The Workmen's Compensation (Supplementation) (Amendment) Scheme 1979

Interpretation Act 1978
Local Government (Miscellaneous Provisions) Act 1976
London Building Act 1930
London Building Acts (Amendment) Act 1939
London County Council (General Powers) Act 1912
London County Council (General Powers) Act 1948

Mines and Quarries (Tips) Act 1969
[Mines and Quarries (Tips) Act 1969]
The Mines and Quarries Acts 1954 to 1971 (Repeals and Modifications) Regulations 1974
[Mines and Quarries (Tips) Act 1969]
The Mines and Quarries Acts 1954 to 1971 (Repeals and Modifications) Regulations 1975
Mines Management Act 1971
[Mines Management Act 1971]
The Mines and Quarries Acts 1954 to 1971 (Repeals and Modifications) Regulations 1974
[Mines Management Act 1971]
The Mines and Quarries Acts 1954 to 1971 (Repeals and Modifications) Regulations 1975
[Northern Ireland Act 1974]
The Clean air (Northern Ireland) Order 1981.
Nuclear Installations Act 1965

[Consumer Protection Act 1961]
The Electrical Equipment (Safety) (Amendment) Regulations 1976

[Consumer Protection Act 1961]
The Electrical Equipment (Safety) Regulations 1975

Consumer Safety Act 1978

[Consumer Safety Act 1978]
The Upholstered Furniture (Safety) (Amendment) Regulations 1983
[Consumer Safety Act 1978]
The Upholstered Furniture (Safety) Regulations 1980

Electricity Supply Regulations 1937
Emergency Laws (Miscellaneous Provisions) Act 1953
Under this Act:
The Control of Explosives Order 1953
The Control of Explosives Order 1954

Employers Liability (Defective Equipment) Act 1969
Employment Act 1980
The Employment Act 1980 (Commencement No. 1) Order 1980
The Employment Act 1980 (Commencement No. 2) Order 1980
Employment Medical Advisory Service Act 1972
Employment of Women, Young Persons, and Children Act 1920
Employment Protection Act 1975

Employment Protection (Consolidation) Act 1978
Under this Act:
The Employment Protection (Variation of Limits) Order 1980

European Communities Act 1972

[Factory and Workshop (Cotton Cloth Factories) Act 1929] The Cotton Cloth Factories Regulations 1929

Fatal Accidents and Sudden Deaths Inquiry (Scotland) Act 1976

Fire Services Act (Northern Ireland) 1969
Fireworks Act 1951
[Gas Act 1948]
The Gas Safety Regulations 1972
Gas Act 1972
[Gas Act 1972] The Gas Safety (Rights of Entry) Regulations 1976
*The Gas Safety (Installations and Use) Regulations 1984
Greater London Council (General Powers) Act 1968
Greater London Council (General Powers) Act 1981
Greater London Council (General Powers) Act 1983

The Petroleum (Regulation) Acts 1928 and 1936 (Fees) Regulations 1983
The Petroleum (Transfer of Licences) Act 1936

[Petroleum (Transfer of Licences) Act 1936]
The Petroleum (Regulations) Acts 1928 and 1936 (Repeals and Modifications) Regulations 1974

Pipe-lines Act 1962
The Pipe-lines Act 1962 (Repeals and Modifications) Regulations 1974

[Pneumoconiosis and Byssinosis Benefit Act 1951]
The Pneumoconiosis, Byssinosis and Miscellaneous Diseases Benefit (Amendment) Scheme 1983

[Pneumoconiosis and Byssinosis Benefit Act 1951]
The Pneumoconiosis, Byssinosis and Micellaneous Diseases Benefit Scheme 1983

Public Health Act 1936 (Ch. 49)
Public Health Act 1961
Public Health (Drainage of Trade Premises) Act 1937
Public Health (Recurring Nuisances) Act 1969

[Public Health (Smoke Abatement) Act 1926]
The Clean Air Enactments (Repeals and Modifications) Regulations 1974

Sex Discrimination Act 1975 [minor amendments to the Health and Safety at Work etc. Act 1974 and the Factories Act 1961]

The Truck Acts 1831 to 1896 (Enforcement) Regulations 1974
Unfair Contract Terms Act 1977

CHAPTER 17

Management of safety and loss prevention

Safety is a management responsibility in the same way that – depending on function – production control, quality control and maintenance are. Ideally safety should be managed in a similar way. This requires[1]:

- A clear, widely publicised statement of company policy on safety.
- Definition of what is required of each level of management to follow that policy.
- Incorporation of the safety policy in the company management systems.
- Procedures for auditing safety performance.

The policy statement should include a general commitment to the provision of a safe and healthy working environment. The responsibilities of the various levels of management and of the workforce should be summarised. The functions of the specialist safety practitioners should be outlined, as should those of the various safety committees and safety representatives.

Excellent guidance on the drafting of safety policies, and for use in the UK detailed interpretation of s.2(3) of the HASAWA, are given in reference 2.

A major area of involvement for middle management is ensuring that safety management systems are devised and applied. Supervisory management may have more clearly defined duties, e.g.:

- Routine supervision of safety systems.
- Investigation of accidents and near-misses.
- Workforce training in safety and loss prevention.
- Drawing-up, and application of, maintenance procedures.

Records are essential, so that action items are not overlooked and safety performance can be measured; these range from minutes of safety committees to accident reports, or vessel/trip system inspection and testing reports, or data from monitoring employee exposure to toxic materials.

While the ultimate responsibility for safety generally rests with line management the safety adviser/manager is concerned with incorporation of the safety policy into the management systems. His responsibility will also include encouraging the introduction of best contemporary practice (e.g. regarding atmosphere monitoring, non-destructive testing) and publicising new safety developments and legislation.

For control it is necessary to obtain an objective assessment of the efficiency of safety management. This requires systems for both internal and external auditing or surveying. (The use of such systems in the design stages, e.g. Hazop and Hazan, is summarised in Ch. 12.)

Systems of work

In addition to safety-conscious design, layout and location (i.e. the provision of safe equipment and a safe place of work) the handling of chemicals requires the establishment, and enforcement, of a proper method of working. It is part of an employer's common law duty to ensure that such 'safe systems of work' are provided. (In the UK, this is also a statutory duty under s.2 of the Health and Safety at Work etc. Act 1974.)

In any event[3]:

> All fires are . . . a demonstration of systems failure, sometimes arising from error or omission in the design stage but invariably associated with human failings in plant operation or maintenance rather than through failure of mechanical plant.

In planning the system of work it is necessary to recognise that workers sometimes become careless or make mistakes, particularly in circumstances where the dangers are obscured by repetition. Lord Oaksey has stated:

> 'It is . . . well known to employers . . . that their workpeople are very frequently, if not habitually, careless about the risks which their work may involve. It is . . ., for that very reason that . . . employers should take reasonable care to lay down a reasonably safe system of work. Employers are not exempted from this duty by the fact that their men are experienced and might, if they were in the position of an employer, be able to lay down a reasonably safe system of work themselves. Workmen are not in the position of employers. Their duties are not performed in the calm atmosphere of a boardroom with the advice of experts. They have to make their decisions on narrow sills and other places of danger and in circumstances where the dangers are obscured by repetition.'[4]

It is not for the workforce to attempt to devise a safe system and precautions, i.e.[4]

> Where a practice of ignoring an obvious danger has grown up I do not think it is reasonable to expect an individual workman to take the initiative in

devising and using precautions. It is the duty of the employer to consider the situation, to devise a suitable system, to instruct his men what they must do and to supply any implements that may be required.'

It is for management to foresee the hazards that can arise in an operation and to devise a system to prevent them causing injury or loss.

As to what may constitute a 'system of work' Lord Green concluded[5]:

'It . . . may include . . . the physical layout of the job – the setting of the stage, so to speak – the sequence in which the work is to be carried out, the provision in proper cases of warnings and notices and the issue of special instructions. A system may be adequate for the whole course of the job or it may have to be modified or improved, to meet the circumstances which arise, such modifications or improvements appear to me equally to fall under the head of system.'

Therefore a system of work normally includes:

- The selection and supervision of personnel.
- The planning and co-ordination of all activities – in particular those involving different sections or departments.
- The training and instruction of the workforce.
- The provision of operating instructions and procedures. These must cover start-up, shut-down, steady-state operation, and procedures for dealing with various breakdowns or emergencies.
- The inspection, testing, maintenance and, where necessary, replacement of equipment.
- The general conditions of work. This covers all the ancillary factors to ensure that, so far as is reasonably practicable, the work is performed safely. Examples are the maintenance of adequate local-exhaust and general ventilation and lighting, the control of access to the site and to specific areas on the plant, fire precautions and numerous measures relating to occupational health. Selections from the above are, of course, covered by specific areas of legislation in the UK.[6]

The combination of measures adopted depends upon the nature and extent of the hazards (e.g. mechanical, toxic, corrosive, dermatitic, fire and explosion), the location, the novelty of process or plant design, the degree of 'sophistication' of the plant, and the qualifications and experience of the workforce. The potential consequences associated with any mistake may also be a factor but, since an ensuing chain of events may be difficult to foresee, an apparently 'safe' consequence should not become an excuse for an inherently faulty procedure.

Clearly no modern factory could function at all without at least a minimum set of systems. Consideration is given below to the major safety systems in plant operation.

Restrictions on access to site; site regulations

Firstly, there is a selection of common-sense restrictions limiting entry to the site to authorised persons, designating specific hazard areas, and generally ensuring that the site is maintained to provide safe access and egress.

A review of current practices for safeguarding industrial facilities against outside intrusion, particularly such as could result in interruption of operations and/or material damage is given in reference 7. The particular measures against arson are reviewed in reference 8.

Secondly, there are simple measures to control non-technical operations on the site. These are summarised in Table 17.1.

These appear rather mundane but are a prerequisite for safe operations within the factory site. Furthermore, management should arrange for monitoring and enforcement of the rules drawn up.

Communication systems

A sound communications system, preferably involving printed or written instructions with built-in checks, is an inherent part of safe working procedures. An indication of the requirements is given in Table 17.2.

This should be a management concern on a day-to-day basis since many incidents have arisen through faulty communications.

> Two different fitters were engaged upon repairing two pumps numbered 1 and 2. The label on the starter for pump 1 was covered with tar and there was no label at all on the starter for pump 2. When he had completed work on pump 1 the fitter decided to test it. In error he pressed the start button for pump 2; it started up and the two men working on it were injured. (In addition to faulty identification this also involved an incorrect procedure in that the starters should have been locked off or the pumps de-fused.)

Table 17.1 Site safety systems

Access systems

Control of access to site. Maintenance of security, e.g. preventing trespass, particularly by children, or vandals.
Control of vehicle speeds and parking.
Recording and accompanying visitors.
Designation of restricted areas, e.g. containing flammable materials, 'eye-protection' zones, 'hearing protection' zones.
Ensuring freedom from obstruction of roads, stairs, gangways, escape routes.

Site restriction systems

Prohibition of eating/drinking except in designated areas.
Prohibit smoking, and carrying matches/cigarette lighters, except in designated areas.
Strict control over the removal of chemicals, 'empty' containers, discarded or used protective clothing or equipment, or contaminated waste from the site.
Restriction on employees permitted to drive factory transport.

Table 17.2 Communications

Maintenance of adequate
Written operating instructions, maintenance procedures, etc.
Log books, recipe sheets, batch sheets.
Identification of vessels, lines, valves, e.g. by clearly visible, and understood, numbers or colour coding.
Inter-shift communication and records.

System for reporting and follow-up of
Plant defects
Process deviations
Hazardous occurrences ('near misses')
Accidents: minor/major injury and/or material loss.
This requires that reports are carried forward from shift to shift, if necessary until dealt with.

Warning notices
Notices relating to specific hazards
Notices relating to temporary hazards (and their subsequent removal)
Notices identifying exits, evacuation zones, emergency equipment locations, etc.
Visual and audible alarms.
(See also permit-to-work systems; see p. 975)

Emergency plans

Where chemical plants, as distinct from factories using relatively limited quantities of chemicals, are concerned it may be necessary to have well-developed and rehearsed measures to reduce the severity of loss from any major incidents.[9] To draw up an emergency plan requires an assessment to be made of the magnitude of the risk on (i) a local, (ii) a plant, and (iii) a community basis; this necessitates knowledge of the identity and quantity of material likely to be involved in any fire, explosion or release, wind direction and areas likely to be affected (i.e. on a map). Established lines of communication must be arranged between key personnel in management, laboratory services, plants supervision, medical services, safety and security services and outside authorities, e.g. police, fire authorities and managements of adjacent factories. What this may involve is summarised in Table 17.3.

The particular case of 'major hazard' installations is discussed in Chapter 15.

Materials storage: disposal of waste

Procedures for the receipt, storage and quality assessment of raw materials is, of course, an inherent part of efficient production management. Control over storage and quality also applies to materials in process and finished products. The proper transfer and handling of these materials is

Table 17.3 Communications in an emergency

Appointment of a central communication co-ordinator.
Clearly identified and agreed duties/responsibilities of each key person.
Identification of persons to be contacted in advance, including those outside, such as in local authorities and neighbouring works.
Arrangements should also be made in advance to cover such contingencies as:

Cancelling signals as soon as possible in the event of false alarms.
Transfer of switch board to night services in the event of evacuation.
Use of neighbouring facilities if the medical centre becomes untenable.
The incident occurring during night time or holidays (there is a 2 in 3 chance of an incident occurring outside the 9 a.m.–5 p.m. period on Mondays to Fridays and a 2 in 7 chance of it occurring at week-ends).
Irrational behaviour under the stressful conditions of an emergency.

also a prime feature of materials and process control. However, the items listed in Table 17.4. are those of particular relevance to safety.

In so far as it tends to be a heterogeneous mixture the systems referring to waste identification, labelling, segregation and disposal are very important when a variety of chemicals are used (see Chs 13 and 14).

Many of the items referred to in Table 17.4 come under 'good housekeeping'. Particular attention is necessary to minimise even small leaks of flammable materials. Leakage into drainage systems, where dense vapours will accumulate, are particularly dangerous since an underground explosion may follow. Leakages which impinge, or accumulate, on hot equipment or pipes may result in spontaneous combustion as described in Chapter 4. When liquid leaks occur at pumps, e.g. due to pump glands failure, the whole pump may subsequently be involved in a fire; an overheated gland or a bearing may serve as the ignition source.

Good housekeeping is important with combustible dusts, i.e. presenting a primary, or secondary, dust explosion hazard. Recommendations include[10]:

- Maintaining clean overhead surfaces, e.g. avoiding accumulations of dust on large ducts or flat surfaces of structural materials (e.g. steel, wood, plasterboard, tiles).
- Using compressed air (cautiously) only for cleaning out small accumulations from relatively inaccessible areas.
- Using central vacuums, portable vacuum cleaners or sweeping to clean up dusts.
- Regular inspection of ducts to avoid dust build-up (these may transmit an explosion otherwise).

Operating and emergency procedures

The systems relating to operation and emergency actions include those listed in Table 17.5.

Table 17.4 Selected systems for materials

Storage
Procedures for ensuring:
- Identification of all raw materials, products, by-products in storage.
- Display of warnings as appropriate.
- Segregation of incompatible materials.
- Good housekeeping.
- Compliance with limits set for stocks of potentially hazardous chemicals.
- Storage, segregation and handling of gas cylinders.
- Supporting of cylinders.

Materials handling
Maintenance of adequate procedures for:
- In bulk, e.g. checks on tanker contents, earthing and bonding, identification of receiving vessel.
- In small containers, e.g. carboy-handling, cylinder unloading.

Dispatch of materials:
- Inspection of tankers, carboys, cylinders, etc., prior to filling.
- Identification and labelling of materials. Driver instructions. Use of earthing clips on portable containers of flammable liquids. Good stacking practice.

'Solid' and waste disposal
Systems for:
- Collection of combustible waste in appropriate containers (e.g. oily rags or other materials subject to spontaneous combustion).
- Identification of wastes. Analytical control procedures. Labelling of containers.
- Segregation of wastes. Identification of potential difficulties due to inadvertent mixing of wastes.
- Selection of appropriate disposal routes, applicable to liquids and solids to comply with acceptable discharge levels.
- Disposal of 'toxic', i.e. hazardous, waste (Table 14.11).

Effluents
Monitoring the adequacy of effluent treatment facilities. Routine checks on discharges. Checks on integrity of surface water drains (freedom from effluents) (Table 14.7).

Emissions
Monitoring atmospheric emissions. Maintenance of pollution control equipment (Tables 6.13 and 14.5).

Precisely written instructions which are kept up to date, and reviewed regularly to incorporate improvements resulting from additional knowledge or experience, are advisable for all operations. These should cover receipt/unloading of materials to dispatch/loading of products. The risks generally appear to be greater on non-standard operations, e.g. unblocking of lines, easing valves, emptying damaged containers or vessels (e.g. cylinders with leaking valves) use of temporary bypasses, changes in operating parameters to cope with significant changes in feed specification, or 'improvisations' to maintain throughput and quality, and non-routine or recycling processes. Procedures for these should be worked out in advance and subjected to a hazard analysis.

Table 17.5 Operating procedures and emergencies

Operating procedures

An adequate system of written and verbal communications (see Communications systems and Table 17.2). These should include:

Provision of clearly written operating instructions, accessible to operators, with reference to numbered/identified plant items covering:

- Start-up and shut-down.
- Normal operation, and dealing with excursions of operating parameters.
- Emergency shut-down, and dealing with mechanical failures or failure of essential services.
- Procedures for non-routine operations, e.g. clearance of blockages, reprocessing of materials, temporary process alteration.

Established procedures for draining, purging, venting, isolating, testing and inspection prior to opening/entering/maintaining plant (see Permit-to-work systems and Maintenance).

Quality control of raw-materials, materials in process, products and wastes.

Materials control to enable losses, over-use, under-use or accumulations of materials to be detected. Control of quantities in storage.

Emergencies

Emergency procedures, following:

- Fire/explosion.
- Toxic release.
- Serious accidents.

Particularly when manning-levels are low, e.g. week-ends.

Major emergency procedures:

- Internal procedures (evacuation, communications and damage control).
- External liaison (fire, police, hospitals, neighbours).

Special operating procedures are required to cover the unsteady-state conditions on start-up and shut-down of equipment or complete plants. If an operation is suspended part way through, e.g. during charging of a precise mixture of chemicals to a vessel, it is important that the procedure includes both a fool-proof system of recording the fact (see Communication systems) and for subsequently drawing it to the attention of the workforce. It is important to maintain continuity of work between shifts, e.g. as to process conditions, equipment status, etc.

In general the steps to be covered on start-up will include an appropriate selection from:

- Preparation for start-up.
- Preparation of services, utilities and ancillary equipment.
- Purging, elimination of air, atmosphere testing.
- Leak testing.
- Introduction of gas/fuel.
- Elimination of water.
- Bringing plant on-stream and up to capacity.

Shut-down instructions may include:

- Depressurising, cooling-down and pumping-out.
- Removal of residual flammable liquids or vapour. Purging.

- Removal of toxic, corrosive or other chemicals.
- Removal of catalyst. Removal and disposal of sludges.
- Drainage and disposal of water-based effluents.
- Isolation, slip-plating, procedures for operating equipment.
- Inspection and testing requirements prior to entry.

Established procedures are needed for operations which involve 'chemicals' handling without normal containment, e.g. transfer to/from portable containers or sampling, or where hoses are used, particularly if the operating position is close by. In the latter case the procedure should allow for a check on the adequacy and integrity of the hose and couplings before use.

The operating instructions must emphasise the hazards in production to reinforce formal training and 'refresher' training, and explain how they are avoided. There should be defined procedures to cope with minor emergency situations, e.g. fires, over-pressures, liquid spillages and personal injury accidents.

Failure of equipment, or a serious deviation from operating conditions, may result in a major leak or spillage of hazardous chemicals. This event will require preplanned, rapid emergency action to bring it under control. Obviously isolation of the source of leakage is a primary objective but depending on the nature of the hazard (i.e. fire, explosion or toxic release), local or general evacuation may be called for.

Generally the instructions will specify the normal operating parameters for the process, e.g. flowrate, temperature, pressure, the tolerance permissible on these and the limits of deviation at which supervision should be informed. Such communication is, of course, in addition to formal recording, logging and reporting systems.

It is particularly important for procedures for use in foreseeable emergencies to be worked out in advance since the need to work under pressure in a limited time-scale is not conducive to good decision making. Thus specific fire-fighting measures are desirable. As an example the action in the case of a fire involving small LPG cylinders should be:

- A small fire at a leak can usually be extinguished using a damp cloth.
- If the cylinder is not directly involved it should be turned off at the main valve, disconnected and removed to a safe place.
- If the cylinder is directly involved it should be cooled with a water spray until the main valve can be closed. The cylinder should then be removed as above.
- If the main valve cannot be turned off (e.g. if damaged), or if gas is issuing from the cylinder itself, cooling with a water spray will reduce the gas flow sufficiently for the fire to be extinguished.
- On no account should the flame at the point of burning be extinguished without turning off the main valve, unless it is certain there is no potential ignition source in the vicinity.

- If the cylinder has been directly heated by the fire for an appreciable time there is a BLEVE hazard (see Ch. 4). Cooling should only then be attempted from behind cover and no approach made to the cylinder until it is certain the explosion hazard has been eliminated.
- The main valve on the cylinder must not be turned on when the cylinder is not vertical (otherwise liquid may emerge instead of gas).

Maintenance

Maintenance operations, particularly any which are non-routine, require a sound system of work. This is based upon an assessment of the risks, e.g. risks arising from toxic or flammable vapours not effectively purged from equipment, or from machinery which is temporarily stopped and unguarded, or the presence of internal pressure when it is 'unexpected'. For this purpose the maintenance staff require an authoritative collection of drawings and technical information; this must include an up-to-date record of all alterations.

Safe plant and machinery maintenance generally depend upon the measures listed in Table 17.6. Some routine inspection and testing requirements are summarised in Table 11.26.

Permit-to-work systems

Cleaning and repair of plant is necessary both during operation and in scheduled shut-downs. All such operations must be carefully controlled based on written procedures and documentation of responsibility.

Operating personnel have the responsibility of ensuring that an area, or item of equipment, is isolated and free from hazards and/or of specifying the precautions necessary. Hand-overs to maintenance/contractors' personnel is controlled by a permit to work (or clearance certificate) which sets out the work to be done and the precautions to be taken. It predetermines a safe procedure and records clearly that all 'foreseeable' hazards have been considered in advance and that all appropriate precautions are defined and followed in the correct sequence.

The following is an example of how double safeguards can be applied to a simple manual operation (after ref. 11):

Operation Repair of a leak from a flanged joint on a caustic line by a craftsman.

Preparation Operator isolates line and washes it out with water until free from caustic, testing with pH paper.

Certificate for craftsman specifies PVC gloves, hood and suit.

Performance Craftsman is aware from certificate that line containing caustic is isolated and washed free. Craftsman wears protective clothing

Table 17.6 Safety in maintenance

- Institution of a programme of regular inspection and examination of all plant, machinery and equipment as a prerequisite of planned maintenance, and to comply with statutory requirements.
- Institution of a system for the issue, control, testing, and repair of lifting gear, ladders, scaffolding, etc.
- Assessment of the possibilities of hazard from fume, dust or other materials – as a risk from fire, explosion, gassing or other kind which may be generated *in situ* as a result of the operations (e.g. vaporisation due to hot work), or from electricity.
- Provision for adequate precautions against any mechanical hazards which might arise from either stationary or moving plant.
- Ensuring that sufficient plant and equipment is provided to create a safe place of work (e.g. temporary lighting, fresh air supply and local exhaust ventilation) and safe access to it (e.g. ladders, temporary scaffolding).
- Provision of efficient tools to do the work required; arrangement for these tools to be regularly inspected and maintained.
- Planning all maintenance procedures and ensuring that responsibility during their performance is clearly understood, i.e. use of permits to work. Such permits may be applicable to:
  Flame cutting/welding (or soldering, brazing).
  Line-breaking (chemicals, pressure, steam, etc.).
  Electrical work – certain classifications.
  Equipment removal, e.g. to workshop.
  Work on roofs. Excavations.
  Work on conveyors, lifts, cranes, etc.
  Entry into confined spaces.
  Work in radio-active areas.
  Work in other unusual/non-routine circumstances.
- Ensuring that personnel are fully instructed and trained to do the work required.
- When special hazards exist, arrangements for the necessary trained personnel (e.g. for fire-fighting, first aid or prevention of gassing) to be present, or on call.
- The keeping of effective records of all maintenance required, work done or remaining to be done (e.g. suspended or incomplete through lack of time, or for other reasons).

specified. Craftsman breaks joint on side opposite so that any residual pressure spurts away from him.

The seven principles embodied in a permit system are[12]:

- The isolation must be adequate.
- The isolation must remain secure.
- Residual hazards must remain controlled.
- Equipment must be correctly and clearly identified.
- 'Maintenance' must be given well-directed instructions.
- Workforce must be on guard against change of intent.
- The system must be monitored.

However, the work can only be performed safely if all the participants follow the permit conscientiously, i.e. without short-cuts/improvements to avoid inconvenience, or to save time under the pressures of production, or through lack of knowledge. Taking 'short-cuts' can be extremely hazardous.

A distillation column had been washed out with water before a shut-down; the water left in had been distilled into the reflux drum, half-filling it, as shown in Fig. 17.1.[13] As the column was warming up a layer of light oil containing LPG was present on top of the water.

Before refluxing started, when an attempt was made to start the water pump it was discovered that a slip-plate (one of several used to isolate the drum for entry) had been left in the suction line. Shutting-down and following the proper procedure to remove this slip-plate would have introduced a 24-hr delay. Therefore, with management's approval, a fitter began to remove the slip-plate hoping to complete the job while only water sprayed out from the flanged joint.

During the operation one of the gaskets was torn and before it was replaced all the water escaped allowing some of LPG to flash off; characteristically this caused ice formation and the attempt to remake the joint had to be abandoned. The vapour spread across the ground towards the associated furnace, some 30 m away, but fortunately there was time for the burners to be shut down manually and no ignition occurred.

Proper identification of equipment is essential.

An experienced contract fitter was killed and an emission of approximately 2 tonnes of ammonia occurred when he removed four top-cap bolts from the wrong plug valve in a liquid anhydrous ammonia system at 6 bar pressure.[14]

Modifications were to be made to four plug valves to enable actuators to be fitted. The bolts on the top cap of each valve were to be replaced, one by one, by stud bolts with nuts. The valves were separated from the ammonia-carrying system by similar valves, acting as stop valves.

Unfortunately the fitter started work on one of the stop valves and, contrary to the safe and general practice, removed all four bolts. When the last bolt was removed the top cap blew off the valve; the fitter was hit in the chest and choked by ammonia.

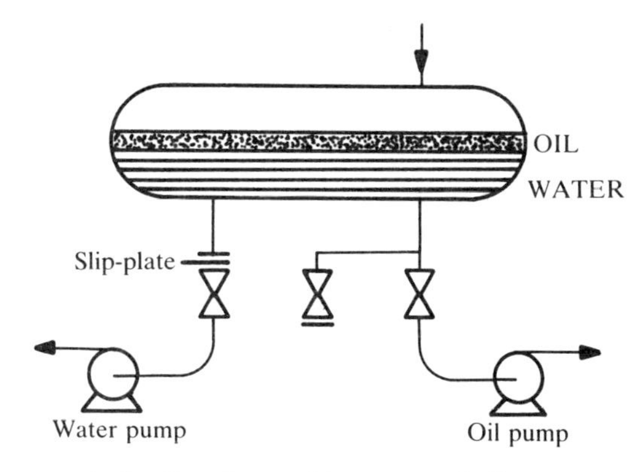

Fig. 17.1 Arrangement of reflux drum with water and oil pumps

So far as pipe-lines and vessels are concerned Table 17.7 illustrates some guidelines. Similar principles apply to mechanical and electrical isolations, i.e. single or double isolation should be used after careful assessment of the hazards.

Only one permit relating to one piece of equipment should be in force over any given period of time. This is to avoid needless confusion over job completion which could result in an accident. In practice it means that *all* the work to be done is listed on one permit irrespective of the tradesmen involved (i.e. fitters, electricians, instrument fitters, etc.) and each job is signed off individually on completion. The system is intended to preclude changes of mind; therefore, the tradesmen and supervisors must appreciate that only the work specified, and the sequence given, on the permit is authorised. If practical difficulties arise which necessitate a change of sequence then a new permit is required. The person authorised to issue permits will normally be the plant superintendent or shift supervisor; he must, of course, be familiar with the system and be qualified, both technically and by experience, to assess the hazards associated with the work. Only one person should be able to issue permits on any plant, or plant area, during a specified period of time.

If a piece of equipment does not have a permanent number clearly visible on it, it should be marked with a numbered tag. The number of the equipment, or the tag number, should be noted on the permit. Accidents are inevitable if reliance is placed upon merely pointing out which equipment is to be isolated, and how it is to be done; the recipient

Table 17.7 Possible vessel/pipeline isolation policy

- Single valve isolation for maintenance work on low-pressure water lines.
- Double-valve isolations with intermediate venting for flashing flammable liquids or flammable gases.
- Single valve isolation of dangerous materials only used where the fitting of slip-plates, blanks, or removal of a section of pipe would take as long, and be as hazardous, as the main job. Protective clothing worn appropriate for the worst hazard likely to be present (probably the isolation valve leaking).
- Isolation by slip-plate, or physical disconnection with the open end leading to the rest of the plant blanked-off, should be the normal standard of isolation before equipment is handed over to maintenance. As further purging and washing can take place between fitting slip-plates and the main job, the standard of protective clothing necessary may be less exacting than when slip-plates are being fitted.
- Where the consequence of error may be serious injury (e.g. for vessel entry) the appropriate standard should be the use of two dissimilar methods of isolation (e.g. valve with slip-plate, valve and physical disconnection, isolator racked out and fuses removed).
  The valve should be considered as part of the isolation system and every effort made to ensure it is tight and secure.

of the instructions may forget them, or be confused by the piping or equipment geometry, or be replaced by a different operator or maintenance worker.

> A permit to work was issued to a fitter to repair a leaking joint on a water line on a pipe-bridge. Access was difficult so the line was pointed out to the maintenance supervisor by the process foreman; subsequently the maintenance supervisor pointed it out to a fitter. In the event the fitter broke a joint in a line carrying carbon monoxide, from which water was dripping; the man was gassed but was fortunately rescued, albeit with some difficulty, by other workmen.[15]

Numbered tags should be fixed to each line at the point it is to be broken.

The adequacy of isolation must be assessed in terms of reasonably foreseeable hazards; this leads to a decision on single versus double isolations and the need for adequate blowdown, venting, purging and washing procedures.

Any permit issued for a job where there is likely to be pyrophoric material should include an instruction such as 'deposit in this line/vessel/equipment is pyrophoric' and list appropriate precautions. Arrangements are also necessary for safe disposal of the pyrophoric material, e.g. by removal to a safe area where ignition may occur without hazard, by controlled oxidation or by burying.

The importance of proper isolation is illustrated by the following examples:

> A large pump was being dismantled for repair; no slip-plates were fitted. When a cover was removed by a fitter, hot oil above its auto-ignition temperature was discharged as the suction valve was open, and a fire ensued.[16]

> The pump shown in Fig. 17.2 was disconnected at the points marked 'broken' in order to repack the glands. The open ends were not blanked. While the maintenance fitters were away from the job an operator arrived to start up the pump. He opened the suction valve on this pump, instead of that on a spare pump, and was affected by escaping ammonia.[17]

Particular problems may arise when work has to be performed on pipelines which inter-connect two plants or processing units. Alternatively, the work may be on the process pipelines or services (e.g. effluent disposal, gas, water, cooling water, inert gas, compressed air, electricity) which are common to, or pass through, other plants. If these plants are not under the same direct management then the tacit approval, or acknowledgement, of each plant superintendent is needed on a clearance certificate.

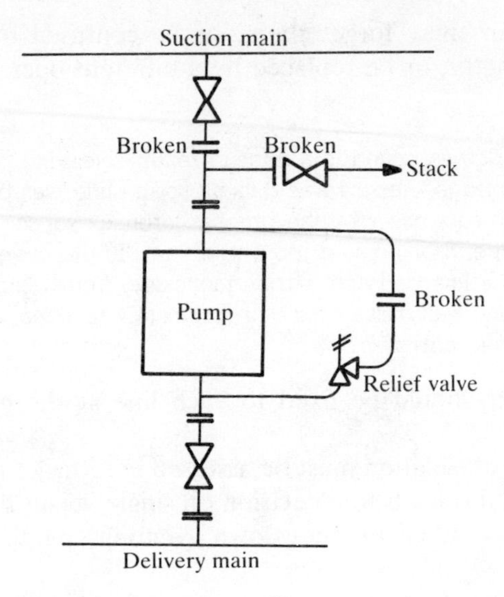

Fig. 17.2 Arrangement of ammonia pump[17]

Each permit should be valid for a limited, clearly specified, period of time. It should be issued just prior to the work being commenced; a permit issued for work to start at some unspecified future time can lead to a dangerous situation as plant conditions and the workforce change. The permit will normally be received by whoever is to take personal charge of, or perform, the work. Otherwise it should be displayed in the area, or on the equipment, involved so that everyone concerned with the work is left in no doubt as to the conditions. A copy should also be kept in a secure place, e.g. in records in a control room, to assist management co-ordination. Discussions between the person issuing the permit and the person receiving it is desirable, in case questions arise which are not covered on the permit, however well devised, and to eliminate any confusion over interpretation.

The importance of strict control including isolation is exemplified by the following incident[18]:

> A piping installation serving two pieces of equipment B and C was as shown diagrammatically in Fig. 17.3. So that filter B could be cleaned by a process worker the valve A was locked shut. At about the same time a permit was issued for maintenance to be carried out on equipment C. Since the lock on valve A isolated C, and the job was of short duration, a slip-plate was not considered necessary.
>
> When work on B was completed valve A was unlocked and opened by the process worker not realising that the lock on valve A also protected C. As a result oil was discharged from the open ends on equipment C.

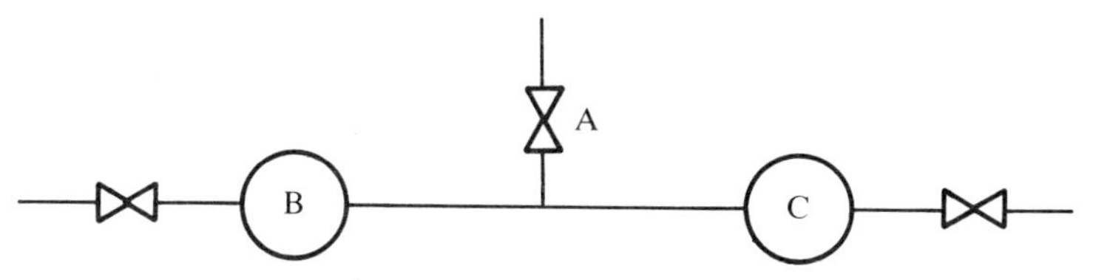

Fig. 17.3 Piping arrangement with common isolation valve

Entry into confined spaces

Safe working in confined spaces where there is likely to be danger from toxic and/or flammable gases, fumes and vapours, or a deficiency of oxygen, depends on a well-devised procedure regulated by a permit-to-work system. 'Confined spaces' include not only reaction vessels, closed tanks and large ducts but also open-topped tanks, vats or pits and medium-sized furnaces with restricted provisions for air circulation.

It is vital that the system of work should not merely exist on paper but should be genuinely followed.

> An inspector first noticed that work was going on inside a chlorine tank because access to the manway was across a single 20 cm plank from the top of an adjacent tank at an elevation of 6 m. No guard rail or handhold was provided. The permit to enter had been given by a manager who had not visited the site, who had been told that the TLV for chlorine was exceeded, and who did not state the length of time for which the permit was valid. However, it was stipulated that breathing apparatus should be worn. This had not been examined for two months and the 46 m airline had never been examined. Earlier in the day the air bottles in use had run out after about 5 to 10 minutes. While the man working in the tank had a safety line attached to him held by a mate outside, neither they nor others on the job had been trained in rescue or in giving artificial respiration. (The work was stopped and the training shortcomings corrected within a month.)[19]

At least 18% oxygen should be present in air in any confined space for entry without wearing breathing apparatus. This is a reasonable minimum, making some allowance for differing personal habits, degree of exertion and physical fitness; however, some procedures recommend a minimum of 19.5%.[20] Therefore, no person should be permitted to enter or remain in any confined space, e.g. vessel, vat, duct, sewer or pit in which the proportion of oxygen in the air is liable to have been substantially reduced unless:

1. Suitable breathing apparatus is worn (i.e. apparatus which enables the wearer to breathe quite independently of the outside atmosphere).
2. The space has been, and remains, adequately ventilated and a responsible person has tested and certified it as safe for entry without breathing apparatus.

The danger of oxygen deficiency is insidious; it tends not to be widely appreciated and therefore, as with many other facets of occupational health, education and instruction of the workforce are all important.

> A fitter's mate . . . was cleaning the inside of a pressure vessel (air–nitrogen receiver) about 6-m long with a 2-m diameter mounted on a cement raft 2.1 m above floor level. The receiver had been used to store compressed air and had not been used for nitrogen for some five weeks. As it was the firm's holiday shut-down period and because the receiver was just over two years old, it was decided to open it up for cleaning in readiness for its first thorough inspection. The man entered the tank to clean it and approximately two hours later a colleague found him lying in the bottom. Another worker became dizzy and collapsed during a rescue attempt. By the time rescue was effected the mate was dead.
>
> On analysis samples of the atmosphere in the vessel were found to be between 11% and 12% oxygen by volume. A detailed examination showed that nitrogen was leaking from the main nitrogen supply into the vessel at 1 m^3/h. Death was due to asphyxia.[21]

Frequently in accidents involving asphyxiation in a confined space there has been a tendency for the first person to find the victim to enter, to attempt a rescue, without wearing breathing apparatus. While instinctive and understandable, this kind of action can result in multiple casualties or fatalities.

> Two employees in a fertiliser plant had to install a float valve in an old 11 m deep water cistern. When the first one dropped on to a wooden platform 2 m below the tank opening he was immediately overcome by hydrogen sulphide gas, and he fell into the water below. His partner went for help and two men entered the tank. They were also overcome, and fell into the water. A passer-by, trying to save the drowning man, jumped into the water; he too was drowned. By this time the fire brigade had arrived and a fire officer wearing breathing apparatus descended to the wooden platform. He removed his face piece for a moment to shout instructions to men outside the tank and he was instantly overcome and died. Thus the original victim and four would-be rescuers lost their lives in one incident.

Clearly, therefore, it is essential that training, to act in a given way in a given set of circumstances, is provided for all employees who may have to enter confined spaces and for the management in contact with them.

It is common but dangerous practice to use oxygen, possibly from the hose of a welding set, to 'sweeten' the atmosphere in a confined space when it is stale or uncomfortable, e.g. due to heat or welding fumes. This is unsafe because a large volume of oxygen can be discharged, even if the valve is open for only a short period, and cause oxygen enrichment (see Ch. 4); in one reported case sparks during subsequent hot-work in a 'sweetened' space resulted in ignition of clothing and fatal injuries.[22] The proper arrangement is to ventilate the working area so as to supply an adequate amount of fresh *air*. The operation of a system for entry into confined spaces is illustrated in Fig. 17.4 and an appropriate certificate is reproduced in Fig. 17.5.[23]

A test in progress above a vessel manway is shown in Fig. 17.6.

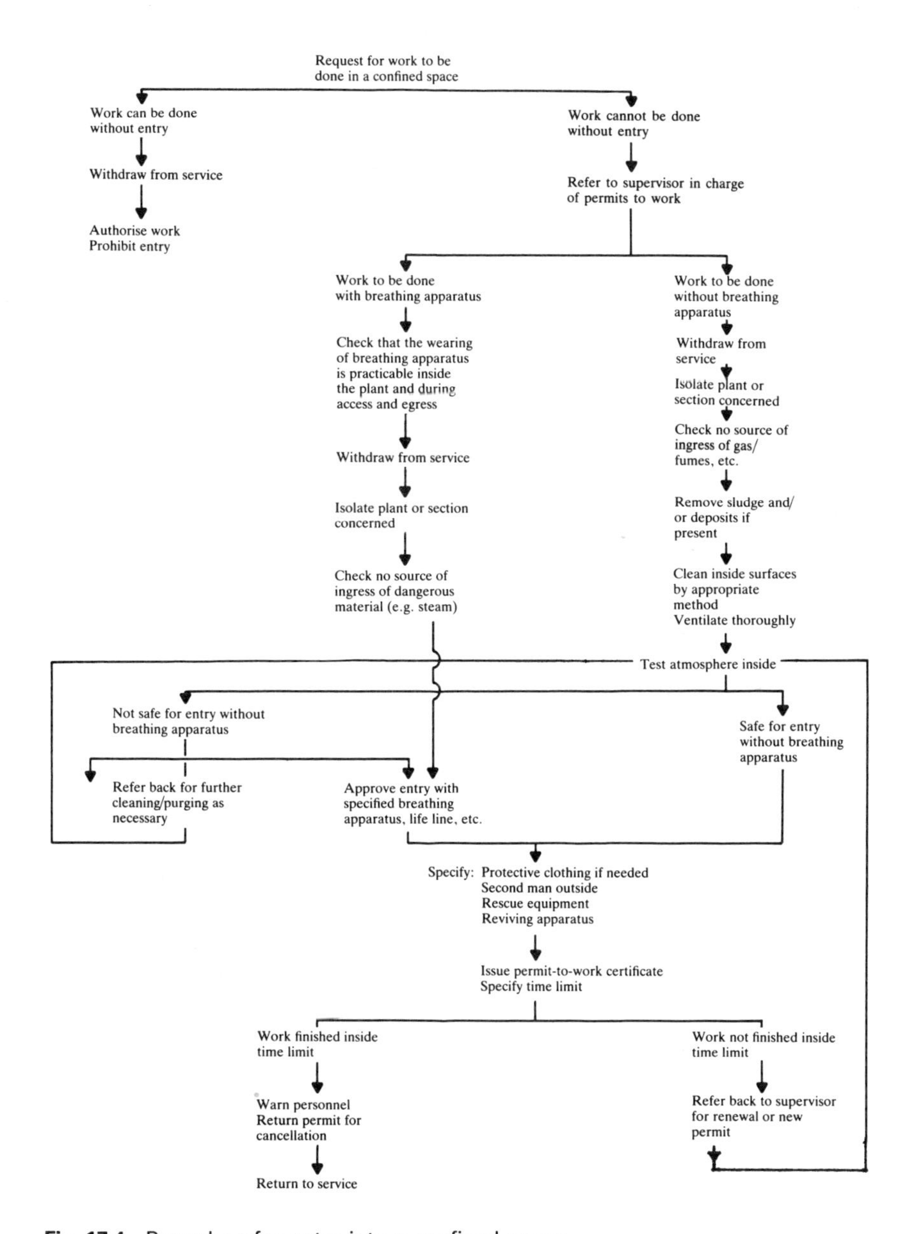

Fig. 17.4 Procedure for entry into a confined space

PLANT DETAILS (Location identifying number, etc.)	
WORK TO BE DONE	
WITHDRAWAL FROM SERVICE	The above plant has been removed from service and persons under my supervision have been informed Signed Date Time
ISOLATION	The above plant has been isolated from all sources of ingress of dangerous fumes, etc. Signed The above has been isolated from all sources of electrical and mechanical power Signed The above plant has been isolated from all sources of heat Signed Date Time
CLEANING AND PURGING	The above plant has been freed of dangerous materials Material(s): Method(s): Signed Time Date
TESTING	Contaminants tested Results Signed Date Time
I CERTIFY THAT I HAVE PERSONALLY EXAMINED THE PLANT DETAILED ABOVE AND SATISFIED MYSELF THAT THE ABOVE PARTICULARS ARE CORRECT * (1) THE PLANT IS SAFE FOR ENTRY WITHOUT BREATHING APPARATUS (2) APPROVED BREATHING APPARATUS MUST BE WORN Other precautions necessary: Time of expire of certificate: Signed Time * Delete (1) or (2) Date	

ACCEPTANCE OF CERTIFICATE	I have read and understood this certificate and will undertake to work in accordance with the conditions in it Signed Date Time
COMPLETION OF WORK	The work has been completed and all persons under my supervision, materials and equipment withdrawn Signed Date Time
REQUEST FOR EXTENSION	The work has not been completed and permission to continue is requested Signed Date Time
EXTENSION	I have re-examined the plant detailed above and confirm that the certificate may be extended to expire at: Further precautions: Signed Date Time
THE PERMIT TO WORK IS NOW CANCELLED A NEW PERMIT WILL BE REQUIRED IF WORK IS TO CONTINUE Signed Date Time	
RETURN TO SERVICE	I accept the above plant back into service Signed Date Time

Fig. 17.5 'Permit-to-Work' certificate for entry into a confined space[23]. A certificate number is added for identification, authenticity checking, etc.

Fig. 17.6 Testing for hazardous vapour at a vessel access port (Courtesy Health and Safety Executive)

Hot-work permits

If flammable materials are handled or may be present in an area a special permit should be issued for:

- Arc- or gas-welding, flame-cutting, use of blow-torches.
- Use of non-classified electrical equipment.
- Chipping concrete.
- Introduction of non-sparkproof vehicles, e.g. fork-lift trucks, cranes.

In the area the hazards to guard against are the presence of flammable vapours/gases in the atmosphere, the presence of residual combustible material, and the presence of residual flammable material in equipment being worked on. Consideration may also have to be given in some cases to the presence of non-flammable vapour subject to thermal degradation,

CERTIFICATE OF CONDITIONS AND RESPONSIBILITY OTHER THAN FOR ENTRY INTO VESSELS OR CONFINED SPACES

Part 1

<table>
<tr><td colspan="2">WORKS</td><td colspan="2">Certificate No.</td></tr>
<tr><td colspan="2">DEPARTMENT</td><td colspan="2">Date of issue   Date ....
Time ....</td></tr>
<tr><td colspan="2">SECTION OF PLANT</td><td colspan="2">Period valid to ....</td></tr>
<tr><td colspan="4">CIRCUMSTANCES</td></tr>
<tr><td colspan="4">NOTE: Strike through where not applicable
(1) The above item of plant is isolated from every (dangerous) source of steam, gas, fume, liquid, motive power, heat and electricity.
Details of isolation:</td></tr>
<tr><td colspan="4">(2) The above item of plant is not isolated from every (dangerous) source of steam, gas, fume, liquid, motive power, heat and electricity.
These special precautions must be taken in addition to (3) and (4) below.</td></tr>
<tr><td>(3) Naked Flame or Other Source of Ignition</td><td>Not Permissible</td><td colspan="2">Permissible: Type:</td></tr>
<tr><td colspan="2">(4) Other hazards which may be encountered</td><td colspan="2">Precautions</td></tr>
<tr><td colspan="2">Hand Over
Issuing Dept .... Date ....
(*Signature*) Time ....
Receiving Dept. ....
(*Signature*)</td><td colspan="2">Hand Back
Issuing Dept .... Date ....
(*Signature*) Time ....
Receiving Dept. ....
(*Signature*)</td></tr>
<tr><td colspan="4">Note 1. This type of certificate may be designed to cover:
Work on roofs, in storage bunkers, on plant containing gas, fume, steam, corrosive, poisonous or radioactive materials, welding and burning in potentially hazardous areas and work being carried out where there is a risk of fire, explosion, electric shock, flooding or high pressure.
Note 2. It is considered that work on Electrical Equipment operating at High Voltage (over 650 volts) should not come within the scope of this pamphlet.</td></tr>
</table>

Fig. 17.7 Hot-work permit[24]: part 1 records conditions and steps required; part 2 authorises work and specifies precautions

e.g. trichloroethylene, or where toxic fumes/vapours may be volatilised by hot-work.

Particular care is needed when welding is performed in confined spaces, for example to guard against excessive carbon monoxide, NO_x and ozone levels in the working atmosphere. Furthermore, the fire/explosion hazard tends to be increased. For example during flame-cutting all the oxygen to the torch may not be consumed and, in some circumstances during normal operation, appreciable quantities of excess oxygen can be released into the atmosphere. In a poorly ventilated space this may result in oxygen enrichment in the vicinity of the operator which can have serious

AUTHORISATION TO WORK
OTHER THAN IN VESSELS OR CONFINED SPACES
Part 2

WORKS	Authorisation No.	
DEPARTMENT	Date of issue	Date .... Time ....
SECTION OF PLANT	Period Valid to	
Certificate No.		
(1) Work to be carried out.		
(2) The following precautions have been taken.		
(3) Naked Flame or Other Source of Ignition	Not Permissible	Permissible: Type:
(4) Precautions to be taken in addition to (2) and (3) above		
Work may start .... (*Signature*)		Date .... Time ....
Work completed .... (*Signature*)		Date .... Time ....

Fig. 17.7 continued

fire/explosion consequences. Clearly, proper ventilation is an essential safeguard.

It is bad practice to allow blowpipes and hoses to be left inside a confined space during, for example a lunch break, or overnight, since a very small leak of fuel gas or oxygen over a period can result in a fire/explosion on resumption of work. Gas cylinders should be located outside confined spaces whenever practicable, and torches and hoses should be moved into the open air during breaks in the work.

A typical hot-work permit is shown in Fig. 17.7.[24] In general, a hot-work permit should not be issued if an atmosphere test indicates the presence of flammable vapour at >20% of the lower flammable limit. If a permit is issued, it is a worthwhile precaution to place portable combustible gas detector alarms at strategic points around the area as described in Ch. 7. In any event, either intermittent or continuous monitoring of the immediate hot-work location is essential since the environment can change due either to some outside event, e.g. spillage or venting of material, or as a result of the hot-work itself (see Chs 3 and 4).

When practicable, it is preferable for equipment which has to be welded/flame-cut to be moved to a safe area; cold-cutting is an alternative.

Equipment on which welding/flame-cutting is to be performed obviously has to be free of flammable liquids or vapour. Special precautions are required when equipment has contained heavy oils, or materials which polymerise or which are solid at normal temperatures, since cracking and/or vaporisation can occur when heat is applied resulting in a flammable vapour–air mixture. Safe procedures with tanks which have contained flammable materials are summarised in references 25 and 26.

Other clearance certificates

Before commencing any work involving excavations a check should be made on site plans to find out whether pipes, sewers or electric cables are nearby. A clearance certificate provides a formal control on such work and the issue of instructions regarding the precautions necessary. A certificate may also be used to control work on roofs.

Equipment which leaves the plant, for repair or servicing at a central workshop or by a contractor, should, whenever possible, be thoroughly cleaned and decontaminated. Otherwise, e.g. if material may have polymerised or become trapped in cavities, a certificate detailing the potential hazard and the appropriate precautions should accompany the equipment.

A special permit to work should be used to control maintenance on radioactive sources. These may be sealed sources, e.g. in level controllers, or radiographic equipment for checking weld integrity. Certificates may also be used:

- For certain classes of electrical work to make known which apparatus is dead and properly isolated from live conductors, that apparatus has been discharged of electricity and is earthed, and that specified apparatus is safe to work on.[27]
- To authorise temporary shut-off of water sprinkler or fire-water mains.
- In locations where accidental or unauthorised starting of plant may endanger workers.

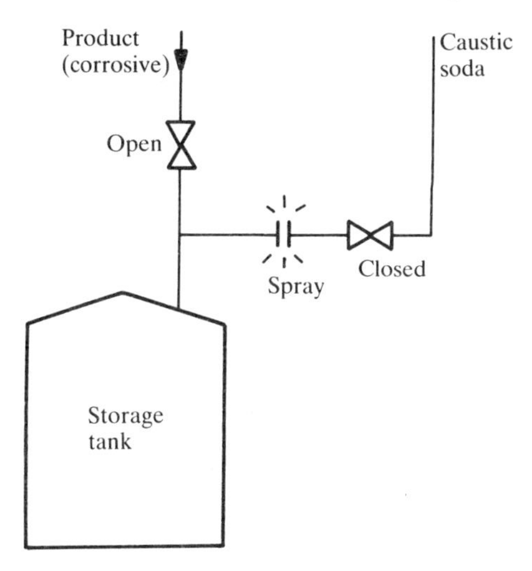

Fig. 17.8 Piping arrangement for caustic soda to a storage tank

- For work on conveyors, lifts, hoists, cranes, etc.
- For work on fragile roofs.
- For work involving the use of high-pressure water-sprays for cleaning.
- In extraordinary/unusual circumstances not covered by routine precautions/instructions.[12]

Clearing of blockages is a special operation which often requires a well-thought-out system of work.

> As illustrated in Fig. 17.8 a permit to work was issued to a fitter to locate and clear a choke in a caustic soda line. He started by breaking the joint shown. When the product line into the tank was blown through with nitrogen, to demonstrate that it was clear, a spray of corrosive product and nitrogen came out of the broken joint.[28]

Monitoring of permit-to-work systems

Whatever the specific system is, it is essential for management to monitor it. This may include:

- A comparison of job-cards or log-books which record the tasks actually performed versus the filed permits to work.
- Random, on the job, checks of permits to work and compliance with the conditions specified.
- Joint consultation about system effectiveness.
- Detailed investigation of any incidents occurring under permit-to-work control.

Modifications (plant or process)

Process and plant modifications are frequently necessary either during operation or in scheduled shut-downs. Even seemingly beneficial modifications can result in an increase in hazard potential unless they are properly thought through first.

Metal was melted in a cupola by means of a thermite reaction. Occasionally the process got out of control and resulted in a spill-over of molten metal from which the operators had to run away.[29] The cupola was therefore relocated over a pit. Accumulation of water in this pit resulted in more serious incidents due to 'steam explosions' on escape of molten metal.

A strict procedure is advisable for the formal proposal, approval, implementation and checking of *all* such modifications to ensure that they do not introduce a new hazard. This involves consideration of all the consequences arising from the changes; a check-list for this type of safety assessment is reproduced as Table 17.8.[30]

Table 17.8 Form for safety assessment of modifications

Safety Assessment

Plant .... *Title* .... *Reg no* ....

Underline those factors which have been changed by the proposal.

Process conditions
temperature
pressure
flow
level
composition
toxicity
flash point
reaction conditions

Operating methods
start-up
routine operation
shut-down
preparation for maintenance
abnormal operation
emergency operation
layout and positioning of controls and instruments

Engineering methods
trip and alarm testing
maintenance procedures
inspection
portable equipment

Safety equipment
fire fighting and detection systems
means of escape
safety equipment for personnel

Environmental conditions
liquid effluent
solid effluent
gaseous effluent
noise

Engineering hardware and design
line diagram
wiring diagram
plant layout
design pressure
design temperature
materials of construction
loads on, or strength of:
foundations
structures
vessels
pipework/supports/bellows
temporary or permanent:
pipework/supports/bellows
valves
slip-plates
restriction plates
filters
instrumentation and control systems
trips and alarms
static electricity
lightning protection
radioactivity
rate of corrosion
rate of erosion
isolation for maintenance
mechanical-electrical
fire protection of cables
handrails
ladders
platforms
walkways
tripping hazard
access for
operation
maintenance
vehicles
plant
fire fighting
underground/overhead:
services
equipment

Table 17.8 (continued)

Within the categories listed below, does the proposal	Yes or no	What problem has been created which affects plant or personnel safety and what action is recommended to minimise it?	Signed & dated
Relief and blowdown			
1 Introduce or alter any potential cause of over/under **pressuring** (or raising or lowering the temperature in) the system or part of it?			
2 Introduce a risk of creating a vacuum in the system or part of it?			
3 In any way affect equipment already installed for the purpose of preventing or minimising over- or under-pressure?			
Area classification			
4 Introduce or alter the location of potential leaks of flammable material?			
5 Alter the chemical composition or the physical properties of the process material?			
6 Introduce new or alter existing electrical equipment?			
Safety equipment			
7 Require the provision of additional safety equipment?			
8 Affect existing safety equipment?			
Operation and design			
9 Introduce new or alter existing hardware?			
10 Require consideration of the relevant Codes of Practice and Specifications?			
11 Affect the process or equipment upstream or downstream of the change?			
12 Affect safe access for pesonnel and equipment, safe places of work and safe layout?			
13 Require revision of equipment inspection frequencies?			
14 Affect any existing trip or alarm system or require additional trip or alarm protection?			
15 Affect the reaction stability or controllability of the process?			
16 Affect existing operating or maintenance procedures or require new procedures?			
17 Alter the composition of, or means of disposal of effluent?			
18 Alter noise level?			

Safety Assessor .... date ....

Checked by ....Plant Manager Checked by ....Engineer

Engineering procedures should be no less strict than those for the original plant. Site work should be regulated by the appropriate permits to work and all obsolete equipment and piping should, when practicable, be isolated or removed from the plant. All drawings, flowsheets, operating instructions or other documents/procedures affected by the modification must be up-dated without delay; all appropriate plant personnel must be informed of the modification and, if necessary, instructed and trained on the new system of work.

The following are some examples of incidents which have arisen because a proper safety check was not made on plant modifications.

A process stream was designed to pass through a series of heat exchangers in series and into a vessel provided with pressure relief as shown in Fig. 17.9. With this arrangement the relief valve served to protect the exchangers from over-pressurisation. During plant construction the isolating valve was added as a useful means of preventing back-flow. Unfortunately the valve was closed during start-up and the heat exchangers were subjected to the full upstream pressure. One burst resulting in a major fire and a delay in start-up.[31]

The piping arrangement in Fig. 17.10 served to provide compressed air to a breathing apparatus. When the compressed air main was renewed the branch to the breathing apparatus network was inadvertently moved to the bottom of the main. Subsequently a man inside a vessel received a faceful of water while wearing a breathing apparatus connected to this system. This was caused by a slug of water in the main filling up the catchpot quicker than it could empty. The purpose of the branch being from the top of the main, i.e. to prevent excessive water draining into the pot, had been overlooked in the modification.[32]

A procedure is necessary to ensure that temporary electrical repairs are carried out properly, and do not become permanent.[33]

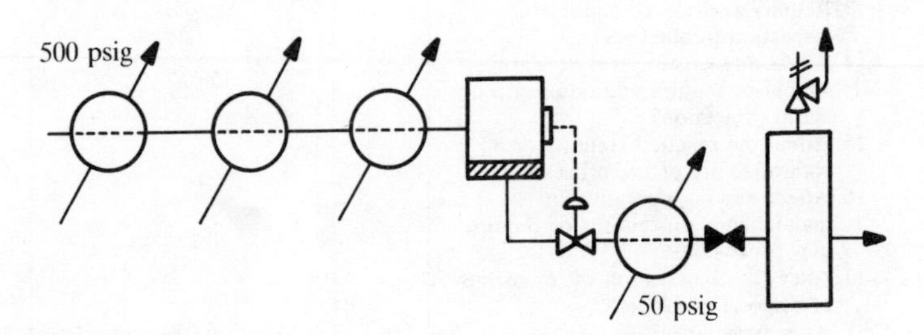

Fig. 17.9 Arrangement allowing isolation of relief valve from heat exchanger train

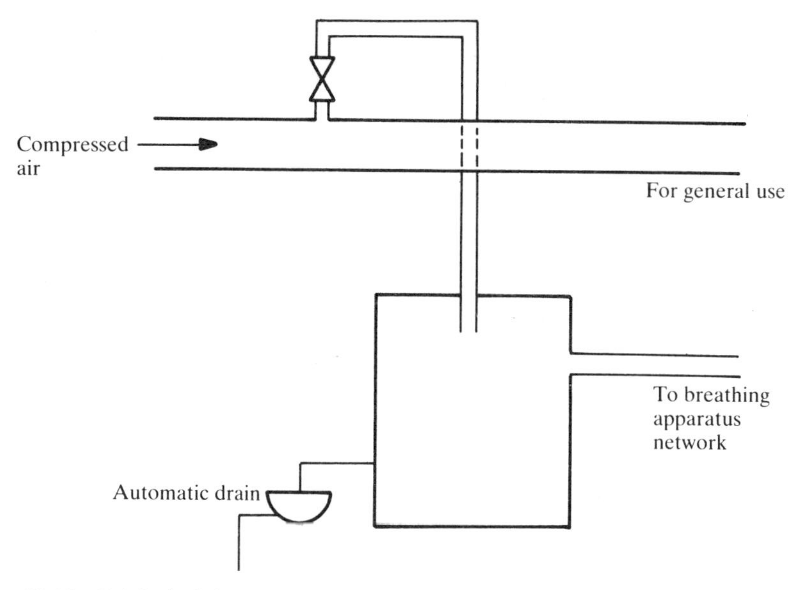

Fig. 17.10 Original piping arrangement to breathing apparatus[32]

Contractors

It is particularly important that non-permanent workers become familiar with plant hazards, rules and safety practices. There should also be a clear delineation of their work and responsibilities.

Therefore, whenever contractors are given access to the site for maintenance, or for construction work, they should work to a strict set of safety rules and procedures.[34] In addition to requiring general observation of rules and safety procedures, and familiarisation of all workers with the fire and evacuation procedures, within the factory these will normally refer to the items in Table 17.9.

Complete dismantling or scrapping of plant handling hazardous substances requires a detailed system of work.

Items to consider include:[38]

- Appointment of a responsible expert e.g. the plant manager.
- Planning the work in single distinct steps.
- Provision of appropriate medical check-ups.
- Isolation, surveillance and monitoring of the area.
- Provision of all appropriate protective equipment.
- Instruction of employees concerning the potential hazards and protective measures.
- Consultation with statutory authorities.

Table 17.9 Control of contractors

Employee control
Provision of an up-dated list of workers on site.
(*This is essential for checks following an emergency evacuation or incident.*)
Requirement for all workers to check in/out of site
Prohibitions on smoking, except in designated areas at approved times.
Prohibition of intoxicating liquor on site.
Limitation of site area to which entry is permitted.

Security
Any work breaching the site fence to require authorisation.
Limitations of vehicles entering the site.
Procedures for authorisation and inspection of vehicles.
Restriction of areas for vehicular access.
Compliance with traffic regulations and site speed limits.

Health and welfare
Provision of toilets and washrooms.
Provision of canteen (and drinking water supply if necessary).
Provision of first-aid kits and first-aid cover.
Procedure for reporting accidents.
Provide copy of Company Safety Policy to contractors at time of agreeing contract.
Give induction training to contractors and provide them with a copy of site safety rules.

Personnel protection
Rules for the provision and use of eye-protection.[35]
Practices to prevent accidents due to falling objects, e.g. use of barriers, notices, helmets and approved methods of working.[36]
Provision of safety-nets, -sheets and -belts.[36]
Maintenance of safe access and egress.

Services
Authorisation required for connection to site services.
Specific authorisation required for work affecting integrity of site safety, e.g. shutting-off/using water, restricting access, etc.
Prohibition on opening valves, breaking lines, opening vessels, etc., without authorisation.

Protection of work, equipment and access
Provision of temporary barriers, warning lights, etc.
Integrity of scaffolding, platforms and equipment, etc., used in the work.[37]
Maintenance of roads, walkways and means of access.
Procedure for maintenance of ladders, tools, lifting gear, equipment, etc.[37]
Procedures for the storage, handling and use of compressed gas cylinders.
Procedure for siting and relocation of cranes.
Provision of fire extinguishers in temporary rooms and working areas.

Work permits and safety permits
Obligation to comply with works permit (unless a safety permit is issued).
Procedures and requirements for safety permits (i.e. permits to carry out potentially hazardous work).

Personal protection

While generally considered only as a back-up to measures for the removal of hazards at source, the provision and regular use of personal protective clothing and equipment is an important system. These should be carefully selected to meet the specific hazards in any particular operation; examples are given in Chapter 6. Procedures are then necessary to allow for proper storage, inspection, cleaning, and repair, or replacement of the clothing and equipment provided.

Protective clothing and apparatus, which can be uncomfortable to wear for extended periods, sometimes falls into disuse, particularly when it is intended to provide protection against long-term hazards (e.g. cumulative exposure to low levels of contamination or to noise, or to risk which the wearer comes to consider as remote). Management has, therefore, to explain the need for items of personal protection, to ensure that replacement items are always readily available, and to admonish the workforce to make full use of them. A sound procedure with regard to eye protection is for it to be provided and used in all operations involving acids, alkalis, dangerous corrosive substances or any substances similarly injurious to the eyes unless, in effect, any foreseeable risk has been 'designed out'.[35] Management should also make provision for the instruction and issue of appropriate protective clothing and apparatus to visitors, etc.

Monitoring, auditing and inspections

Monitoring of physical plant and of control procedures and formal inspection and testing programmes are very important. Therefore, they appear under several headings above. Surveys of plant-operating procedures, the hardware and the environment by qualified personnel are invaluable. In particular they may identify unsafe practices which have developed with time or deviations from established procedures.

Furthermore, they are an essential part of the appraisal, analysis and measurement of safety performance.[39]

Safety audits[40]

In a safety audit each area of a company's activity is subjected to a systematic critical examination with the object of minimising loss. Each component of the total system is covered, e.g.:

- Management policy
- Attitudes training
- Process and design features
- Plant construction and layout
- Operating procedures
- Emergency plans
- Standards of personal protection
- Accident records.

The aim is to discover the strengths and weaknesses and the main areas of vulnerability, or risk, so that the action necessary to minimise hazards can be determined *before* personal injury, or damage, occurs.

However, even audits of this type may not be completely effective.

The fire which followed the explosion at Flixborough was fuelled from the tank farm where flammable liquids were stored under licence from the local authority for 38,600 litres but there was an actual quantity of 197,300 litres. This had not been revealed by the special inspection.[41]

Ideally an audit should cover the examination and qualitative assessment of all facets of safety and every activity. Such activities include:

- Research and development
- Design
- Occupational health and hygiene
- Environmental control
- Product and public safety (including storage, packaging, labelling and transportation)
- Production (including technical operation, maintenance, clearance certificates, emergency procedures, job descriptions, operating instructions, training, housekeeping, personal attitudes).

Audit techniques, i.e. a formalised, systematic, critical approach can usefully be applied in the development of a new process or plant or in major alterations. An example is proposed in Table 17.10.[40]

Several design safety check-lists are available in the literature.[42–44] Whenever an audit is applied, questions may result which require policy decisions and proposals for the authorisation of capital expenditure. Therefore the company's senior management should be seen:

1. To be the authority for the formal audit.
2. To be prepared to commit manpower and resources to implement agreed changes.
3. To be directly involved in the review of audit reports leading to an action plan and in subsequent progress reviews.

Inspections will, in addition to covering the specified objectives of the audit, aim to promote contact with the workforce. This serves to publicise

Table 17.10 Audits during plant design, construction and commissioning

Stage 1	Pre-authorisation study
Stage 2	In-depth study – carried out when process package plan is complete.
Stage 3	Insurance study – usually undertaken as soon as agreement on stage 2 is reached.
Stage 4	Follow-up studies – carried out at regular intervals during construction.
Stage 5	Pre-start-up study – carried out within one month of start-up.
Stage 6	Review studies – carried out within six months of start-up and subsequently regular audits of defined areas and activities as with existing plants.

management's concern for safety and loss prevention, to gain the involvement of the workforce and to encourage suggestions relating to safety. The co-operation of the plant operators, maintenance workers, foremen, etc., is extremely important in order to discover conditions and practices which need correction to bring a plant/works up to an approved safety standard.

The frequency of auditing is clearly related to the type of activities. In general, on chemical plants inspections should be on an annual basis with more frequent inspections for specific activities or plant areas. The latter may be identified from records of injury and damage accidents or may be 'obvious' high-risk activities or locations, e.g. LPG-handling facilities.

The general principles of an audit system involve:

- Identification of possible loss-producing situations.
- Assessment of potential losses associated with these risks.
- Selection of measures to minimise losses.
- Implementation of these measures within the organisation.
- Monitoring of the changes.

For any on-site audit it is usual to examine first the accident records and to prepare and submit a questionnaire or check-list. An example is given in Table 17.11. A well-constructed questionnaire, with a full list of matters to be covered, will reveal many of the areas requiring attention. The answers should be discussed between the auditors and management before on-site work begins and the areas and activities to be inspected agreed. There is, of course, no limit to items which may be investigated; some possible additions are listed in Table 17.12.

Table 17.11 Plant safety audit check-list

Statutory requirements
Methods of process operation. Hazards of the materials, unit operations and processes
Material handling
Tools, machinery, maintenance equipment
Permit-to-work system. Schedules for regular inspection of emergency equipment
Personal protective equipment; condition, care and suitability
Plant tidiness, condition of floors, stairs, walkways, cleanliness of toilets and washing facilities, environmental factors, waste disposal
Fire prevention and protection, alarm systems, emergency exits. Flammable material storage
Unsafe practices
Arrangements for treating injuries. Condition of safety showers, eye-wash facilities, resuscitation equipment, first-aid boxes
Involvement of employees in safety activities, their knowledge of the safety organisation, attitudes. Condition of displays and notices.

Table 17.12 Additional items which can be covered by check-list (or be the subject of specific audits or surveys)

Electro-static hazards
Alarms and trip systems and testing
Pressure vessels and relief valve inspections
Cable protection
Radiation hazards
Operation and maintenance of internal and external transport
Occupational health and hygiene standards
Electrical equipment maintenance
Major emergency arrangements (updating of plans and procedures)
New projects and processes
Adherence to Codes of Practice (internal and external)
Equipment faults (type, incidence, reporting and correction)
Labelling of products
Adherence to approved work and operational procedures (i.e. systems)
Training procedures/programmes and their effectiveness
Adequacy of safety procedures in office, administrative and ancillary buildings
Methods of communication
Receipt, storage and transportation of chemicals and other materials
Extensions/alterations to existing production and storage facilities
Pollution/environmental control methods
Liaison with contractors
Safety standards in research, development and laboratory work
Consideration of safety and loss prevention at the design stage
Security arrangements covering the storage and issue of harmful substances
Safety advisory service to customers, covering products, their handling and use

The type of detailed check-list varies with company size, activities, commitment and nationality; some include a rating form by means of which standards may be compared within a company.[45]

The audit 'team' may in a small unit consist of one person, e.g. the works manager or engineer; if there are several units the audit may be carried out by a 'visiting' works manager. Competent consultants may be used. On larger chemical plants a team, e.g. including an engineer, chemist and a safety committee representative, together with the safety officer and relevant specialists may be used. Some training in auditing is advantageous.

The check-list will be completed by the audit team by:

- Incorporating basic information supplied by management.
- Personal observation.
- Inspection of records.

- Discussions with local management, technical staff, supervisors and workers in the area.
- Verification that established procedures and standards are in fact followed, and that emergency arrangements are known and understood.

Any unsafe conditions or practices which may be revealed, and which require immediate action, should be drawn to the attention of local management; the situation should still, however, be noted in the audit report.

An audit report is finally prepared according to an agreed timetable. This is discussed with the authorising company executive to generate minuted actions from the recommendations. A monitoring system is required to ensure that the work or modifications required are in fact implemented within the allotted time-scale. It is also important to communicate to appropriate personnel (including the safety committee and any appropriate works technical committees):

- A summary of the audit report.
- Agreed recommendations for action and the time-scale.
- The reasons for any inaction on any matters referred to or recommended in the audit report.

To exemplify the type of check-list which can easily be drawn up just one topic from Table 17.11 – fire prevention and protection – because of its particular relevance to chemicals handling, is dealt with in Table 17.13 (after ref. 46). This is a general factory list for expansion in specific cases (see Ch. 4).

The adequacy and limitations of insurance cover may also be audited. In a complementary activity – safety surveying – instead of examining all the equipment and activities in a plant, one particular type is chosen and examined in depth. For example testing (e.g. routine testing of trips and alarms), inspecting, or one particular type of equipment may be selected for thorough investigation.[47]

Other inspections

Within the UK union-appointed safety representatives are legally entitled to carry out inspections,[48] viz.

- Of the workplace on a regular basis, for hazard spotting.
- Following a notifiable accident[49] to discover the cause.
- Of any part of the workplace which has housed a dangerous occurrence.[50]
- Of relevant parts of a workplace in which there have been substantial changes in work conditions, e.g. introduction of new machinery, new substances or new processes.
- Of the workplace where a notifiable disease has been contracted.[51]

Table 17.13 Fire safety check-list

Control of ignition sources

- Instruction of all relevant employees in use, maintenance, and hazard detection with electrical equipment.
- Instruction of employees regarding smoking rules. Provision of adequate arrangements for publicity and compliance.
- Instruction of all relevant employees in use, maintenance, and hazard detection with all equipment which can generate heat mechanically or sparks.
- Permit system for all 'hot-work'.
- Training of staff to recognise, and to avoid, dangerous actions likely to result in a fire.
- Adequate arrangements for inspection and maintenance of boilers, gas- or oil-heating plant.
- Adequate arrangements to prevent arson or fires started by children.[100]
- Adequate arrangements, including a designated area, for rubbish burning.
- Routine investigation of all new processes for potential fire hazards.

Combustion prevention

- Regular clearance of waste and rubbish.
- Segregation, and spacing, of waste and rubbish from potential ignition sources and other combustible materials.
- Adequate coverings for waste bins.
- Control of use of combustible materials in building structure and fittings.
- Selection of electrical insulation to appropriate specification and regular inspections.
- Spacing of heating and lighting appliances from combustible materials.
- Clearances at the rear of radiators and heating units.
- Strict control of any portable heating units.
- Segregation, and spacing, of raw materials, process materials and products from potential ignition sources.
- Control of all substances liable to spontaneous combustion.
- Removal and disposal of all packaging/wrapping materials from any non-flammable materials stored close to potential ignition sources.
- Installation of a sufficient number of the correct type of fire-fighting appliances for the fire risks in the area.
- Performance of morning and evening checks (for combustion hazards, equipment left running, etc.).
- Performance of periodic audits.
- Location of fire-fighting appliances in the correct position and unobstructed.
- Regular inspection, testing and maintenance of all fire-fighting appliances.
- Provision of automatic sprinkler systems or automatic roof vents (in addition to portable extinguishers).

Limitation of fire spread and effects

- Possibility of toxic products due to pyrolysis or burning.
- Rapidity of fire discovery.
- Adequacy of installed fire detectors (smoke, radiation, flammable gas, heat).
- Regular inspection, maintenance and testing of detectors.
- Spacing of storage from detectors and sprinklers.

Table 17.13 (continued)

- Spacing of large inventories of combustible materials from employee activities, or provision of fire-separation walls.
- Unless isolated, spacing of storage areas at an adequate distance from production areas.
- Adequate arrangements for workplace ventilation.
- Adequacy of provisions for containing leakages/spillages of flammable liquids, oils, fats, etc.
- Design and construction of buildings to prevent rapid spread of fire.
- Provision for dealing with leaks, and clearing up, spillages rapidly.
- Selection of building construction materials, roof linings, ceiling and wall panels, etc., of appropriate fire resistance.
- Arrangement of work-flows to allow openings between areas to be fitted with self-closing, fire-resistant doors (to remain closed except during use).
- Control of airborne materials and deposits.
- Periodic audits; morning and evening checks (on correct storage, access, fire doors, etc.).
- Operability of all fire doors, escape doors, etc.
- Provision of adequate fire hydrants, clearly marked and unobstructed.

Systems of work

- Adequate fire alarms, visible/audible to all employees.
- Regular inspection, maintenance and testing of all alarms.
- Provision of clear escape routes; clearly marked, and unobstructed.
- Provision of clearly marked assembly points.
- Adequate training of employees in fire prevention, fire drills and selection/use of appropriate portable fire-fighting equipment.
- Training of security staff in fire detection, fire-fighting, and in spotting fire hazards during patrols.
- Provision of clear, comprehensive instructions for normal and emergency procedures.
- Clear allocation of responsibilities in the event of a fire.
- Efficient evacuation procedures.
- Familiarity of all employees with main and alternative escape routes and assembly points.
- Arrangements for rapid roll-call on assembly and for searching every part of the premises.
- Performance of fire drills at least twice a year.

- Of that part of a workplace about which new information relevant to hazards is published by the HSC or HSE.
- To investigate complaints from any employee they represent relating to that employee's health, safety or welfare at work.
- Under their own initiative.

Table 17.14 Possible equipment, facilities and information requirements for safety representative's inspections[46]

Facilities and information	Equipment
Access to data banks	Audiometric devices
Access to industrial engineer	Camera
Access to occupational hygienist	Danger tags to attach to equipment that should not be used
Access to occupational physician	Flashlights – insulated
Access to safety engineer	Goggles – appropriate to conditions
Access to specialists in chemicals, e.g. toxicologist, chief chemist, research chemist, etc.	Hard hats
Accommodation for library	Light meter
Address book and indexing system	Measuring tape
Blackboard and chalk	Micrometer
Explanation of information which is vague or not understood	Multi-pocket shoulder bag
Filing cabinet	Notebooks, pencils and coloured pens
Identification of, and access to, any source of relevant knowledge or information not in possession of organisation, e.g. geologist, construction engineer, design engineer, etc.	Padlocks for locking out switches when inspecting certain machines or cranes
Office and desk	Plumb line
Photocopying services	Portable instruments for testing the atmosphere for toxic, inflammable or explosive substances
Services of external specialist consultants	Pressure gauges
Services of independent testing organisations	Respiratory protective equipment
Sight of relevant published material	Revolution counter for checking speeds of grinding wheels, pulleys, shafting, flywheels and saws
Stationery and other office materials	Safe and snug-fitting protective clothing (loose sleeves, flowing ties, jewellery, etc., should be protected or divested)
Telephone	Safety shoes – appropriate to conditions
Typing services	Small tape cassette recorder
Wall charts	Spirit level
	Stop watch
	Thermometer
	Velometer for testing air movement

- Of documents which the employer is required to keep under the Factories Act or other statutory rules, e.g. records of tests on specified pieces of equipment, or data sheets under s.6 of the HASAWA, etc.

Full details of the rights of employers and statutory safety representatives are given in reference 46. However, guidance on the type of equipment, facilities and information that may be required on such inspections is summarised in Table 17.14.[46]

Products and wastes management

As is clear from Chapters 13 and 14, management's responsibilities for safety are not constrained by site boundaries.

Thus the activities related to marketing and, as mentioned earlier, wastes disposal also need auditing for safety and loss prevention. For example it follows that items to be covered in a product safety audit will include:

- Product testing, limitations on use, precautions needed in use and disposal.
- Quality control; adequacy to prevent off-specification/defective product being marketed.
- Packaging and labelling adequacy.
- Supply of information to packagers, shippers, warehouse staff, distributors, etc.
- Supply of information to users. Any restrictions on sale. Records of users.
- Adequacy of system for feedback of user queries, complaints, problems, etc.
- Up-dating of information supply with increased knowledge, variation in product formulation, etc.
- Provisions for product recall in extreme cases.

Rapid ranking of hazards

In one method of rapid ranking,[52] to set priorities for loss control resources, the hazards are categorised by their severity in relation to:

- The unit and its workforce.
- Installation or plant.
- Business (i.e. interruption or loss).
- Public (i.e. property damage, personal injury, environmental effects).

Typically, each of five or six categories is assigned a guide frequency decreasing with severity; it is sufficiently precise to agree the guide frequencies in decades. For example, once per year for hazard category 1, decreasing to one in 10^4 years for the very severe consequences included in category 5. Table 17.15 shows one hazard categorisation scheme used in locations, ranging from oil and petrochemical plants to warehouses and distribution depots. The expected frequencies for the hazards to be ranked are evaluated by a combination of:

- Historical data from the site, industry or generic data.
- Consensus of estimates from site personnel.
- Simple fault-tree analysis down to levels where the frequency of events can be verified.

The hazards are ranked A to D; A requires the highest priority of loss-control effort and those ranked D may not justify effort to reduce their expected frequency. Table 17.6 sets out the rules for ranking identified hazards, by comparing their expected frequency with the guide-frequencies set out in Table 17.15 for the category into which the hazard falls.

Table 17.15 The principles of hazard categorisation for consequences

Area at risk	Description of risk	Hazard category				
		1	2	3	4	5
Plant	Damage	Minor < £10^3	Appreciable < £10^4	Major < £10^5	Severe < £10^6	Total destruction
	Effect on personnel	Minor injury only	Injuries	1 in 10 chance of a fatality	Fatality	Multiple fatalities
Works	Damage	None	None	Minor	Appreciable	Serious
Business	Business loss	None	None	Minor	Severe	Total loss of business
Public	Damage	None	Very minor	Minor	Appreciable	Widespread
	Effects on people	None (smells)	Minor	Some hospitalisation	1 in 10 chance of public fatality	Fatality
	Reaction	None/mild	Minor local outcry	Considerable local and national press reaction	Severe local and considerable national press reaction	Severe national (pressure to stop business)
Guide values of publicly acceptable frequencies		1/yr	1/10 yr	1/100 yr	1/10^3 yr	1/10^4 yr

Table 17.16 Hazard ranking rules

Hazard category (from Table 17.15)	Expected frequency compared with guide frequency			
	Smaller	Same	Greater	Uncertain U
1	Such incidents do not fall within the scope of this method and many will remain unidentified. Such as are identified are recorded thus:			
	D	D	D/C at team's discretion	
2	D	Normally C, but if upper end of frequency/potential could be raised to B at team's discretion	Equally damaging hazard as those below A, but if lower end of frequency/potential could be lowered to B at team's discretion	B Frequency estimate should not be difficult at this category; may be a lack of fundamental knowledge which requires research.
3	C	B	A Major hazard	A/B at team's discretion. Such potential should be better understood.
4 & 5	B/C at team's discretion	B but can be raised to A at team's discretion	A Major hazard	A Such potential should be better understood.

If a frequency cannot be estimated with sufficient precision the 'uncertain' column in Table 17.16 may be used to rank these hazards. Evaluation of these uncertain frequencies may be deferred until their hazards become the highest ranked.

The following example illustrates how hazards are given a category rating[53]:

Fire in a pipe trench[54] Ignition of oil, spilled during the removal of a slip-plate, causes a fire in a pipe trench which requires that throughput be reduced on a process unit for 1 hr with a 1-in-10 chance that the worker removing the slip-plate may receive fatal burns. Referring to Table 17.15 this hazard is category 3, because of the personnel risk. The scenario suggests minor disruption to the process unit, negligible effect on the plant and the loss/damage in the pipe trench should not exceed £10,000. The effect on equipment of this hazard would place it in category 2 but, unless the chance of a fatality is discounted, it must remain in category 3. (When this hazard is ranked, the expected frequency of pipe-trench fires will be scaled down by the probability of ignition while the worker is removing the slip-plate *and* where it is 10% certain that he will be trapped by the fire.)

A fire occurred while a new pipe was being laid in a trench when welding was in progress 20 m from the flange where the slip-plate was to be removed. Vapour from a limited spillage of light oil will normally not spread 20 m, but there was water in the trench and the assistant foreman did not inspect the work area before signing the permit to work nor did the foreman who endorsed this permit and that for the welding, issued earlier that day. The worker who was removing the slip-plate sustained fatal burns.

The following data were used to estimate the annual frequency of the pipe-trench hazard[53]:

- Number of slip-plates inserted, or removed, in the pipe trenches of this plant per year (from work permit records). = 52
- Estimated proportion of these lines which carry volatile oils. = 50%
- Time to swing a slip-plate ≃ 20 min. = 1/26,000 year
- Frequency of pipe-trench fires (ignition source unknown) on this site. = 1 in 2 yr = (1/2 per yr)
- Duration of welding in pipe trenches – 3 weeks per year during day-shift only. = 1/52 yr
- Estimated proportion of pipe-trench area covered by water. = 20%
- Proportion of slip-plate work-permits issued without inspection of the area for ignition sources. = 10%

Two frequencies can be calculated for the pipe-trench fire hazard with a 1-in-10 chance of the operator being caught in the fire.

- Exposure time of operators to volatile oil spills in any one year =

$$\frac{52 \times 0.5}{26{,}000} = 10^{-3} \text{ yr/yr.}$$

- Frequency of pipe-trench fires anywhere on site while an operator is carrying out a slip-plate change on a volatile oil line = $1/2 \times 10^{-3}$ per year.

  For a 1-in-10 chance of the operator being caught, 9 out of 10 of these fires must not be related to the operator's current task.

- Estimated frequency due to pipe-trench fires where the operator has a 1-in-10 chance of not being trapped = $1/2 \times 10^{-2}$ per year.

  If the swinging of slip-plates is uniformly distributed throughout a year, the number of such operations carried out at the same time as welding = $52 \times 0.5 \times 1/52 = 0.5$ per year.

  Probability of the above operation causing ignition of oil spreading on water in the pipe trench = $0.2 \times 0.1 = 1/50$.

- Estimated frequency where the operator will be caught by fire spreading on water = $0.5/50 = 1/100$ year.

  The aggregate of frequencies calculated above is 1.5×10^{2} per year, which exceeds the guide frequency of 10^{-2} per year set in Table 17.15 for category 3 harzards.

  There were two hazard scenarios:

1. Pipe-trench fire causes up to £10,000 damage and reduction of throughput on a process unit for up to 1 hr: category = 2. Expected frequency = 5 in 10 years. Using Table 17.16 this is ranked A by the rules, and cannot be lowered to B because its frequency is 5 times the guide-frequency.
2. As for (1) but a worker who is swinging a slip-plate receives fatal burns: category = 3. Expected frequency = 1.5 in 100 years. This is ranked A, but with only 1.5 times more than the guide-frequency.

By improving the permit-to-work system, and ensuring that foremen issuing permits inspect the site immediately before issues, the proportion of slip-plate permits which are still issued without inspection could drop to 1 in 50. The aggregate expected frequency for hazard (2) would be reduced to (1/5 + 1/2) in 100 years. From Table 17.16 this hazard would then be ranked C, which is acceptable.

Hazard (1) is ranked A, owing to the high frequency of pipe-trench fires on this site. By good housekeeping and adequate drains, coupled with stringent inspection and control of welding, the frequency of fires causing £10,000 damage could be reduced to 1 in 10 years (ranked C) and the scale of pipe-trench fires would be reduced to those causing less than £1,000 worth of damage (category 1, ranked D).

(Incidentally this is not an isolated type of incident; rapid spread of flame across a thin layer of organics on top of water in a pit was one factor in a fire which destroyed a cobalt refinery utilising solvent extraction in 1972.)[55]

Accident investigation, analysis and reporting

The purpose of accident investigation is to establish the cause, e.g. the chain of events resulting in the incident; this may in the case of some 'chemicals-related' incidents involve considerable technical detail. Having established the cause it is usually necessary to assess the losses, e.g. personal injury, damage to plant and equipment, and third party losses, consequential losses, etc.

The conclusions from an accident investigation may be required for any, or all, of the following actions:

1. To prevent a recurrence on the same plant or other plants using the same, or similar, processes or operations.
2. To establish liability, e.g. whether failure of equipment was due to an inherent fault or to lack of care by installers, contractors, etc.
3. To settle, or initiate legal proceedings for compensation, claims.

As noted on page 999, safety representatives may become involved in relevant inspections or the investigation. Of course investigations may also be instituted by statutory authorities (e.g. HSE in the UK) but these should be considered separate from those carried out in-company or on behalf of insurers.

In fact, the interests in loss and potential loss investigation vary widely as exemplified in Table 17.17 (after ref. 56).

Table 17.17 'Interests' in loss investigation

	Interests
Owner	Liability for safety of workplace and the value of property.
Management	Sales, work processes, general liability. Business organisation, direction and control. General interest from a moral standpoint. Desire to limit future losses.
Suppliers	Product liability relating to materials or equipment provided.
Insurers	Obligations to pay benefits, damage repairs, business interruption settlements.
Supervisors	Management of day-to-day work tasks.
Workers	Personal health and welfare. Compensation/benefits for any 'injury' sustained.
Union	As for workers, plus 'workers rights'.
Community	Hazards to residents. Economic and environmental impacts on the local community.
Government	Compliance with regulations, codes, standards. Any national impact.

Ideally every accident, dangerous incident and 'near miss' situation should be properly investigated, since the difference between a minor accident and a serious one is often a matter of chance. However, the thoroughness of investigation tends inevitably, in practice, to be related to the seriousness of the *actual* consequences. The technical and administrative aspects of arson investigations are summarised in reference 100.

Different types of report forms may be used for accidents involving injury to personnel, e.g. as in Fig. 17.11 or 'dangerous occurrences', e.g. an incident in which property was, or may have been, damaged or where personal injury may have occurred.[57]

INJURY REPORT

Report No. ....
Date ....

Plant Name ....

"A" TO BE COMPLETED BY THE SURGERY — NOTE: – USE BLOCK LETTERS

Surname — Christian Names — No. — Age

1. Name of Injured Person ....
2. Home Address ....
3. Employed by GRC ...... Yrs. ...... Mths. ...... On Present Job ...... Yrs. ...... Mths. ......
4. Location of Accident ....
   Date of Injury .... Time of Injury .... a.m./p.m.
   Date Reported to Surgery .... Time Reported to Surgery .... a.m./p.m. ....
5. Nature of Injury ....
6. Witness — Name .... Dept. ....
   Name .... Dept. ....
7. Details of Accident as Reported by Employee ....
8. Hour Commenced Work .... a.m./p.m. .... Date and Time Left Work ....
   Date and Time Returned .... Signature ....

"B" SUPERVISOR'S REPORT (Complete and Forward Within 24 Hours)

1. How did the Accident Occur? ....
2. If Unsafe Condition What Have You Done to Rectify? ....
3. Had the Injured Person been Instructed in a manner that should have Prevented this Accident? ....
4. Material or Equipment Damage ....
5. Loss of Production Time ....
   Date .... Signature ....

"C" SUPERVISOR OF PERSON COMPLETING "B" ....

Signature .... Date ....

"D" DEPARTMENTAL MANAGER	"E" LOSS CONTROL MANAGER
Signature .... Date	Signature .... Date ....

Fig. 17.11 Injury report[57]

One recommended standard practice[56] requires any employee to immediately report any accident involving 'personal injury' (i.e. involving fractures, work restriction, suturing, medical treatment of the eyes, lost time) or – of course – any fatal accident to his immediate supervisor. Any property damage above a specified minimum must also be reported. All injuries should be treated in the surgery, with the immediate supervisor being informed either before or afterwards depending on the emergency. The employee is expected to provide all possible information regarding causation.

Where personal injury is involved the supervisor is required immediately to conduct a complete investigation with all persons concerned, with such interdepartmental co-operation as is necessary. He should then submit a report within 24 hours; he is also expected to visit an injured worker throughout the recovery period.

All property damage accidents should be similarly investigated and reported. The Safety Department should be notified and assist in the investigation. Evidence should not be disturbed pending its completion; a service request to repair damages is then initiated by the supervisor – with clear indications that it is 'accidental damage' for accounting purposes.

In the case of any 'major accident', i.e. involving lost-time injury, or fatality, or property damage in excess of a significant cost, the relevant superintendent is required to conduct a meeting within 48 hours. Managers are expected to follow this up promptly.

The Safety Department is expected to assist in complete investigations as required, make periodic checks that remedial action/repairs have been made, audit repair work orders and collect costs, and complete part of the Accident Investigation Report. This department should also initiate investigations by the supervisor of any accidental damage which has not been reported.[56] (Note that this is US recommended practice and would vary in the UK and elsewhere in factories where line management possibly plays a lesser role in safety management.)

Accident analysis

Accident analysis is a systematic study of the factors and events leading up to an accident, and of the different factors which determined its consequences, e.g. damage or personal injury. It is illustrated in Fig. 17.12.

In the induction phase events occur which lead up to an accident. A point of no return is precipitated by a triggering incident or final event (e.g. the cyclohexane vapour cloud released at Flixborough finding a source of ignition). Up to this point an accident may be avoided (e.g. the cloud may have drifted, been diluted beneath the LEL and dispersed harmlessly); once past it an accident is inevitable.

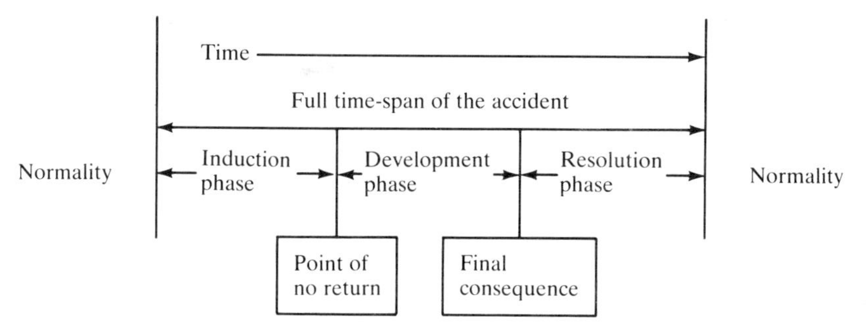

Fig. 17.12 Accident analysis

In the development phase events and situations interact to determine the consequences of the accident. Finally, in the resolution phase resources must be expended to return to normality.

Site investigation

The following guidelines will assist effective site investigation bearing in mind that once the resolution phase commences valuable evidence may be lost.

Preparation

- Collect any available documents of relevance, e.g. operating or maintenance instructions, manufacturer's instructions.
- Obtain plans, elevations, flowsheets, instrument diagrams, etc., as necessary.
- If statements have been taken from witnesses (anyone who saw the accident, knows something of the events leading up to it, or who came on the scene soon afterwards and can describe what he or she found) collect and read them.

On-site

- Determine the correct orientation of the accident scene relative to drawings.
- Interview technical/managerial personnel. (This is best done by going through events in chronological order.)
- Collect any internal reports or notifications of the accident, any reports of previous similar incidents, any descriptions of process/plant modifications of relevance.
- Interview witnesses (if not done already).
- Take, or have taken, a full photographic record including several views, detailed views of all items to figure in the report, and close-up views of specific items.

- Collect, label (as to source and drawings/photographs) any materials for evidence or technical examination or chemical analysis.

Conflicting descriptions or histories are often obtained from witnesses following the confusion associated with an accident. However, in very many cases the site technical management will be able to pin-point the cause with reasonable certainty.

If material or specimens are to be sent to independent laboratories to assist with the investigation, it is important to brief them properly as to the purpose for which their specialist assistance is required. For example, full chemical analyses may be quite superfluous if all that is required is confirmation of the presence of a particular contaminant. Any samples must be representative and, as mentioned earlier, be from a clearly identifiable source. Duplicate samples should be retained whenever possible.

Reporting

The nature of the report will vary with the intended readership and objectives. However, the following sections are typical:

1. Background.
   Site and process description.
   Normal mode of operation.
2. Sequence of events in the induction phase (with reference to chronology and supporting evidence from witnesses – and to plant records).
3. Development phase.
4. Final event. Consequences/losses. Injuries sustained. Emergency actions.
5. Resolution to date of reports.
6. Conclusions as to causation, liabilities, actions to minimise similar events in future, etc.
7. Appendices. Drawings, sketches, extracts from log-books, record-charts. Summary of chemical analyses. Collected statements from witnesses.

Occupational diseases

The investigation of cases of occupational disease, often involving prolonged working in an environment over a significant time-span, follows a similar pattern to accidents but on-site enquiries must also include:

- Identification, and if possible quantification, of the chemicals or physical environment involved. (These may have changed considerably over the time-span involved.)

- Assessment of the conditions on inspection, noting changes/improvements over the time-span. Evaluation of the personal protective equipment, again noting changes made over the period concerned.
- Identification of any similar or related cases.
- Recording the working history and exposures involved.
- If necessary arranging for measurements to be made of exposures, etc. (see Ch. 7).
- Noting previous occupational history before the current employment, and any relevant observations from pre-employment medicals.

Education, training and supervision

The training and instruction of the workforce, covering all grades and skills, is an essential part of any system of work. In particular management has a duty to instruct operators, maintenance workers, etc., in safe procedures and of any hazards associated with their work and to provide adequate training.

The following discussion relates to safety training as an integral part of job training. The importance of this is made clear in the UK in the HASAW etc. Act, s.2.2(c), which imposes an obligation on employers to provide:

> 'such information, training and supervision as is necessary to ensure, as far as is reasonably practicable, the health and safety at work of their employees'.

In general, safety training can be identified under the main headings[58]:

- Legal requirements
- Hazards at work
- Safe working methods
- Accident prevention.

However, the training needs obviously vary between managers, supervisors, specialist staff, designers, operators and support personnel, engineering staff, laboratory staff, etc. Additional training is necessary for safety representatives[59] regarding technical hazards of the workplace, appropriate precautions, methods of work and company organisation and systems.

In fact, training dcsign and the preparation of operating manuals, or other job-aids, depend upon a clear understanding of what the specific person is supposed to do. The technique of task analysis[60,61] may be applied to ascertain what workers do when performing tasks, i.e. how they plan their activities, how they respond to different cues in the working environment (e.g. feedback from the plant) and what actions they actually carry out. (Incidentally, this also enables a check to be made that suitable controls/equipment and adequate information are provided for the workers to carry out their tasks properly and reliably.)

Some guidance is available on ways of reducing human error in process operation – by using good ergonomics practice to propose improvements in working methods, and indeed in equipment.[62]

Induction training

On commencement of a new job basic safety information and training should include, as a minimum:

- Fire and emergency procedures
- First-aid and hygiene facilities
- Company safety policy and its implications
- Basic safety rules and procedures
- Major hazards and risks
- Accident reporting procedure
- Identity and locations of supervisor, safety advisor and safety representatives.

Here 'an emergency' relates not only to major incidents, as discussed earlier and in Chapter 15, but to actions to be taken[63]:

- If an employee is seriously injured or develops severe symptoms of illness.
- To evacuate the building (and when) and assemble in designated locations.
- To raise a fire alarm, to report a fire (e.g. to the local brigade) and to use portable extinguishers and any other fire-fighting equipment.
- To shut off electricity and other services, e.g. gas, and other special hazards (and when).

The topics may be covered in a formal course or be dealt with by supervision, e.g. using a check-list. The detail in which each item is covered will vary with the age and experience of the new employee and the nature of the job.

Common human failings in fires, which training should attempt to eliminate, include:

- Failing to detect a fire rapidly.
- Inability – due to unfamiliarity, lack of practice or excitement – to use a fire extinguisher.

A man ran for a portable, soda-acid type, water extinguisher to deal with a small fire. He carried it over his shoulder upside down and, as the soda and acid mixed, it expelled the water. It formed a pool on the floor in which he slipped, injuring his back.[64]

- Entering areas to recover personal possessions – ignoring the hazard from accumulating heat and smoke.
- Following normal routes out of areas ignoring fire escapes.

It is also important that employees are told to 'size-up' any fire, i.e. to assess whether it is likely to be extinguished, or contained, by the fire extinguisher to hand.

Elements of a health and safety induction training programme, which can be adapted to specific needs of a company and employees, are given in reference 58. The normal job instructions should also include reference to safe working methods, the obligation to follow safety precautions, and the legal duties of the employee (e.g. under the HASAW etc. Act and other relevant statutory legislation). Finally, the supervisor/manager should be satisfied that the employee is capable of safe and effective performance before permitting 'unsupervised' work.

Training programmes

Managers

Managers will normally be responsible for safety within their department, for the implementation of the company's safety policy, and for ensuring that all relevant statutory legislation is complied with.

How much knowledge and skill a manager requires with regard to health and safety matters will obviously vary with the post and its responsibilities. However, items which may need to be covered in the training programme, possibly as 'up-dating', are listed in Table 17.18.[58]

Supervisors

Supervisors generally require extensive basic job training and, as with other grades, safety and loss prevention should form an integral part of their training and experience. Training will probably be required under the various headings listed in Table 17.19.[58]

The recognition of hazardous situations is a very important skill for the supervisor, together with the knowledge and ability to take preventive action. For example, in the near-disaster at Three Mile Island:[65]

> It seems clear . . . when the relief valve on the steam pressuriser failed to reset, the plant operators did not appreciate what had happened. They thought there was too much water in the system, not too little, and therefore took several inappropriate actions that made the situation worse. . . . the operators were dealing with a situation that had not been foreseen and for which they had no relevant training.

This universal problem of training 'panel operators' in fault diagnosis, and improving their ability to deal with abnormal and possibly dangerous situations, has led to some large companies developing programmes using

Table 17.18 Items for inclusion in a training programme for managers

- The Health and Safety at Work etc. Act 1974, including appreciation of personal statutory obligations.
- Acts, Regulations and Codes of Practice relevant to the company.
- The working of the Health and Safety Commission and Executive and other enforcing authorities.
- The difference between legal liabilities under criminal and civil/common law.
- Joint consultation and the involvement of the industrial relations machinery in health and safety.
- The principles and practice of communication as applied to health and safety, including liaison with enforcing authorities.
- The basic causes of accidents, the importance of preventing them and the responsibility for doing so.
- The importance of investigating and reporting accidents or dangerous occurrences.
- The cost of accidents and their effect on morale.
- The principles of loss prevention.
- The hazards of the machinery, plant and processes within the business and the safety precautions needed.
- Appreciation of hazard and operability studies and hazard analysis and their implications in terms of resource requirements.
- The importance of safety being incorporated in the design of new machinery, plant and equipment.
- Appreciation of the interdependence of managers, safety advisers and technical staff within the operational sphere.
- The company's health and safety responsibilities to the public.
- Principles of fire prevention and fire-fighting.
- Fire and evacuation drills.
- The firm's health and safety policy, with all its implications.
- How to apply the results of safety audits and safety surveys.
- The need for people to be trained effectively in safe methods of working and the awareness of job hazards.
- The need to delegate responsibility and authority to subordinates.
- How to evaluate the effectiveness of the firm's health and safety policy.
- How to control and manage health and accident prevention in their own company.
- The contribution of training to improving the company's health and safety performance.
- The need to consider hazards and procedures in adjacent areas.
- The correct use and care of protective equipment.
- The role of the safety adviser.
- The role of the medical adviser.

simulated control panels.[65] The use of computerised simulators can now provide realistic experience covering a range of start-up, shut-down parameter excursion and equipment 'failure' scenarios.[66]

Laboratory staff

The multitude of hazards which may arise in laboratories were summarised in Chapter 11. A training programme is required which results in the employee being able[58]:

- To understand relevant laws, regulations and Codes of Practice in health and safety at work and the company safety policy.
- To understand his own responsibility for safety.
- To work safely in laboratory and plant and use equipment safely.

Table 17.19 Elements of a training programme for supervisors

Company safety policy
- The policy itself, with all its implications.
- The supervisor's safety responsibilities.
- The agreed bounds of his/her safety authority.
- An understanding between management and supervisor of the concept of accountability.
- Company policy on safety representatives and safety committees.

Company systems
- Fire and hazard precautions.
- Fire and evacuation drills.
- Channels for communication on safety.
- Accident reporting and investigation procedures.
- Safe working procedures relevant to their area.
  Standard operating procedures.
- Permit-to-work procedures.
- Protection and protective clothing.
- The functions of safety specialists.
- The role of the safety committee and how it works.
- The functions of safety representatives, and the supervisor's relations with them.

Hazards of the working place
- To understand and recognise the hazards which are present or may arise in the section for which they are responsible, caused by the nature of the work or process, the materials in use, movement of persons, material or machines and general hazards due to poor housekeeping, poor maintenance or other causes.
- To know the systems and procedures for liaising with safety representatives.
- To know how to eliminate or guard against such hazards, as far as is reasonably practicable for them.
- To recognise the limit of their authority (including financial limits) in this context.
- To understand when and how to tell their superior of action they have taken or consider necessary to secure a safe and healthy working place.
- To be able to recognise individual training needs (such as for lifting, fire-fighting, safe working methods).
- To appreciate the techniques of hazard surveys and safety sampling and the concept of the safety audit in as far as they relate to their working area.
- To know the systems and procedures for liaising with maintenance staff.
- To know about the use and abuse of protective equipment and methods of enforcing these.

General
- The Health and Safety at Work etc. Act 1974 (general appreciation and their personal responsibilities).
- Other safety legislation relevant to their working area.
- Communication and industrial relations skills needed for consultation with management, workers, safety representatives and inspectors of relevant enforcing authorities for passing information.
- In many cases, instructional techniques training.

- To identify hazards of chemicals and other materials and understand correct handling and storage procedures.
- To understand emergency procedures, including fire precautions, use of fire-fighting equipment, evacuation and fire drills and accident procedures.

The programme will obviously vary with the type of laboratory, particularly with regard to hazardous emissions, but a basic list is given in Table 17.20.[58] Numerous texts are available on laboratory safety and the trainee should become familiar with them[67–69] and indeed with the contents of Chapter 11. Further training advice is given in references 70 and 71.

Table 17.20 Elements of a training programme for laboratory staff

Statutory and company regulations affecting health, safety and hygiene
Health and Safety at Work etc. Act 1974, other relevant legislation and company safety policy: the need to observe regulations; personal responsibility for safety.

Personal hygiene and clothing
The need for cleanliness; removal of contaminants from clothes and skin, e.g. use of emergency showers; dangers of wearing rings, loose clothing, long hair; reasons for not preparing or consuming food in the laboratory.

Good housekeeping
The need for tidiness; disposal of waste; correct storage of all equipment and materials; dealing with breakages.

Protective clothing and personal equipment
Correct use and care of gloves, footwear, goggles and overalls; the use of masks, types and uses of respirators; barrier cream.

Accidents
Procedure in the event of accidents; initial action in case of bleeding, burns, gassing, falls, electric shock and splashes in eyes; artificial respiration.

Combustion
Types of fires; extinguishers, types and uses, hazards involved; practice in their use, fire alarms and fire drill; emergency services.

Common types of chemicals
Handling of chemicals and procedures for diluting acids and alkalis; flammable, corrosive, toxic and other dangerous materials; precautions in storage and handling; labelling of containers; precautions in dealing with seized bottle stoppers and corroded containers; antidotes to toxic materials.

Flammable chemicals and other materials
Sources of ignition; flash points, explosive mixtures, spontaneous combustion, dust explosion risks; handling and storage of flammable liquids; ethers and other oxidisable liquids; formation of peroxides.

Toxic and corrosive solids, liquids and gases
Effect of inhalation, ingestion and absorption; fume cupboards and hoods, with particular emphasis on purposes; identification, use, handling and hazards of gas cylinders.

Disposal of toxic and flammable materials
Methods used including neutralisation and detoxification; special dangers e.g. cyanides, phosgene, antidotes; disposal of hazardous materials.

Emissions
Hazards likely to be encountered in the laboratory, e.g. ultraviolet and infrared radiations, laser beams and ionising radiations.

Process operators and support staff

Good systems of operation on modern chemical plants cannot be enforced by authority alone; operators expect to understand the reasons for their actions and be convinced that what they are doing is right.[72] Thus in UK petrochemical plants new operators are carefully selected, receive six to eight weeks' induction, followed by on-the-job training; nine to twelve months is required for a new recruit to become fully competent.[72]

Follow-up training includes practice and discussion of emergency procedures including fire fighting, hazard workshops and work group discussions; particular attention is paid to preparation for maintenance and trip testing.

Mechanical equipment
Care in operation etc., guarding of belts, drives etc., operation under special conditions.

Electrical equipment
Limits of authority in operation and remedy of faults; high and low voltage equipment; connection to mains voltage.

Permits to work
Those used to permit craftsmen to maintain and repair equipment; statement of hazards and limitation of work; special precautions in flame-free areas.

Lifting and handling
Safe lifting and handling techniques.

Consultation
The roles and functions of safety representatives, safety committees, safety advisers and medical advisers.

A selection of items to be covered in job safety training programmes for operators and support staff (e.g. internal transport workers, cleaners, effluent treatment operators, warehousemen) is given in Table 17.21.[58] Clearly there will be overlap between the items there and in Tables 17.20 and 17.22, and it is for management to select an appropriate programme which is economical in time but effective.

However, so far as the response to a fire is concerned the operator must never be in any doubt[63]:

- When and how to signal an alarm.
- How to determine whether extinguishment should be attempted, and if so how to go about it.
- What action to take in the event of an alarm, e.g. turning off supplies, equipment, services.

Table 17.21 Elements of a training programme for operators and support staff

Safety policy

- Company health and safety policy – explanation and check on understanding.
- The work of safety committees and safety representatives.
- Relevant safety legislation, emphasising the individual's personal responsibilities.
- The safety of others at the workplace.

Hazards of the workingplace

- Toxic and fire hazards of chemicals (explained in terms which the operator can understand).
- Hazards of clothing, rings, jewellery, long hair.
- Hazards created by badly placed materials.
- Hazards created by defective or inappropriate tools.
- Specific short-term and long-term health hazards of the job.
- Safe use of fork-lift trucks and other potentially dangerous equipment.
- Relevant basic electrical safety rules.
- Specific fire hazards of the job and the precautions to be taken.

Company systems

(Some of which are legal requirements)

- Explanation of the reasons for using guards on machines.
- The correct use of guards and other protective devices.
- The correct safe methods of setting up and operating machines.
- Pre-operating tests where necessary.
- Safety procedures for cleaning plant, machinery or other equipment.
- Safe handling and use of chemicals, particularly those which are corrosive, flammable, toxic or irritant.
- Protective clothing or equipment to be used, with limitations of protection.
- Maintenance of housekeeping standards.
- Maintenance of clear access and egress.
- Safe stacking of material in specified areas.
- Correct handling, lifting and carrying of materials.
- Observance of the safety procedures in other departments or sites visited in the course of the job.
- Action in the event of plant or machine fire or other safety-related malfunction.
- Types of fire extinguisher to be used on various parts of the machine and how to operate each.
- Location of fire extinguishers.
- Placing and use of emergency facilities (e.g. first-aid box, eye washes, antidotes).
- First-aiders – who they are and where to find them.
- Machines to be operated only when authorised to do so and under adequate supervision.
- Immediate reporting of any faults on plant, machine or other equipment.
- Safety rules and procedures specific to the job.
- The permit-to-work system – what it is and how it operates.

- How to evacuate the workplace.

Further information is given in references 73 and 74.

Regular fire drills are essential for all grades of staff with the basic objectives of familiarising them with the sound of the alarm, actions to take (e.g. shutting-down, or leaving their workplace safe, rapidly), escape routes and the locations of assembly areas. Ideally, in order to overcome apathy, some elements of surprise should be introduced into the series of drills (e.g. drills at different times of day or night, drills with certain escape routes blocked-off).

Table 17.22 Elements of a training programme for engineering staff

General requirements:
- A working knowledge of the relevant provisions of the Health and Safety at Work etc. Act 1974 and other appropriate safety legislation.
- An understanding of company safety policy.
- A clear understanding and acceptance of personal safety responsibilities, including those to other employees.
- Relevant company safety rules, procedures, standard operating procedures and Codes of Practice.
- Emergency procedures, including fire drills and evacuation drills.
- For craftsmen who will instruct apprentices and other trainees on or off the job, instructional techniques training and a sound appreciation of the safety aspects of the apprentices' own training.
- Company accident prevention, reporting procedures and lines of communication.
- The role of the safety committee, safety adviser and safety representative.
- Hazard recognition techniques and the craftsman's contribution to workplace safety inspections.

Craft requirements (all trades):
- Electrical isolation/locking off.
- Permit-to-work systems.
- Safety precautions when drilling, welding, burning or grinding; entry into vessels/areas, etc., atmosphere tests before/during operations; spark-proof tools; filing.
- Use of barrier creams, etc.
- Protective clothing, i.e. special overalls, safety footwear, goggles, gloves, etc.
- Use and abuse of solvents.
- Good housekeeping practices.
- Safe place of work and safe means of access (e.g. working from ladders and other temporary workplaces).
- Screening.
- Job responsibilities in emergencies.
- Earthing (static electricity).

Craft requirements (fitter):
- Product line isolation/drain/purge.
- Trapped material (product).
- Breaking joints/opening vessels.
- Adjustments to moving machinery.
- Chemicals risks.
- Permit-to-work systems.

Craft requirements (electrician):
- Fuse withdrawal
- Permits-to-work: soldering; burning; hacksawing; temporary lighting; lone operator working.
- Working near moving machinery.

Engineering staff

The specific safety training programmes for engineering staff (fitters, electrical, instrument engineers, etc.) within any company can be based upon Table 17.22.

The identification, planning and implementation, and – an important feature – the review, of training are outside the scope of this text but first-class guidance is given in combined references 58, 59, 65–71, 73, 74. The depth of training is a matter which can only be assessed by professionals with specialist knowledge of the specific hazards. For example, in the US *Training Requirements of OSHA Standards* OSHA 2254 has been produced for use in determining the individual training needs of each company but, as is clear from the specimens reproduced in Table 17.23,[75] the detail required is a matter of judgement. All relevant extracts from OSHA 2254 (pp. 8–31) are published in reference 63.

Table 17.23 Specimen OSHA Safety and Health Requirements for General Industry (ex OSHA 2254 from Part 1910 of the Code of Federal Regulations)

1910.252(b)(1) (iii) ***Welding, cutting, and brazing***	(iii) *Instruction*. Workmen designated to operate arc welding equipment shall have been properly instructed and qualified to operate such equipment as specified in subparagraph (4) of this paragraph.
1910.252(c)(1) (iii)	(iii) *Personnel*. Workmen designated to operate resistance welding equipment shall have been properly instructed and judged competent to operate such equipment.
1910.252(c)(6)	(6) *Maintenance*. Periodic inspection shall be made by qualified maintenance personnel, and records of the same maintained. The operator shall be instructed to report any equipment defects to his supervisor and the use of the equipment shall be discontinued until safety repairs have been completed.
1910.252(d)(2) (xii)(c)	(xii) *Management*. (c) Insist that cutters or welders and their supervisors are suitably trained in the safe operation of their equipment and the safe use of the process.
1910.1017 (j)(1) (i) thru (ix) ***Vinyl chloride***	(j) *Training*. Each employee engaged in vinyl chloride or polyvinyl chloride operations shall be provided training in a program relating to the hazards of vinyl chloride and precautions for its safe use. (1) The program shall include: (i) The nature of the health hazard from chronic exposure to vinyl chloride including specifically the carcinogenic hazard; (ii) The specific nature of operations which could result in exposure to vinyl chloride in excess of the permissible limit and necessary protective steps; (iii) The purpose for, proper use, and limitations of respiratory protective devices; (iv) The fire hazard and acute toxicity of vinyl chloride, and the necessary protective steps; (v) The purpose for and a description of the monitoring program; (vi) The purpose for, and a description of, the medical surveillance program; (vii) Emergency procedures; (viii) Specific information to aid the employee in recognition of conditions which may result in the release of vinyl chloride; and (ix) A review of this standard at the employee's first training and indoctrination program, and annually thereafter.

Specific training requirements

Within the UK there are legal obligations, in addition to those generally, arising under the HASAWA, to provide training for particular tasks. Examples arise in woodworking,[76] the preparation of power presses for use,[77] the mounting of abrasive wheels,[78] and in the training of first-aiders

Table 17.24 Special training requirements – lift truck drivers

- All prospective lift truck drivers should attend a training course covering:
  1. The basic skill and knowledge to operate a lift truck safely and efficiently, and to carry out routine daily checks.
  2. Specific job training tailored to the employer's needs, but including:
     - knowledge of the operating principles and controls of the lift truck to be used and any special attachments, and routine inspection and servicing in accordance with manufacturer's instructions.
     - training and practice in its use in conditions to be met on the job, e.g. gangways, slopes, loading dock, etc.
     - training and practice in the work to be undertaken, e.g. loading and unloading vehicles; stacking and de-stacking; familiarisation with loads and materials of the type normally handled, including weight assessment.
  3. Familiarisation training at the workplace under supervision, e.g. the lift truck to be used, site layout, company safety rules, the features of the work, emergency procedures, etc.
- Supervisors and operators should be familiar with the following information (i.e. equipment specification):

  Name of the manufacturer of the truck; type truck; serial number; unloaded weight; load capacity; load centre distance; and maximum lift height.

for construction sites.[79] Operatives requiring special skills are also recipients of specialist training, e.g. if required to check scaffolding,[80] to work with ionising radiations,[81] to perform 'gas-freeing' associated with the issue of permits to work, drivers of internal transport; and competent persons to inspect, test and examine ventilation provisions.[82] Clearly first-aiders also require specialist training[83] and there are numerous other examples. As an illustration, the special training requirements for lift truck drivers are listed in Table 17.24.

'Incidents' as training material

When an 'incident' occurs within, or is reported from outside, the company the maximum practical use should be made of it in training activities.[84] Therefore, circulation of information of an incident should ideally extend beyond managers and safety engineers to operating levels. The preparation of an abbreviated investigation report, with simplified flow diagrams and sketches, will generally provide a more suitable document for general distribution and training sessions. If still or cine (or video) pictures can be obtained during an incident then, in addition to assisting investigation, they will add visual impact to subsequent emergency training.

As well as being used to highlight equipment or system of work or human error causes, the incident report can be useful in illustrating problems of emergency response, e.g. process emergencies or fire-fighting, or human reaction under stress. Some may be adapted for use as hazard workshop modules[85,86] or for emergency training programmes.

Numerous sources of well-written case histories[87–91] can be abstracted and adapted to a company's needs or used to highlight particular hazards. For example, 'system-of-work failures':

- An operator was charging sulphur (a Class 1 explosive dust) down a chute into a vessel. While awaiting another pallet-load of sulphur he decided to charge another ingredient on the 'charge sheet'. As he pushed the drum over the charge grid a sheet of flame erupted from the chute. This self-extinguished in a few seconds and there was no damage.

  The cause was deduced to be ignition of a cloud of sulphur residues, dispersed as dust, due to a spark caused by the drum hitting the grid.[92]
  Lesson: Understanding instructions.
  The operator did not appreciate that the reason the 'charge-sheet' specified swilling down of the charge chute after sulphur powder charging was not just good housekeeping but an essential safety feature.
- A road tanker was filled during the night and a 'filling certificate', a very small piece of paper, was completed and slipped in the set of dispatch papers.

  The tanker was not collected before the end of the shift and the morning-shift operator was unsure whether or not it had been filled, since the record sheets had been sent to records office and when he shook the dispatch papers no filling certificate fell out. It was in fact caught up in the other papers and when he started to 'fill' the tanker it overflowed.[93]
  Lesson: The need for a good handover between shifts.

Alternatively good design practice can be emphasised:

- Five reactors were installed in parallel with two gas-feed connections which were cross-connected as shown in Fig. 17.13. Oxygen was also fed to the reactors via separate lines (not shown).

  While only two reactors, No. 1 and No. 4, were on line, an operator believed valve B was open and therefore closed valve A; this effectively shut-off the gas supply to reactor No. 1. The oxygen flow was normally regulated by a ratio controller but, since this was 'out of zero', a small flow continued.

  The operator realised he had made a mistake and restored the gas flow to reactor No. 1, which contained excess oxygen. An explosion occurred in the downstream waste-heat boiler causing four fatalities.[94]

  Lesson: Design plant, or initial protective systems, so that the accidental operation of a valve cannot result in an explosion or runaway reaction.
- If, when an alarm sounds, a man has to go outside and close a valve, he may forget to do so (or close the wrong valve):

  1 in 10 times if he works in a busy control room.

  1 in 100 times if he works in a quiet control room.[95]

  Lessons: Either an occasional mistake has to be accepted – if the consequences are unlikely to be serious – or the possibility of error should be designed out, or the probability of error should be reduced by installing an alarm or introducing a supervisory check.

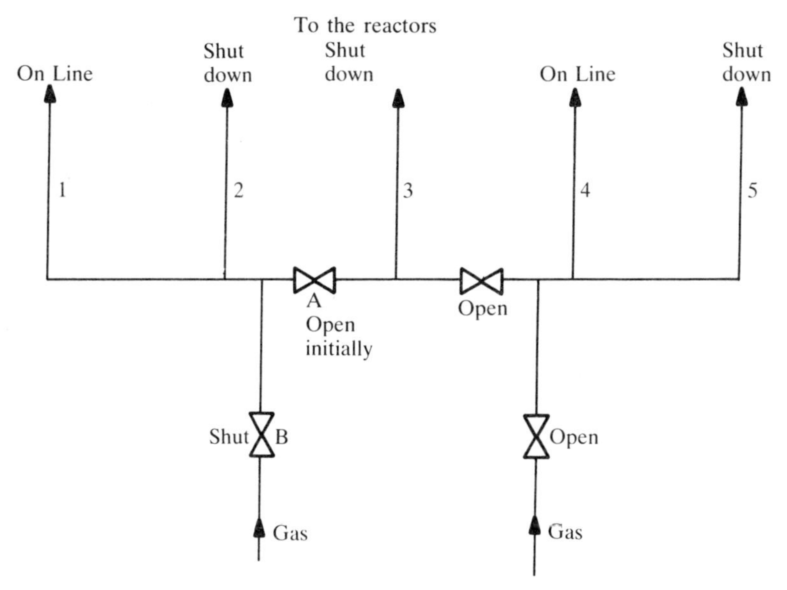

Fig. 17.13 Gas-feed arrangement to five reactors (oxygen lines not shown)

Alternatively, case studies may serve as a reminder of basic physico-chemical principles (see Ch. 3) which the audience know, or knew of, and of their relevance to safety.

- A vessel which was 13.7 m high was swept out with nitrogen and handed over to maintenance for a slip-plate to be installed in an overhead line, as shown in Fig. 17.14, at a distance of 12.8 m from the top of the vessel.
  A small quantity of naphtha was drained from the vessel via the bottom drain; it was realised that some might be left inside but, because of the size of the vessel and distance to the flanges to be broken, this was not believed to constitute a hazard. The drain valve near this flange was opened, but the slight pressure of 20 millibar had not blown off before the fitter broke the flange. A mixture of nitrogen and naphtha was emitted and, following ignition by a hot pipe, resulted in a fire.[96]
  Lesson: Mixtures of hydrocarbon vapour and nitrogen will burn above a critical oxygen content (see Ch. 4).
- Two chlorine drums were changed-over in a water processing plant, a manoeuvre performed safely many times before. Although the main safety valve was closed a significant leak of chlorine was observed from the connecting pipe. A fitter and several colleagues attempted to rectify the leak wearing canister-type respirators. In the attempt a large emission of chlorine occurred via a faulty valve and all the men were badly gassed.[97]
  Lessons: Limitations of personal protection (self-contained breathing apparatus would have prevented the gassings). Need for regular maintenance and replacement of equipment in toxic gas service.

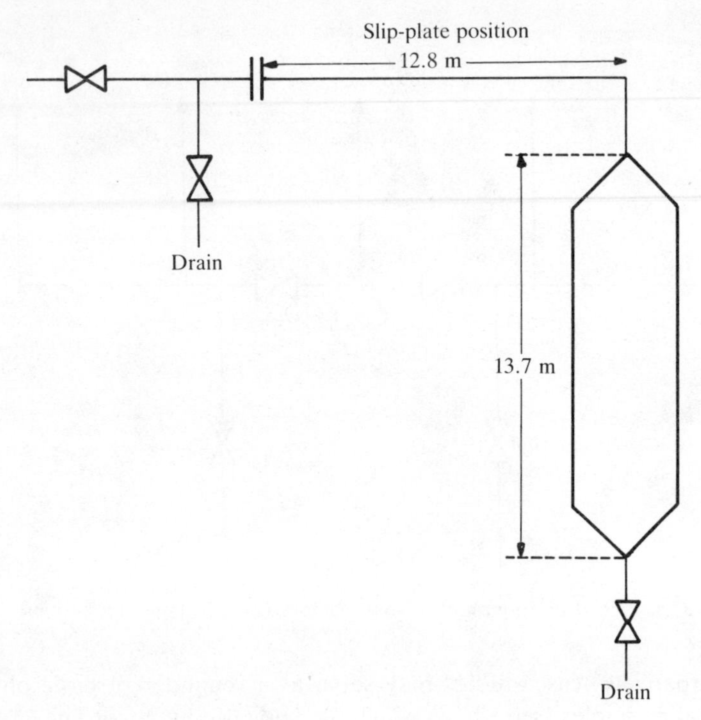

Fig. 17.14 Slip-plate installations job on overhead line

- Because of a leak in a carbate heat exchanger cooling water containing chromate ion contaminated the brine fed to a mercury cell. The cell should have produced only chlorine but hydrogen was consequently evolved in large quantities. The hydrogen burned with the chlorine in an electrostatic precipitator resulting in damage to this and adjacent equipment.[98]
  Lesson: Flammable gases or vapours will burn in chlorine, albeit less vigorously than in air.

The groupings of the above examples was clearly arbitrary since such incidents involve a combination of factors/events. Human error is often a contributory factor, but it is to be expected that when a dangerous process or operation is continually repeated there will be mistakes or accidental slips due to wavering attention or the urgency of completing the work; thus management should 'have in mind not only the careful man but also the man who is inattentive to such a degree as can normally be expected'.[99]

For impact a series of incidents can be grouped under a common theme.[85] The presenter may then prompt the audience/reader to draw analogies with plant/procedures/materials within his area of responsibility.

Refresher training

As with fire drills, it is recognised that refresher training is necessary for those tasks which are not undertaken regularly. Relevant areas include the use of breathing apparatus, action in local emergencies (e.g. rescues from confined spaces, fire fighting) and first aid.

References

(*All places of publication are London unless otherwise stated*)

1. Petersen, D., *Techniques of Safety Management* (2nd edn). McGraw Hill, 1978.
2. Egan, B., *Safety Policies* (New Law Guidance Series No. 11). New Commercial Publishing 1979.
3. Bird, R. K., *Fire Eng. Jnl.*, 1981, **41**, 123, 29.
4. Oaksey, Lord, *General Cleaning Contractors Ltd* v. *Christmas* [1953] A. C. 180, 1952, 2 All E. R. 1110.
5. Greene, Lord, *Speed v. Swift (Thomas) Co. Ltd* [1943] K. B. 557, [1943] 1 All E. R. 539.
6. Health and Safety at Work etc. Act 1974; Factories Act 1961; Legislation made under the Factories Act 1961, e.g. Chem. Works Regs. HMSO.
7. Atallah, S, *Chem. Engr.*, Oct. 10 1977
8. Carson, P. A., Mumford, C. J. & Ward, R., *Loss Prevention Bulletin*. Institution of Chemical Engineers, 1985 (065), 1.
9. Chemical Industries Assn, *Recommended Procedures for Handling; Major Emergencies* (2nd edn). CIA 1976.
10. Cocks, R. E., *Chemical Engineering*, 5 Nov. 1979.
11. Bond, J. & Bryans. J. W., *Chem. Engr.*, Apr. 1975, 296, 245–247.
12. Chemical and Allied Products Industry Training Board, *Permit to Work Systems*. 1977.
13. Kletz, T. A., *Plant/Operations Progress*, **3**, 1 Jan. 1984, 1–3.
14. Friedrichsen, F. G., *Plant/Operations Progress*, 1983, **2**, 2, 122–3.
15. Imperial Chemical Industries, *Accident Case History 105*; and Institution of Chemical Engineers, *Hazard Workshop Module 004*, 7, 'Preparation for Maintenance'.
16. Imperial Chemical Industries, *Case History No. 21; Accidents Illustrated*. I.C.I. Ltd.
17. Institution of Chemical Engineers, *Case History, Hazard Workshop Module 004*, 3. Instn of Chem. Engrs, 1981.
18. Imperial Chemical Industries, *Safety Newsletter* 19, 1.
19. Health and Safety Executive, *Health and Safety: Industry & Services*. HSE 1975, 15.
20. 'Confined Space Entry', *Loss Prevention Bulletin 053*, Institution of Chemical Engineers, Oct. 1983, 15–26.
21. *Annual Report of H.M. Chief Inspector of Factories 1967*, 101. HMSO.
22. Health and Safety Executive, *Health and Safety: Industry & Services*. HMSO 1975, 33.

23. H.M. Factory Inspectorate, *Entry into Confined Spaces: Hazards and Precautions*, Technical Data Note 47.
24. British Chemical Industry Safety Council, *Safety in Inspection and Maintenance of Chemical Plant*.
25. Health and Safety Executive, *The Safe Cleaning, Repair and Demolition of Large Tanks for Flammable Liquids*, Technical Data Note 18 (Rev.).
26. Health and Safety Executive, 'Repair of drums and small tanks: Explosion and fire risk', *Health & Safety at Work*, 32. HMSO.
27. Hooper, E., *Beckinsale's Safe Use of Electricity*. RoSPA 1981.
28. Institution of Chemical Engineers, *Case History 104: Accidents Illustrated, Preparing for Maintenance*, ICI Ltd & Hazard Workshop Module 004, 5.
29. Webster, T. J., 'Loss Prevention and Safety Promotion in the Process Industries, *Proceedings of 2nd International Symposium on Loss Prevention*, Dechema, 1978, 1–15.
30. Kletz, T. A., *Chem. Eng. Prog., Loss Prevention*, 1976, **10**, 91.
31. Imperial Chemical Industries, 'Must plant modifications lead to accidents?', *Safety Newsletter 83*, Jan. 1976, 2.
32. Imperial Chemical Industries, *Case History No. 9/2: Fires caused by plant modifications, Accidents illustrated*; and Henderson, J. M. and Kletz, T. A., (Instn Chem. Engrs Symp. Series, 47.)
33. Marshall, G., *Manuf. Chemist*, Feb 1983, 54(2), 45.
34. Bergtraun, E. M., *Chem. Engr*, Jan. 30; Feb. 27, 1978
35. The Protection of the Eyes Regulations 1975. HMSO.
36. The Construction (General Provisions) Regulations 1961. HMSO.
37. The Construction (Lifting Operations) Regulations, 1961, and The Construction (Working Places) Regulations 1966. HMSO.
38. Oberhansberg, J., 'Loss Prevention and Safety Promotion in the Process Industries, *Proceedings of 2nd International Symposium on Loss Prevention*, Dechema, 1978, 11–77.
39. DeReamer, R., *Modern Safety and Health Technology*. Wiley, Chichester, 1980.
40. Chemical Industries Association, *Safety Audits, A Guide for the Chemical Industry*. CIA Chemical Industry Safety and Health Council 1977.
41. Warner, F., Sir, 'Engineering Safety and the Environment'. The Eleventh Maurice Lubbock Memorial Lecture, 6 May 1976.
42. Webb, H. E., *Chem. Engr*, 1 Aug. 1977.
43. Yelland, A. E. J., *Chem Engr*, 201, 1966, 214.
44. Wells, G. L., *Safety in Process Plant Design*. George Godwin, 1980, 239–57.
45. Kletz, T. A., *Chemical Processing*, Sept. 1974.
46. Egan, B., *Safety Inspections* (New Law Guidance Series No. 9). New Commercial Publishing 1979.
47. Haddaway, L., Hunt, E. S. & Kletz, T. A., paper presented at Safety Audits – Essential for Every Factory Conference, Institution of Chemical Engineers, 27 Jan. 1977.
48. The Safety Representatives and Safety Committees Regulations 1977. HMSO.
49. Factories Act 1961, s.80. HMSO.
50. Factories Act 1961, s.81 and the Dangerous Occurrences (Notification) Regulations, 1947. HMSO.

51. Factories Act 1961, s.82 and Various Orders relating thereto. HMSO.
52. Gillett, J. E., Unpublished work. ICI Pharmaceuticals Division.
53. Mumford, C. J. & Lihou, D. A., *Loss Prevention Bulletin*. Institution of Chemical Engineers 1984 (059), 6–14.
54. Institution of Chemical Engineers, *Case History No. 10: Fires & Explosions*, Hazard Workshop Module 003.
55. Hoy-Petersen, R., 'Fire prevention in solvent extraction plants', *Proceedings of 1st International Loss Prevention Symposium*, Hague, 28–30 May 1974. Elsevier.
56. Kuhlman, R. L., *Professional Accident Investigation*. Institute Press, USA, 1977.
57. Ling, K. C., 'Process Industry Hazards: Accidental Release, Assessment, Containment and Control', *Instn Chem. Engrs Symp. Series*, No. 47, 1976, 109–118.
58. Chemical and Allied Products Industry Training Board, *Assessing Safety Training Needs* (Supplement 16D), 1980.
59. Egan, B., *Appointment and Functions of Safety Representatives*. New Commercial Publishing, 1978.
60. Duncan, K. D., 'Analytical techniques in training design', in E. Edwards & F. P. Lees (eds), *The Human Operator in Process Control*. Taylor & Francis, 1974.
61. Annett, J., Duncan, K. D., Stammers, R. B. & Gray, M. J., *Task Analysis*, Training Information No. 6, 1971. HMSO.
62. Safety and Reliability Directorate, *The Guide to Reducing Human Error in Process Operation*, SRD 347.
63. ReVelle, J. B., *Safety Training Methods*. Wiley, 1980.
64. King, R. & Magid, I., *Industrial Hazard and Safety Handbook*. Newnes Butterworth, 1979.
65. Marshall, E. C., Scanlon, K. E., Shepherd, A. & Duncan, K. D., *Chem. Engr* 1981, 365, 66–69.
66. TecQuipment International Ltd UK, *Simulator CH30*.
67. Bretherick, L (ed.), *Hazards in the Chemical Laboratory*, (3rd edn). 1981.
68. Pieters, H. A. J. & Creyghton, J. W., *Safety in the Chemical Laboratory*. Butterworth.
69. Gaston, P. J., *The Care, Handling and Disposal of Dangerous Chemicals*. Northern Publ. (Aberdeen), 1970.
70. Chemical and Allied Products Industry Training Board, *Laboratory Safety Management*, I.P. 24
71. Chemical & Allied Products Industry Training Board, *Training of Scientific Laboratory Technicians*, T. R. 17.
72. Barker, G. F., Kletz, T. A. & Knight, H. A., *Chem. Eng. Prog.*, Sept. 1977, 4–8.
73. Chemical and Allied Products Industry Training Board, *Training Operators*, T. R. 12
74. Chemical and Allied Products Industry Training Board, *Health and Safety at Work – Everybody's Concern*, I.P. 16.
75. OSHA 2254 *Training Requirements of OSHA Standards*, Part IS10.
76. The Woodworking Machines Regulations 1974, 13. HMSO.
77. The Power Press Regulations 1965, 4 and Schedule. HMSO.

78. The Abrasive Wheels Regulations 1970, 9. HMSO.
79. The Construction (Health and Welfare) Regulations 1966, 7. HMSO.
80. The Construction (Working Places) Regulations. HMSO.
81. The Ionising Radiations (Sealed Sources) Regulations 1961, 7. HMSO.
82. Asbestos Regulations 1969. HMSO.
83. Health and Safety (First-Aid) Regulations 1981. HMSO.
84. Searson, A. H., 'Learning from incident experience, safety and loss prevention', *Instn of Chem. Engrs Symp. Series*, No. 73, 1982, B37–B40.
85. Lihou, D. A. & Beveridge, G. S. C., *An Introduction to Hazard Workshop Training Modules*. Institution of Chemical Engineers, 1981.
86. Institution of Chemical Engineers, *Hazard Workshop Modules*.
87. *Safety in Air and Ammonia Plants, Subsequently Ammonia Plant Safety (and Related Facilities) (Chem. Eng. Prog. Tech. Manual Series* 1–19).
88. *Loss Prevention Bulletins*. Institution of Chemical Engineers.
89. *Loss Prevention* (*Chem. Eng. Prog. Techn Manuals*).
90. Imperial Chemical Industries, Petrochemicals Division *Newsletters* ICI plc.
91. Symposia on Chemical Process Hazards with Special Reference to Plant Design. Institution of Chemical Engineers.
92. Chemical Industries Association, *Chemical Safety Summary*, 54, 216, 499.
93. Imperial Chemical Industries *Safety Newsletter*, 105, 3.
94. Imperial Chemical Industries *Safety Newsletter*, 86, 2.
95. Kletz, T. A., Hazop & Hazan, *Notes on the Identification and Assessment of Hazards*. Institution of Chemical Engineers, 1983.
96. Imperial Chemical Industries *Safety Newsletter, 31*, 1.
97. Atherley, G. R.C., *Occupational Health and Safety Concepts*. Applied Science, 1978, 338.
98. Anon *Loss Prevention Bulletin*. Institution of Chemical Engineers, 1975 (003).
99. *Smith* (*or Westwood*) v. *National Coal Board*, 1967, 2 All E. R. 593.
100. Carson, P. A. & Mumford, C. J., *Loss Prevention Bulletin*. Institution of Chemical Engineers, 1986 (070), 15.

CHAPTER 18

Chemical safety information

It is clear from the foregoing chapters that a prerequisite for the safe handling of chemicals is a knowledge of their inherent chemico-physical and toxicological properties. In addition to any moral obligation, strict legal requirements exist to inform intending users of chemicals of any hazardous properties together with recommendations for correct handling precautions. The UK scene has been described in Chapters 13 and 16. Advice must be given on all aspects of dealing with the chemicals including their transportation, storage, use and disposal, together with arrangements for coping with spillages, fires and other emergencies including first-aid measures. Although details of legislation differ from country to country, in general this duty to notify extends to employees, contractors, customers, government bodies, etc., and is encompassed by a wide variety of laws.

Interestingly, the Robens Committee[1] observed that:

> 'It is apparent that a great deal of research and advisory literature is provided. What it is lacking is an effective means of ensuring that the information is always available to those who need to know.'

This Chapter concentrates on identifying useful sources of information on health and safety with bias towards chemicals; the factors which influence the effectiveness of the communication of the data are discussed in reference 2.

Many sources of health and safety information are available and indeed in some large organisations librarians and information scientists specialise in the storage and retrieval of such information. In the majority of smaller establishments, employing no such information specialist or qualified health and safety adviser, knowing how to obtain data can in itself present a major problem for the person charged with the responsibility of promoting safety in individual chemical processes or for providing information to prospective users. Experience suggests that information retrieval carries with it an element of chance.[3]

Since it is impossible to anticipate all future information needs, and because it is important to keep abreast of changing legislation and recent announcements of newly discovered hazards on materials hitherto considered as safe, it is necessary to know how to obtain information when required and how to maintain an awareness of the most recent publications and discoveries.

The large number of sources of information on chemical safety includes collated works such as books, codes, reviews, manuals, data sheets, proceedings of symposia, etc.; abstracts; computerised data bases; trade and professional journals; academic institutions and other organisations such as government bodies and professional, trade, industry, insurance and fire-protection associations, private consultants, etc. This chapter serves as an introduction to the subject of sources of safety and health information with a few hundred key examples. A valuable review of the subject is given in reference 4.

Legislation

There are many laws and regulations relevant to occupational health and safety, as discussed in Chapter 16. This contains an appended list of hundreds of relevant examples. Further guidance can be obtained from the Health and Safety Executive and the Health and Safety Commission, which publish approved Codes of Practice (Table 18.1), Guidance Notes (Table 18.2) and other booklets and pamphlets on the subject (Table 18.3). A publications catalogue is issued annually by the Health and Safety Executive Library and Information Service listing all publications produced by the HSC/HSE since 1974 plus earlier publications by the separate inspectorates. Also new publications are announced in the *HSC Newsletter* which is issued six times a year. Other publications by HSC/HSE include an annual report, technical papers, toxicology reviews, translations, and discussion and consultative documents.

Table 18.1 Approved Codes of Practice published in the UK by the HSC

COP 3	Working with Asbestos Insulation and Asbestos Coating
COP 4	Health and Safety (First Aid) Regulations 1981
COP 5	Classification of Dangerous Substances for Conveyance in Tankers and Tank Containers
COP 6	Petroleum Spirit (Plastic Containers) Regulations 1982
COP 7	Principles of Good Laboratory Practice (Notification of New Substances Regulations 1982)
COP 8	Methods of Determining Ecotoxicity (Notification of New Substances Regulations 1982)
COP 9	Methods for the Determination of Physico-chemical Properties (Notification of New Substances Regulations 1982)
COP 10	Methods for the Determination of Toxicity (Notification of New Substances Regulations 1982)

Table 18.2 Guidance Notes published in the UK by the HSE

General Series

GS 1	Fumigation using methyl bromide
GS 2	Metrication of construction safety regulations
GS 3	Fire in the storage and industrial use of cellular plastics
GS 4	Safety in pressure testing
GS 5	Entry into confined spaces
GS 6	Avoidance of danger from overhead electrical lines
GS 7	Accidents to children on construction sites
GS 8	Articles or substances for use at work – guidance for designers, manufacturers, importers, suppliers, erectors and installers
GS 9	Road transport in factories
GS 10	Roofwork prevention of falls
GS 11	Whisky cask racking
GS 12	Effluent storage on farms
GS 13	Reporting of accidents to pupils and students
GS 14	Provision of sanitary conveniences and washing facilities in agriculture
GS 15	General access scaffolds
GS 16	Gas flooding fire extinguishing systems: toxic hazards and precautions
GS 17	Safe custody and handling of stock bulls on farms and at artificial insemination centres
GS 18	Commercial ultra-violet tanning equipment
GS 19	General precautions aboard ships being fitted out or under repair
GS 20	Fire precautions in pressurised workings
GS 21	Assessment of radio frequency ignition hazards

Chemical Safety Series

CS 1	Industrial use of flammable gas detectors
CS 2	The storage of highly flammable liquids
CS 3	Storage and use of sodium chlorate
CS 4	The keeping of LPG in cylinders and similar containers
CS 5	The storage of LPG at fixed installations
CS 6	The storage and use of LPG on construction sites

Plant and Machinery Series

PM 1	Guarding of portable pipe-threading machines
PM 2	Guards for planing machines
PM 3	Erection and dismantling of tower cranes
PM 4	Safety at high temperature dyeing machines
PM 5	Automatically controlled steam and hot water boilers
PM 6	Dough dividers
PM 7	Lifts
PM 8	Passenger carrying paternosters
PM 9	Access to tower cranes
PM 10	Tripping devices for radial and heavy vertical drilling machines
PM 13	Zinc embrittlement of austenitic stainless steel
PM 14	Safety in the use of cartridge operated tools
PM 15	Safety in the use of timber pallets
PM 16	Eyebolts
PM 17	Pneumatic nailing and stapling tools
PM 18	Locomotive boilers
PM 19	Use of lasers for display purposes
PM 20	Cable-laid slings and grommets
PM 21	Safety in the use of woodworking machines
PM 22	Mounting of abrasive wheels
PM 23	Photo-electric safety systems
PM 24	Safety at rack and pinion hoists
PM 25	Vehicle finishing units: fire and explosion hazards

Table 18.2 (continued)

PM 26 Safety at lift landings
PM 27 Construction hoists
PM 28 Working platforms on fork lift trucks
PM 29 Electrical hazards from steam/water pressure cleaners etc
PM 30 Suspended access equipment
PM 31 Chain saws
PM 32 The safe use of portable electrical apparatus
PM 33 Safety of bandsaws in the food industry
PM 34 Safety in the use of escalators

Medical Series
MS 3 Skin tests in dermatitis and occupational chest disease
MS 4 Organic dust surveys
MS 5 Lung funtion
MS 6 Chest X-rays in dust disease
MS 7 Colour vision
MS 8 Isocyanates – medical surveillance
MS 9 Byssinosis
MS 10 Beat conditions and tenosynovitis
MS 12 Mercury – medical surveillance
MS 13 Asbestos
MS 15 Welding
MS 16 Training of offshore sick-bay attendants ('rig-medics')
MS 17 Biological monitoring of workers exposed to organophosphorous pesticides
MS 18 Health surveillance by routine procedures
MS 20 Pre-employment health screening

Environmental Hygiene Series
EH 2 Chromium – health and safety precautions
EH 4 Aniline – health and safety precautions
EH 5 Trichloroethylene – health and safety precautions
EH 6 Chromic acid concentrations in air
EH 7 Petroleum based adhesives in building operations
EH 8 Arsenic – health and safety precautions
EH 9 Spraying of highly flammable liquids
EH 10 Asbestos – control limits and measurement of airborne dust concentrations
EH 11 Arsine – health and safety precautions
EH 12 Stibine – health and safety precautions
EH 13 Beryllium – health and safety precautions
EH 14 Level of training for technicians making noise surveys
EH 16 Isocyanates: toxic hazards and precautions
EH 17 Mercury – health and safety precautions
EH 18 Toxic substances: a precautionary policy
EH 19 Antimony – health and safety precautions
EH 20 Phosphine – health and safety precautions
EH 21 Carbon dust – health and safety precautions
EH 22 Ventilation of buildings: fresh air requirements
EH 23 Anthrax – health hazards
EH 24 Dust accidents in malthouses
EH 25 Cotton dust sampling
EH 26 Occupational skin diseases: health and safety precautions
EH 27 Acrylonitrile: personal protective equipment
EH 28 Control of lead: air sampling techniques and strategies
EH 29 Control of lead: outside workers
EH 30 Control of lead: pottery and related industries
EH 31 Control of exposure to polyvinyl chloride dust

Table 18.2 (continued)

EH 32	Control of exposure to talc dust
EH 33	Atmospheric pollution in car parks
EH 34	Benzidine based dyes
EH 38	Ozone: health hazards and precautionary measures
EH 40	Occupational exposure limits 1986
EH 42	Monitoring strategies for toxic substances
EH 43	Carbon monoxide
EH 44	Dust in the workplace: general principles of protection
EH 45	Carbon disulphide: control of exposure in the viscose industry

Table 18.3 Leaflets and booklets published in the UK by the HSE

Agriculture Series	
AS 3	Bulls
AS 4	Respiratory disease in the mushroom industry
AS 5	Farmer's lung
AS 6	Crop spraying
AS 7	Guns
AS 8	Noise
AS 10	Children
AS 11	Circular saws
AS 12	Prevention of accidents
AS 13	Employment of children and young persons on farms
AS 14	Cyanide gassing powders
AS 15	Tree-felling, hauling and scrubland clearance
AS 16	Checking tractor-trailer brakes for safety
AS 17	Electricity on the farm
AS 18	Storage of pesticides on farms
AS 19	Hydrofluoric acid
AS 20	Chain saws
AS 21	First aid in agriculture
AS 22	Prevention of tractors overturning
AS 23	Lifting and carrying
AS 24	Power take-off and power take-off shafts
Health and Safety at Work Series	
1	Lifting and carrying
2	Canteens, messrooms and refreshment services
3	Safety devices for hand and foot operated presses
4	Safety in the use of abrasive wheels
5	Cloakroom accommodation and washing facilities
6A	Safety in construction work: general site safety practice
6B	Safety in construction work: roofing
6C	Safety in construction work: excavations
6D	Safety in construction work: scaffolding
6E	Safety in construction work: demolitions
6F	Safety in construction work: system building
8	Dust and fumes in factory atmospheres
9	Safety in the use of machinery in bakeries
10	Fire-fighting in factories
11	Guarding of hand-fed platen machines
12	Safety at drop forging hammers
13	Ionising radiations – precautions for industrial users

Table 18.3 (continued)

14	Safety in the use of mechanical power presses
15	Dry cleaning plant: precautions against solvent risks
16	The structural requirements of the Factories Act
17	Improving the foundry environment
18	Industrial dermatitis: precautionary measures
19	Safety in laundries
20	Drilling machines: guarding of spindles and attachments
21	Organisation of industrial health services
22	Dust explosions in factories
23	Hours of employment of women and young persons
24	Electrical limit switches and their applications
25	Noise and the worker
26	Safety in the use of biscuit-making machinery
27	Precautions in the use of nitrate salt baths
28	Plant and machinery maintenance
29	Carbon monoxide poisoning: causes and prevention
30	The bulk storage of liquefied petroleum gas at factories
31	Safety in electrical testing
32	Repair of drums and small tanks: explosion and fire risk
33	Safety in the use of guillotines and shears
34	Guide to the use of flame arresters and explosion reliefs
35	Basic rules for safety and health at work
36	First aid in factories
37	Precautions in the handling, storage and use of liquid chlorine
38	Electric arc welding
39	Lighting in offices, shops and railway premises
40	Means of escape in case of fire in offices, shops and railway premises
41	Safety in the use of woodworking machines
42	Guarding of cutters of horizontal milling machines
43	Safety in mechanical handling
44	Asbestos: health precautions in industry
45	Seats for workers in factories, offices and shops
46	Evaporating and other ovens
47	Safety in the stacking of materials
48	First aid in offices, shops and railway premises
49a	Safety in the cotton and allied fibres industry: opening processes
49b	Safety in the cotton and allied fibres industry: cardroom processes
49c	Safety in the cotton and allied fibres industry: spinning, winding and sizing
49d	Safety in the cotton and allied fibres industry: fabric production
50	Welding and flame-cutting using compressed gases
HSE Series	
4	Short guide to employer's liability (Compulsory Insurance Act 1969)
5	An introduction to the Employment Medical Advisory Service
7	Health and safety for young workers
8	Fires and explosions due to the misuse of oxygen
9	Area office information services
11	Reporting an accident (The Notification of Accidents and Dangerous Occurrences Regulations 1980)
12	Control of dust in the cotton textile industry
HS(G) Series	
1	Polyurethane foam
2	Poisonous chemicals on the farm
3	Highly flammable materials on construction sites
4	Highly flammable liquids in the paint industry

Table 18.3 (continued)

5	Hot work
6	Life trucks
7	Container terminals
8	Fabric production
9	Spinning, winding and sizing
10	Cloakroom accommodation and washing facilities
11	Flame arresters and explosion reliefs
12	Offshore construction
13	Electrical testing
14	Opening processes: cotton and allied fibres
16	Evaporating and other ovens
17	Safety in the use of abrasive wheels
18	Portable grinding machines: control of dust
19	Safety in working with power-operated mobile work platforms
20	Guidelines for occupational health services
22	Electrical apparatus for use in potentially explosive atmospheres
HS(R) Series	
1	Packaging and labelling of dangerous substances
2	A guide to agricultural legislation
3	A guide to Tanker Marking Regulations
5	The notification of accidents and dangerous occurrences
6	A guide to The Health & Safety at Work Act
7	A guide to The Safety Signs Regulations 1980
8	A guide to the Diving Operations at Work Regulations 1981
9	A guide to the Woodworking Machines Regulations 1974
11	First aid at work
12	A guide to the Health and Safety (Dangerous Pathogens) Regulations 1981
13	A guide to the Dangerous Substances (Conveyance by Road in Road Tankers and Tank Containers) Regulations 1981
14	A guide to the Notification of New Substances Regulations 1982
15	Administrative Guidance on the Implementation of EEC 'Explosive Atmospheres' Directive (76/117 EEC and 79/196/EEC)
16	A guide to the Notification of Installations Handling Hazardous Substances Regulations 1982
17	A guide to the Classification and Labelling of Explosives Regulations 1983
18	Administratives guidance on the application of the EEC 'Low Voltage' Directive (73/23/EEC)
19	A guide to the Asbestos (Licensing Regulations) 1983
MS (A) Series	
1	Lead and you
2	Vinyl chloride and you
MS (B) Series	
1	Ulceration of the skin and inside the nose caused by chrome
3	Anthrax
4	Skin cancer caused by pitch and tar
5	Skin cancer caused by oil
6	Occupational (industrial) dermatitis
7	Poisoning by pesticides
MA Series	
1	The medical examination of divers
3	The control of lead at work: medical surveillance
4	Asbestos (Licensing) Regulations 1983: medical surveillance

Table 18.3 (continued)

IND (G) Series	
3L	First aid provisions in small workplaces
4P	First aid at work
5L	Code of safety at fairs
6L	The EINECS Inventory
7C	Press brakes
8C	Power presses
12C	Cold degreasing solvents
IND (S) Series	
3L	Dust and fume control in the rubber industry
4L	Preventing falls to window cleaners
HSC Series	
2	The Act outlined
3	Newsletter supplement. Advice to employers
4	HASAWA–Advice to the self-employed
5	Advice to employees
6	Guidance notes on employers' policy statements for health and safety at work
7	Newsletter supplement. Regulations and Approved Codes of Practice and guidance literature
8	Safety committees; Guidance to employers whose employees are not members of recognised independent trade unions
9	Time off for the training of safety representatives
Methods for the Detection of Toxic Substances in Air	These series are covered on page 441
Methods for the Determination of Hazardous Substances	
Notes on best practicable means	
5	Hydrochloric acid works
6	Mineral works (sintered aggregates)
8	Mineral works (plaster)
9	Aluminium (secondary) works
10	Amines works
11	Chemical incineration works
12	Iron works and steel works. Electric arc furnaces
13	Mineral works (roadstone plants)
Occasional papers	
1	The problem drinker at work
2	Microprocessors in industry
3	Managing safety
5	Electrostatic ignition
6	Underground cable damage survey
8	Training for hazardous occupations
Research papers	
9	Vibration injuries of the hand and arm
10	Human factors – aspects of visual display units
Sampling and analysis of emission to the air from scheduled works	
1	An introduction
2	The sampling of gaseous emissions
3	Total acidity
4	Hydrogen chloride

Table 18.3 (continued)

Technical reports	
1	Carcinogens in the workplace
Testing memoranda	
2	Test and approval of explosives for use in coal mines
3	Certification of breathing apparatus
Toxicity reviews	
1	Styrene
2	Formaldehyde
3	Carbon disulphide
4	Benzene
5	Pentachlorophenol
6	Trichloroethylene
7	Cadmium and its compounds
9	1,1,1-Trichloroethane

Collated sources

One UK company[5] provides an indexed library of over 81,000 pages of full-text health and safety information on microfiche. The data are automatically updated three times a year to enable subscribers to keep abreast of new developments. The system is aimed primarily at satisfying the needs of safety officers. It is extremely valuable as an introduction to key publications on all aspects of industrial safety particularly legislation, standards, codes, etc., although deeper searches of the literature are often required, e.g. for most recent toxicology information on a specific chemical.

Table 18.4 lists a selection of standard texts on chemical safety, divided into those dealing with general aspects of the subject, those covering fire and explosion hazards and their control, and those sources biased towards occupational health, hygiene and toxicology. Some books cover specific topics in detail (e.g. *Dust Extraction Technology*), while others are very general (e.g. *Industrial Hazard and Safety Handbook*). Texts are included which are solely compilations of data sheets (e.g. *Toxic and Hazardous Industrial Chemicals Manual, Environmental and Industrial Health Hazards*, etc.) while others adopt a mixed approach with chapters on specific subjects together with resumés of the hazardous properties of many hundreds of substances (e.g. *Hazards in the Chemical Laboratory, Dangerous Properties of Industrial Materials,* etc.). For the layman with little knowledge of chemistry *Hazardous Materials* by Schieler and Pauze serves as an elementary text. It contains an introductory section on the fundamentals of science with chapters describing the different types of hazardous chemicals presented in simple form.

Table 18.4 Selection of safety books on chemicals

General

Aerosol Guide (C.S.M.A. Incorp) 7th edn., 1981
The Care, Handling and Disposal of Dangerous Chemicals; Gaston P J (Northern Publ. (Aberdeen) Ltd), 1970
Chemical Safety Supervision; Guelich J (Reinhold Publ. Co.)
Cryogenics Safety Manual, British Cryogenics Council, London, 1970.
Code of Practice Against Radiation Hazards (Imperial College of Science and Technology), 1973.
Code of Practice for the Protection of Persons Exposed to Ionising Radiation in Research and Teaching (HMSO), 1968
Dangerous Properties of Industrial Materials; Sax N I (Van Nostrand Reinhold Co.), 6th edn., 1984
Dangerous Substances (Croner Publications) 1984
Disposal of Process Wastes, Symposium at the ACHEMA, Frankfurt, Germany 1964
Encyclopaedia of Chemical Technology; eds Kirk R E and Othmer D F (Interscience Publ.) 3rd edn., 1984
Experimental Cryophysics; Hoare F E, Jackson L C and Kurti N (Butterworth Inc.)
Guide for the Perplexed Organic Experimentalist: Loewenthall, H J E (Heyden & Son Ltd), 1978
A Guide to Laboratory Law; Cooke A J D (Butterworths), 1976
Guide to Safety in the Chemical Laboratory (Van Nostrand Reinhold Co.), 2nd edn., 1972
Guide to Safety in Aerosol Manufacture (British Aerosol Manufacturers Assoc.), 1977
Handbook of Radiological Protection (Dept. of Employment), 1971
Handbook of Reactive Chemical Hazards; Bretherick L (Butterworths), 1979
Handbook of Toxic and Hazardous Chemicals; Sittig, M (Noyes Publications), 1981
Handling Chemicals Safely (Dutch Chemical Ind. Assoc.), 1980
Hazard Control Policy in Britain; Chicken J C (Pergamon Press), 1975
Hazards in the Chemical Laboratory; Bretherick L (The Chemical Soc.), 3rd edn., 1981
Hazardous Chemicals Data Book; ed. Weis G. (Noyes Data Corpn.), 1980
Handbook of Laboratory Safety; Steere N V (CRC Press Inc.), 2nd edn., 1979
Hazardous Materials; Schieler L and Pauze D (Van Nostrand Reinhold Co.) 1976
Hazardous Materials; Isman, W E and Carlson, G P (Glencoe Publ. Co.) 1980
Hazardous Materials Handbook; Meidl J H (Glencoe Press), 1972
Hazardous and Toxic Materials: Safe Handling & Disposal; Fawcett, H H (John Wiley & Sons), 1984
High Temperature Technology; IPUA (Butterworth Inc.)
Industrial Hazard and Safety Handbook; King R W and Magid J (Newnes-Butterworths) 1979
Industrial Safety Handbook; Handley W (McGraw), 1977
Industrial Waste Disposal; Ross E D (Van Nostrand Reinhold), 1980
An Introduction to Radiation Protection; Martin A and Harbison S A (Science Paperbacks), 1972
Inert Atmospheres; White P A F and Smith S E (Butterworth Inc.)
Laboratory Management and Techniques; Edwards J A (Butterworth Inc.)
Laboratory Planning; Munce J F (Butterworths)
Laboratory Waste Disposal Manual (MCA) 2nd edn., 1969
Laboratory Safety – Theory and Practice; ed. Fuscaldo, A *et al.* (Academic Press), 1980
Loss Prevention in the Process Industries; Lees F P (Butterworths), 1980
Major Loss Prevention in the Process Industry (Institution of Chemical Engineers), 1971
Manual of Hazardous Chemical Reactions (Boston NFPA), 1968
Matherson Gas Data Book (East Rutherford), 5th edn., 1971
Modern Safety and Health Technology; De Reamer R (Wiley-Interscience), 1980
Plastics Safety Handbook (Soc. of the Plastics Ind., USA)
Proceedings of the Seminar on Safety of Chemicals in the Environment (Harwell), 1979
Process Industry Hazards (Institution of Chemical Engineers), 1976

Table 18.4 (continued)

Prudent Practices for Handling Hazardous Chemicals in Laboratories (National Academy Press), 1981
Radiological Protection Tests for Products Which Can Lead to Exposure of the Public to Ionising Radiation (National Radiological Protection Board)
Safe Storage and Handling Dyestuffs (Ciba-Geigy)
Safe Storage of Laboratory Chemicals; ed. Pipitone D, A. (John Wiley & Sons Ltd.), 1984
Safety and Accident Prevention in Chemical Operations; eds Fawcett H H and Woods W S, 2nd edn. (John Wiley & Sons Ltd), 1982
Safety and Health in Wastewater Systems (Water Pollution Control Federation), 1983
Safety at Work; ed. Ridley J (Butterworths), 1983
Safety in Biological Laboratories; eds Hartree E and Booth V (Biochemical Soc. Special Publications), 1977
Safety in the Chemical Laboratory, Pieters H A J and Greyghton J W (Butterworths), 2nd edn., 1957
Safety in Process Plant Design; Wells G L (George Godwin Ltd.), 1980
Safety in Working with Chemicals; Green M E and Turk A (McMillan Publ. Co.), 1979
Safe Use of Solvents; eds Collings A J and Luxon S G (Academy Press), 1982
Solvent Guide; Marsden C (Clever-Hume), 1963
Storage and Handling of Petroleum Liquids – Practice and Law; J R Hughes, (Griffin), 1967
Toxic and Hazardous Industrial Chemicals Safety Manual (The International Technical Information Institute, Tokyo), 1976
Techniques of Safety Management; Peterson, D (McGraw-Hill Book Co.), 2nd edn., 1978
UMIST Safety Manual (University of Manchester Institute of Science and Technology), 1973

Fire and explosion
Bulk Storage of Highly Inflammable Liquids (BP Chemicals)
Combustion, Flames and Explosions of Gases; Lewis B and Von Elbe G (Academic Press), 1961
Dust Explosions and Fires; Palmer K N (Chapman and Hall), 1973
Explosives; Meyer R (Verlag Chemie), 2nd edn., 1981
Flammable Hazardous Materials, 2nd edn., James H Meidl (Glencoe Publishing Co. Inc.), 1978
Flammability Characteristics of Combustible Gases and Vapours; Zabetakis M G (US Bureau of Mines)
Flash Points (BDH)
Flash Point Index of Trade Name Liquids (Boston NFPA)
Fire and Explosion – Safety and Loss Prevention Guide (Amer. Chem. Eng.), 1973
Fire Investigation; Dennett, M E (Pergamon Press), 1980
Fire Protection Guide on Hazardous Materials (Boston NFPA), 1972
Fire Protection Manual for Hydrocarbon Processing Plants; Vervalin G (Gulf Publ. Co.), 1973
Guide to Safety with Electrical Equipment (FPA)
Handbook of Industrial Fire Protection and Security (Trade and Technical Press Ltd), 1977
Hydrocarbon Propellants (Chemical Speciality Manufacturers' Assoc.), 1979
Intrinsic Safety: The Safe Use of Electronics in Hazardous Locations; Redding R J (McGraw-Hill), 1979
Manual of Firemanship, Parts 1–7, Home Office (Fire Department), HMSO, 1980
Properties of Flammable Liquids (Boston NFPA)
Standard on Fire Protection for Laboratories Using Chemicals (Boston NFPA)

Occupational health, hygiene and toxicology
Air Pollution; Stern A C (Academic Press), 1977
Air Pollution Manual (American Industrial Hygiene Association)
Air Pollution Reference Library (American Conference of Governmental Industrial Hygienists)

Table 18.4 (continued)

Air Sampling Instruments Manual (ACGIH)
Analytical Methods Manual (American Conference of Governmental Industrial Hygienists)
Cancer Causing Chemicals; Sax, N I (Van Nostrand Reinhold), 1981
Casarett and Doull's Toxicology: the basic science of poisons; eds Doull *et al.* (MacMillan Publishing Co. Inc.), 1980
Carcinogenic and Chronic Toxic Hazards of Aromatic Amines; Scott T S (Butterworths Inc.)
Chemical Carcinogenisis and Cancers; Husper W C and Conway W D (Charles C Thomas), 1964
Chemical Carcinogens; ed Searle C (Amer. Chemical Soc.), 1977
Chemical Detection of Gaseous Pollutants; Ruch W E (Ann Arbor Publications Inc.), 1966
Clean Air – Law and Practice; Garner J F and Crow R K (Shaw & Sons Ltd.), 1976
Clinical Toxicology of Commercial Products; Gleason M N (Van Nostrand Reinhold Co.) 3rd edn., 1969
Community Air Quality Guides (American Industrial Hygiene Association)
Design of Industrial Exhaust Systems; Alden J L and Kane J M (Industrial Press Ltd.), 2nd edn., 1970
Determination of Toxic Substances in Air, A Manual of ICI Practice; Hanson N W (Hoffer), 1965
Diseases of Occupations; Hunter D (The English Universities Press Ltd), 5th edn., 1975
Documentation of TLV's for Substances in Workroom Air (American Conference of Governmental Industrial Hygienists), 4th edn., 1980
Dust Extraction Technology; Batel W (Technicopy Ltd.), 1976
Effects of Exposure to Toxic Gases – First Aid and Medical Treatment; Braker W (Matheson Gas Products), 1970
Encyclopaedia of Instrumentation for Industrial Hygiene; Yaffe C D *et al.* (Ann Arbor)
Encyclopaedia of Occupational Safety and Health (International Labour Office, Geneva), 3rd edn., 1983
Environmental and Industrial Health Hazards; Trevethrick, R A (Heinemann Medical Books), 1978
Fundamentals of Industrial Toxicology; Anderson K and Scott R (Ann Arbor Science), 1981
Fundamentals of Industrial Hygiene (National Safety Council), 2nd edn., 1979
Halogenated Hydrocarbons: Health & Ecological Effects; Khan & Stanton (Pergamon), 1981
The Halogenated Hydrocarbons of Industrial and Toxicological Importance; Von Oettingen W F (Butterworths Inc.)
Handbook of Air Pollution Analysis; eds Perry and Young (Chapman and Hall), 1977
Handbook of Occupational Hygiene; ed. Harvey, B (Kluwer Publications), 1980–84
Health Effects of Environmental Pollutants; Waldbott G L, 2nd edn. (The C V Mosby Co.), 1978
Industrial Control Equipment for Gaseous Pollutants, Buonicore A J and Theodore L (CRC Press, Ohio), 1975
Industrial Hygiene Aspects of Plant Operations; eds Cralley, L V and Cralley, L L (MacMillan Publishing Co. Inc.), 1983/84
The Industrial Environment – Its Evaluation and Control (US Dept. of Health, Education and Welfare), 1973
Industrial Toxicology and Dermatology in the Production and Processing of Plastics; Malten K E and Zielhuis R L, (Elsevier), 1964
Industrial Ventilation (Amer. Conference of Governmental Industrial Hygienists), 16th edn., 1980
Isocyanates in Industry, Code of Practice (British Rubber Manufacturers' Assoc.)
Lecture Notes on Occupational Medicine; Waldron H A (Blackwell), 1976

Table 18.4 (continued)

Monitoring Toxic Gases in the Atmosphere for Hygiene and Pollution Control; Thain W (Pergamon Press), 1980
Occupational Diseases (NIOSH), 1977
Occupational Health and Safety Concepts; Atherley G (Applied Science Publ.), 1978
Occupational Health Practice; Schilling, R S F (Butterworths), 2nd edn., 1981
Occupational Hygiene; eds Waldron H A and Harrington, J M (Blackwell Scientific Publications), 1980
Patty's Industrial Hygiene and Toxicology; eds Clayton, G D and Clayton F E (J Wiley & Sons), 3rd edn., 1982
Plant and Process Ventilation; Hemeon W (Industrial Press Inc.)
The Prevention of Occupational Cancer (ASTMS)
The Protection Handbook of Pollution Control; Sutton P (A Osbourne & Associates), 1975
Registry of Toxic Effects of Chemical Substances (NIOSH) – including sub-file of Suspected Carcinogens (1980)
Safe Handling of Chemical Carcinogens, Mutagens, Teratogens, and Highly Toxic Substances; Walters D B (Ann Arbor Science), 1980
The Social Audit Pollution Handbook; Frankel M (MacMillan Press Ltd.), 1978
Suspected Carcinogens; Fairchild, E J (Castle House Publications Ltd.), 1978
Toxicity and Metabolism of Industrial Solvents; Browning E (Elsevier Publishing Co.) 1965
Toxicity of Industrial Metals; Browning E (Butterworths), 1969
Toxicity of Industrial Organic Solvents; (Medical Research Council)
Toxicity of Industrial Solvents and Plasticizers; Browning E (Elsevier Publ. Co.)
Waste Recycling and Pollution Control Handbook; Bridgwater A V and Mumford C J (George Godwin Ltd), 1979

Journals/abstracts

Periodicals are valuable for keeping abreast of the latest publication titles, safety techniques and equipment, and for changes in philosophy, legislation, etc., and for the review articles often featured. A list of relevant periodicals is given in Table 18.5, arranged so as to identify their prime subject area and including both mainstream safety journals along with a selection of non-safety publications which frequently contain items of interest on chemical safety. The list includes both journals and current awareness bulletins produced by the many abstracting services now available, some of which are accessible on-line (see below). The International Occupational Safety and Health Information Centre in Geneva, one of the best safety data bases in the world, currently searches around 1,500 journals and contains a computerised index-searching service. A current awareness bulletin, *CIS Abstracts*, is available which covers all aspects of industrial safety worldwide.

The chemical literature has been abstracted by *Chemical Abstracts* since 1907. This service, provided by the American Chemical Society, monitors nearly 14,000 periodicals from more than 150 countries, plus

Table 18.5 A selected list of relevant periodicals

General
Ambio (The Royal Swedish Academy of Sciences and Pergamon Press)
Accident Analysis and Prevention (Pergamon Press)
Accident Prevention (IAPA, Canada)
Australian Safety
BSI News (British Standards Institution)
Archives des maladies professionelles de médicine du travail et de sécurité sociale
Chemical Abstracts
CIS Abstracts (International Labour Office)
Chemical Hazards (American Chemical Soc.)
Hazards Bulletin (British Soc. for Social Responsibility in Science)
Health and Safety Information Bulletin (Industrial Relations Services)
Health and Safety Commission Newsletter
Health and Safety at Work (MacLaren Publ. Ltd)
Health and Safety Monitor (Monitor Press)
Industrial Health and Safety Bulletin (Amalgamated Union of Engineering Workers)
Industrial Health and Safety (Engineers Employers Federation)
Industrial Safety (United Press Ltd)
Journal of Hazardous Materials (Elsevier)
Journal of Occupational Accidents (Elsevier)
Journal of the American Soc. of Safety Engineers (Amer. Soc. Safety Engineers)
Lawlines (British Safety Council)
Loss Prevention (Amer. Inst. of Chemical Engineers)
Loss Prevention Bulletin (Instit. Chem. Engineers)
Monthly Newsletters (National Safety Council)
Occupational Hazards (PO Box 5746–Cleveland, Ohio 44115)
Occupational Safety and Health (Royal Soc. for the Prevention of Accidents)
Occupational Safety and Health Reporter (Bureau of National Affairs)
Professional Safety (Amer. Soc. of Safety Engineers)
RAPRA Communications (Rubber and Plastics Research Association)
RoSPA Bulletin (Royal Society for the Prevention of Accidents)
Safety (British Safety Council)
Safety Science Abstracts (Cambridge Abstracts, Maryland)
The Safety Practitioner (Inst. of Occupational Safety and Health)
Travail et sécurité (Institut National de Recherche et de Sécurité pour la Prevention des Accidents du Travail et des Maladies Prof., Paris, France)
Safety Surveyor (Victor Green Publ. Ltd.)

Fire and explosion
Fire Journal (National Fire Protection Assoc.)
Fire Research Abstracts and Reviews (National Academy of Sciences)
Fire Technology (Soc. of Protection Engineers)
Fire Prevention (Fire Protection Assoc.)
Fire Protection Journal (Fire Protection Assoc.)
Fire Technology (Fire Protection Assoc.)

Occupational health, hygiene and toxicology
American Industrial Hygiene Assoc. Journal. (Amer. Ind. Hygiene Assoc.)
Annals of Occupational Hygiene (British Occupational Hygiene Soc.)
Archives of Environmental Health (Amer. Medical Assoc.)
British Journal of Industrial Medicine (British Medical Assoc.)
British Medical Journal
Environmental Control and Safety Management (Morristown New Jersey)
Environmental Health (Environmental Officers Assoc.)
International Environment and Safety (Labmate Ltd.)
Industrial Hygiene Digest (Industrial Hygiene Foundation)

Table 18.5 (continued)

Industrial Medicine and Surgery (Industrial Medicine Publ. Co.)
Industrial Hygiene News Report (Flourney and Assoc.)
Journal of Occupational Medicine (Harper and Row)
Journal of the Society of Occupational Medicine (J. Wright & Sons)
Occupational Health (Macmillan Journals Ltd)
Occupational Safety and Health Abstracts (International Labor Office)

Journals regularly containing items on chemical safety
Analyst
American Institute of Chemical Engineers Journal
Chemie Ingenieur Technik
Analytical Chemistry
Chemical Age
Bibra Information Bulletin
Chemical Engineering
Chemical Engineering Progress
Chemical and Engineering News
Chemistry and Industry
Chemistry in Britain
Chemical Processing
Control and Instrumentation
Education in Chemistry
Environmental Engineering
Hydrocarbon Processing
Industrial and Engineering Chemistry
Laboratory Practice
Lancet
Loss Prevention Bulletin, (Institution of Chemical Engineers)
Journal of Chemical Education
Journal of Hazardous Materials
New Scientist
Official Journal of the European Communities
Process Engineering
The Chemical Engineer, Institution of Chemical Engineers

patents, conference proceedings, etc. Citations cover biochemistry, organic chemistry, macromolecular chemistry, physical and analytical chemistry, applied chemistry and chemical engineering. It is the largest bibliographic chemistry file in the world with 5 million references and 420,000 updates annually. Information relevant to a particular chemical, or class of chemicals, is indexed under specific headings for that material or class. These headings include flammability, fires, combustion, explosions, explosibility, detonation, heat of explosion, health hazards, etc. Also, since 1974 the inclusion of the key-word 'safety' has increased the value of *Chemical Abstracts* and *Chemcon* (the computer readable chemical abstracts which currently dates back only to 1969) in searching for chemical safety information. Current awareness bulletins based on *Chemical Abstracts* are also available and include *CA Reviews* and *CA Selects*. The latter includes a fortnightly bulletin on chemical hazards. It covers articles dealing with radiological hazards or chemically hazardous substances which might affect the health and safety of personnel working with the substances where the author did not emphasize chemical safety. The UK agent is UKCIS.

Computerised data bases

It is estimated that over 200 on-line data bases are commercially available for information retrieval worldwide covering a wide range of subjects and totalling 60 million references. Hardware requirements for accessing these data bases usually includes a VDU, keyboard/microcomputer, printer and modem (to transmit and receive data via the public telephone system) or a direct link to a data network system such as Euronet, Telenet, etc. In addition to chemical abstracts the selection of on-line data bases in Table 18.6 contain information on safety and health aspects of chemicals. Other data bases include, in the UK, Harwell's data base on industrial waste information; it also publishes an *Industrial Wastes Information Bulletin*. Some private companies also possess data bases which can be used for external enquiries. One such example is the biological data base at ICI's Central Toxicological Laboratory in the UK.

While advantages of on-line searching of the literature include speed and reliability, in general the data bases are limited to information published during the last 10–15 years.

Organisations

Valuable information on chemical safety can be obtained from a host of organisations. These include government departments, trade associations, professional bodies/learned societies, universities, independent consultants, agencies preparing standards. The compilation of organisations in Table 18.7 includes those culled from the literature (4,6,7) plus additions by the authors.

Information available in the UK from the HSC/HSE is mentioned on page 1032. The Federal Register is useful for keeping abreast of regulatory developments in the USA, and the *Official Journal of the European Committees* identifies developments within the European Community.

In the UK the Chemical Industries Association, through the Chemical Industry Safety Health Council, has published a wide variety of documents on subjects ranging from a hygiene manual to Codes of Practice for the safe handling of specific chemicals and advice on major emergencies. A list of publications is given in Table 18.8. Similarly in the USA the Manufacturing Chemists' Association (MCA) has, for over quarter of a century, compiled manuals, safety guides and data sheets, as illustrated

Table 18.6 A selection of computerised data bases of information relating to chemical safety.

Biosis	Contains citations from Biological Abstracts (BA) Biological Abstracts/Reports, Reviews, and Meetings (BA/RRM) and Bioresearch Index. BA alone contains about 165,000 accounts of original research from over 9,000 primary journals. BA/RRM encompasses an additional 125,000 citations per year from meetings, reviews, books, etc. Biosis files date back to 1969 and cover such subjects as all life sciences, toxicology and environmental biology.
Chemline	Contains the records of more than 250,000 chemicals identified by chemical abstracts service (CAS) Registry Number. About 550,000 chemical names can be used to access the records. Though Chemline is primarily an on-line dictionary file which typically one would use to find the CAS Registry Number prior to undertaking e.g. a Toxline search, (see later) itself it gives an indication of whether the substance is a candidate under the Toxic Substances Control Act.
Chemical Exposure	A comprehensive data base of chemicals identified in tissues and body fluids, set up because of concern by Government Agencies of toxic chemical effects. Data compiled by the Chemical Effects Information Centre at the Oak Ridge National Laboratory identifies body burdens which reflects exposure to contaminants in air, food and water (as well as drugs). Subject coverage includes chemical identity and chemical properties, methods of analysis, toxicology, health effects, etc. It is accessible back to 1974 with 14,370 records citing about 1,000 unique substances.
Chemical Regulations and Guidelines	An authorative index to US federal regulatory material relating to the control of chemical substances covering federal statutes, promulgated regulations and available federal guidelines, standards and supporting documents. Subjects span disposal, manufacture, occupational health, product registration, transportation, use etc. Records are accessible from 1981 with about 1,000 new records added monthly.
Compendex	Covers all aspects of engineering for all industries. It is particularly useful for agricultural, aerospace, automotive, bioengineering, chemical, civil, electrical, electronic, food, geological, mechanical, mining, and nuclear engineering, and dates back to 1970.
HSE Line	Produced in the UK by HSE. It provides access to a major source of references on the different aspects of health and safety. Source material consists of documents acquired and produced by HSE and includes report literature, legislation, monographs and conference proceedings. In addition about 250 periodicals are scanned. Subject coverage includes science, manufacturing industries, production, agriculture, mining, explosives, workplace pollution, occupational medicine, toxic substances, risk assessment and environmental and industrial hazards. HSE Line is also accessible (together with other material published by the HSE Library and Information Service) on Prestel (the British Telecom Public Videotext system).
Medline	The on-line version of Medlars. It is produced by the US National Library of Medicine and is one of the major sources for biomedical literature. It is accessible back to 1966 and contains over 2 million citations from about 3,200 journals with about 250,000 records added each year. Of the many subject areas covered, Toxicology is of particular relevance.

Table 18.6 (continued)

NTIS	Produced by the National Information Service of the US Department of Commerce. This database consists of government sponsored research, development and engineering plus other reports prepared for the government and covers over 300 government agencies. Subject areas include Atmospheric Sciences, Medicine and Biology, Chemistry, Environmental Pollution and Control, Material Sciences, and Transportation. It dates back to 1964 and is updated biweekly.
RTECS	The Registry of Toxic Effects of Chemical Substances is compiled by the National Institute of Occupational Safety and Health. It contains toxicity data for over 55,000 substances with listings of toxicological effects, Threshold Limit Values, recommended air standards, aquatic toxicity and other related information.
Toxline	Toxline and its associated backfile, Toxback, is made up from eleven separate subfiles, viz. Toxicity Bibliography, Chemical and Biological Activities, Pesticides Abstracts, Health Aspects of Pesticides Abstracts Bulletin, International Pharmaceutical Abstracts, Abstracts on Health Effects of Environmental Pollutants, a special collection of material collected by Hayes, Environmental Mutagen Information Centre, Toxic Materials Information Centre, Teratology, and Environmental Teratology Information Centre. Though Toxline itself contains 410,000 records from primary journals from 1974 onwards only Toxback contains older material consisting of approximately 380,000 records.

Table 18.7 List of organisations

UK Organisations	
Asbestos Information Centre	Sackville House, 40 Picadilly, London W1V 9PA
Asbestos Removal Contractors Assoc.	45 Sheen Lane, London SW14 8AB.
Asbestosis Research Council	PO Box 40, Rochdale, Lancs, OL12 7EQ
ASLIB	3 Belgrave Square, London SW1X 8PL
Association of Scientific, Technical and Managerial Staffs (ASTMS)	Health and Safety Office, Whitehall Office, Dane, O'Coys Road, Bishops Stortford, Herts.
Aston University	Health and Safety Unit, Dept of Mech. & Prod. Eng. Aston Triangle, Birmingham B4 7ET
BDH Chemicals Ltd	Broom Road, Poole, Dorset BH12 24N
British Adhesive Manufacturers Association	2a High St. Hythe, Southampton SO4 6YW
British Agrochemicals Association Ltd	Alembic House, 93 Albert Embankment, London SE1 7TU
British Automatic Sprinkler Assoc.	PO Box 207, 128 Queen Victoria St., London, EC4.
British Chemical and Dyestuffs Trades Assoc.	126 Westminster Palace Gardens, Artillery Row, London SW1P 1RL

Table 18.7 (continued)

British Medical Assoc.	BMA House, Tavistock Square, London WC1H 9JP
British Occupational Hygiene Soc.	c/o Dept. of Occupational Health, London School of Hygiene & Tropical Medicine, Keppel St., London WC1
British Materials Handling Fed.	192–109 Vauxhall Bridge Road, London SW1V 1DX
British Paper & Board Industry Federation	3 Plough Place, London EC4
British Plastics Federation	5 Belgrave Square, London SW1X 8PH
British Printing Industries Federation	11 Bedford Row, London WC1R 4DX
British Resin Manufacturers Assoc.	Queensway House, Queensway, Redhill, Surrey RH1 1QS
British Rubber Manufacturers Assoc. Ltd	90–91 Tottenham Court Road, London W1P 0BR
British Soc. for Social Responsibility in Science	9 Poland Street, London W1B 3DG
Chemical Industries Assoc.	Alembic House, 93 Albert Embankment, London SE1 7TU
Employment Medical Advisory Service	Atlantic House, Farrington Street, London EC4
Fire Extinguishing Trades Assoc.	48A Eden St., Kingston-upon-Thames, Surrey
Fire Protection Assoc	Aldermary House, Queen Street, London EC4N 1TJ
Fire Research Station	Borehamwood, Herts, WD6 2BL
Harwell, Hazardous Materials Service	Toxic and Hazardous Materials Group, Harwell Laboratory, Oxfordshire OX11 0RA
Health and Safety Executive	Baynards House, 1 Chepstow Place, Westbourne Grove, London W2
and	The Triad, Stanley Road, Bootle, Merseyside L20 3PG
Health and Safety Executive Occupational Medicine and Hygiene Laboratory	403/405 Edgeware Road, London NW 6LN
Health and Safety Executive Laboratory, Safety in Mines Research Establishment	Red Hill, Sheffield, S3 7HQ
Health and Safety Executive British Approval Service for Electrical Equipment in Flammable Atmospheres (BASEEFA)	Harpur Hill, Buxton, Derbyshire SK17 9JN
Heating and Ventilating Contractors Assoc.	34 Palace Court, London W2 4JG

Table 18.7 (continued)

Hydrocarbon Solvents Assoc.	19 The Boundary, Langton Green, Tunbridge Wells, Kent TN3 0YA
ICI Central Toxicology Laboratory	ICI Ltd., Alderley Park, Manchester SK10 4TJ
Industrial Safety (Protective Equipment) Manufacturers Assoc.	69 Cannon Street, London EC4N 5AB
Institute of Occupational Medicine	9 Roxburgh Place, Edinburgh EH8 9SU
Institute of Petroleum	61 New Cavendish St., London W1M 8AR
Institution of Chemical Engineers	165–171 Railway Terrace, Rugby. Warwickshire, CV21 3HQ
Institution of Environmental Health Officers	10 Grosvenor Place, London SW1X 7HU
Institution of Occupational Safety & Health	222 Uppingham Road, Leicester LE5 0QG
Laboratory of the Government Chemist	Cornwall House, Stamford St., London
Liquefied Petroleum Gas Industry Technical Assoc.	17 Grosvenor Crescent, London SW1X 7ES
National Assoc. of Waste Disposal Contractors	Suite 1, 14 Uxbridge Road, Ealing, London WS 2BP
National Radiological Protection Board	Harwell, Didcot, Berks.
Oil & Chemical Plant Constructors Assoc.	Suites 41–48 Kent House, 87 Regents St., London, W1R 7HF
Paint Research Assoc.	Waldegrave Road, Teddington, Middlesex TW11 8LD
Paint Makers' Assoc.	Alembic House, 93 Albert Embankment, London SE1 7TY
Plastics and Rubber Institute	11 Hobart Place, London SW1W 0HL
Poisons Reference Service	New Cross Hospital, London.
Royal Soc. of Chemistry	30 Russell Square, London WC1B 5DT or Burlington House, London W1V 0BN
Royal Society for the Prevention of Accidents (RoSPA)	Cannon House, The Priory, Queensway, Birmingham B4 6BS
Rubber and Plastics Research Assoc.	Shawbury, Shrewsbury, SY4 4NR
Safety Equipment Distributors Assoc.	Gateway House, 50 High St., Birmingham B4 7SY
Science Reference Library	10 Porchester Gardens, London W2 4DE
The Environmental Health Officers Assoc.	19 Grosvenor Place, London, SW1X 7HU
Trades Union Congress	Congress House, 23–28 Great Russell St., London WC1B 3LS
TUC Centenary Institute of Occupational Health	London School of Hygiene and Tropical Medicine, Keppel St., London WC1E 7HT

Table 18.7 (continued)

University of Birmingham	Institute of Occupational Health, PO Box 363, Birmingham, B15 2TT
University of Manchester	Radiological Protection Service, Manchester M13 9PL
University of Surrey	Institute of Industrial and Environmental Health and Safety University of Surrey, Guildford, Surrey GU2 5XH
UKCIS	The University, Nottingham NG7 2RD
Welding Institute	Abington Hall, Abington, Cambridge, CB1 6AL
Warren Spring Laboratory	Gunnels Wood Road, Stevenage, Herts, SG1 2BX

US Organisations

Air Pollution Control Assoc., 4400 Fifth Avenue, Pittsburgh, Pa 15213
Amer. Academy of Industrial Hygiene, 475 Wolf Ledges Parkway, Akron, Ohio, 44311
Amer. Chemical Soc., 1155 16th Street, N.W. Washington DC, 20036
Amer. Conference of Governmental Industrial Hygienists, PO Box 1937, Cincinnati, Ohio 45201
Amer. Industrial Hygiene Assoc., 475 Wolf Ledges Parkway, Akron, Ohio, 44311
Amer. Institute of Chemical Engineers, 345 47th Street, New York, NY, 10017
Amer. Institute of Chemists, 7315 Wisconsin Ave., Washington, DC, 20014
American Petroleum Institute, 2101 L Street, N.W., Washington DC, 20037
American Petroleum Refiners Association, 1110 Ring Building, Washington DC., 20036
Amer. Society of Safety Engineers, 850 Busse Highway, Park Ridge, Ill. 60068
Bureau of Radiological Health, FDA, 5600 Fishers Lane, Rockville, Md 20852
Compressed Gas Assoc., 500 5th Avenue, New York, NY., 10036
Consumer Product Safety Commission, 111 Eighteenth St., NW., Washington DC, 20237
Department of Labor, 200 Constitution Ave., NW., Washington, DC, 20210
Environmental Protection Agency, 401 M Street, SW, Washington DC, 20460
Health Physics Soc., PO Box 156 East Weymouth, Mass., 02189
Institute of Environmental Science, 940 East Northwest Highway, Mount Prospect, Ill. 60056
National Centre for Toxicological Research, FDA, 5600 Fishers Lane, Rockville, Md., 20857
National Fire Protection Association, 470 Atlantic Ave., Boston, Mass., 02210
National Fire Prevention and Control Administration, Dept. of Commerce, Washington, DC., 20230
National Institute for Occupational Safety and Health, Parklawn Building, 5600 Fishers Lane, Rockville, Md., 20852
National Safety Council, 444 North Michigan Ave., Chicago, Ill. 60611
National Technical Information Service, Dept. of Commerce, 5285 Port Royal Rd., Springfield Va. 22151
National Transportation Safety Board, Dept. of Transportation, Washington DC, 20591
Occupational Safety and Health Administration, US Dept. of Labor, 200 Constitution Ave., NW Washington, DC., 20210
Soap and Detergent Assoc., 475 Park Avenue South, New York, NY 10016
Society of Toxicology, 475 Wolf Ledges Parkway, Akron, Ohio, 44311
Underwriters Laboratories Inc., 207 East Ohio St., Chicago, Ill. 60611

Table 18.7 (continued)

Organisations in Other Countries

Belgium
Comite Securite-Hygiene de la Federation des Industries, Chimique de Belgique, 1040 Brussels.

Canada
Canadian Centre for Occupational Health and Safety, 250 Main Street East, Hamilton, Ontario, L8N 1H6
Ontario Ministry of Labour, Library, 400 University Ave., Toronto, Ontario, M7a 1T7

Finland
Tyoterveyslaitos, Haartmaninkatu 1, 00290 Helsinki 29

France
Institut National de Recherche et de Securite Pour la Prevention des Accidents du Travail et des Maladies Professionelles, 30 rue Olivier Noyer, 75680 Paris Cedex 14
International Radiation Protection Association, 3 Sq. Albin Cochot, 75-Paris

Germany
Abteilung Fur Arbeitsschutz, Gewerbeaufsicht und Arbeitsmedizin,
Rheinland-Pfalz Ministerium fur Soziales, Gesundheit und Sport, Bauhofstrasse 4, Mainz
Abteilung Hygiene und Arbeitsmedizin, Technische, Hochschule, 51000 Aachen, Lochnerstrasse 4–20
Bundesanstalt Fur Arbeitsschutz und Unfallforschung, Vogelpothsweg 50–52, 4600 Dortmund-Dorstfeld
Hauptverband Der Gewerblichen Berufsgenossenchaften, Langwartweg 103, Postfach 5040, 5300 Bonn
Niedersachsisches Landesvervaltungsamt – Arbeitsmedizin und Gewerbehygiene, Bertastrasse 4–6, Hannover
Technische Hochschule Darmstadt Institut fur Arbeitswissenchaft, Petersenstrasse 30, 6100 Darmstadt

Netherlands
Veiligheidsinstituut, PO Box BUS 5665, 1007 AR, Amsterdam

Norway
Institute of Occupational Health, Gydasvei 8, Bolks 81, 49 Dept., Oslo 1

Sweden
Arbetarskyddsstyrelsen, S-171 84 Solna

Switzerland
International Labour Office, International Occupational Safety and Health Information Centre, 1211 Geneva 22 (UK Office: Marsham St., London) World Health Organisation, H 1211 Geneva 27

Table 18.8 List of health, safety and related publications available from the Chemical Industries Assoc. Ltd

Health and safety
Acrylonitrile in the Atmosphere
Allergy to Chemicals at Work
An approach to the Categorisation of Process Plant Hazard and Control Building Design

Table 18.8 (continued)

Benzene Guidelines
Cancer in Modern Mortality
Chemical Safety Summary
Codes of Practice for Chemicals with Major Hazards – (7 books) Acrylonitrile; Anhydrous Hydrogen Chloride; Chlorine; Ethylene Dichloride; Hydrogen Fluoride; Phosgene; Vinyl Chloride
Determination of formaldehyde in the workplace atmosphere
Determination of Vinyl Chloride
EINECS – A Guide to Completion
Employment and Reproductive Health
Exposure to Gases & Vapours – Notes
Fire Retardants Directory
Formaldehyde Exposure in Industry
Guide to Fire Prevention in the Chemical Industry
Guide for Storage of Highly Flammable Liquids
Guide to the Evaluation & Control of Toxic Substances in the Work Environment
Guidelines for Bulk Handling of Chlorine at Customer Installations
Guidelines for Bulk Handling of Ethylene Oxide
Guidelines for Safe Warehousing
Hazard & Operability Studies
Induction of Vomiting – Card
Isocyanates – Incidents Documents
Isocyanates – Questions & Answers
Major Emergencies
Major Hazards
Measurement of Formaldehyde in the Atmosphere
Phenol Splashes
Protection of the Eyes
Reporting for EINECS – CIA Seminar Package
Safe and Sound
Safe use of VDUs
Safety Audits
Safety Pennants
Treatment Labels
Tumours of the Bladder in the Chemical Industry
Vinyl Chloride Monomer

Distribution
Ammonia Aid Emergency Scheme Manual
Black and White Marking – Low Hazard Chemicals
Chemicals on the Move
Chemsafe Manual
Chlor-Aid Emergency Scheme Manual
Code of Practice for Safe Handling and Transport of Anhydrous Ammonia by Rail
Code of Practice for Safe Handling and Transport of Anhydrous Ammonia by Road
Code of Practice for Storage of Fully Refrigerated Anhydrous Ammonia
Code of Practice for Storage of Anhydrous Ammonia Under Pressure in the UK – Spherical and Cylindrical Vessels
Design and Construction of Vehicles
Drivers and the Hazardous Load – Package
Drivers and the Hazardous Load – Tankers
Proceedings of Distribution Conference 1981
Road Transport of Hazardous Chemicals
UN Numbers 1983 (Alphabetical and Numerical)

Table 18.8 (continued)

CEFIC publications
CEFIC Annual Report
CEFIC Criterion Document on Benzene
CEFIC Introducing Occupational Epidemiology
CEFIC Tremcards Group Texts
CEFIC Tremcards Reference Edition
CEFIC Tremcards Reference Edition Supplement 1982
CAPITB publications
(Training Recommendation Number – Tr Rec No)
Tr Rec No 11
Industrial Training Associated with Standing Courses (Science & Engineering) January 1971
Tr Rec No 12
Training Operators A-1 March 1978
Tr Rec No 14
Engineering Craftsmen: Electrical Intermediate Training & Further Training March 1971
Tr Rec No 15
Engineering Craftsmen: Mechanical Intermediate & Further Training August 1971
Tr Rec No 17
Training of Scientific Laboratory Technicians February 1972
Tr Rec No 17 Supplement C
Training for Scientific Laboratory Technicians Surface Coating July 1973
Tr Rec No 17 Supplement D
Training for Scientific Laboratory Technicians Plastics Processing July 1973
Tr Rec No 20
Training for Engineering Technicians May 1972
Tr Rec No 20 Supplement A
Training for Engineering Technicians Specific Training & Experience for Instrument Technicians May 1972
Tr Rec No 20 Supplement B
Training for Engineering Technicians Specific Training & Experience for Electrical Technicians July1973
Tr Rec No 20 Supplement C
Training for Engineering Technicians Specific Training & Experience for Mechanical Technicians September 1973
Tr Rec No 28
Training for Training Staff June 1973
Information Paper No 13
Basic Course for Adult Entrants into Chemical Process Operations & an Example Training Unit Distillation August 1972
Logical Approach to Fault-Finding March 1978

by Table 18.9. Other organisations able to help with information on chemicals' safety including safety data sheets are: the Fire Protection Association, FPA (UK); the National Fire Protection Association, NFPA (USA); the National Safety Council, NSC (USA); and the American Industrial Hygiene Association, AIHA (USA). The quality and content of the data sheets varies with the source. Thus, MCA sheets are comprehensive multi-page documents covering properties, hazards, engineering-control methods, employee safety, techniques for handling and storage, procedures for cleaning and repair of equipment and tanks, methods for waste disposal and medical arrangements and first-aid procedures. The AIHA data sheets tend to be single-page documents recommending maximum atmospheric concentrations and control measures; they comment on the degree of hazard, significant properties, means of recognition, methods for evaluation of exposure, along with specific procedures for first aid and biological monitoring. Data sheets presented in *Environmental and Industrial Health Hazards* were compiled by an occupational physician and therefore tend to be medically and first-aid orientated and of no value, for example, in identifying reactive hazards. The NFPA sheets contain a description of the material and outline fire and explosion hazards, life hazards, fire-fighting appliances, usual shipping containers, storage arrangements and some miscellaneous remarks. FPA data sheets tend to concentrate on fire hazards and their control. Some industrial associations issue current awareness abstract publications to members, e.g. *Regulatory Alert* published by the Chemical Specialities Manufacturers Association.

An important source of information on occupational safety and health is 'standards'. UK government has enhanced the status of standard specifications[8] and the EEC has issued a policy of harmonisation of standard specifications. Agencies issuing standards are listed in Table 18.10 while Table 18.11 identifies British Standards with relevance to safety. The BSI represents the UK in the International Organisation for Standardisation, in the International Electrotechnical Commission, and in West European organisations concerned with harmonisation of standards. Scope of standards includes glossaries of terms, definitions, quantities, units and symbols; specifications for quality, safety, performance or dimensions; preferred sizes and types; Codes of Practice. The BSI is also concerned with certification and assessment of products as complying with standards and international aspects of these. A *BSI Yearbook* lists all British Standards in numerical order. Announcements relating to new, revised, amended or withdrawn standards are made monthly in the *BSI News* or *BSI Sales Bulletin*.

With regards to the setting of hygiene standards for environmental pollutants, the Environmental Protection Agency (EPA), the National Institute for Occupational Safety and Health (NIOSH) and the American Conference of Government Industrial Hygienists (ACGIH) all feature prominently in the USA. Furthermore, many countries adopt USA values as a basis for setting their own values.

In the UK two relevant learned societies are the Royal Society of Chemistry and The Institution of Chemical Engineers. A list of their publications is given in Table 18.12

Table 18.9 Publications by the Manufacturing Chemists Assoc.

1. Chem-Cards (Transportation Emergency Guides).

2. Chemical Safety Data Sheets for the following substances:

Substance	Year	Data Sheet
Acetaldehyde	(1952)	SD-43
Acetic Acid	(1973)	SD-41
Acetic Anhydride	(1962)	SD-15
Acetone	(1962)	SD-87
Acetylene	(1957)	SD-7
Acrolein	(1961)	SD-85
Acrylonitrile	(1974)	SD-31
Allyl Chloride	(1973)	SD-99
Aluminium Chloride	(1956)	SD-62
Ammonia Anhydrous	(1960)	SD-8
Ammonia Aqua	(1947)	SD-13
Ammonium Dichromate	(1952)	SD-45
Aniline	(1963)	SD-17
Antimony Trichloride (Anhydrous)	(1957)	SD-66
Arsenic Trioxide	(1956)	SD-60
Benzene	(1960)	SD-2
Benzoyl Peroxide	(1960)	SD-81
Benzyl Chloride	(1974)	SD-69
Bromine	(1968)	SD-49
Butadiene	(1974)	SD-55
n-Butyllithium in Hydrocarbon Solvents	(1966)	SD-91
Butyraldehydes	(1960)	SD-78
Calcium Carbide	(1967)	SD-23
Carbon Disulfide	(1967)	SD-12
Carbon Tetrachloride	(1963)	SD-3
Caustic Potash	(1968)	SD-10
Caustic Soda	(1974)	SD-9
Chloroform	(1974)	SD-89
Chlorosulfonic Acid	(1968)	SD-33
Chromic Acid	(1952)	SD-44
Diethylamine	(1971)	SD-97
Diethylenetriamine	(1959)	SD-76
Dimethyl Sulfate	(1966)	SD-19
Dinitrotoluenes	(1966)	SD-93
Ethyl Acetate	(1972)	SD-51
Ethyl Chloride	(1953)	SD-50
Ethyl Ether	(1965)	SD-29
Ethylene	(1973)	SD-100
Ethylene Dichloride	(1971)	SD-18
Ethylene Oxide	(1971)	SD-38
Formaldehyde	(1960)	SD-1
Hydrochloric Acid	(1970)	SD-39
Hydrocyanic Acid	(1961)	SD-67
Hydrofluoric Acid	(1970)	SD-25
Hydrogen Peroxide	(1969)	SD-53
Hydrogen Sulfide	(1968)	SD-36
Isopropyl Alcohol	(1972)	SD-98
Isopropylamine	(1959)	SD-72
Lead Oxides	(1956)	SD-64
Methanol	(1970)	SD-22
Maleic Anhydride	(1974)	SD-88
Methyl Acrylate and Ethyl Acrylate	(1960)	SD-79
Methylamines	(1955)	SD-57
Methyl Bromide	(1968)	SD-35
Methyl Chloride	(1970)	SD-40
Methylene Chloride	(1962)	SD-86
Methyl Ethyl Ketone	(1961)	SD-83
Mixed Acid	(1974)	SD-65
Naphthalene	(1956)	SD-58
Nitric Acid	(1961)	SD-5
Nitric-Sulfuric Acid		
Nitrocellulose (Wet Types)	(1970)	SD-96
Ortho-Dichlorobenzene	(1974)	SD-54
Paraformaldehyde	(1974)	SD-6
Perchloroethylene	(1971)	SD-24
Perchloric Acid Solution	(1965)	SD-11
Phenol	(1964)	SD-4
Phosgene	(1967)	SD-95
Phosphoric Acid	(1958)	SD-70
Phosphoric Anhydride	(1974)	SD-28
Phosphorus, Elemental	(1976)	SD-16
Phosphorus Oxychloride	(1968)	SD-26
Phosphorus Pentasulfide	(1958)	SD-71
Phosphorus Trichloride	(1972)	SD-27
Phthalic Anhydride	(1956)	SD-61
Propylene	(1974)	SD-59
Sodium Chlorate	(1952)	SD-42
Sodium Cyanide	(1967)	SD-30
Sodium, Metallic	(1974)	SD-47
Sodium and Potassium Dichromates	(1952)	SD-46
Styrene Monomer	(1971)	SD-37
Sulfur	(1959)	SD-74
Sulfur Chlorides	(1960)	SD-77
Sulfur Dioxide	(1953)	SD-52
Sulfur Trioxide	(1976)	SD-101
Sulfuric Acid	(1963)	SD-20
Tetrachloroethane	(1949)	SD-34
Toluene	(1956)	SD-63
Toluene Diisocyanate	(1971)	SD-73
Toluidine	(1961)	SD-82
1,1,1-Trichloroethane	(1965)	SD-90

Cresol .... (1952) SD-48 Mixtures .... (1974) SD-65 Trichloroethylene .... (1956) SD-14
Cyclohexane .... (1957) SD-68 paraNitroaniline .... (1966) SD-94 Vinyl Acetate .... (1970) SD-75
o-Dichlorobenzene .... (1974) SD-54 Nitrobenzene .... (1967) SD-21

3. Manuals as exemplified by the following:

Guide to Precautionary Labelling of Hazardous Chemicals (Sixth Edition – 1961) L-1

Tank Cars – ICC Spec. 103B, Rubber-lined – Unloading when filled with nuriatic acid, phosphoric acid, or other authorised liquids TC-2

Tank Cars – Unloading when filled with liquid caustic soda or caustic potash (Revised 1946, 1950, 1952) TC-3

Tank Cars – Unloading when filled with flammable liquids (Revised, 1952) TC-4

Tank Cars – Unloading when filled with Phenol (Revised, 1959) TC-6

Tank Car Approach Platforms TC-7

4. Laboratory Safety:

– A Film entitled 'Safety in the Chemical Laboratory'
– *A Guide for Safety in the Chemical Laboratory*, Van Nostrand & Co., Inc., Princeton, NJ.

5. Chemical safety guides as illustrated below:

Health Factors in the Safe Handling of Chemicals SG-1
Housekeeping in the Chemical Industry SG-2
Flammable Liquids – Storage and Handling of Drum Lots and Smaller Quantities SG-3
Emergency Organisation for the Chemical Industry SG-4
Plastic Foams – Storage, Handling and Fabrication SG-5
Forklift Operations SG-6
Guide for Storage and Handling of Shock and Impact Sensitive Materials SG-7
Electrical Switch Lockout Procedure SG-8
Disposal of Hazardous Waste SG-9
Entering Tanks and Other Enclosed Spaces SG-10
Off-The-Job Safety SG-11
Public Relations in Emergencies SG-12
Maintenance and Inspection of Fire Protection Equipment SG-13
Safety in the Scale-up and Transfer of Chemical Processes SG-14
Training of Process Operators SG-15
Liquid Chemicals; Sampling of Tank Car and Tank Truck Shipments SG-16
Fire Protection in the Chemical Industry SG-17
Identification of Materials SG-18
Electrical Equipment in Hazardous Areas SG-19

Table 18.10 Agencies providing standards

Standard	Agency Address
ANSI	The American National Standard, 1430 Broadway, New York NY 10018, USA.
ASTM	American Society for Testing and Materials, 1916, Race Street, Philadelphia, Pa. 19103, USA.
BS	British Standards Institution, 2 Park Street, London W1A 2BS, United Kingdom
BASEEFA	British Approvals Service for Electrical Equipment in Flammable Atmospheres, HSE, Harpur Hill, Buxton, Derby SK17 9JN
DIN	Deutsches Institut fur Normung, Beuth-Verlag GmbH Burggrafenstr. 4–10, D-1000 Berlin, 30, Germany
ISO	International Organisation for Standardisation, 1, rue de Varembi, CH-Geneva 20, Switzerland
JIS	Japanese Standards Association, 1–24, Akasaka 4, Minato-ku, Tokyo 107, Japan
NEN	Nederlands Normalisatie Instituut, Polakweg 5, Rigswigk (ZH), Netherlands
NF	L'Association Francaise de Normalisation, Tour Europe, 92400 Courbevoie, France
PTB Method	Physikalisch-Technische Bundesanstalt, Bundesallee 100, D-3300 Braunschweig, Germany

Table 18.11 A selection of British Standards with relevance to safety

BS No.	Subjects
Respiratory protection	
DD 54	Methods of the sampling and analysis of fume from welding (and allied processes)
341	Valve fittings for compressed gas cylinders
679	Filter for use during welding and similar industrial operations
1747	The measurement of air pollution
2091	Respirators for protection against harmful dusts, gases and scheduled agricultural chemicals
4001	Recommendations for the care and maintenance of underwater breathing apparatus
4275	Recommendations for the selection, use and maintenance of respiratory protective equipment
4400	Sodium chloride particulate test for respirator filters
4555	High efficiency dust respirators
4558	Positive pressure powered dust respirators
4667	Breathing apparatus
4771	Positive pressure powered dust hoods and blouses
5343	Gas detector tubes
6016	Specification for filtering facepiece chest respirators

Table 18.11 (continued)

BS No.	Subjects
Clothing	
PAS 18	Workwear materials
1547	Flameproof industrial clothing (materials and design)
1771	Outdoor uniform clothes
2653	Protective clothing for welders
3119	Method of test for flameproof materials
3120	Performance requirements of flameproof materials for clothing and other purposes
3121	Performance requirements of fabrics described as of low flammability
3314	Specification for protective aprons for wetwork
3546	Coated fabrics for water resistant clothing
3783	X-ray lead–rubber protective aprons for personal use
3791	Clothing for protection against intense heat for short periods
4170	Specification for waterproof protective clothing
4679	Protective suits for construction workers and others in similar arduous activities
4724	Method of tests for resistance for air-impermeable clothing materials to penetration by harmful liquids
5426	Specification for workwear
6249	Materials and material assemblies used in clothing for protection against heat and flame
6357	Assessment of resistance of materials used in protective clothing to molten metal splash
6408	Clothing made from coated fabrics for protection against wet weather
Cranes and lifting tackle	
302	Wire ropes for cranes, excavators and general engineering purposes
327	Power driven derrick cranes
330	Standard wire ropes for haulage purposes
357	Power-driven travelling jib cranes
462	Wire rope grips
466	Power driven overhead travelling cranes, semi-goliath and goliath cranes for general use
1290	Wire rope slings and sling legs for general lifting purposes
1663	High tensile steel chain grade 40 for lifting purposes
1757	Power-driven mobile cranes
2573	Permissible stresses in cranes and design rules
2799	Power-driven tower cranes for building and engineering construction
2837	Steel links and strap assemblies for lifting attachments for packing cases
2902	High tensile steel chain slings
2903	High tensile steel hooks for chains, slings, blocks, and general engineering purposes
CP 3010	Safe use of cranes (mobile cranes, tower cranes and derrick cranes)
3125	Power-driven mast hoists for materials

Table 18.11 (continued)

BS No.	Subjects
3243	Hand-operated chain pulley blocks
3458	Alloy steel chain slings
3481	Flat lift slings
4018	Pulley blocks for use with wire ropes
4278	Eye bolts for lifting purposes
4465	Electrical hoists for passengers and materials
4898	Chain lever hoists
4942	Short link chains for lifting purposes
5323	Scissor lifts
5744	Safe use of cranes
6166	Rating of lifting gear for general purposes
6210	The safe use of wire rope slings for general lifting purposes
Electrical	
88	Cartridge fuses for voltages up to and including 1000 V a.c. and 1500 V d.c.
89	Direct acting indicating electrical measuring instruments and their accessories
196	Plugs, socket-outlets, cable-couplers and applicance-couplers
229	Flame proof enclosures of electrical apparatus
279	100-amperes flame-proof plugs and sockets (restrained type)
415	Safety requirements for mains operated and related apparatus for household and general purpose use
587	Motor starters and controllers
638	Arc welding power sources, equipment, and accessories
646	Cartridge fuse-links
741	Flame-proof electric motor for conveyors, coalcutters, loaders and similar purposes for use in mines
775	Contactors
787	Mining type flame-proof gate-end boxes
816	Requirements for electrical appliances and accessories
889	(1982) Flame-proof electric light fittings
921	Rubber mats for electrical purposes
CP 1017	Distribution of electricity on construction and building sites
1090	Flame-proof hand-held electric drilling machines primarily for use in mines
1259	Intrinsically safe electrical apparatus and circuits for use in explosive atmospheres
1361	Cartridge fuses for a.c. circuits
1362	General purpose fuse links for domestic and similar purposes (primarily for use in plugs)
1363	13 A plugs, switched and unswitched socket-outlets and boxes
1538	Intrinsically-safe transformers primarily for bell-signalling circuits
1539	Moulded electrical insulating materials for use at high temps.
1540	Moulded electrical insulating materials for use at radio frequencies

Table 18.11 (continued)

BS No.	Subjects
2044	Laboratory tests for resistivity of conducting and anti-static rubbers
2754	Memorandum construction of electrical equipment for protection against electric shock
2769	Portable electric motor-operated tools
2950	Cartridge fuse-links
3036	Semi-enclosed electric fuses
3101	Intrinsically-safe remote-control circuits associated with restrained plugs and sockets for use in coal mines
3187	Electrically conducting rubber flooring
3395	Electrically bonded hose and hose assemblies for fuel dispensers
3454	Bolted flame-proof cable couplers and adaptors primarily for use in mines
3492	Electrically bonded road and rail tanker hose and hose assemblies
3535	Safety isolating transformers for industrial and domestic purposes
3681	Electrical safety ot office machines
3861	Electrical safety of office machines
3905	Bolted flame-proof cable-couplers and adaptors for use in mines
4066	Tests on electric cables under fire conditions
4137	Guide to the selection of electrical equipment for use in Division 2 areas
4343	Industrial plugs, socket-outlets and couplers for a.c. and d.c. supplies
4363	Specification for distribution units for electricity supplies for construction and building sites
4444	Electrical earth monitoring
4683	Electrical apparatus for explosive atmospheres
4743	Safety requirements for electronic measuring apparatus
4794	Control switches (electrical)
4934	Specification for safety requirements for electric fans and regulators
5067	Flame-proof transformers for use in mines
5125	50 A flame-proof restrained and bolted plugs and sockets (primarily for use in mines)
5126	Mining type flame-proof supply and control units
5175	(1982) Safety of commercial electrical appliances using microwave energy for heating foodstuffs
5266	Emergency lighting
5311	AC circuit breakers
5345	Selection, installation and maintenance of electrical apparatus for use in potentially explosive atmospheres. Basic requirements for all parts of the code
5405	Maintenance of electrical switchgear (formerly CP 1008)
5415	Safety of electrical motor-operated industrial cleaning appliances
5451	Electrically conducting and antistatic rubber footwear
5458	Safety requirements for indicating and recording electrical measuring instruments
5463	AC switches

Table 18.11 (continued)

BS No.	Subjects
5501	Electrical apparatus for potentially explosive atmospheres
5620	Flame-proof restrained and bolted plugs and sockets (primarily for mines)
5724	Safety of medical electrical equipment
5784	Safety of electrical commercial catering equipment
5958	Control of undesirable static electricity
6007	Rubber-insulated cables for electric power and lighting
6020	Instruments for the detection of combustible gases
6387	Performance requirements for cables required to maintain circuit integrity under fire conditions
6396	Electrical systems in office furniture and office screens
6423	Maintenance of electrical switchgear and control gear for voltages up to and including 650 V
Eyes and face protection	
679	Filters for use during welding and similar industrial operations
1542	Equipment for eye, face and neck protection against non-ionising radiation arising during welding and similar operations
2092	Industrial eye protectors
2724	Filters for protection against intense sunglare (for general and industrial use)
4110	Eye protection for vehicle users
4031	X-ray protection lead glasses
Fire prevention	
DD 58	Tests of the ignitability of upholstered seating
138	Portable fire extinguishers of the water type (soda acid)
336	Fire hose couplings and ancillary equipment
459	Fire-check flush doors and frames
476	Fire tests on building materials and structures
738	The non-ignitable and self-extinguishing properties of solid electrical insulating materials
750	Underground fire hydrants and surface box frames and covers
1382	Portable fire extinguishers of the water type (gas pressure)
1641	Cast iron pipe fittings for sprinklers and other fire protection installation
1689	Galvanised mild steel fire buckets
1721	Portable fire extinguishers of the halogenated hydrocarbon type
2000	Petroleum and its products P34 Flash Point by Pensky-Martens closed tester
2740	Simple smoke alarms
2788	Fireguards for solid fuel fires
2839	Flashpoint of petroleum products by Pensky-Martens closed tester
3116	Automatic fire alarms systems in buildings
3119	Method of test for flameproof materials
3165	Rubber suction hose for fire fighting purposes
3169	Rubber reel hose for fire fighting purposes

Table 18.11 (continued)

BS No.	Subjects
3251	Indicator plates for fire hydrants and emergency water supplies
3326	Portable carbon dioxide fire extinguishers
3442	Flashpoint by the Abel apparatus
3709	Portable fire extinguishers of the water type (stored pressure)
4056	Ignition temperature of gases and vapour
4422	Glossary of terms associated with fire-phenomenon of fire
4547	Classification of fires
4688	Method for determination of flashpoint (open) and fire point of petroleum products by the Pensky-Martens apparatus
4689	Method for determination of flash and fire points for petroleum products and other liquids by the Cleveland open cup
4992	Protection against ignition and deterioration initiated by radio frequency radiation
5041	Fire hydrant systems equipment
5274	Fire hose reels for fixed installations
5306	Fire extinguishing installations and equipment on premises
5364	Manual call points for electric fire alarm systems
5396	Stainless steel CO_2 containers for fixed fire-fighting installations on ships
5423	Portable fire extinguishers
5445	Components of automatic fire detection systems
5446	Components of automatic fire alarm systems for residential premises
5588	Code of Practice for fire precautions in the design of buildings
5839	Fire detection and alarm systems in buildings
5909	Code of Practice for fire precautions in chemical plant
6165	Specification for small disposable fire extinguishers of the aerosol type
6266	Fire protection for electronic data processing installations
6327	Fire protection of reciprocating internal combustion engines
6336	Development and presentation of fire tests and their use in hazard assessment
Footwear	
953	Methods of test for safety and protective footwear
1870	Safety footwear
2723	Firemen's leather boots
4676	Gaiters and footwear for protection against burns and impact risks in foundries
4972	Women's protective footwear
5145	Lined industrial rubber boots
5451	Electrically conducting and antistatic rubber footwear
5462	Lined rubber boots with protective (penetration resistant) midsoles
6159	Polyvinyl chloride boots
General	
10	Flanges and bolting for piping, valves and fittings
CP 153	Windows and rooflights

Table 18.11 (continued)

BS No.	Subjects
341	Valve fittings for compressed gas cylinders
759	Valves, gauges and other safety fittings for application to boilers and to piping installations for and in connection with boilers
767	Centrifuges of the basket and bowl types for use in industrial and commercial applications
848	Fans for general purposes
893	Measurement of the concentration of particulate material in ducts carrying gases
1113	Water-tube steam generating plant
1123	Safety valves, gauges and other safety fittings for air receivers and compressed air installations
1319	Medical gas cylinders, valves and yoke connections
1500	Fusion welded pressure vessels for general purposes
1640	Steel butt-welding pipe fittings for the petroleum industry
1756	Methods for the sampling and analysis of flue gases
1821	Class 1 oxy-acetylene welding of ferritic steel pipework for carrying fluids
1831	Recommended common names for pesticides
1979	Safety requirements for radio transmitting equipment
2595	Environmental cleanliness in enclosed spaces
2654	Vertical steel welded storage tanks with butt-welded shells for the petroleum industry
2718	Gas cylinder trolleys
2831	Air filters used in air conditioning and general ventilation
2881	Hospital cupboards (wall fixing) for poisons and dangerous drugs
CP 3003	Lining of vessels and equipment for chemical processes
3044	Design of office chairs and tables
3202	Recommendations on laboratory furniture and fittings (includes fume cupboards)
3405	Measurement of particulate emission including grit and dust
3601	Steel pipes and tubes for pressure purposes
3602	Steel pipes and tubes for pressure purposes
3604	Steel pipes and tubes for pressure purposes
4089	Rubber hose and hose assemblies for liquefied petroleum gas lines
4163	Recommendations for health and safety in workshops of schools and colleges
4390	Portable pneumatic grinding machines
4434	Refrigeration safety
4947	Test gases for gas appliances
5045	Transportable gas containers
5073	Storage of goods in freight containers
5289	Visual inspection of fusion welded joints
5295	Environmental cleanliness in enclosed spaces
5384	The selection and use of control systems for heating, ventilating and air-conditioning installations

Table 18.11 (continued)

BS No.	Subjects
5395	Stairs
5429	Safe operation of small-scale storage facilities for cryogenic liquids
5430	Periodic inspection, testing and maintenance of transportable gas containers (excluding dissolved acetylene containers)
5607	Safe use of explosives in the construction industry
5667	Continuous mechanical handling equipment
5720	Mechanical ventilation and air conditioning in buildings
5725	Emergency exit devices
5726	Microbiological safety cabinets
5925	Design of buildings: ventilation principles and designing for natural ventilation
5997	British Standard codes of practice for building services
6133	Safe operation of lead–acid stationary cells and batteries
6164	Safety in tunnelling in the construction industry
6283	Safety devices for use in hot water systems
6399	Design loading for buildings
6759	Safety Valves
Hand protection	
PAS 16	Gel type hand cleaner
697	Rubber gloves for electrical purposes
1651	Industrial gloves
1884	Rubber post-mortem gloves
2606	X-ray protective gloves for medical diagnostic purposes up to 150 kV peak
Harness, prevention of falls and access	
CP 93	The use of safety nets on constructional works
470	Access and inspection openings for pressure vessels
1129	Portable timber ladders, steps, trestles and lightweight stagings
1139	Metal scaffolding
1247	Manhole step irons
1397	Specification for industrial safety belts, harnesses and safety lanyards
2037	Aluminium ladders, steps and trestles
2482	Timber scaffold boards
2655	Lift, escalators, passenger conveyors and paternosters
2830	Specification for suspended safety chairs and cradles for use in the construction industry
3049	Pedestrian guard rails
3367	Specification for fire brigade and industrial ropes and rescue lines
3572	Specification for access fittings for chimneys and other high structures in concrete or brickwork
3678	Specification for access hooks for chimneys and other high structures in steel
3913	Industrial safety nets
4211	Specification for steel ladders for permanent access
5062	Self-locking safety anchorages for industrial use

Table 18.11 (continued)

BS No.	Subjects
5507	Falsework equipment
5531	Safety in erecting structural frames
5845	Permanent anchors for industrial safety belts and harnesses
5973	Access and working scaffolds and special scaffold structures in steel
5974	Temporarily installed suspended scaffolds and access equipment
5975	Falsework
6037	Permanently installed suspended access equipment
6180	Protective barriers in and about buildings
Head protection	
2495	Protective helmets for vehicle users
3864	Firemen's helmets
4033	Industrial scalp protectors (light duty)
5240	General purpose industrial safety helmets
5361	Protective helmets for vehicle users
Hearing protection, noise and vibration	
2042	An artificial ear for the calibration of earphones
2475	Octave and one-third octave band-pass filters
2497	A reference zero for the calibration of pure-tone audiometers
3045	Expression of physical and subjective magnitude of sound or noise in air
3383	Normal equal-loudness contours for pure tones and normal threshold of hearing
3425	Measurements of noise emitted by motor vehicles
3489	Specification for sound level meters (industrial grade)
3539	Sound level meters
3593	Preferred frequencies for acoustical measurements
4142	Method of rating noise affecting mixed residential and industrial areas
4196	Methods of determination of sound power levels of noise sources
4197	Specification for a precision sound level meter
4198	Method for calculating loudness
4718	Silencers for air distribution systems
4813	Method of measuring noise from machine tools: excluding testing in anechoic chambers
5108	Method of measurement of attenuation of hearing protectors at threshold
5228	Noise control on construction and open sites
5330	Estimating the risk of hearing handicap due to noise exposure
5944	Measurement of airborne noise from hydraulic fluid power systems and components
5969	Sound level meters
6055	Measurement of whole-body vibration of the operators of agricultural wheeled tractors and machinery
6056	Measurement of transformer and reactor sound levels
6177	Selection and use of elastomeric bearings for vibration isolation of buildings

Table 18.11 (continued)

BS No.	Subjects
6294	Measurement of body vibration transmitted to the operator of earth-moving machinery
6344	Industrial hearing protectors
Industrial vehicles	
1761	Single bucket excavators
3318	Earth-moving machinery – method for locating the centre of gravity
3726	Counterbalanced lift trucks, Stability – basic tests
4063	Requirements and testing of protective cabs and frames for agricultural wheeled tractors
4338	Specification for rated capacities of fork lift trucks
4430	Safety of powered industrial trucks
4436	Reach and straddle fork lift trucks – stability tests
5453	Anchorages for seat belts in protective cabs and frames on agricultural tractors
5526	Specification for falling-object, protective structures on earth moving machinery
5527	Specification for roll-over protective structures on earth-moving machinery
5528	Specification for operator's controls on excavators used for earth-moving
5777	Methods of test for verification of stability of pallet stackers and high lift platform trucks
5778	Methods of test for verification of stability of industrial trucks operating in special conditions of stacking with mast tilted forward
5933	Specification for overhead guards for high-lift rider trucks
6218	Specification for performance requirements for seatbelts and seatbelt anchorages for earth-moving machinery fitted with roll-over protective structures
Machinery and guarding	
1123	Safety valves, gauges and other safety fittings for air receivers and compressed air installations
3042	Standard test fingers and probes for checking protection against electrical mechanical and thermal hazards
3417	Agricultural power take-off shafts and guards
4402	Mechanical safety requirements for laboratory centrifuges
4581	Dimension of flanges for the mounting of plain grinding wheels
4640	Metal working machine tools
4644	Recommendations for safety of office machines and data processing equipment
4999	General requirements for rotating electrical machines (enclosures)
5304	Safeguarding of machinery
5498	Safety of hand-operated paper-cutting machines
5667	Specification for continuous mechanical handling equipment safety requirements
5850	Safety of electrically energized office machines
5924	Safety requirements for electrical equipment of machines for resistance welding and allied processes
5945	Guards and shields for earth-moving machinery
6204	Safety of data processing equipment

Table 18.11 (continued)

BS No.	Subjects
Radiation	
DD 66	Methods of leak testing for sealed radioactive sources
1542	Equipment for eye, face and neck protection against radiation
2606	X-ray protective gloves
3232	Medical treatment lamps
3510	A basic symbol to denote the actual or potential presence of ionising radiation
3664	Film badges for personal radiation monitoring
3783	X-ray lead – rubber protective aprons
3890	The testing, calibration and processing of radiation monitoring films
3895	Guide to the design, testing and use of packaging for the safe transport of radioactive materials
3909	Ingot lead for radiation shielding
4094	Data on shielding from ionising radiation
4247	The assessment of surface materials for use in radioactive areas.
4803	Protection of personnel against hazards from laser radiation
5243	Sampling airborne radioactive materials
5288	Sealed radioactive sources
5566	Installed exposure rate meters, warning assemblies and monitors for gamma radiation of energy between 80 keV and 3 MeV
5650	Apparatus for gamma radiography
Signs and colours	
DD 48	Identification of fire extinguishers
349	Identification of the contents of industrial gas containers
381	Colours for identification, coding and special purposes
1635	Graphical symbols and abbreviations for fire protection drawings
1710	Identification for pipelines
2474	Recommended names for chemicals used in industry
2560	Exit signs (internally illuminated)
3351	Piping systems for petroleum refineries and petrochemical plants
4218	Self-luminous exit signs
4580	Number designation of organic refrigerants
4610	Colours for high visibility clothing
4765	Safety signs for radio frequency or other non-ionising radiations
4964	Symbols and displays for control markings for agricultural tractors and machinery
5252	Framework for colour co-ordination for building purposes
5378	Safety signs and colours
5383	Materials marking and colour coding of metal pipes and piping systems components in steel, nickel alloys and titanium alloys
5499	Fire safety signs, notices and graphic symbols
6034	Specification for public information symbols
6309	Symbols for operator controls and for controls other than operator controls for use on earth-moving machinery

Table 18.12 Examples of publications from learned bodies

Relevant publications available from The Royal Society of Chemistry

Air pollution damage to vegetation, ACS Advances in Chemistry Ser. No. 122 (1973)
Assessment and management of chemical risks, ed Rodericks, JV, ACS Symposium series No. 239
Biological activities of polymers, ACS Symposia Series No. 156 (1982)
Biological effects of non-ionizing radiation, ACS Symposia Series No. 157 (1981)
Chemical carcinogens, ACS Monograph No. 173 (1976)
Chemical hazards in industry (Monthly Current Awareness Bulletin)
Chemical hazards in the Workplace, ACS Symposia, Series No. 149 (1981)
Chemistry and the environment, ACS, 1976
Cleaning our environment, ACS, 1978
Cotton Dust: controlling an occupational health hazard
Environmental Chemistry, Brown, HJM
Environmental Monitoring, ACS, 1976
Handbook of particle sampling and analysis methods, ed Murphy, C H, Verlag Chemie
Hazards in the chemical laboratory, ed Bretherick, 3rd edn. 1981
Health and safety in the laboratory – where do we go from here, publication 51 (1984)
Heavy metals as contaminants of the human environment, Bryce-Smith, D, 1975 (Cassette and workbook)
Industry and the environment in perspective, ed Hunter, R E (1983)
Laboratory hazards bulletin (monthly current awareness periodical)
Maximum concentrations at the workplace and biological tolerance values for working with materials, ed The Commission for Investigation of Health
Hazards of Chemical Compounds in the Work Area, 1983
Molecular aspects of toxicology, Hathway, D E, 1984
Monitoring toxic substances, ACS Symposia Series No. 94 (1979)
The Pesticide chemist and modern technology, No. 160, (1981)
Pollution: causes, effects, and control, ed Harrison, R M, Publication No. 44 (1983)
Risk assessment at waste sites, ACS Symposia Series No. 24 (1982)
Safety in academic chemistry laboratories, ACS Monograph, 1979
Toxic chemical and explosive facilities, ACS Symposia Series No. 96 (1979)
TSCA's impact on society and chemical industry, ACS Symposia Series No. 213 (1983)

Relevant publications from the Institution of Chemical Engineers

A First Guide to Loss Prevention, 1981
The Assessment of Major Hazards, Symposium Series No 71, 1982
The Assessment and Control of Major Hazards, Symposium Series No 93, 1985
Cheaper Safer Plants or Wealth and Safety at work, Kletz T A, 1984
Chemical Pollution – A General Survey of Research, Jaff D and Walters J K, 1975
Chemical Process Hazards with Special Reference to Plant Design Vols 5 and 6, Symposium Series Nos 39 and 49
The Control of Sulphur and Other Gaseous Emissions, Symposium Series No 57, 1979
Effluent Treatment in the Process Industries, Symposium Series No 77, 1983
Ergonomics Problems in Process Operations, Symposium Series No 90, 1984
Flowsheeting for Safety, 1976
Guide to Dust Explosion Prevention and Protection. Part 1-Venting, Schofield D, 1984
Guide to Safety in Mixing Operations, 1982
Hazop and Hazan, Kletz T A, 1985
Hazard workshop Training Modules:
  Hazards of Over and Under Pressurising of Vessels
  Hazards of Plant Modification
  Fires and Explosions
  Preparation for Maintenance
  Furnace Fires and Explosions
  Work Permit Systems
ISGHO Guide Notes on
  Safe Application of Oxygen Analysers to Hydrocarbon Oxidation Reactions in Chemical Process Plant, 1983

Table 18.12 (continued)

Safe Use of Stainless Steel in Chemical Process Plant, 1978
Use of Acoustic Emission Testing in Chemical Plants
4th International Symposium on Loss Prevention and Safety Promotion in the Process Industries, Symposium Series Nos 80, 81 and 82
Loss Prevention Bulletin
Myths of the Chemical Industry, Kletz T A, 1984
Noise in Chemical Plant, Mather J S B
Nomenclature for Hazard and Risk Assessment in the Process Industries, 1985
Offshore and Onshore Engineering Practices Compared, 1984
Preparation of Plant for Maintenance, 1980
Preventing Emergencies in the Process Industries, (Film/Video), 1985
Process Industry Hazards, Symposium Series No 47, 1976
The Protection of Exothermic Reactors and Pressurised Storage Vessels, Symposium Series No 85, 1984
User Guide to Dust and Fume Control, 1985
User Guide to Fire and Explosion Hazards in the Drying of Particulate materials, 1977
Venting Gas and Dust Explosions – A Review, Lunn G, 1984

References

(*All places of publication are London unless otherwise stated*)

1. Department of Employment, *Report of the Roben's Committee*. HMSO 1972.
2. Carson, P. A. & Jones, K., *J. Hazardous Materials*, 1984 (9), 305.
3. Bretherick, L., 'Prevention of Risks in the Chemical Industry'. Paper presented for the International Social Security Association Symposium, Frankfurt, June 21–23, 1976.
4. Pantry, S., *Health and Safety: A Guide to Sources of Information*. Capital Planning Information: Edinburgh 1983.
5. Barbour Index Ltd, New Lodge, Drift Road, Windsor, Berks. SL4 4RQ.
6. Churchley, A., *Chem. and Ind.*, 1977, 624.
7. Olishifski, J. B. (ed.)., *Fundamentals of Industrial Hygiene* (2nd edn). National Safety Council 1979.
8. Department of Trade, *Standards, Quality and International Competitiveness*, HMSO 1982.

Index